Chromosome Engineering in Plants: Genetics, Breeding, Evolution

Chromosome Engineering in Plants: Genetics, Breeding, Evolution

Editor:

Carlos Tello Lacal, Ph.D

Delve Publishing
www.delvepublishing.com

Chromosome Engineering in Plants: Genetics, Breeding, Evolution

Editor: Carlos Tello Lacal, Ph.D

708 3rd Avenue, 6th Floor
New York, NY 10017
United States of America

www.delvepublishing.com

ISBN: 978-1-68095-897-3

Library of Congress Control Number: 2016951500

About the Editor

Carlos Tello Lacal, Ph.D

Carlos Tello (Valencia, Spain, 1981) studied Agricultural Engineering with specialization in Biotechnology at the Universidad Politécnica de Valencia, where he graduated with a BSc thesis on the characterization of an Arabidopsis mutant with increased tolerance to seed ageing and salt stress. He moved afterwards to Sevilla to complete his MSc and PhD at the Instituto de Recursos Naturales y Agrobiología, where he studied the mechanisms employed by plants in their adaptation to abiotic stress and more specifically the role of the Arabidopsis Na^+/H^+ antiporter SOS1 in the regulation of Na^+ and K^+ homeostasis. Since 2014, he is based in Zürich and develops several freelance science-related jobs. He has recently retrained himself in the field of cell culture in bioreactors by attending an advanced training course at the Zürcher Hochschule für Angewandte Wissenschaften.

Contents

List of Contributors

Linda Zamariola
Department of Plant Production, Faculty of Bioscience Engineering, University of Ghent, Ghent, Belgium

Choon Lin Tiang
Department of Plant Breeding and Genetics, Cornell University, Ithaca, NY, USA

Nico De Storme
Department of Plant Production, Faculty of Bioscience Engineering, University of Ghent, Ghent, Belgium

Wojtek Pawlowski
Department of Plant Breeding and Genetics, Cornell University, Ithaca, NY, USA

Danny Geelen
Department of Plant Production, Faculty of Bioscience Engineering, University of Ghent, Ghent, Belgium

Shawn R. Carlson
Chromatin, Chicago, Illinois, United States of America

Gary W. Rudgers
Chromatin, Chicago, Illinois, United States of America

Helge Zieler
Chromatin, Chicago, Illinois, United States of America

Jennifer M. Mach
Chromatin, Chicago, Illinois, United States of America

Song Luo
Chromatin, Chicago, Illinois, United States of America

Eric Grunden
Chromatin, Chicago, Illinois, United States of America

Cheryl Krol
Chromatin, Chicago, Illinois, United States of America

Gregory P. Copenhaver
Chromatin, Chicago, Illinois, United States of America
Department of Biology, The University of North Carolina at Chapel Hill, Chapel Hill, North Carolina, United States of America
Carolina Center for Genome Sciences, The University of North Carolina at Chapel Hill, Chapel Hill, North Carolina, United States of America

Daphne Preuss
Chromatin, Chicago, Illinois, United States of America
Department of Molecular Genetics and Cell Biology, The University of Chicago, Chicago, Illinois, United States of America

Joseph F. Petolino
Dow AgroSciences, 9330 Zionsville Rd., Indianapolis, IN USA

Abdellah Barakate
Cell and Molecular Sciences, The James Hutton Institute, Dundee, UK

Jennifer Stephens
Cell and Molecular Sciences, The James Hutton Institute, Dundee, UK

Riccardo Aversano
Department of Soil, Plant, Environmental and Animal Production Sciences, University of Naples Federico II, Via Università 100, Portici 80055, Italy

Maria Raffaella Ercolano
Department of Soil, Plant, Environmental and Animal Production Sciences, University of Naples Federico II, Via Università 100, Portici 80055, Italy

Immacolata Caruso
Department of Soil, Plant, Environmental and Animal Production Sciences, University of Naples Federico II, Via Università 100, Portici 80055, Italy

Carlo Fasano
Department of Soil, Plant, Environmental and Animal Production Sciences, University of Naples Federico II, Via Università 100, Portici 80055, Italy

Daniele Rosellini
Department of Applied Biology, University of Perugia, Borgo XX Giugno 74, Perugia 06121, Italy

Domenico Carputo
Department of Applied Biology, University of Perugia, Borgo XX Giugno 74, Perugia 06121, Italy

Ruvini Ariyadasa
Leibniz Institute of Plant Genetics and Crop Plant Research (IPK), Corrensstraße 3, 06466 Gatersleben, Germany

Nils Stein
Leibniz Institute of Plant Genetics and Crop Plant Research (IPK), Corrensstraße 3, 06466 Gatersleben, Germany

Dominika Idziak
Department of Plant Anatomy and Cytology, Faculty of Biology and Environmental Protection, University of Silesia, Jagiellonska 28, 40-032 Katowice, Poland

Alexander Betekhtin
Department of Plant Anatomy and Cytology, Faculty of Biology and Environmental Protection, University of Silesia, Jagiellonska 28, 40-032 Katowice, Poland

Elzbieta Wolny
Department of Plant Anatomy and Cytology, Faculty of Biology and Environmental Protection, University of Silesia, Jagiellonska 28, 40-032 Katowice, Poland

Karolina Lesniewska
Department of Plant Anatomy and Cytology, Faculty of Biology and Environmental Protection, University of Silesia, Jagiellonska 28, 40-032 Katowice, Poland

Jonathan Wright
Department of Cell and Developmental Biology, John Innes Centre, Colney Lane, Norwich NR4 7UH, UK

Melanie Febrer
Department of Cell and Developmental Biology, John Innes Centre, Colney Lane, Norwich NR4 7UH, UK

Michael W. Bevan
Department of Cell and Developmental Biology, John Innes Centre, Colney Lane, Norwich NR4 7UH, UK

Glyn Jenkins
Institute of Biological, Environmental and Rural Sciences, Aberystwyth University, Edward Llwyd Building, Penglais, Aberystwyth, Wales SY23 3DA Ceredigion, UK

Robert Hasterok
Department of Plant Anatomy and Cytology, Faculty of Biology and Environmental Protection, University of Silesia, Jagiellonska 28, 40-032 Katowice, Poland

Jinxia Wu
Biotechnology Research Institute/National Key Facility for Genetic Resources and Gene Improvement, The Chinese Academy of Agricultural Sciences, Beijing, China

Zhiguo Zhang
Biotechnology Research Institute/National Key Facility for Genetic Resources and Gene Improvement, The Chinese Academy of Agricultural Sciences, Beijing, China

Qian Zhang
Biotechnology Research Institute/National Key Facility for Genetic Resources and Gene Improvement, The Chinese Academy of Agricultural Sciences, Beijing, China

Xiao Han
Biotechnology Research Institute/National Key Facility for Genetic Resources and Gene Improvement, The Chinese Academy of Agricultural Sciences, Beijing, China

Xiaofeng Gu
Biotechnology Research Institute/National Key Facility for Genetic Resources and Gene Improvement, The Chinese Academy of Agricultural Sciences, Beijing, China

Tiegang Lu
Biotechnology Research Institute/National Key Facility for Genetic Resources and Gene Improvement, The Chinese Academy of Agricultural Sciences, Beijing, China

Yufang Guo
Department of Horticulture, University of Georgia -Tifton Campus, 2360 Rainwater Rd, Tifton, GA 31793-5766, USA

Brian Abernathy
Center for Applied Genetic Technologies, University of Georgia, 111 Riverbend Road, Athens, GA 30602, USA.

Yajuan Zeng
Department of Horticulture, University of Georgia -Tifton Campus, 2360 Rainwater Rd, Tifton, GA 31793-5766, USA

Peggy Ozias-Akins
Department of Horticulture, University of Georgia -Tifton Campus, 2360 Rainwater Rd, Tifton, GA 31793-5766, USA

Yukio Kaneko
Laboratory of Plant Breeding, Faculty of Agriculture, Utsunomiya University, 350 Minemachi, Utsunomiya, Tochigi 321-8505, Japan

Sang Woo Bang
Laboratory of Plant Breeding, Faculty of Agriculture, Utsunomiya University, 350 Minemachi, Utsunomiya, Tochigi 321-8505, Japan

Anne B. Britt
Department of Plant Biology, University of California, Davis, Davis, CA, USA

Sundaram Kuppu
Department of Plant Biology, University of California, Davis, Davis, CA, USA

Tobias Würschum
State Plant Breeding Institute, University of Hohenheim, Stuttgart 70593, Germany.

Matthew R Tucker
ARC Centre of Excellence for Plant Cell Walls, University of Adelaide, Waite Campus, Adelaide, Urrbrae SA 5064, Australia.

Jochen C Reif
State Plant Breeding Institute, University of Hohenheim, Stuttgart 70593, Germany

Hans Peter Maurer
State Plant Breeding Institute, University of Hohenheim, Stuttgart 70593, Germany

Karen S Aitken
[1]CSIRO Plant Industry, Queensland Bioscience Precinct, 306 Carmody Rd, St Lucia, Brisbane, QLD 4067, Australia

Meredith D McNeil
CSIRO Plant Industry, Queensland Bioscience Precinct, 306 Carmody Rd, St Lucia, Brisbane, QLD 4067, Australia

Paul J Berkman
CSIRO Plant Industry, Queensland Bioscience Precinct, 306 Carmody Rd, St Lucia, Brisbane, QLD 4067, Australia

Scott Hermann
BSES Limited, 50 Meiers Road, Indooroopilly, Brisbane, QLD 4068, Australia

Andrzej Kilian
Diversity Arrays P/L, 1 Wilf Crane Crescent, Yarralumla, Canberra, ACT 2600, Australia

Peter C Bundock
Southern Cross University, Ctr Plant Conservation Genetics, Lismore, NSW 2480, Australia

Jingchuan Li
CSIRO Plant Industry, Queensland Bioscience Precinct, 306 Carmody Rd, St Lucia, Brisbane, QLD 4067, Australia

Grit Winterfeld
Institute of Biology, Martin Luther University Halle-Wittenberg, Neuwerk 21, 06099 Halle, Germany

Alexandra Wolk
Institute of Biology, Martin Luther University Halle-Wittenberg, Neuwerk 21, 06099 Halle, Germany

Martin Roser
Institute of Biology, Martin Luther University Halle-Wittenberg, Neuwerk 21, 06099 Halle, Germany

Jing Hou, Ning Ye
The Southern Modern Forestry Collaborative Innovation Center, Nanjing Forestry University, Nanjing 210037, China

Defang Zhang
The Southern Modern Forestry Collaborative Innovation Center, Nanjing Forestry University, Nanjing 210037, China

Yingnan Chen
The Southern Modern Forestry Collaborative Innovation Center, Nanjing Forestry University, Nanjing 210037, China

Lecheng Fang
The Southern Modern Forestry Collaborative Innovation Center, Nanjing Forestry University, Nanjing 210037, China

Xiaogang Dai
The Southern Modern Forestry Collaborative Innovation Center, Nanjing Forestry University, Nanjing 210037, China

Tongming Yin
The Southern Modern Forestry Collaborative Innovation Center, Nanjing Forestry University, Nanjing 210037, China

Edelín Roque
Institute of Molecular and Cell Biology of Plants Higher Council for Scientific Research & Polytechnic University of Valencia (CSIC-UPV), Polytechnic City of Innovation, Edf. 8E, C / Ingeniero Fausto Elio s / n, E 46011 Valencia, Spain

Mario A. Fares
Institute of Molecular and Cell Biology of Plants Higher Council for Scientific Research & Polytechnic University of Valencia (CSIC-UPV), Polytechnic City of Innovation, Edf. 8E, C / Ingeniero Fausto Elio s / n, E 46011 Valencia, Spain

Lynne Yenush
Institute of Molecular and Cell Biology of Plants Higher Council for Scientific Research & Polytechnic University of Valencia (CSIC-UPV), Polytechnic City of Innovation, Edf. 8E, C / Ingeniero Fausto Elio s / n, E 46011 Valencia, Spain

Mari Cruz Rochina
Institute of Molecular and Cell Biology of Plants Higher Council for Scientific Research & Polytechnic University of Valencia (CSIC-UPV), Polytechnic City of Innovation, Edf. 8E, C / Ingeniero Fausto Elio s / n, E 46011 Valencia, Spain

Jiangqi Wen
Plant Biology Division, The Samuel Roberts Noble Foundation, 2510 Sam Noble Parkway, Ardmore, OK 73401, USA

Kirankumar S. Mysore
Plant Biology Division, The Samuel Roberts Noble Foundation, 2510 Sam Noble Parkway, Ardmore, OK 73401, USA

Concepción Gómez-Mena
Institute of Molecular and Cell Biology of Plants Higher Council for Scientific Research & Polytechnic University of Valencia (CSIC-UPV), Polytechnic City of Innovation, Edf. 8E, C / Ingeniero Fausto Elio s / n, E 46011 Valencia, Spain

José Pío Beltrán
Institute of Molecular and Cell Biology of Plants Higher Council for Scientific Research & Polytechnic University of Valencia (CSIC-UPV), Polytechnic City of Innovation, Edf. 8E, C / Ingeniero Fausto Elio s / n, E 46011 Valencia, Spain

Luis A. Cañas
Institute of Molecular and Cell Biology of Plants Higher Council for Scientific Research & Polytechnic University of Valencia (CSIC-UPV), Polytechnic City of Innovation, Edf. 8E, C / Ingeniero Fausto Elio s / n, E 46011 Valencia, Spain

Leandro Quadrana
Biology Institute of the Ecole Normale experience greater than, National Center for Scientific Research, National Institute of HEALTH and Research Me'- of association, Ecole Normale experience greater than, Paris, France

Amanda Bortolini Silveira
Biology Institute of the Ecole Normale experience greater than, National Center for Scientific Research, National Institute of HEALTH and Research Me'- of association, Ecole Normale experience greater than, Paris, France

George F Mayhew
Roche NimbleGen, Inc, Madison, United States

Chantal LeBlanc
Department of Molecular, Cellular and Developmental Biology, Yale University, New Haven, United States

Robert A Martienssen
Watson School of Biological Sciences, Howard Hughes Medical Institute, Gordon and Betty Moore Foundation, Palo Alto, United States
Cold Spring Harbor Laboratory, Cold Spring Harbor, United States

Jeffrey A Jeddeloh
Roche NimbleGen, Inc, Madison, United States

Vincent Colot
Biology Institute of the Ecole Normale experience greater than, National Center for Scientific Research, National Institute of HEALTH and Research Me'- of association, Ecole Normale experience greater than, Paris, France

Maria Cuacos
School of Biosciences, University of Birmingham, Birmingham, UK

F. Chris H. Franklin
School of Biosciences, University of Birmingham, Birmingham, UK

Stefan Heckmann
School of Biosciences, University of Birmingham, Birmingham, UK

Preface

The generic term "chromosome engineering" encompasses a set of biotechnological procedures in which chromosome features are manipulated to change their structure and mode of genetic inheritance. The employment of these technologies can enormously facilitate and accelerate the mapping of plant genomes, the analysis of gene functions and ultimately the development of plant breeding programs. Additionally, the study of chromosome biology and genome structure allows a broadened view of plant species evolution. In this book, a wide range of peer-reviewed research papers fully covering the above-mentioned topics have been compiled.

Chapter 1 focuses on plant chromosome segregation during meiosis and describes the events taking place, as well as the molecular factors involved. In Chapter 2, the de novo construction of a maize minichromosome is reported and its potential applications in plant biotechnology are discussed. In Chapters 3 and 4, genome editing in plants through the respective employment of zinc finger nucleases and the CRISPR/Cas9 System are analyzed. In the following block of five chapters, we review the application of chromosome engineering technology to genome analysis and functional genomics. Chapter 5 examines the main tools allowing an in-depth analysis of polyploid genome structure and gene expression. In Chapter 6, the construction of BAC libraries and their application in genome physical mapping are described. Chapter 7 depicts the genome structure and evolution analysis of three Brachypodium distachyon cytotypes by employing the chromosome painting technique. Chapter 8 describes the use of the enhancer trapping system in gene identification, illustrated by the characterization of the rice oscdm1 mutant. Chapter 9 shows the development of a TILLING by sequencing method in peanut in order to identify mutations in selected loci. The next four chapters investigate practical applications of chromosome engineering in crop breeding. In Chapter 10, the obtaining of new Brassicaceae amphidiploid, alloplasmic, gene integration, chromosome addition and chromosome substitution lines by chromosome engineering methods is presented. In Chapter 11, the manipulation of the Arabidopsis histone CENH3 to produce haploid plants is shown. Similarly, Chapter 12 features the generation of doubled haploid triticale plants for their application in modern breeding programs through microspore embryogenesis. Chapter 13 describes a comparative mapping study between sugarcane and sorghum providing new information on the sugarcane genome structure and revealing the presence of chromosome rearrangements between both species. In Chapter 14, the homoploid hybridization and polyploidy events that led to the evolution of the endemic alpine Helictotrichon parlatorei group are reviewed. Chapter 15 sheds some light on the evolution of sex chromosomes by comparing their divergence in the genera Salix and Populus. The role played by gene duplication in plant genome evolution is illustrated in Chapter 16 through a comparative functional study of the Medicago truncatula transcription factor PISTILLATA and its paralogs MtPI and MtNGL9. In Chapter 17, the role of transposable elements in plant genome evolution is investigated through the genome sequencing data of 211 Arabidopsis thaliana accessions. Finally, Chapter 18 reviews the role of plant centromere biology in evolution, with focus on atypical centromeres, such as holo- and neocentromeres. As an approach to the exciting field of chromosome engineering, the present book intends to be beneficial for college students, researchers and teachers.

Editor
Carlos Tello Lacal, Ph.D

INTRODUCTION

Chromosomes are highly condensed rods of Deoxyribonucleic Acid (DNA), the genetic material which contains the building blocks of life. DNA stores important information about the structure of an animal or plant, and it helps direct the organism as it grows and manages daily tasks. Chromosomes serve as the storage for this important material, periodically dividing along with cells and replicating to make copies of the DNA they contain. Chromosomes are also very important in sexual reproduction, as they allow an organism to pass genetic material on to descendants.

In organisms with cell nuclei, known as eukaryotes, chromosomes are found inside the nucleus. Most of these organisms have a set of chromosomes which come in pairs. In structural cells, each cell retains a complete set of chromosomes, in what is known as diploid form, referring to the fact that the chromosome set is complete. In cells for sexual reproduction like eggs or sperm, each cell only has half of the parent organism's genetic material, stored in haploid form, ensuring that the parent passes down half of its genes.

Each end of a chromosome is capped by a telomere, a string of repetitive DNA which protects the chromosome from damage. Scientists sometimes look at telomeres to gain important information about an organism, as they appear to change over time and they may be associated with aging. When chromosomes divide to make copies, the telomere also ensures that all of the important genetic material is copied.

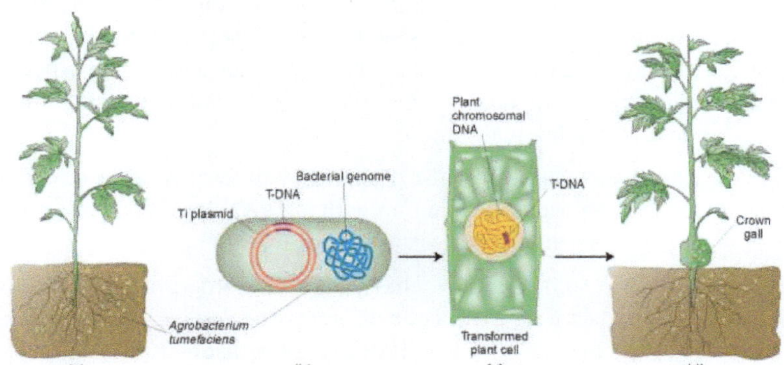

Every time a cell divides, the chromosomes inside are replicated. In mitosis, normal cell division, the chromosomes make copies of themselves which later pair up, so that at the end of the cell division process there are two cells with a set of diploid chromosomes each. When cells divide to create reproductive material, a process called meiosis, each division yields four cells, each with a haploid set of chromosomes. These cells are known as gametes, and when they meet, they contain enough genetic material to create an entirely new organism.

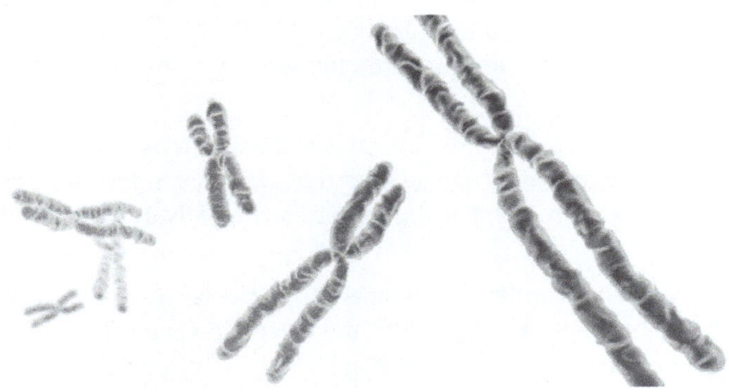

In humans, the normal number of chromosomes is 46, appearing in 23 pairs. Each pair of chromosomes stores distinct information, and any damage to a chromosome can cause serious problems for the parent organism. Errors usually occur during cell division, creating gaps in the genetic material in the chromosome. In some cases, an abnormal number of chromosomes appears, in a condition known as aneuploidy. Aneuploidy in reproductive cells can be a major issue, as it will cause birth defects.

What is a Y chromosome?

A Y chromosome is one of the sex chromosomes in humans and other mammals, with genetic code determining whether an organism will be genetically male or female. The presence of a Y chromosome will usually cause someone to be male, except in the case of certain genetic disorders. The complementary chromosome is the X chromosome; women generally have two X chromosomes in their karyotype, while men inherit an X and a Y.

Not all organisms have sex chromosomes, and those that do may have a complicated system of inheritance when it comes to sex. This article focuses specifically on the role of the Y chromosome in humans. It is believed that initially, ancestors of humans inherited two basically identical sex chromosomes, and over time they diverged into an X and Y chromosome. The Y chromosome is somewhat shorter than the X chromosome, and primarily contains genetic material related to male sex characteristics.

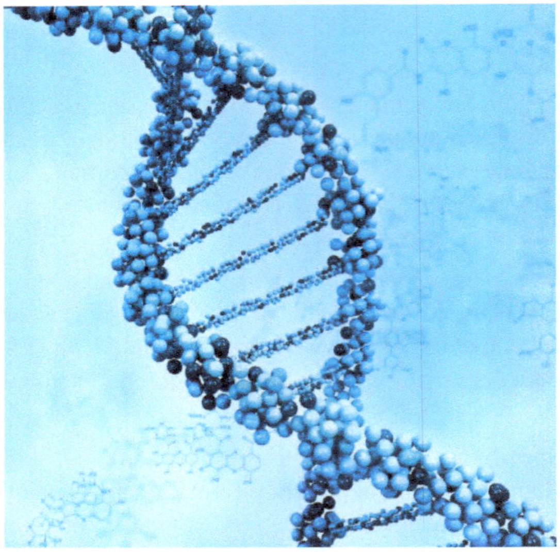

One of the most important genes among the 58 million base pairs on the Y chromosome is the SRY gene, which determines the formation of testes. When the testes form, a cascading series of reactions happens as they produce sex hormones like testosterone, contributing to the development of secondary sex characteristics like chest hair and a deep voice. The genetic material in the chromosome makes up about 2% of the genetic material in any given cell, a relatively small amount, and historical evidence suggests it was once larger and is actually shrinking with time.

Genetic disorders involving the chromosome usually cause problems with fertility or the development of sex characteristics. In some cases, people inherit an extra X chromosome, ending up with an XXY karyotype and a condition called Klinefelter's syndrome. The subject is male, but may have fertility problems and other issues.

Men pass their Y chromosomes on to their sons, and if they have genetic disorders on the Y chromosome, their sons will have these disorders as well. In contrast, if they are carriers for genetic disorders seen on the X chromosome and they have daughters, their daughters will also become carriers and may be at risk of developing the disorder if it is dominant or if they inherit another copy of the defective gene from their mothers. One example of a sex-linked genetic disorder is hemophilia, a condition primarily seen in men who inherit a single defective X chromosome and lack a corresponding allele on the Y chromosome to inhibit the expression of the deleterious gene.

What is Genetic Research?

Genetic research is the term used by scientists for the study of how heredity affects living beings. Genes contain all of the information that is needed to create a new being, and they are housed in DNA. Scientists study these genes to determine how diseases are inherited as well as how beneficial traits are passed along. While we often think of genetic research as solely a human pursuit, plants and animals are studied as well.

Perhaps the most widespread use of this research field in plants is the development of seeds that are resistant to herbicides. These widely available seeds are genetically modified so that after they are planted the farmer can spray them with herbicide to kill weeds without damaging the plant. Genetic research made this product readily available as well as affordable for the average farmer.

Genetic research is used when trying to determine treatments for particular diseases. Researchers study various genes that are associated with a condition. They may try to develop a way to switch the gene off, produce a medication that will shut the gene down, or change the process that a gene goes through that leads to the development of a genetic disease.

Genetic research is extremely complicated. Genes are only one factor in many cases of disease. Environmental issues are an important component as well. Some people may carry the gene for a particular genetic disease but never develop any health problems, while others may be stricken aggressively. Doctors are unable to look at different people carrying the same gene and determine who will be affected and who will not.

Determining who will develop a genetic disease is one of the more difficult aspects of genetic research. For instance, a young woman that carries the gene for an aggressive form of breast cancer may want to undergo a double mastectomy in order to reduce her risk of developing breast cancer. While this is understandable, there is no guarantee that she would develop the cancer, and the surgery is very aggressive. In cases like this, genetic research often generates as many questions as it does answers.

Couples that are older or having difficulty getting pregnant or maintaining a pregnancy are often referred to a genetic counselor. The counselor can conduct tests to determine if the potential parents are carrying genes that may increase the likelihood of a child having a genetic disease. While there are definitive tests that can be conduct on the fetus in utero, many people prefer to have this information before they pursue a pregnancy.

Genetic Engineer

An individual who wishes to become a genetic engineer will likely have a strong interest in science. The path to genetic engineering is often fairly specific and includes many years of education. Specific skills may be required, such research skills and knowledge of specialized equipment. Current knowledge of scientific literature in the field is also often necessary to become a genetic engineer.

Genetic engineers are usually research scientists who work with animal, human, or plant genes. These individuals manipulate genes and produce living organisms. They may look for ways to fight or cure diseases. Genetic engineers often specialize in a particular scientific area, such as animal husbandry, and work in that field for their entire careers.

After completing high school, someone who wishes to become a genetic engineer should attend an accredited four year institution. Students should earn undergraduate degrees in a science-related field. Biology, chemistry, zoology, botany, or general science are all relevant options and will prepare the student for advanced study, which is the next step in the process of becoming a genetic engineer.

Generally, the advanced field of study should be chosen before applying to graduate schools. Some options for advanced study include genetics, microbiology, biochemistry, medical science, or biosciences. Courses in immunology and advanced genetics may also be included in the course of study.

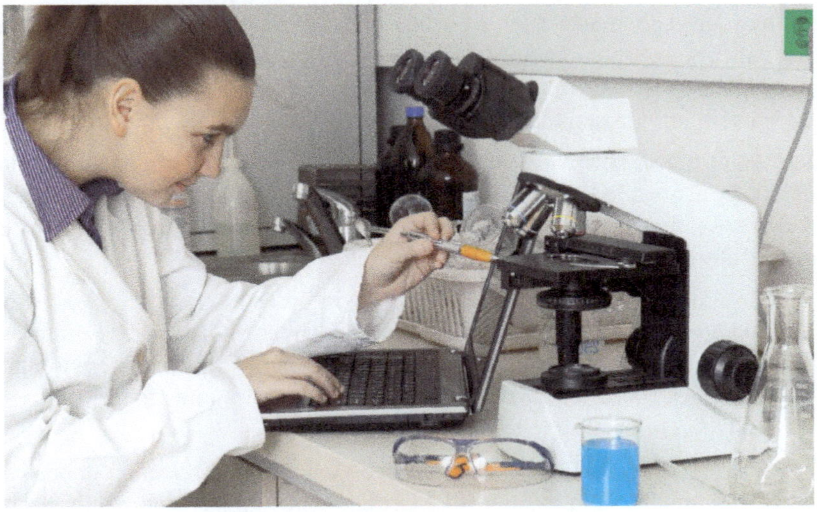

Some students decide to spend an additional three to four years pursuing a doctoral degree in genetic engineering. Many individuals spend much of their doctoral study performing research or assisting in a lab in their specialized field of study. Concurrently, they will also draft and complete a dissertation that details their research or study.

Practical experience is important for individuals who wish to become a genetic engineer; in addition to work completed during the course of doctoral study, students may often benefit from lab courses. Professors or other contacts developed during study can often be good resources when study is complete and full time work is sought. Networking in the field while taking courses may improve the chances of obtaining a genetic engineering job.

In addition to formal education, genetic engineers need to be familiar with certain lab techniques and tools; for example, some work may require the use of scanners and microscopes. Analytical skills are important for interpreting results of lab experiments. Research and writing, in order to keep up-to-date in the field, are also important skills to have in order to become a genetic engineer.

Transgenic Plants

Transgenic plants are those that have been genetically engineered to contain gene sequences that do not naturally occur within their species. These gene sequences can come from plants of a different species, and are introduced in order to try to change some fundamental characteristics of the plant. Some plants that most commonly undergo this process are food crops, which can ideally be made more efficient and more productive by the introduction of new genetic material.

Some of the desirable traits that can be bred into these transgenic plants include resistances to disease and pests, higher yield, higher quality fruits, vegetables, or flowers, and an increased tolerance to weather conditions. Until the invention of inserting new genetic material artificially, plants were bred to accent these characteristics by simply taking the best examples from the same species and crossing them in the hopes of developing the most impressive offspring. This process can be made more efficient with the aid of science.

One of the first steps is determining which genes are to be replaced. Every section of deoxyribonucleic acid (DNA) governs a different part of the plant, whether it is responsible for how many petals are on a flower or how long the cells develop. Genetic experts must determine what gene controls each specific process and then also determine which portion of what plant it will be replaced with.

In their native environments, plants receive new genetic material through the process of pollination. This new information is inserted artificially in processes that can be done in several ways in transgenic plants. Boilistics, a term that combines the words biology and ballistics, is the process by which new DNA is injected directly into the plant cells through the cell walls. This is the favored process when implanting a monocot, or plant with seedlings with only one seed leaf.

When it comes to creating transgenic dicots, the agrobacterium method has met with the most success. In this process, a soil-based species of bacteria called Agrobacterium tumefaciens is used as a carrier. Injected with the new, desired strain of DNA, the bacteria are then introduced to the soil the plant is rooted it. This unique strain of bacteria then invades the plant and uses the plant's own cells to reproduce itself, introducing the new genetic strain.

Creating a successful group of transgenic plants depends on factors such as the ability of the plant to pass its new genetic sequence on to future generations. Once the gene has been successfully introduced and inherited, biologists must continue to study the new plant to make sure that there are no unforeseen complications that arise from the new genetic material. Transgenic plants are also known as genetically modified (GM) plants.

Plant breeding is the purposeful manipulation of plant species in order to create desired genotypes and phenotypes for particular purposes, such as food production, forestry, and horticulture. Breeders have tried to introduce, into crop plants, traits that will increase the quality and yield of the crop; enhance their ability to tolerate harsh environmental conditions including salinity, high temperatures, and water shortage; enhance their ability to resist infection by viruses, fungi and bacteria, and enhance their ability to withstand insect pests and to tolerate herbicides.

Plant breeding is an artificial version of natural evolution, involving artificial selection of desired plant characteristics and artificial generation of genetic variation; it complements other farming innovations (such as introduction of new crops, grafting, changed crop rotations and tillage practices, irrigation, and integrated pest management) for improving crop productivity and land stewardship. Since 1930, crop breeding, in concert with these other innovations, has led to spectacular increases in crop yields especially with cereal grains. Plant breeding has been practiced since the beginning of human civilization, but rapid advances since 1900 have been driven by concepts and techniques developed in the science of genetics. This scientific plant breeding has contributed about 50% of the improvement in crop yield seen in the major cereal crops since 1930.

Plant breeding is practiced by both government institutions and commercial enterprises. International development agencies believe that breeding new crops is important for ensuring food security and for developing sustainable agriculture by developing crops that minimize agriculture's impact on the environment.

Genetics and crop breeding intertwined

Plant breeders created the earliest artificial hybrids between different species (such as the feed cereal Triticale) in the 19th century, and progress in the science of genetics has been intertwined with crop breeding ever since. In 1903, W. L. Johannsen defined the heritable component of biological variability in experiments with a self-pollinated bean. In 1908, US plant breeder G. H. Scull discovered substantial crop yield improvements in interspecies hybrids of maize (now called 'hybrid vigor' or heterosis) with dramatic consequences for the development of a successful commercial corn-seed breeding industry, first in the USA and later elsewhere. Plant geneticists subsequently realized that hybridization can allow fuller exploitation of genetic diversity available in a plant population, and developed widely used interspecies hybrids for pearl millet, sorghum, rice and rapeseed (canola) crops.

Barbara McClintock developed fundamental concepts about chromosome behavior and cytogenetics with maize in the 1930s. Chromosome and genome relationships both within and between crops species have been the conceptual keystones to much successful crop breeding. It is now realized that most crops have undergone duplication of the minimal diploid set of chromosome at some stage, either as ancient duplications (as in the case of maize), or by hybridization between different species (as in allopolyploids such as wheat)

Protecting major crops against disease has rested on alien gene transfer into the crop variety using chromosomal engineering. This technique relies on manipulation of chromosome pairing at meiosis. In 1956, chromosomal engineering enabled Ernest Sears to translocate a segment of goatgrass (Agilops umbellulata Zhuk) chromosome that conferred resistance to leaf rust onto wheat chromosome 6B, and this achievement was followed by numerous successes in transfers of alien genes to crop plants by chromosome manipulation (the so-called wide-crosses; see also wheat.)

Chromosomal engineering is facilitated by using X-ray radiation to promote chromosome breakage, plant tissue culture, and mutants such as Ph-1 of wheat that relax normal strict pairing of homologous chromosomes in meiosis. It can be valuable to artificially promote doubling of chromosome number using colchicine in inter-species and inter-generic cross hybrid lines as a step to enable transfer of alien chromosome segments into crop germplasm.

The expansion of biotechnology after 1980 led to many advances in genetics which are accelerating the usually slow process of plant breeding and leading to important discoveries that improve plant breeding, such as molecular markers, plant genomics, RNA interference, and numerous gene characterizations such as the genes affecting abiotic stress responses and pest and pathogen tolerance.

Strategies for scientific plant breeding

The starting point for plant breeding is the identification of novel, useful traits in populations of parental organisms. The use of chemically or radiation induced mutation can also exploited to increase the range of useful genetic variability, and plant breeding exploits genetic recombination to generate novel combinations of traits. The end-result of lengthy selective breeding programs using these methods is "elite" varieties of high performing germplasm for major crops that are used widely in broad-acre farming.

Plant breeding strategies used by the International Maize and Wheat Improvement Center (CIMMYT) encapulate the way in which modern plant breeding should be pursued. CIMMYT wheat breeding strategies aim to achieve an optimal combination of the best genotypes, in the right environments, under appropriate crop management, and appropriate to the needs of the people who must implement and manage them.

Useful traits

Techniques for increasing the available heritable variation in the initial population include introduction of new germplasm from distant geographical regions or from seed-bank collections, cross-pollination, either within the species, or between related species and genera - including wide-crosses using wild relatives of domesticated plants to introduce pest resistant traits needed in domesticated varieties, and creation of mutations by irradiation or chemical treatment.

'Hope'

In 1930, resistance to stem rust was transferred from T. turgiduminto wheat by producing the variety 'Hope'. In 1979, it was determined that stem rust resistance in 'Hope' was largely controlled by a single gene on the short arm of chromosome 3B. This gene was named Sr2 and assumed to have come from T. turgidum. 'Hope' was used in Mexico in the 1940s as the donor for developing the stem rust resistant wheat cultivar 'Yaqui 48'. Since then, the Sr2 gene has been employed widely in wheat improvement programs and distributed it to many wheat production regions of the world. The gene confers durable, broad-spectrum rust resistance effective against all isolates of P. graminis, and has been described as one of the most important disease resistance genes deployed in modern plant breeding.

The dwarfing trait confers a change in plant stature that provides resistance to lodging and nitrogen fertilizer responsiveness, and exploitation of this trait in wheat and rice was responsible for many of the successes of the Green Revolution, which revolutionized wheat and rice production in Asia, averting mass-famine.

In 1963, scientists discovered the gene opaque-2, which improves the nutritional quality of maize by increasing its lysine and tryptophan content. At first, farmers initially showed little interest in opaque-2 maize because of its low yields, chalky-looking grain, and other defects. But thirty years of breeding effort finally led to releases of Quality Protein Maize (QPM) varieties, which are now widely used in regions where maize is a staple food.

Fusarium head blight is a devastating disease of durum wheat, and a resistant gene to fungus that causes head blight has been transferred into durum wheat germplasm from wheatgrass Lophopyrum elongatum by chromosome engineering.

Classical plant breeding

One major technique of plant breeding is selection, the process of selectively propagating plants with desirable characteristics and eliminating or "culling" those with less desirable characteristics.

Another technique is the deliberate interbreeding (crossing) of closely or distantly related individuals to produce new crop varieties or lines with desirable properties. Plants are crossbred to introduce traits/genes from one variety or line into a new genetic background. For example, a mildew-resistant pea may be crossed with a high-yielding but susceptible pea, the goal of the cross being to introduce mildew resistance without losing the high-yield characteristics. Progeny from the cross would then be crossed with the high-yielding parent to ensure that the progeny were most like the high-yielding parent, (backcrossing). The progeny from that cross would then be tested for yield (selection, as described above) and mildew resistance and high-yielding resistant plants would be further developed. Plants may also be crossed with themselves to produce inbred varieties for breeding. Pollinators may be excluded through the use of pollination bags.

Classical breeding relies largely on homologous recombination between chromosomes to generate genetic diversity. The classical plant breeder may also make use of a number of in vitro techniques such as protoplast fusion, embryo rescue or mutagenesis (see below) to generate diversity and produce hybrid plants that would not exist in nature.

Traits that breeders have tried to incorporate into crop plants include:

- Improved quality, such as increased nutrition, improved flavor, or greater beauty
- Increased yield of the crop
- Increased tolerance of environmental pressures (salinity, extreme temperature, drought)
- Resistance to viruses, fungi and bacteria
- Increased tolerance to insect pests
- Increased tolerance of herbicides
- Longer storage period for the harvested crop

Theories of Evolution

Evolutionary theory has come a long way since Charles Darwin published On the Origin of Species in 1859. Prior to its publication, and in direct competition, Jean-Bapiste Lamarck was promulgating a well-known incorrect variant of evolutionary theory: Lamarckism or Lamarckian evolution. Lamarck thought that changes to the body over the course of a lifetime could be passed down to ancestors as long as they were common to both sexes of the species.

For instance, his explanation for why a giraffe has a long neck is because the giraffe had a history of predecessors that tried harder and harder to reach tall leaves on a tree, and accordingly the offspring were born with larger necks. This theory of evolution has been demonstrated to be false - the genetic difference between parents and offspring can be explained entirely by their genes, and requires no reference to the daily lives of the parents.

The most important theory of evolution is known as the modern synthesis or neo-Darwinian synthesis, which is the combination of Darwinian natural selection with Mendelian genetics and Thomas Morgan's chromosome theory of inheritance. This theory continues to be predominant to this day, with a few minor differences.

Mendel, a German monk who lived around the same time as Darwin but whose work was not rediscovered until 1900, experimented extensively with plants and found out the concepts of alleles (different versions of the same gene), dominant and recessive traits, and that the expression of different traits are exclusive to one another. Darwinian evolution and natural selection is just the idea that life evolves based on random mutations and selective pressures on individuals trying to reproduce.

In the 1960s, it became necessary to introduce the "gene-centric" view of natural selection, not really an independent theory but more of a footnote on the modern synthesis. Around this time, some biologists mistakenly were thinking that individuals evolve traits for the good of their species or the group.

To correct this, it was necessary to put forth a gene-centric view of evolution, where organisms as viewed as the "survival machines" of genes, only constructed for the sole purpose of passing those genes on. For example, a gene can opt for an organism with a shorter lifespan and faster reproductive cycle if that's what evolution selects for. The genes do not care about the well-being of their shells, the organism. This is probably the most sophisticated and recent iteration in the ongoing evolution of evolutionary theory.

CHAPTER 1 CHROMOSOME SEGREGATION IN PLANT MEIOSIS

Linda Zamariola, Choon Lin Tiang, Nico De Storme, Wojtek Pawlowski, and Danny Geelen

Keywords: Chromosome segregation, cohesion, kinetochore, meiosis, plant, recombination, spindle, synapsis

INTRODUCTION

Meiosis is a specialized cell division that generates four haploid daughter cells from a diploid parent cell after a single round of DNA replication and two consecutive rounds of nuclear division. In the first nuclear division, homologous chromosomes segregate (reductional cell division), and in the second one, sister chromatids segregate (equational cell division). As such, each daughter cell carries half the amount of the parental genetic material. The accurate segregation of chromosomes during meiosis is essential for the formation of haploid gametes. Failure in the proper execution of chromosome segregation inevitably leads to the formation of imbalanced gametes and aneuploid or polyploid progeny. In plants, aneuploidy is more tolerated than in animals and viable aneuploid plants have been observed, especially among the progeny of triploid individuals (Henry et al., 2005,2010). Despite being affected in growth and reproduction (Birchler et al., 2001), aneuploids may have an evolutionary role, serving as a bridge to euploid polyploid plant formation through repeated generations of selfing (Ramsey and Schemske, 1998; Henry et al., 2005). Polyploid plants generated through aneuploids or by the polyploidization events of somatic doubling and unreduced gametes, are considered as a prominent driving force in plant genome evolution (Ramsey and Schemske, 2002; Adams and Wendel, 2005; Comai, 2005; Otto, 2007).

To ensure the correct completion of the meiotic cell division program, a sequence of coordinated steps must take place during the two phases of meiosis. In meiosis I, homologous chromosomes must pair and synapse and physically exchange genetic material through recombination. The resulting points of crossing-over, also termed chiasmata, form links between the two homologs in the bivalent configuration and ensure proper positioning of the bivalent relative to the division spindle and balanced segregation of homologs in anaphase I. Additionally, to achieve this, sister kinetochores from each homolog must attach to microtubules emanating from the same spindle pole, a process called monopolar kinetochore attachment, and cohesion must be lost in a stepwise manner. More specifically, at anaphase I, cohesion is released at chromosome arms but not at sister centromeres, allowing homologs to segregate without affecting the physical connection between both sister chromatids. In meiosis II, chromosome segregation in the two resulting haploid interphase nuclei occurs in an equational manner and hence strongly resembles the dynamics of a mitotic cell division. Cohesion at centromeres is retained until anaphase II to ensure bipolar attachment of sister kinetochores to microtubules and equational segregation of chromatids into four haploid daughter cells. Progression through the meiotic cell division is regulated at determined checkpoints by the activity of CDK (Cyclin-Dependent Kinase) - cyclin complexes and the Anaphase Promoting Complex/Cyclosome (APC/C) (Harper et al., 2002; Cooper and Strich, 2011). In particular, the Spindle Assembly Checkpoint (SAC) acts during the transition between metaphase and anaphase of the two meiotic cell divisions to ensure correct kinetochore-microtubule attachments and faithful chromosome segregation (Malmanche et al., 2006; Yamamoto et al., 2008). In plants, checkpoints appear to be less stringent compared to yeast and animals, since completion of meiosis is achieved in several meiotic mutants creating imbalanced gametes (Wijnker and Schnittger, 2013).

Most of the knowledge on the molecular biology of mitotic and meiotic chromosome segregation comes from studies in yeast (reviewed in Marston, 2014). However, the

Citation: Linda Zamariola, Choon Lin Tiang, Nico De Storme, Wojtek Pawlowski and Danny Geelen, Chromosome segregation in plant meiosis, dx.doi.org/10.3389/fpls.2014.00279.

mechanisms of chromosome segregation are conserved in eukaryotes, including plants (Dawe, 1998; Bhatt et al., 2001). In the last decade, the increasing availability of genomic tools and the development of*Arabidopsis thaliana*, but also maize (*Zea mays*) and rice (*Oryza sativa*), as model systems, have led to the identification of a large number of conserved meiotic genes (Mercier and Grelon, 2008). Phenotypic and cytogenetic analyses of the corresponding mutants, have unraveled the function of several molecular factors required for proper chromosome segregation in plants (Bhatt et al., 2001;Ma, 2006). Therefore, the focus of this review will be on the major cellular processes that take place to ensure accurate chromosome segregation in plant meiosis and the related genes that have been yet identified in *Arabidopsis*, maize and rice. Other factors having an effect on chromosome segregation in plant meiosis, such as environmental stresses and changes in ploidy level have been described in recent reviews (Comai, 2005; Madlung and Wendel, 2013; De Storme and Geelen, 2014). After mentioning the importance of homologous chromosome pairing and recombination, two subjects extensively discussed in other reviews (Hamant et al., 2006; Edlinger and Schlögelhofer, 2011; Osman et al., 2011; Tiang et al., 2012; Da Ines et al., 2014), we describe the relevance of cohesion, focusing on the roles of the cohesin complex and on the cohesion dynamics (e.g., loading, release and protection) during meiotic cell division. Next, we discuss the role of centromeric and kinetochore proteins in establishing proper spindle attachment during meiosis I and II, and additionally describe what is currently known on the checkpoint control mechanisms acting at kinetochores. Finally, we report the molecular mechanisms underlying microtubule organization and we focus on the relevance of spindle orientation in plant meiosis.

HOMOLOGOUS PAIRING AND RECOMBINATION AS A BASIS FOR REDUCTIONAL CELL DIVISION IN MEIOSIS I

Homologous Chromosome Pairing and Synapsis

To ensure accurate segregation, chromosomes must first recognize their homologous partners and pair with them during early meiotic prophase I. This process leads to the formation of bivalents, which ensures correct bipolar attachment of homologous centromeres to the division spindle at metaphase I in a way that each of the chromosomes in the bivalent moves to a different pole at anaphase I. Bivalent formation is also required for proper positioning of chromosomes at the metaphase plate. Consequently, mutants with chromosome pairing problems exhibit chromosome segregation defects (Bozza and Pawlowski, 2008).

It is assumed that chromosome homology recognition is based on their DNA sequence. Although mechanisms that bring homologous chromosomes together have yet to be fully elucidated, studies in a variety of species, including plants, have shown that chromosome pairing is strongly dependent on their dynamics in early meiotic prophase as well as the initiation and progression through the early stages of the recombination pathway. Chromosome dynamics in prophase I is largely controlled by the behavior of telomeres, blocks of highly conserved repetitive DNA sequence at the ends of chromosomes (Siderakis and Tarsounas, 2007). Telomeres attach to the nuclear envelope before the onset of chromosome pairing, and gather on a small region, forming a unique structure that resembles a flower bouquet, the so called telomere bouquet (Bass et al., 2000; Golubovskaya et al., 2002; Harper et al., 2004; Richards et al., 2012). The bouquet arrangement has been observed in most eukaryotes (Klutstein and Cooper, 2014). The exact role of the bouquet is still being debated. However, mutants defective in bouquet formation are frequently also defective in chromosome pairing, which implies a role of the bouquet in this process (Harper et al., 2004;Klutstein and Cooper, 2014). One example of such mutant is *plural abnormality of meiosis 1*(*pam1*) in maize, which exhibits significant reduction in homologous pairing (Golubovskaya et al., 2002). In this mutant, telomeres attach to the nuclear envelope but fail to cluster. The bouquet formation has been, therefore, suggested to promote homologous paring by bringing chromosome ends together (Harper et al., 2004).

Alternative chromosome interaction mechanisms have been described in several species, including*Caenorhabditis elegans* and *Arabidopsis* (Armstrong et al., 2001; Phillips and Dernburg, 2006). In*C. elegans*, telomeres do not form the bouquet but pairing centers, short chromosome segments recognized by specific zinc-finger proteins, that attach to the nuclear envelope during early prophase I, also bringing homologous chromosomes together (Phillips and Dernburg, 2006). In*Arabidopsis*, telomeres cluster in meiotic interphase on the nucleolus rather than the nuclear envelope (Armstrong et al., 2001). Subtelomeric regions of *Arabidopsis* chromosomes start to pair before telomeres dissociate from the nucleolus, suggesting that the clustering on the nucleolus may play a role similar to that of the canonical bouquet. *Arabidopsis* telomeres establish their connections with the nuclear envelope during leptotene and zygotene, although without an obvious bouquet formation (Armstrong et al., 2001).

Interestingly, the connections used to attach chromosomes to the nuclear envelope in *C. elegans*and *Arabidopsis* are homologs of the same transmembrane proteins that are used in other species to tether telomeres to the nuclear envelope during bouquet formation. SUN domain proteins, identified in yeast, mammals, *C. elegans*, maize, as well as *Arabidopsis*, cross the inner nuclear membrane (Chikashige et al., 2007; Schmitt et al., 2007; Penkner et al., 2009; Sato et al., 2009;Graumann et al., 2010; Murphy et al., 2010). They interact at their N-termini with telomere binding proteins while their C-termini bind transmembrane proteins containing a conserved KASH domain that cross the outer membrane and interact with the cytoskeleton (Miki et al., 2004; Zhou et al., 2012). The commonality of the structures attaching telomeres to the nuclear envelope reinforces the notion that the telomere-nuclear membrane attachments in *C. elegans* and*Arabidopsis* may be functionally similar to the presence of the canonical bouquet.

It has been shown in several species that the cytoskeleton acts through the telomere-nuclear membrane attachments to induce dynamic motility of chromosomes (Bhalla and Dernburg, 2008;Koszul et al., 2009; Sheehan and Pawlowski, 2009; Woglar and Jantsch, 2013). The chromosome movements are thought to help the chromosomes to engage in finding their pairing partners as well as resolving their entanglements.

Another process, which is required for proper chromosome segregation, and closely follows chromosome pairing, is synapsis. Synapsis is installation of a proteinaceous structure, the synaptonemal complex (SC), between the paired homologous chromosomes, which stabilizes the pairing interactions. The SC consists of two lateral elements (LEs) which reside at the base of the chromosome loops and are held together in parallel by transverse filament proteins. In most eukaryotes, the LEs are derived from the axial elements (AEs) loaded on the chromosomal axis before synapsis. Installation of the synaptonemal complex is also closely linked with the formation of crossovers (see the following section), and so synapsis also affects chromosome segregation through its role in crossover formation. *Arabidopsis* mutants defective in synaptonemal complex formation exhibit univalents at metaphase I and improper chromosome segregation at anaphase I (Ross et al., 1997; Higgins et al., 2005).

Meiotic Recombination

Meiotic recombination affects segregation of chromosomes in at least two ways. First, studies in many species, including plants, mammals, and fungi, have indicated that homologous chromosome pairing is closely connected to meiotic recombination (Pawlowski and Cande, 2005). Second, crossovers, reciprocal chromosome segment exchanges formed as a result of meiotic recombination, form physical connections, known as chiasmata, between homologous chromosomes in each bivalents. Chiasmata keep bivalents together to ensure proper orientation and segregation of chromosomes during the first meiotic division.

Recombination in meiosis is initiated by the formation of double strand breaks (DSBs) in chromosomal DNA, triggered by Spo11, a conserved topoisomerase type-II-like protein (Keeney et al., 1997). The MRN complex (MRE11/RAD50/NBS1) then resects the breaks creating single-stranded DNA overhangs (Borde, 2007), which then invade appropriate regions on the homologous chromosomes. This process is promoted by two recombination proteins, Rad51 and Dmc1 (Masson and West, 2001). Rad51 is solely responsible for the repair of DNA breaks using sister chromatids as templates. However, this process is restrained and replaced by repair via the homologous chromosome when Dmc1 is localized to meiotic DNA break sites together with Rad51 (Bishop et al., 1992; Niu et al., 2009). In *Arabidopsis*, mutating *Rad51* results in chromosome fragmentation (Li et al., 2004). However, fragmentation is not observed in the *dmc1* mutant (Couteau et al., 1999). These observations suggest that the function of Dmc1 is distinct from Rad51, as Dmc1 promotes interhomolog recombination rather than intersister recombination (Kurzbauer et al., 2012; Pradillo et al., 2012).

Meiotic recombination results in formation of crossovers and non-crossovers (which include gene conversions). The number and location of crossovers are tightly regulated. In most plant species, only one to four crossovers are formed per bivalent (Crismani and Mercier, 2012). At least one crossover must be formed per bivalent to ensure correct chromosome segregation at anaphase I. However, the number of crossovers per chromosome is limited by crossover interference, a mechanism that prevents formation of crossovers next to each other (Jones, 1984). A group of proteins called ZMM, which contains Zip1, Zip2, Zip3, Zip4, Msh4, Msh5, and Mer3, have been identified as essential for the formation of interference-dependent crossovers in yeast (Börner et al., 2004). Homologs of several of these proteins have been studied in *Arabidopsis* and found to play similar roles in crossover formation (Higgins et al., 2004, 2005, 2008; Chen et al., 2005;Mercier et al., 2005; Chelysheva et al., 2007). Loss of MSH4 in *Arabidopsis*, results in a reduction in crossover frequency to 15% of the wild-type level (Higgins et al., 2004). Similar effect was shown in the *Arabidopsis mer3* mutant (Chen et al., 2005; Mercier et al., 2005). Interestingly, the ZMM group includes proteins that are primary components of the synaptonemal complex, such as ZIP1. This interdependence indicates a link between crossover formation and synapsis. Overall, about 85% of *Arabidopsis* crossovers arise from the interference-dependent pathway (Higgins et al., 2004). The remaining crossovers are interference-independent, and are generated by a distinct group of proteins including MUS81 and EME1/MMS4 (Berchowitz et al., 2007).

Recombination events, including crossovers are not distributed randomly along chromosomes. Instead they tend to appear at certain chromosomal locations known as recombination hotspots (Drouaud et al., 2006). In plant species with large genomes, such as maize, barley, or wheat, crossovers are predominantly present in chromosome regions close to the telomeres (Akhunov et al., 2003; Gore et al., 2009). Crossover distribution affects the positions of chiasmata and may have implications for bivalent stability and chromosome segregation. However, neither mechanisms that control crossover distribution nor implications of crossover distribution for chromosome behavior in meiosis are well understood.

Early Defects in Chromatin Structure have an Impact on Homologous Chromosome Segregation: ASK1

ASK1 (*Arabidopsis* SKP1-like1) encodes one of the 21 predicted *Arabidopsis* homologs of the yeast and human Skp1 proteins (Yang et al., 1999; Zhao et al., 2003a,b). Skp proteins are an essential component of the Skp1-Cullin-F-box (SCF) complex, that belongs to a class of E3 ubiquitin ligases that target a variety of proteins for ubiquitin-mediated degradation via the 26S proteasome pathway (Petroski and Deshaies, 2005). ASK1 is the Skp homolog that has been best characterized in *Arabidopsis*. *ask1-1* mutants display defects in plant growth, flower development and male fertility (Yang et al., 1999; Zhao et al., 2001, 2003b). Male sterility arises from meiotic defects in prophase I that lead to erroneous homologous chromosome segregation in meiosis I and sister chromatid segregation in meiosis II, and to the subsequent formation of unbalanced spores. During prophase I, chromosomes maintain a leptotene-like structure with long and thin threads that do not synapse, as demonstrated by the absence of the typical SC structure (Wang et al., 2004). FISH experiments using a centromeric probe showed the presence of more than 5 signals in *ask1-1*meiocytes during pachytene, confirming lack of homologous pairing and bivalents formation (Zhao et al., 2006). The localization of the α-kleisin subunit of the cohesin complex SYN1 (described in the next paragraph) was also found to be altered in *ask1* meiocytes from zygotene to anaphase I. These observations together with a premature sister chromatid detachment detected by FISH in anaphase I, suggest that *ask1* mutation alters cohesin distribution and function, which is necessary for proper pairing and synapsis (Zhao et al., 2006). The abnormalities detected in *ask1* seem to derive from early defects in meiotic chromatin structure and chromosome reorganization in leptotene that cause a prolonged attachment of chromosomes to the nuclear membrane and the nucleolus, alterations in rDNA structure, prolonged attachment of the telomeres to the nucleolus, and defects in histone 3 acetylation, overall leading to the absence of homologous chromosome pairing (Yang et al., 2006). Hence, ASK1 is most likely required for chromosome conformation and remodeling of meiotic chromosomes by controlling the release of chromatin from the nucleolus and nuclear membrane starting from leptotene (Yang et al., 2006). Several hypotheses have been currently proposed to explain the potential role of ASK1 in meiosis, consistent with the meiotic defects observed in the mutant and the homology of ASK1 to Skp proteins (Yang et al., 2006; Zhao et al., 2006). ASK1 may control the degradation of a protein which inhibits the leptotene to zygotene transition, so that the alterations observed in chromatin structure and organization would be a consequence of the block of this transition. Alternatively, ASK1 might regulate the interaction of chromosomes to the nuclear membrane by degrading one or more proteins that link chromatin to the nuclear matrix, thus allowing a nuclear reorganization during leptotene and zygotene. ASK1 may also control chromatin structure by regulating chromatin remodeling proteins, as suggested by the alterations detected in histone 3 acetylation. However, the specific function of ASK1 in male meiosis is not yet defined.

SISTER CHROMATID COHESION IS ESSENTIAL FOR FAITHFUL CHROMOSOME SEGREGATION

The Cohesin Complex

Sister chromatids must be held together from the moment of their synthesis in S-phase until their separation in anaphase to ensure correct attachment of chromosomes to the spindle and accurate chromosome segregation in dividing cells. Cohesin is the multi-subunit protein complex that mediates sister chromatid cohesion in meiosis and mitosis by physically trapping them in a tripartite ring structure (Haering et al., 2008). The complex is highly conserved in eukaryotes and is composed of a core of four evolutionary conserved proteins, extensively studied in yeast and animals. In mitosis, the cohesin complex is composed of two members of the SMC family (structural maintenance of chromosomes), SMC1 and SMC3, and two auxiliary SCC subunits (sister chromatid cohesion), the α-kleisin RAD21/SCC1 and SCC3. In meiosis, the structure of the cohesin complex is highly similar, except for the RAD21/SSC1 component, which is replaced by its counterpart Rec8 (Stoop-Myer and Amon, 1999; Watanabe and Nurse, 1999). SMC1 and SMC3 consist, in their folded configuration, of a globular head and a hinge domain, connected by a long anti-parallel coiled coil.

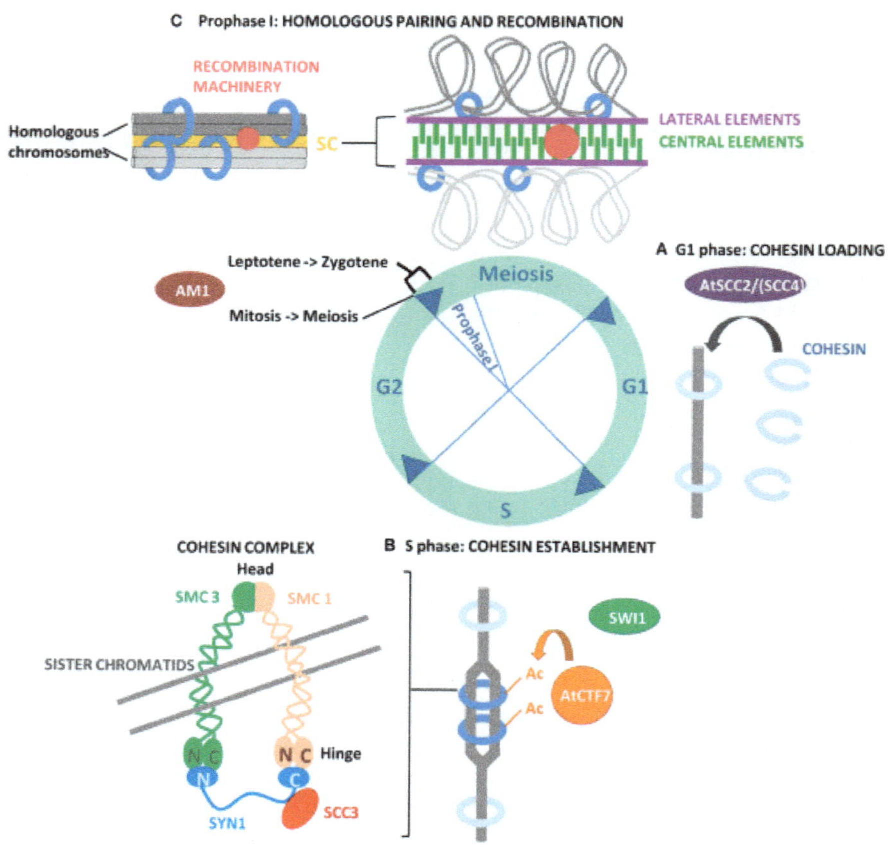

Figure 1: Overview of the events that allow the establishment of the cohesin complex on chromosomes, necessary for the successive steps of chromosome segregation, including homologous pairing and recombination in meiosis I. (A) Loading of cohesin on chromosomes requires the SCC2/SCC4 complex, only AtSCC2 has been characterized in *Arabidopsis*. **(B)** The establishment of chromosome cohesion takes place during DNA replication in S phase when Eco1/CTF7 acetylates SMC3 residues, effectively closing the cohesin ring. AtCTF7 has acetylatransferase activity *in vitro* and is required for the establishment of chromosome cohesion in *Arabidopsis*. SWI1 is an*Arabidopsis* protein with a role

in cohesin establishment even if the mechanism of action is not yet known. Its maize homolog AM1 is required for the switch from mitosis to meiosis and for a putative checkpoint between leptotene and zygotene in prophase I. **(C)** Cohesion is required for SC elongation and polymerization and for meiotic recombination in prophase I.

The proposed model of action of cohesin, the embrace model, requires the connection of the SMC hinge domains to form a SMC1/SMC3 heterodimer with a V-shaped structure, that can bind across sister chromatids and close, forming a ring, through a physical connection of the α-kleisin subunit to the C-terminal domain of SMC1 and the N-terminal domain of SMC3 (Gruber et al., 2003). The complex is stabilized by recruitment of SCC3 by the α-kleisin subunit (Figure 1) (for reviews on cohesin complex: Nasmyth and Haering, 2005; Onn et al., 2008;Peters et al., 2008).

SMC PROTEINS

Similar as in yeast and animals, the sister chromatid connection in plants is also established through the cohesin complex. Homologs of the cohesin complex have been identified in some plant species and major progress on the understanding of their function has been achieved in the model plant Arabidopsis thaliana, in which all the components have been described. The Arabidopsis genome contains single copies of SMC1 and SMC3 cohesin subunits. Genetic studies revealed that loss of AtSMC1 or AtSMC3 functionality causes seedling lethality, hence impairing functional characterization (Liu et al., 2002). Localization studies using a specific antibody revealed that AtSMC3 is present in the cytoplasm and nucleus, on chromosomes and in the nuclear matrix of meiotic and mitotic cells, indicating a function in both types of cell divisions (Lam et al., 2005). At meiotic prophase, AtSMC3 localizes along sister chromatids to axial elements and lateral elements, similar to the Arabidopsis α-kleisin subunit SYN1. This observation confirms the conserved role of the cohesin complex in sister chromatid cohesion but also supports an additional function in SC formation, as proposed in yeast and mammals (Klein et al., 1999; Eijpe et al., 2000). By metaphase I, AtSMC3 localizes only to chromosome centromeres and, in addition, co-localizes to the spindle at metaphase I and anaphase I and II. The spindle localization is independent of SYN1 functionality and suggests that AtSMC3 might play an additional role as spindle associated protein, distinct from its conserved role in sister chromatid cohesion. This novel localization pattern is also conserved in mitosis and could be related to a role of AtSMC3 in spindle assembly and/or in the chromosome association with the spindle (Lam et al., 2005). A similar novel function has been suggested in human mitosis for the entire cohesin complex (Gregson et al., 2001).

Immunolocalization studies in tomato meiocytes (Solanum lycopersicum) revealed that SMC1 and SMC3 show a similar localization pattern as AtSMC3. In prophase I, SMC1 and SMC3 antibodies display a signal along AEs of the SC from leptotene to diplotene and a weak and diffuse signal on chromosomes at metaphase I and telophase II (Lhuissier et al., 2007). However, no localization to the spindle was documented, suggesting that the novel spindle function might be specific for Arabidopsis AtSMC3 and not conserved in other plant species.

REC8 AND SCC3

More intensive studies have been undertaken on the role of the meiotic α-kleisin subunit Rec8 in Arabidopsis (named SYN1 but also DIF1 and AtRec8), maize (AFD1), and rice (OsRad21-4/OsRec8) meiosis. In Arabidopsis, the homolog of Rec8, SYN1, is required for sister chromatid cohesion in meiosis (Cai et al., 2003; Chelysheva et al., 2005). SYN1 fully co-localizes with AtSCC3 at pachytene and is necessary for its proper loading on sister chromatids, confirming that they are indeed part of a complex (Chelysheva et al., 2005). FISH (fluorescence *in situ* hybridization) analysis on syn1 meiocytes using chromosome arm and centromeric probes show defective sister chromatid arm and centromere cohesion in meiosis I, confirming that SYN1 functions in cohesion (Cai et al., 2003).

In addition, SYN1 is required for synapsis of homologous chromosomes, being necessary for SC polymerization and elongation (Chelysheva et al., 2005). In syn1 meiocytes, synapsis is blocked and chromosome condensation and pairing are almost completely absent, leading to the presence of univalents at metaphase I (Bai et al., 1999; Bhatt et al., 1999). Localization of ASY1, a protein required for chromosome synapsis, recombination and SC assembly and widely used as a marker for chromosome axes in meiosis (Sanchez-Moran et al., 2008), is impaired in syn1 mutants, confirming the requirement of SYN1 for AE polymerization and elongation but not for their formation (Chelysheva et al., 2005). Synapsis is known to be closely related to meiotic recombination. Therefore, it is not surprising that SYN1 plays also a role in recombination, specifically in DSBs repair. Indeed, chromatin bridges and chromosome fragmentation are observed in syn1 meiosis I. They are suppressed by introducing into the syn1 mutant background the Atspo11 mutation, which abolishes DSBs formation and prevents recombination, confirming that SYN1 is required for DSBs repair (Chelysheva et al., 2005). Involvement of the cohesin complex in homologous chromosome pairing, assembly of the SC, and in meiotic recombination has been shown previously in other organisms. In yeast, Rec8 and SMC3 are required for SC formation and for repairing DSBs (Klein et al., 1999). In mouse, loss of Rec8 affects homologous recombination but does not affect SC formation and assembly. However, synapsis occurs between sister chromatids instead of homologous chromosomes, suggesting that Rec8 might define the chromosome unit and limit the SC binding sites to one single chromosome surface of a sister-chromatid pair in mammals (Xu et al., 2005).

Support for an additional role of Rec8 in homologous pairing and recombination in plants comes from studies on the maize α-kleisin subunit AFD1 and the rice OsRad21-4/OsRec8. A study on different afd1 alleles has revealed that AFD1 is required for AE installation, affecting the deposition of the recombination machinery on chromosomes (Golubovskaya et al., 2006). The rice OsRec8 regulates AE formation and may have a role in DNA DSBs repair, since localization of PAIR2 (homolog of Arabidopsis ASY1), ZEP1 (ZYP1 homolog), and MER3, involved in the formation of crossovers, is affected in Osrec8 mutants. As a consequence, no proper homologous pairing occurs (Zhang et al., 2006; Shao et al., 2011). Moreover, defective telomere bouquet formation is observed in Osrec8 and afd1 mutants, also preventing proper pairing of homologous chromosomes. Hence, OsRec8 regulates

AE formation, homologous recombination and synapsis by affecting downstream proteins PAIR2, ZEP1, and MER3 (Shao et al., 2011).

Rec8 has a crucial role in the determination of kinetochore geometry for monopolar orientation in fission yeast, since rec8 mutants display loss of monopolar orientation at meiosis I and chromosome segregation defects (Yokobayashi et al., 2003; Sakuno et al., 2009). Similarly, Arabidopsis SYN1 is necessary for the monopolar attachment of sister kinetochores in meiosis I, as indicated by the observation of bipolar sister kinetochore attachment in meiosis I in the double syn1 Atspo11 mutant, in which syn1 chromosome fragmentation is suppressed allowing a clearer observation of chromosome segregation. However, the same defect in kinetochore orientation is observed for the other SCC cohesin subunit, AtSCC3, indicating that SYN1 is not sufficient for monopolar kinetochore orientation or, most likely, is inactive when the other members of the complex are not present. These data suggests that Rec8-containing cohesin complex is responsible for defining kinetochore geometry in meiosis I in plants, as proposed in yeast, Drosophila and mammals (Chelysheva et al., 2005; Watanabe, 2012).

AtSCC3 is the sole SCC3 homolog investigated in plants so far. It is required for normal plant growth and fertility and has a conserved role in proper sister chromatid cohesion, confirmed by the combination of univalents and bivalents observed in Atscc3 mutants (Chelysheva et al., 2005). However, in contrast to SYN1, AtSCC3 is not required for AE formation, since ASY1 localization in Atscc3 is normal and synapsis does not show major defects in the mutant. Moreover, only a low level of fragmentation is observed in Atscc3 and recombination is not defective, suggesting that the two SCC subunits, although being part of the same complex, may fulfill different additional functions (Chelysheva et al., 2005).

While AtSCC3 has no paralogs in the Arabidopsis genome, three α-kleisin homologs, SYN2, SYN3, and SYN4, are present that share about 38 % sequence similarity at their N-termini and 20 % at their C-termini with SYN1, and could partially compensate for each other (Schubert et al., 2009a). Two observations raise the hypothesis that the α-kleisin paralogs may be involved in cohesion in meiosis. First, in the syn1 Atspo11 double mutant, sister chromatid cohesion is only lost at anaphase I, suggesting that other homologs of the SYN1 family might be responsible for cohesion before that stage (Chelysheva et al., 2005). Second, SYN1 localization is only observed along chromosome axes but not at the core centromeres at metaphase I and metaphase II (Chelysheva et al., 2005; Cromer et al., 2013; Zamariola et al., 2013). It is known that SYN1, SYN2 and SYN4 may partially compensate for each other whereas SYN3 is required for plant viability, it localizes to the nucleolus and might have evolved a role in rDNA transcription and/or processing. A specific function in DNA repair in somatic cells has been suggested for SYN2, while SYN4 is required for centromere cohesion in mitosis (Schubert et al., 2009b). However, the role of the different paralogs is, at this time, not clear, and the creation of double or triple mutants might help unravelling their specific functions (Schubert et al., 2009b).

Loading and Establishment of Chromosome Cohesion

The loading of the cohesin complex onto chromosomes starts at telophase in humans and at the end of G1 in yeast and requires the evolutionary conserved SCC2/SCC4 complex (for reviews seeUhlmann, 2009; Ocampo-Hafalla and Uhlmann, 2011). Cohesin loading has been shown to be enriched at centromeric and pericentromeric regions promoting high fidelity chromosome segregation (Eckert et al., 2007). Recent studies in budding yeast have revealed that the observed enrichment is defined by the presence of the kinetochore subcomplex Ctf19, that promotes SCC2/SCC4 centromere association (Fernius et al., 2013). Also in Angiosperms, interphase nuclei show a preferred alignment of sister chromatids at centromeres, which might facilitate kinetochore bipolar orientation in mitosis, essential for correct chromosome segregation (Schubert et al., 2007). In plants, only the *Arabidopsis* homolog of the adherin SCC2 has been described. AtSCC2 is essential for plant viability and *Atscc2* plants show defects in embryogenesis and endosperm development (Schubert et al., 2009b; Sebastian et al., 2009). Using an inducible RNAi (RNA interference) system, Sebastian et al. (2009) demonstrated that AtSCC2 is required for sister chromatid cohesion and loading of the cohesin complex in meiosis, as indicated by defects in AtSCC3 localization. Furthermore, *Atscc2* mutants show an irregular localization of ASY1 and chromosome fragmentation, indicating that AtSCC2 is required for axial development and most likely for repair of DNA DSBs, supporting the notion that sister chromatid cohesion is a prerequisite for axial development and DSBs resolution (Sebastian et al., 2009).

The loading of cohesin is the first step through the establishment of sister chromatid cohesion that takes place during DNA replication. After the loading, cohesin is unstable due to the activity of the Wapl-Pds5 complex that promotes cohesin dissociation (Rowland et al., 2009). In yeast, cohesion is established during S-phase by the Eco1/CTF7 protein, that acetylates the SMC3 residues, effectively closing the cohesin ring (Rowland et al., 2009). In *Arabidopsis*, AtCTF7 exhibits acetyltranferase activity *in vitro* like its yeast and human homologs (Jiang et al., 2010). *Atctf7*homozygous mutants display a dwarf phenotype and aberrant microsporogenesis due to defects in chromosome segregation in mitosis and PMCs (pollen mother cells). FISH performed with a centromeric and a chromosome 4 arm probes on male meiocytes of *Atctf7* and AtCTF7 RNAi plants, revealed that the protein is required for both centromere and arm cohesion in meiosis (Bolaños-Villegas et al., 2013; Singh et al., 2013). Furthermore, localization of the cohesin complex subunits AtSMC3, AtSYN1 and AtSCC3 is impaired in *Atctf7* male meiosis, indicating that AtCTF7 is necessary for association of cohesin on chromatin in meiosis (Bolaños-Villegas et al., 2013; Singh et al., 2013). In addition, the level of expression of genes required for DNA repair is significantly altered in *Atctf7* mitotic and meiotic tissues, and the mutant plants show a lower ability to repair DNA double strand breaks *in vivo* in mitotic cells (Bolaños-Villegas et al., 2013). Taken together, these observations suggest that AtCTF7 is also required for DNA repair in *Arabidopsis*, as shown for Eco1 in yeast mitosis (Lu et al., 2010).

SWITCH1/DYAD (SWI1/DYAD) is an *Arabidopsis* protein with an essential role in the establishment of sister chromatid cohesion during early meiosis (Mercier et al., 2001, 2003). Different allelic mutations have been investigated for the *SWI1/DYAD* gene, all of them showing an impact on fertility due to different mechanisms affecting megasporogenesis (*swi1-1* and *dyad*;Motamayor et al., 2000; Siddiqi et al., 2000; Mercier et al., 2001; Agashe et al., 2002) or both mega and microsporogenesis (*swi1-2* and *dsy10*; Mercier et al., 2003; Boateng et al., 2008). *Swi1-1*and *swi1-2* alleles have been shown to have an effect on the female mitosis-meiosis switch, so that meiosis is converted into a mitotic cell division (Motamayor et al., 2000; Mercier et al., 2001). However, analysis of the *dyad* allele by Agashe et al. (2002) and Siddiqi et al. (2000) with a meiotic marker, provided evidence that the female megaspore enters the meiotic programme but does not progress into further meiotic divisions. Detailed studies of male meiosis for *swi1-2* and *dsy10*alleles, have shown that the mutants loose cohesion in a stepwise manner already in meiosis I, leading to the presence of 20 chromatids at metaphase I which segregate randomly in meiosis II, forming polyads (Mercier et al., 2001). Furthermore, the mutant lacks AE formation, leading to incorrect pairing and synapsis, and does not initiate recombination. These defects probably all derive from defective establishment of cohesion before the initiation of meiosis, since the protein is expressed exclusively in meiotic G1 and S phase (Mercier et al., 2003). Specifically, the localization of SYN1 in *swi1-2* meiocytes, indicates that SWI1 performs its function after the loading of the cohesin complex (Mercier et al., 2003). However, its specific function in chromosome cohesion is not yet understood.

Maize AM1 and rice OsAM1 are proteins closely related to SWI1. Mutants in *AM1* and *OsAM1* genes show defective sister chromatid cohesion, absence of homologous pairing and synapsis, and lack of homologous recombination (Pawlowski et al., 2009; Che et al., 2011). However, while *Arabidopsis swi1* mutants affect meiotic processes downstream of meiotic initiation and do not affect entrance in meiosis, maize *am1* mutants show typical features of mitotic division in the early steps of meiosis, indicating that AM1 is required for the transition from the mitotic cell cycle into meiosis. Meiocytes of a specific *am1* allele arrest during early meiotic prophase at the transition between leptotene and zygotene, suggesting the presence of a novel checkpoint in maize required for progression through prezygotene (Pawlowski et al., 2009). Similarly, in rice, OsAM1 is also likely involved in a checkpoint mechanism that regulates the transition from leptotene to zygotene (Che et al., 2011).

A schematic overview of the processes of cohesin loading and establishment and homologous chromosome pairing and recombination, is shown in Figure 1.

Release of Chromosome Cohesion: Separase

Cleavage of the α-klesin subunit occurs in a stepwise manner during meiosis. In meiosis I, Rec8 is cleaved at chromosome arms, allowing the resolution of chiasmata and homologous chromosome segregation in meiosis I, whereas in meiosis II cohesin is released at centromeres, enabling sister chromatid separation (Nasmyth, 2001). Cleavage of Rec8 is performed by the cysteine protease Separase, which is conserved in various organisms, including yeast and vertebrates (Kitajima et al., 2003; Kudo et al., 2009). Separase function is inhibited by a protein called Securin, which is degraded at the onset of anaphase by ubiquitylation by the APC/C (Uhlmann, 2001). Homologs of separase are present in many plant species. However, the studies undertaken so far have only focused on the *Arabidopsis* separase AESP (Liu and Makaroff, 2006). AESP is an essential gene but RNA interference of AESP under the control of the meiotic DMC1 promoter, and the finding of the temperature permissive mutant *rsw4* (radially swollen 4), have allowed to investigate AESP function in meiosis (Liu and Makaroff, 2006; Wu et al., 2010; Yang et al., 2011). *Aesp* and *rsw4*mutants display defective chromosome segregation in meiosis I, in which entangled chromosomes and chromosome fragments are observed, and in meiosis II, where bivalents are still present, indicating persistence of cohesion (Liu and Makaroff, 2006; Yang et al., 2011). In support of this, SYN1 and SMC3 signals persist on *aesp* and *rsw4* chromosomes at later stages after metaphase I, demonstrating that AESP is responsible for removal of the cohesin complex from chromosomes. The creation of a double mutant between *aesp* and *ask1*, in which homologous chromosomes prematurely separate in meiosis I due to defects in homologous synapsis, showed that sister chromatids did not separate in meiosis II. This observation confirms that AESP is responsible for sister chromatid separation also in anaphase II (Yang et al., 2009). In *Arabidopsis*, a large amount of cohesin is released from chromosome arms in prophase I and the residual arm cohesin is released at anaphase I (Cai et al., 2003). While AESP is required for the release of cohesin at anaphase I and in meiosis II, it does not participate in the first step of release in prophase I, suggesting that a separase-independent mechanism might exist at early stages in *Arabidopsis*, similar to budding yeast, in which the condensin complex SMC2/SMC4 and a Polo kinase are responsible for cohesin removal at chromosome arms before metaphase I (Sumara et al., 2002; Yu and Koshland, 2005; Liu and Makaroff, 2006).

Separase is a multifunctional protein that in various organisms possesses additional roles to sister chromatid separation mechanistically less understood, such as proteolytic cleavage of other target proteins in yeast and spindle assembly in humans (Moschou and Bozhkov, 2012). Also in*Arabidopsis* additional functions of separase have been reported (Yang et al., 2009, 2011). *Aesp*mutants show alterations in non-homologous centromere associations at zygotene, suggesting that AESP might play a role in the control/release of the transient centromere associations that occur during zygotene in *Arabidopsis* (Armstrong et al., 2001). Furthermore, in *aesp* male meiocytes the radial microtubule array (RMA) is disturbed at telophase II and phragmoplast-like structures are observed, suggesting that AESP might have a function in microtubule organization or cell polarity (Yang et al., 2009). Absence of AESP also causes the formation of multinucleate microspores as a consequence of defective RMA (Yang et al., 2009). In contrast to yeast, where separase is required for normal meiotic spindle formation (Jensen et al., 2001; Baskerville et al., 2008), in *Arabidopsis*only RMA formation is defective while AESP might be required for the proper interaction of microtubules with the nuclear envelope at the tetrad stage (Yang et al., 2009).

Protection of Centromere Cohesion: Shugoshin and Patronus

In meiosis, sister chromatid cohesion is controlled in a time- and space-dependent manner, with chromosome arm cohesion release at the start of anaphase I, and maintenance of centromeric cohesion up till anaphase II. Meiosis-specific protection of Rec8 at pericentromeric regions from anaphase I to anaphase II is performed by Shugoshin (Sgo), a protein first described in *Drosophila*(MEI-S332; Kerrebrock et al., 1995), and successively identified in yeast, mammals and plants (Yao and Dai, 2012). Studies from yeast and vertebrates have elucidated the mechanism of action of Sgo, which is recruited at pericentromeric heterochromatin regions where it associates with the phosphatase PP2A to dephosphorylate Rec8 and prevent its cleavage in meiosis I (Lee et al., 2008a;Xu et al., 2009). In yeast, Sgo1 localizes at centromeres until the end of anaphase I (Kitajima et al., 2004), whereas in vertebrates SGOL2 persists on the chromosomes also in meiosis II (Lee et al., 2008a). Currently, two hypotheses are postulated to explain the dynamic association of Shugoshin with centromeres. On the one hand, Sgo function may be controlled by microtubule attachment and deactivated by a spatial change of its localization in the peri-centromeric domain in response to a change in microtubule tension (Lee et al., 2008a). Alternatively, a PP2A inhibitor may block dephosphorylation thereby conferring loss of protection of centromeric cohesion in meiosis II (Chambon et al., 2013). Flies and budding yeast possess a single copy of Sgo, while fission yeast, mammal and plant genomes have two Sgo paralogs, Sgo1 and Sgo2. In *Drosophila*, yeasts and plants, Sgo1 is responsible for the protection of centromere-specific sister chromatid cohesion in meiosis I, while in mammals SGOL2 performs the function of protector (Gutiérrez-Caballero et al., 2012). Though they are homologs, Sgo genes share limited sequence similarity and display in the different organisms somewhat different functions which have been acquired during evolution (for a recent review on the Shugoshin protein family and the additional roles of Shugoshin see Clift and Marston, 2011; Gutiérrez-Caballero et al., 2012). The *Sgo1* paralog *Sgo2* possesses different properties depending on the species examined. In fission yeast, Sgo2 plays a role in chromosome segregation in mitosis (Kitajima et al., 2004), in particular it has been shown to control the localization of the CPC, a protein complex that senses lack of tension between kinetochores and microtubules (Kawashima et al., 2007; Vanoosthuyse et al., 2007; Tsukahara et al., 2010). In addition, fission yeast Sgo2 also plays a role in meiosis, as *Sgo2* deletion leads to a modest increase in non-disjunction of homologs at meiosis I (Kitajima et al., 2004). In humans, hSGOL1 protects centromeric cohesion in mitosis (Salic et al., 2004; McGuinness et al., 2005), whereas hSGOL2 is dispensable for sister chromatid cohesion in mitotic cell division but is essential for correcting erroneous kinetochore attachments by recruiting the microtubule depolymerase MCAK to the centromeres (Huang et al., 2007), a role that is consistent with the one shown for fission yeast Sgo2 (Kawashima et al., 2007).

In plants, the role of Sgo as protector of centromere cohesion in meiosis has been described for the maize ZmSGO1, the rice OsSGO1 as well AtSGO1 and AtSGO2 of *Arabidopsis* (Hamant et al., 2005;Wang et al., 2011; Cromer et al., 2013; Zamariola et al., 2013, 2014). FISH analysis performed on*sgo1* meiocytes with a centromeric probe revealed a premature detachment of sister chromatid centromeres in anaphase I, resulting in random chromosome segregation in meiosis II. However, monopolar orientation of sister kinetochores in meiosis I is not affected in the mutants and chromosomes normally segregate in the reductional division, indicating that SGO proteins are required for protection of cohesion at anaphase I but not for monopolar orientation of sister kinetochores. In fission yeast and mammals, which possess two Sgo homologs, one copy is generally required for protection of sister chromatid cohesion in meiosis, while the other has evolved additional roles, as previously mentioned. So far, no function in somatic cells has been described for any of the plant Sgo proteins. *Arabidopsis* is the only species in which the role of both Sgo paralogs has been investigated. Single mutants show no vegetative phenotype and a meiotic phenotype is detected exclusively for *Atsgo1*. However, *Atsgo1 Atsgo2* double mutants reveal a partially redundant role for the two SGOs, opposite to yeast and vertebrate (Cromer et al., 2013;Zamariola et al., 2014). Immunolocalization of ZmSGO1 and OsSGO1 has revealed that SGO1 is loaded on chromosomes at leptotene, earlier than in other organisms such as yeast or mammals in which loading occurs during late prophase I or at diplotene, respectively (Kitajima et al., 2004;Gómez et al., 2007). Thus, plant SGO proteins might have a function in prophase I. In support of this hypothesis, ZEP1 localization is defective in *Ossgo1* mutants in about 21% of meiocytes, indicating that OsSGO1 may be required for the timely assembly of the SC, even if not for its initial assembly (Wang et al., 2011). In contrast, *Arabidopsis* ZYP1 localizes normally in *Atsgo1* mutants (Zamariola et al., 2013).

Recently, a novel protein involved in the protection of sister chromatid cohesion during meiosis II has been identified in *Arabidopsis*, named PATRONUS (PANS1) (Cromer et al., 2013; Zamariola et al., 2014). PANS1 is a plant specific protein that shares homology with genes belonging to the Eudicots family. *Pans1* meiocytes show a premature release of sister chromatid cohesion at metaphase II but not at meiosis I, indicating that the protein is required for protection of cohesion during interkinesis, at a later stage than SGOs. Moreover, similar to SGOs, PANS1 is not required for monopolar attachment of sister kinetochores in meiosis I. TAP-TAG and Y2H experiments have revealed that PANS1 may be a regulator of the APC/C complex because of the interaction with some of the APC/C subunits. In addition, the presence of two destruction boxes in the PANS1 sequence may indicate that PANS1 is at the same time also targeted by the APC/C complex. Currently, three hypotheses have been suggested to explain how PANS1 maintains sister chromatid cohesion at interkinesis: (1) by protecting SGOs from destruction by the APC/C; (2) by protecting sister chromatid cohesion from Separase independently of SGOs, in the case SGOs are no longer present after anaphase I, and (3) by inhibiting via APC/C regulation the Wapl-dependent process of cohesin release, which is usually activated at the end of mitosis/G1 phase to allow dynamic cohesin renewal and that could be present also at the end of meiotic telophase I (Cromer et al., 2013). At the moment, AtSGOs and PANS1 localization, that could help unraveling the function of PANS1 in meiosis and the relation among the protectors, is lacking. Besides its role as protector of cohesion, PANS1 has also been shown to be required for spindle organization in meiosis since *pans1*meiocytes display defective spindles starting from telophase I. Defective spindles is probably the cause of the formation of an aberrant internuclear organelle band at interkinesis, detected in 7% of*pans1* meiocytes. Taken together, these phenotypes and the premature separation of sister chromatids observed in meiosis II, suggest a function of PANS1 in ensuring the coordinate organization of the cell organelles in accordance with the meiotic cell cycle phase and chromosome cohesion (Zamariola et al., 2014), which is in agreement with the interaction of PANS1 with the APC/C.

CENTROMERES AND KINETOCHORES

Role of Centromeres and Kinetochores in Chromosome Segregation

Centromeres are DNA-protein structures necessary to direct chromosome movement in cell division. Centromere DNA sequences are fast evolving and highly variable among species. However, centromeric regions in most plant species encompass mainly two domains. One is the core centromere, which contains satellite tandem repeats, usually 150–180 bp long, and specialized nucleosomes in which histone H3 is replaced by a centromere-specific H3 histone variant, CENH3. This region is required for the assembly of the kinetochore, a protein structure that binds to spindle microtubules allowing faithful chromosome segregation. The core centromere is flanked by pericentromeric heterochromatin domains containing retroelements and other transposons. In yeast the pericentromeric domains have been shown to have mainly a role in the recruitment of Shugoshin (Pidoux and Allshire, 2005; Yamagishi et al., 2008). In addition, epigenetic mechanisms may be involved in the specification of centromeric chromatin and propagation of centromeres (Houben and Schubert, 2003; Ekwall, 2007; Torras-Llort et al., 2009; Wang et al., 2009).

The specific centromeric variant Histone 3, CENH3, was first identified in human as CENP-A and subsequently in all eukaryotic model systems (De Rop et al., 2012), including *Arabidopsis* (also called HTR12; Talbert, 2002). Despite its essential and conserved role in ensuring proper chromosome segregation, CENH3 proteins are highly variable in their sequences and fast evolving, especially their N-terminal tail domain and a loop 1 region at the C-terminal domain, which are necessary for CENH3 localization to centromeres in *Arabidopsis* (Ravi et al., 2010; Moraes et al., 2011). The C-terminal part of the protein is sufficient for the centromeric localization of CENH3 in mitotic cells even when the N-terminal part is absent (Lermontova et al., 2006). In meiosis, a different loading mechanism for CENH3 is present, in which the N-terminal tail plays a critical role. *Arabidopsis* plants transformed with a N-terminally truncated YFP-CENH3(C) protein show meiotic defects and partial sterility and the YFP signal cannot be detected in meiotic nuclei (Lermontova et al., 2011). Similarly, the replacement of the N-terminal tail with a GFP tagged variant, GFP-tailswap, causes sterility due to defects during sporogenesis (Ravi et al., 2011). In GFP-tailswap plants, meiosis is disturbed starting from metaphase I, in which bivalents align on the division plate but are not subjected to tension from the spindle, which is confirmed by decreased interkinetochore distance and by defective spindles (Ravi et al., 2011). CENH3 protein signal is reduced or not detected in GFP-tailswap meiocytes and is again detected after meiosis on mitotic chromosomes at the microspore stage, indicating the existence of distinct mechanisms for CENH3 loading in meiosis and mitosis (Ravi et al., 2011). The work of Lermontova et al. (2011) also suggests a different loading mechanism in meiosis and mitosis, since the YFP-CENH3(C) variant is deposited to the centromeres in mitosis but not in meiotic nuclei.

Recently, the *Arabidopsis* homolog of KNL2 has been identified. It represents one of the components of the Mis18 complex, responsible for the initiation of CENH3 deposition at the centromeres in humans (Hayashi et al., 2004), *C. elegans* (De Rop et al., 2012) and fission yeast (Hayashi et al., 2004). In *Arabidopsis*, KNL2 is associated with centromeres at all stages of the cell cycle except from metaphase to mid-anaphase. *Arabidopsis* KNL2 knockout mutants show defects in mitosis and meiosis and reduced CENH3 loading at the centromeres (Lermontova et al., 2013). Furthermore, CENH3 gene expression is decreased in *knl2* mutants but KNL2 expression is stable in CENH3 RNAi transformants, indicating that KNL2 acts upstream of CENH3 and has a function in the assembly of CENH3 at the centromeres (Lermontova et al., 2013). Moreover, KNL2 is co-expressed with H3K9 histone methyltransferases genes, whose expression is reduced in *knl2*mutants. Also DNA methylation levels are lower in *knl2* mutant plants. The requirement of KNL2 for CENH3 expression and for DNA methylation, suggests that KNL2 may interact with methyltransferases to allow the maintenance of DNA methylation, in order to control the epigenetic status of centromeric chromatin and to control CENH3 loading (Lermontova et al., 2013).

Sister kinetochores must behave differently in meiosis I and II: in meiosis I are oriented toward the same pole (mono-orientation) to allow homologous chromosomes segregation, while in meiosis II they face opposite poles (bi-orientation) (Brar and Amon, 2009). The tension exerted at the kinetochores by microtubules during division, and the kinetochore geometry, defined in meiosis and mitosis by cohesion, are fundamental for stabilizing the monopolar attachment in MI and the bipolar in MII (for review see Watanabe, 2012). In contrast to the high variability of centromeric sequences, more than 20 active kinetochore proteins are conserved between humans and yeasts (Lampert and Westermann, 2011), and are specific either for the inner kinetochore, where they directly recognize and bind DNA, or for the outer kinetochore, being responsible for the interaction with microtubules (Santaguida and Musacchio, 2009; Wang et al., 2009). However, to date, only 7 kinetochore proteins have been reported to be conserved in *A. thaliana*, the majority of which has not been yet functionally characterized (Murata, 2013). The inability to identify homologs of many human and yeast kinetochore proteins in plants, may suggest the existence of different kinetochore structure in plants (Murata, 2013).

Kinetochore functionality depends on the presence of a functional centromere in meiosis. Indeed MIS-12, a kinetochore protein which co-localizes with CENH3 at the centromere regions (Sato et al., 2005), does not do it in *Arabidopsis* GFP-tailswap meiocytes. In contrast, CENP-C, another kinetochore protein which localizes at the centromeres in mitotic cells (Ogura et al., 2004), is not affected in CENH3 RNAi transformants, suggesting that its localization does not depend on the presence of a functional CENH3 (Lermontova et al., 2011).

In maize, kinetochore proteins have been more thoroughly investigated. CENPC is part of the inner kinetochore and interacts at one side with the DNA repeats located at the centromeric regions, and, on the other side, with the members of the outer kinetochore (Dawe et al., 1999; Zhong et al., 2002). At the outer kinetochore NCD80 and MIS12 are present. Homologs of these two proteins are known to be parts of the KMN (KNL-1/Mis12/Ndc80) complex that constitutes the core microtubule-binding site of the kinetochore in *C. elegans* (Cheeseman et al., 2006). NDC80 is a constitutive kinetochore protein which localizes at kinetochores in all meiotic and mitotic stages (Du and Dawe, 2007). It does not bind DNA directly and interacts with MIS12, which is also present at kinetochores during all stages of the cell cycle. NCD80 and MIS12 form at metaphase I a bridge structure that links sister kinetochores, while CENH3 and CENPC appear at the inner side of sister kinetochores as two distinct signals (Li and Dawe, 2009). MIS12 has an important role in sister chromatid connection at meiosis I and is required for the initiation of reductional division (Li and Dawe, 2009). Knock-down of MIS12 by RNAi leads to a weakening of the MIS12-NCD80 bridge and aberrant chromosome segregation in meiosis I, where in 30% of the cells sister kinetochores separate and segregate in an equational division instead of reductional (Li and Dawe, 2009). In MIS12 RNAi cells, the signal of the centromere protector ZmSGO1 does not weaken (Li and Dawe, 2009). The protein lies between sister kinetochores but cannot restore kinetochore co-orientation, confirming that Shugoshin is not required for the monopolar orientation of kinetochores (Hamant et al., 2005; Li and Dawe, 2009). A model, in which axial elements and cohesin hold sister chromatids together during prophase I and create the base for fused sister kinetochore formation promoted by the MIS12-NCD80 bridge has been proposed (Li and Dawe, 2009). This structure would cooperate with Shugoshin to induce reductional segregation by co-orienting sister kinetochores (Figure 2). MIS12 and NCD80 are thought to be similar to the monopolin complex, which promotes sister kinetochore co-orientation in budding yeast (Corbett and Harrison, 2012).

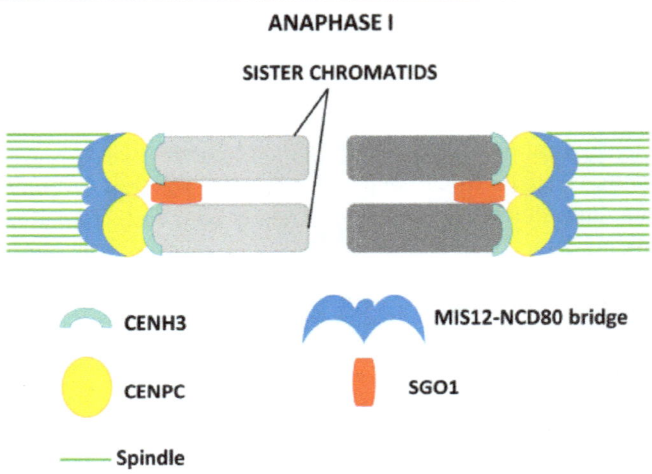

Figure 2: Model proposed in maize by Li and Dawe (2009) for reductional segregation in meiosis I. Sister kinetochores are fused in meiosis I by formation of the MIS12-NCD80 bridge that, together with SGO1, allows monopolar attachment of sister chromatids to the spindle pole. The inner kinetochore proteins CENPC and CENH3 are visualized as two distinct signals.

Checkpoint Mechanisms Acting at Kinetochores

In eukaryotes, checkpoint mechanisms are present in meiosis and mitosis to prevent chromosome mis-segregation that would result in aneuploidy or apoptosis (Murray, 1994). The SAC is a conserved protein complex that controls proper attachment of microtubules to kinetochores in the metaphase to anaphase transition. In case of lacking or improper kinetochore-microtubule attachment, SAC creates a “wait anaphase” signal that stops anaphase progression. This response is promoted by APC/C together with its co-activator Cdc20 protein (for reviews see Peters, 2006;Vader et al., 2008; Musacchio, 2011). When all kinetochores are properly attached to microtubules, APC/C targets the destruction of Securin, the inhibitor of Separase as well as other cyclins, promoting chromosome segregation and exit from meiosis or mitosis. Evolutionary conserved proteins of SAC are MAD1, MAD2 (mitotic arrest deficient), Bub1, Bub3 (budding unhibited by benomyl), BubR1 kinase (bub-related1, Mad3 in yeast), and Mps1 (Monopolar kinase1) (May and Hardwick, 2006). The SAC proteins BubR1, Bub3, and MAD2 are also members of the Mitotic Checkpoint Complex (MCC), which is the effector of SAC that physically inhibits APC/C by binding to its co-activator Cdc20 until the moment all chromosomes are properly attached to kinetochores (Sudakin et al., 2001). SAC function has been investigated in depth in mitosis. However, a similar control mechanism is active also during meiosis (Malmanche et al., 2006; Sun and Kim, 2012).

Homologs of SAC proteins have been described in plants. MAD2 was first identified in maize where it localizes to the outer kinetochore in prometaphase I and II of meiosis, next to the inner kinetochore protein CENPC (Yu et al., 1999). During meiosis I and II, microtubule attachment is not sufficient for MAD2 dissociation from kinetochores, and the dissociation might occur in response to tension applied to the kinetochores (Yu et al., 1999). This hypothesis is supported by the concomitant staining of the MAD2 and 3F3/2 antibodies in maize meiosis.

3F3/2 recognizes a kinetochore phosphoepitope that is known to disappear in animal cells when tension is applied to the kinetochore (Nicklas et al., 1995). MAD2 homologs have been identified also in wheat and*Arabidopsis*, and their roles have been mainly investigated in mitotic checkpoint control. In wheat, intense MAD2 signal was observed at all centromeres in colchicine treated cells but not in untreated cells, confirming the function of MAD2 in the spindle checkpoint (Kimbara et al., 2004). In *Arabidopsis*, MAD2 localization was studied together with BubR1 and Bub3.1, the others SAC proteins identified in the model plant (Caillaud et al., 2009). Interactions between the three proteins were observed in the nuclei of tobacco cells using bimolecular fluorescence complementation (Caillaud et al., 2009). During normal mitosis, localization of the SAC proteins to the kinetochores was not detected. However, by application of microtubule destabilizing drugs or of the proteasome inhibitor MG132, MAD2, BubR1, and Bub3.1 localized at the kinetochores, suggesting that SAC proteins are only recruited at kinetochores in case of defective spindle assembly in *Arabidopsis* (Caillaud et al., 2009). In contrast, a study by Ding et al. (2012) on the*Arabidopsis* MAD2 protein, showed that MAD2-GFP localizes at kinetochores also during normal mitotic progression from prophase to metaphase, as shown in maize. Moreover, AtMAD2 binds to AtMAD1, which interacts with the nucleoporin NUA, showing that SAC components interact with the nuclear pore. This interaction has been found in several other organisms, and it seems that the presence of SAC proteins at the nuclear pore mediates mitotic spindle checkpoint (Lee et al., 2008b).

Mps1 (Monopolar kinase 1) is also required for SAC function in the mitotic checkpoint in several eukaryotes and has been shown to be responsible for the recruitment of Mad1 and Mad2 at kinetochores in humans (Hewitt et al., 2010). Mps1 *Arabidopsis* homolog has conserved motifs which could mediate its interaction with MAD2 but also with cyclins, the APC/C and MAPK (mitogen-activated protein kinases), however, proof of its biological role in the checkpoint mechanism is still required (De Oliveira et al., 2012).

In most organisms, SAC is controlled by the chromosome passenger complex (CPC). In general, the CPC consists of the core enzyme Aurora B kinase, and three non-enzymatic subunits that control the targeting, enzymatic activity and stability of Aurora B: inner centromeric protein (INCENP), borealin and survivin (for review see Ruchaud et al., 2007). The major role of CPC is sensing incorrect kinetochore-microtubule attachments and generating, in response, unattached kinetochores, which allows new rounds of attachment until the correct configuration is obtained. The presence of unattached kinetochores activates the SAC that blocks the progression of cell divisions until all chromosomes are under tension. In plants, little is known about the role of the CPC in meiosis and few components of the complex have been identified. Like animals,*Arabidopsis* possesses three Aurora kinase homologs, which share a similar structure to the ones of other species (Kawabe et al., 2005). AtAurora1 and AtAurora2 display similar localization dynamics to Aurora B kinase in *Arabidopsis* mitosis, suggesting that they could function as chromosomal passenger proteins (Demidov et al., 2005). AtAurora1 interacts with SAC proteins BubR1 and MAD2 *in vivo* and phosphorylate them *in vitro*, which suggests that it functions in checkpoint mechanisms (Demidov D., personal communication). Furthermore, deregulation of AtAurora kinases activity, either by mutagenesis or by chemical treatment, results in defects in microsporogenesis and generation of polyploid and aneuploid progeny, suggesting that AtAurora may regulate correct chromosome segregation in*Arabidopsis* meiosis (Demidov D., personal communication).

A putative ortholog of the CPC subunit INCENP, WYR, has been identified in *Arabidopsis*(Kirioukhova et al., 2011). WYR shares with the INCENP homolog proteins a characteristic C-terminal domain, a coiled coil domain and a IN-box at the C-terminus, required for the binding of Aurora kinase. WYR is an essential gene with a role in cell cycle control and, independently, in cell fate and differentiation in *Arabidopsis*, since is required for both female and male gametogenesis. Similar functions have been reported also for the orthologs of INCENP in animals (Ruchaud et al., 2007). However, further genetic and biochemical analyses on WYR and Aurora kinases are required to establish the role of CPC proteins in plants.

MICROTUBULE ORGANIZATION AND SPINDLE DYNAMICS

In all eukaryotic cells, faithful chromosome segregation is accomplished by microtubule-based movement and requires a bipolar structure, the spindle, which consists of an antiparallel array of microtubules. The microtubules have their minus-end anchored at the spindle pole and their plus-end toward the chromosomes (Wittmann et al., 2001). They are highly dynamic polar polymers of noncovalently bound α and β tubulin heterodimers and represent the major components of the cytoskeleton in eukaryotic cells (Nogales, 2000). They rapidly polymerize and depolymerize while being continually translocated toward the poles. In animal and yeast cells, microtubules nucleate from microtubule-organizing centers (MTOC), such as the centrosome and the spindle pole body, which are responsible for the organization of the cortical astral arrays in interphase and mitotic spindles during cell division (Pereira and Schiebel, 1997; Jaspersen and Winey, 2004). γ-tubulin is enriched at the nucleation centers where it is recruited as a ring-shaped complex together with associated proteins, enhancing the nucleation of microtubules (O'Toole et al., 2012). In contrast to animals and yeast, plant microtubules lack conspicuous organizing centers. However, they are organized into ordered arrays that are associated with a growth pattern of the plant cell and relocate in a cell-cycle specific manner (Azimzadeh et al., 2001). During cell division, a succession of microtubule arrays is identified: radial arrays from the nuclear surface and cortical arrays of interphase, preprophase bands, spindles, and phragmoplasts (Wasteneys, 2002; De Storme and Geelen, 2013a). Like in animal and yeast, γ-tubulin is also required for microtubule nucleation in plants, being essential for the organization of the microtubule structures in interphase and cell division (Canaday et al., 2000; Shimamura et al., 2004; Pastuglia et al., 2006).

Microtubule motor proteins have an essential role in spindle assembly in both centrosomal and acentrosomal systems (Walczak et al., 1998). The best studied class of microtubule motor proteins are kinesins, proteins that participate in a variety of biological processes, including transport of vesicles, chromosomes or organelles, and organization of spindle microtubules, and chromosome segregation (Woehlke and Schliwa, 2000). They move unidirectionally along microtubules toward their plus or minus-ends. They use energy derived from ATP hydrolysis, in a processive or non-processive way, depending on their capacity of moving cargo long or only short distances before detaching from the microtubules. Several kinesins are known to be required for the structure, assembly and positioning of the mitotic and meiotic spindles in animals and fungi (Endow, 1999;Sharp et al., 2000). The *Arabidopsis* genome contains 61 predicted kinesins, one-third of them belonging to the kinesin-14 family that includes minus end-directed motor proteins (Reddy and Day, 2001). ATK1 is a member of this family and has been shown to support microtubule movement in an ATP-dependent manner and to be a non-processive, minus-end motor protein (Marcus et al., 2002). ATK1 has a specific role in male meiosis, in which *atk1-1* meiocytes display defective chromosome alignment and segregation in meiosis I and II due to aberrant formation of metaphase and anaphase spindles, leading to spore and pollen abortion and decreased plant fertility (Chen et al., 2002). ATK1 is involved in the assembly of the meiotic spindle and is needed for organizing microtubules at the two poles at metaphase and anaphase I and II, but not for the organization of microtubules for other structures, such as the interzonal microtubule array formed at telophase I (Chen et al., 2002). Studies in yeast and *Drosophila* have suggested that minus and plus-ended motor proteins could produce counteracting forces within the spindle to maintain its structure (Sharp et al., 1999, 2000). Thus, ATK1 might have a similar function in plant male meiosis, by producing inward-acting forces necessary for the assembly and maintenance of a bipolar spindle (Chen et al., 2002). The creation of a double heterozygote mutant between ATK1 and its homolog ATK5 (also named AtKIN14a and AtKIN14b, respectively), has shown that both proteins are required for proper chromosome segregation in female and male meiosis and for normal spindle morphogenesis in male meiosis (Quan et al., 2008). In addition to its male meiotic function, ATK1 localizes to the midzone of the mitotic spindle from metaphase through anaphase, suggesting a function also in the mitotic spindle apparatus (Liu et al., 1996).

AtPRD2/MPS1 (Multi-polar spindle1) is a putative *Arabidopsis* coiled-coil protein with homologs only among Embryophytes. Although having been identified as AtPRD2, an essential protein for DSBs formation, due to the presence of univalents in *Atprd2* mutant meiosis (De Muyt et al., 2009), the protein has also been found to be required for spindle organization and determination of spindle polarity in male meiosis (MPS1; Jiang et al., 2009). *Mps1* meiocytes display multiple focused spindles at metaphase I, indicating that spindle assembly is not defective, in contrast to*atk1* and *atk1/atk5* mutants, but spindle bipolarity is compromised in meiosis I and II, and chromosome segregation results more affected than in the kinesin mutants. This observations suggest that MPS1, ATK1, and ATK5 play a role in different mechanisms in plant meiosis. It has been proposed that MPS1 might guide microtubule minus-end migration in meiosis, maybe through binding to an unknown MAP (microtubule associated proteins) or, alternatively, could be a component of the spindle pole transmitting the signal to attract the minus-end of the spindle microtubules before spindle assembly (Jiang et al., 2009). However, whether the spindle defects observed in *mps1* meiocytes correspond to a primary function of the protein in spindle organization and polarity, or to a secondary effect caused by univalents formation in meiosis I, is not clear since conflicting observations on the relationship between unpaired chromosomes and spindle aberrations have been reported (Chan and Cande, 1998; Dawe, 1998).

In rice, a Kinesin-1-like protein, Pollen Semisterility 1 (PSS1), has been shown to have microtubule-stimulated ATPase activity and to be required for proper chromosome alignment and segregation in meiosis. However, spindle morphology is only slightly affected in *pss1* mutants, indicating that PSS1 might have a minor and not essential role in the formation of the meiotic spindle or alternatively might be involved in the regulation of chromosome movements along the spindles, as suggested by the delayed chromosomes observed in meiosis in *pss1* (Zhou et al., 2011).

Recently, the identification of a MATH-BTB domain protein, MAB1 (MATH-BTB1) in maize has been reported. This protein is required for organizing microtubule spindles and nuclei positioning in meiosis II and in the first mitotic division in both male and female germlines. Since no direct interaction between MAB1 and the spindles has been observed, it has been proposed that MAB1 may act through the control of a spindle apparatus regulator(s) (Juranič et al., 2012). Six MATH-BTB proteins have been currently identified in the *Arabidopsis* genome, however, no similar function has been reported (Weber and Hellmann, 2009).

The correct orientation of spindles in the second meiotic division is an essential requirement for faithful chromosome segregation. Alterations in the orthogonal configuration of the division planes in meiosis II lead to co-orientation of the spindles producing unreduced gametes, that represent the major route to polyploidization in plants (Brownfield and Köhler, 2011). Co-orientation can lead to the formation of three types of MII spindle defects which usually occur together in cells: parallel, tripolar or fused (Conicella et al., 2003; De Storme and Geelen, 2013b). This phenomenon only takes place in PMCs (pollen mother cells) of plants with simultaneous cytokinesis. In this type of cytokinesis, as opposed to the successive type, no cell plate is formed at the end of meiosis I and the two sets of chromosomes stay in the same cytoplasm and need to be perpendicularly oriented to create the tetrahedral configuration observed at the end of meiosis II (De Storme and Geelen, 2013a). They have been documented in many plant species, however, the molecular mechanisms behind their occurrence are still largely unknown. Two proteins involved in spindle orientation specifically in male meiosis II have been identified in *Arabidopsis*: AtPS1 and JASON (D'Erfurth et al., 2008; Erilova et al., 2009; De Storme and Geelen, 2011). Mutations in these genes produce at the end of meiosis II a high number of unreduced gametes (i.e., dyads and triads) instead of normal haploid gametes, leading to diploid pollen formation and triploid offspring. The biological mechanism causing 2n gamete formation in the mutants has been elucidated by tubulin immunostainings, which have shown the formation of parallel, tripolar and fused spindles in meiosis II. The defective spindles lead to 2n spores that retain parental heterozygosity at the centromeres, indicative of a FDR-type (first division restitution) of meiotic restitution (D'Erfurth et al., 2008; De Storme and Geelen, 2011). The introduction of *Atsp1* or *jason* mutations into the*Atspo11* mutant background has confirmed the model of 2n gametes formation through co-oriented spindles, since the unbalanced segregation caused by *Atspo11* at meiosis I is nullified by parallel spindles in meiosis II, leading to the formation of mainly balanced dyads as result of meiosis in the double mutants. *Atps1* and *jason* meiocytes lack the characteristic interzonal microtubule array (IMA) observed in simultaneous PMCs at telophase I, which physically separates the two new formed nuclei. They mostly show fused nuclei at metaphase II. In potato, the absence of IMA has also been proposed to cause alterations in cell polarity and the formation of fused spindles (Conicella et al., 2003), suggesting that also in the *Arabidopsis* mutants depending on the total, partial, or unipolar loss of IMA fused, parallel or tripolar spindles are formed (De Storme and Geelen, 2013b).

AtPS1 is a protein conserved in the plant kingdom (Cigliano et al., 2011), which contains two conserved domains in its structure: an N-terminal Forkhead-associated (FHA) domain required for phosphoprotein interaction in many signaling pathways (Li et al., 2000) and a PINc domain that has RNA-binding properties associated with RNAse activity, and which is generally found in proteins involved in RNAi and in nonsense-mediated mRNA decay (NMRD) (Clissold and Ponting, 2000). JASON encodes a protein of unknown function and no known domains that is conserved in plants (Erilova et al., 2009). Expression analysis have demonstrated that *JASON* controls the *AtPS1*transcript level specifically in meiotic flower buds, suggesting the existence of a regulatory mini-network for the control of spindle orientation in meiosis II (De Storme and Geelen, 2011).

Defects in spindle orientation in the second meiotic division have been also reported in mutants in one of the *Arabidopsis* formins, AFH14 (Li et al., 2010). Formins are a class of proteins known to regulate the microfilament cytoskeleton (Blanchoin and Staiger, 2010), but have been recently shown to have also a prominent role in microtubule regulation and in the crosstalk between actin filaments and microtubules in higher eukaryotes (Bartolini and Gundersen, 2010). Indeed, microtubules and microfilaments have been shown to co-distribute and interact in the meiotic spindle and in the phragmoplast in maize (Staiger and Cande, 1991). AFH14 co-localizes with MTs and MFs arrays during cell division in *Arabidopsis* suspension cells and with MTs in meiotic cells, affecting their arrangement during microsporogenesis. *Afh14* mutants display abnormal MTs structures including defective RMS at telophase I, parallel spindles at metaphase II and the absence of phragmoplast structures at late cytokinesis. AFH14 has been shown to preferentially bind MTs and to link MTs and MFs *in vitro*, thus playing a key role in cytoskeletal dynamics and organization required for cell division, including MII spindle orientation (Li et al., 2010; De Storme and Geelen, 2013b).

An overview of the process of chromosome segregation between metaphase I and anaphase II, and of the molecular factors playing an essential role in *Arabidopsis* chromosome segregation, is displayed in Figure 3.

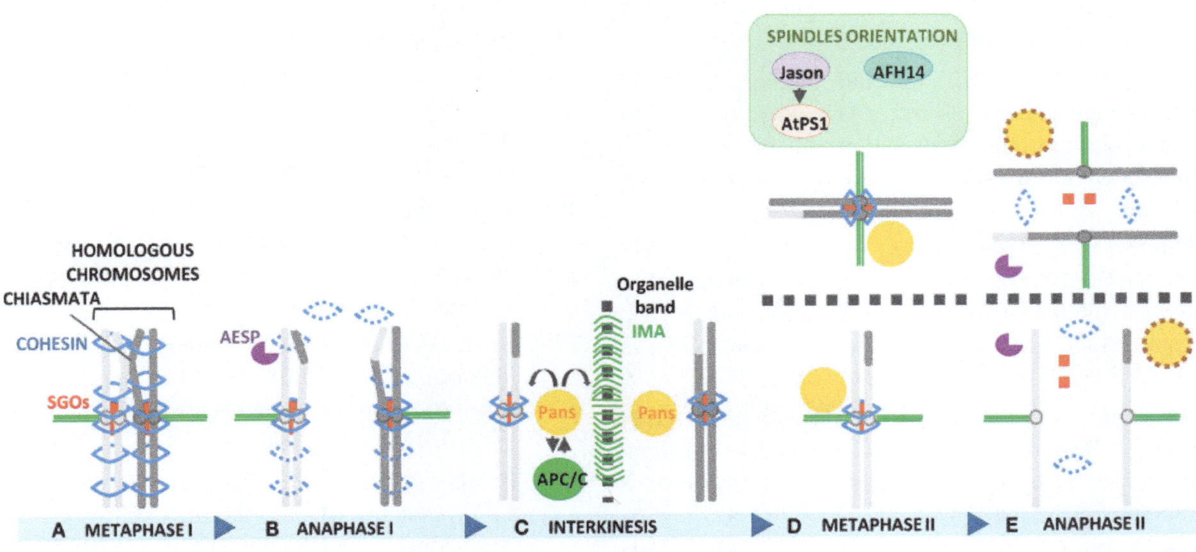

Figure 3: Overview of chromosome segregation in *Arabidopsis* **meiosis. (A)** At metaphase I, homologous chromosomes are connected by chiasmata and SHUGOSHINs (AtSGOs) are present at the centromeres. **(B)** At anaphase I, Separase AESP is activated and cleaves the cohesin at chromosome arms but not at centromeres, allowing resolution of chiasmata and homologous chromosomes segregation by monopolar attachment to the spindles. **(C)** At interkinesis, an internuclear microtubule array (IMA) is formed at the site of the organelle band, to physically separate homologous chromosomes. PANS1 is active and protects centromere cohesin, probably in conjunction with SGOs. PANS1 also interacts with the APC/C, and it is probably also an APC/C target. In addition, PANS1 plays a role in spindle organization from telophase I to telophase II. **(D)** At metaphase II, the chromosomes orient perpendicularly to the metaphase plate through the perpendicular orientation of spindles regulated by Jason and AtPS1. Also the formin AFH14 influences spindle orientation by linking MTs and MFs. **(E)** Releasing or degradation of SGOs and PANS allows cleavage of centromeric cohesin by separase and sister chromatids segregation.

CONCLUSIONS AND PERSPECTIVES

In the past 15 years the identification and characterization of plant meiotic genes has seen a remarkable acceleration due to the forward and reverse genetics strategies used in the model plants*Arabidopsis*, maize and rice. In addition, investigation of the molecular mechanisms regulating meiosis in other kingdoms has enormously contributed to the development of plant research in this field. The coordinate events leading to accurate chromosome segregation have been elucidated in budding yeast and studies in plants have confirmed the conserved role of many proteins in the steps of meiotic chromosome segregation, such as cohesin and the dynamics of cohesin removal and protection, the machinery of homologous pairing and recombination, and the function of kinetochores and microtubules. However, even if the main genes have been identified and their function in chromosome segregation confirmed, not much is known about their regulation in accordance with the cell cycle. Further research should focus on investigating the molecular mechanisms regulating protein functions and the interaction between the proteins to define their role in the broader context of chromosome segregation.

CONFLICT OF INTEREST STATEMENT

The authors declare that the research was conducted in the absence of any commercial or financial relationships that could be construed as a potential conflict of interest.

REFERENCES

1. Adams, K. L., and Wendel, J. F. (2005). Polyploidy and genome evolution in plants. *Curr. Opin. Plant Biol.* 8, 135–141. doi: 10.1016/j.pbi.2005.01.001
2. Agashe, B., Prasad, C. K., and Siddiqi, I. (2002). Identification and analysis of DYAD: a gene required for meiotic chromosome organisation and female meiotic progression in Arabidopsis. *Development* 129, 3935–3943.
3. Akhunov, E. D., Goodyear, A. W., Geng, S., Qi, L.-L., Echalier, B., Gill, B. S., et al. (2003). The organization and rate of evolution of wheat genomes are correlated with recombination rates along chromosome arms. *Genome Res.* 13, 753–763. doi: 10.1101/gr.808603
4. Armstrong, S. J., Franklin, F. C. H., and Jones, G. H. (2001). Nucleolus-associated telomere clustering and pairing precede meiotic chromosome synapsis in Arabidopsis thaliana. *J. Cell Sci.* 114, 4207–4217.
5. Azimzadeh, J., Traas, J., and Pastuglia, M. (2001). Molecular aspects of microtubule dynamics in plants. *Curr. Opin. Plant Biol.* 4, 513–519. doi: 10.1016/S1369-5266(00)00209-0
6. Bai, X., Peirson, B. N., Dong, F., Xue, C., and Makaroff, C. A (1999). Isolation and characterization of SYN1, a RAD21-like gene essential for meiosis in Arabidopsis. *Plant Cell* 11, 417–430. doi: 10.1105/tpc.11.3.417
7. Bartolini, F., and Gundersen, G. G. (2010). Formins and microtubules. *Biochim. Biophys. Acta* 1803, 164–173. doi: 10.1016/j.bbamcr.2009.07.006
8. Baskerville, C., Segal, M., and Reed, S. I. (2008). The protease activity of yeast separase (esp1) is required for anaphase spindle elongation independently of its role in cleavage of cohesin. *Genetics* 178, 2361–2372. doi: 10.1534/genetics.107.085308
9. Bass, H. W., Riera-Lizarazu, O., Ananiev, E. V, Bordoli, S. J., Rines, H. W., Phillips, R. L., et al. (2000). Evidence for the coincident initiation of homolog pairing and synapsis during the telomere-clustering (bouquet) stage of meiotic prophase. *J. Cell Sci.* 113, 1033–1042.
10. Berchowitz, L. E., Francis, K. E., Bey, A. L., and Copenhaver, G. P. (2007). The role of AtMUS81 in interference-insensitive crossovers in *A.* thaliana. *PLoS Genet.* 3:e132. doi: 10.1371/journal.pgen.0030132
11. Bhalla, N., and Dernburg, A. F. (2008). Prelude to a division. *Annu. Rev. Cell Dev. Biol.* 24, 397–424. doi: 10.1146/annurev.cellbio.23.090506.123245
12. Bhatt, A. M., Canales, C., and Dickinson, H. G. (2001). Plant meiosis: the means to 1N. *Trends Plant Sci.* 6, 114–121. doi: 10.1016/S1360-1385(00)01861-6
13. Bhatt, A. M., Lister, C., Page, T., Fransz, P., Findlay, K., Jones, G. H., et al. (1999). The DIF1 gene of Arabidopsis is required for meiotic chromosome segregation and belongs to the REC8/RAD21 cohesin gene family. *Plant J.* 19, 463–472. doi: 10.1046/j.1365-313X.1999.00548.x
14. Birchler, J. A., Bhadra, U., Bhadra, M. P., and Auger, D. L. (2001). Dosage-dependent gene regulation in multicellular eukaryotes: implications for dosage compensation, aneuploid syndromes, and quantitative traits. *Dev. Biol.* 234, 275–288. doi: 10.1006/dbio.2001.0262
15. Bishop, D. K., Park, D., Xu, L., and Kleckner, N. (1992). DMC1: a meiosis-specific yeast homolog of E. coli recA required for recombination, synaptonemal complex formation, and cell cycle progression. *Cell* 69, 439–456. doi: 10.1016/0092-8674(92)90446-J
16. Blanchoin, L., and Staiger, C. J. (2010). Plant formins: diverse isoforms and unique molecular mechanism. *Biochim. Biophys. Acta* 1803, 201–206. doi: 10.1016/j.bbamcr.2008.09.015
17. Boateng, K. A., Yang, X., Dong, F., Owen, H. A., and Makaroff, C. A. (2008). SWI1 is required for meiotic chromosome remodeling events. *Mol. Plant* 1, 620–633. doi: 10.1093/mp/ssn030
18. Bolaños-Villegas, P., Yang, X., Wang, H.-J., Juan, C.-T., Chuang, M.-H., Makaroff, C. A., et al. (2013). Arabidopsis CHROMOSOME TRANSMISSION FIDELITY 7 (AtCTF7/ECO1) is required for DNA repair, mitosis and meiosis. *Plant J.* 75, 927–940. doi: 10.1111/tpj.12261
19. Borde, V. (2007). The multiple roles of the Mre11 complex for meiotic recombination. *Chromosome Res.* 15, 551–563. doi: 10.1007/s10577-007-1147-9
20. Börner, G. V., Kleckner, N., and Hunter, N. (2004). Crossover/noncrossover differentiation, synaptonemal complex formation, and regulatory surveillance at the leptotene/zygotene transition of meiosis. *Cell* 117, 29–45. doi: 10.1016/S0092-8674(04)00292-2
21. Bozza, C. G., and Pawlowski, W. P. (2008). The

cytogenetics of homologous chromosome pairing in meiosis in plants. *Cytogenet. Genome Res.* 120, 313–319. doi: 10.1159/000121080

22. Brar, G. A., and Amon, A. (2009). Emerging roles for centromeres in meiosis I chromosome segregation. *Nat. Rev. Genet.* 9, 899–910. doi: 10.1038/nrg2454
23. Brownfield, L., and Köhler, C. (2011). Unreduced gamete formation in plants: mechanisms and prospects. *J. Exp. Bot.* 62, 1659–1668. doi: 10.1093/jxb/erq371
24. Cai, X., Dong, F., Edelmann, R. E., and Makaroff, C. A. (2003). The Arabidopsis SYN1 cohesin protein is required for sister chromatid arm cohesion and homologous chromosome pairing. *J. Cell Sci.* 116, 2999–3007. doi: 10.1242/jcs.00601
25. Caillaud, M.-C., Paganelli, L., Lecomte, P., Deslandes, L., Quentin, M., Pecrix, Y., et al. (2009). Spindle assembly checkpoint protein dynamics reveal conserved and unsuspected roles in plant cell division. *PLoS ONE* 4:e6757. doi: 10.1371/journal.pone.0006757
26. Canaday, J., Stoppin-Mellet, V., Mutterer, J., Lambert, A. M., and Schmit, A. C. (2000). Higher plant cells: gamma-tubulin and microtubule nucleation in the absence of centrosomes. *Microsc. Res. Tech.* 49, 487–495. doi: 10.1002/(SICI)1097-0029(20000601)49:5<487::AID-JEMT11>3.0.CO;2-I
27. Chambon, J.-P., Touati, S. A., Berneau, S., Cladière, D., Hebras, C., Groeme, R., et al. (2013). The PP2A inhibitor I2PP2A is essential for sister chromatid segregation in oocyte meiosis II. *Curr. Biol.* 23, 485–490. doi: 10.1016/j.cub.2013.02.004
28. Chan, A., and Cande, W. Z. (1998). Maize meiotic spindles assemble around chromatin and do not require paired chromosomes. *J. Cell Sci.* 111, 3507–3515.
29. Che, L., Tang, D., Wang, K., Wang, M., Zhu, K., Yu, H., et al. (2011). OsAM1 is required for leptotene-zygotene transition in rice. *Cell Res.* 21, 654–665. doi: 10.1038/cr.2011.7
30. Cheeseman, I. M., Chappie, J. S., Wilson-Kubalek, E. M., and Desai, A. (2006). The conserved KMN network constitutes the core microtubule-binding site of the kinetochore. *Cell* 127, 983–997. doi: 10.1016/j.cell.2006.09.039
31. Chelysheva, L., Diallo, S., Vezon, D., Gendrot, G., Vrielynck, N., Belcram, K., et al. (2005). AtREC8 and AtSCC3 are essential to the monopolar orientation of the kinetochores during meiosis. *J. Cell Sci.* 118, 4621–4632. doi: 10.1242/jcs.02583
32. Chelysheva, L., Gendrot, G., Vezon, D., Doutriaux, M.-P., Mercier, R., and Grelon, M. (2007). Zip4/Spo22 is required for class I CO formation but not for synapsis completion in Arabidopsis thaliana. *PLoS Genet.* 3:e83. doi: 10.1371/journal.pgen.0030083
33. Chen, C., Marcus, A., Li, W., Hu, Y., Calzada, J.-P. V., Grossniklaus, U., et al. (2002). The Arabidopsis ATK1 gene is required for spindle morphogenesis in male meiosis. *Development* 129, 2401–2409.
34. Chen, C., Zhang, W., Timofejeva, L., Gerardin, Y., and Ma, H. (2005). The Arabidopsis ROCK-N-ROLLERS gene encodes a homolog of the yeast ATP-dependent DNA helicase MER3 and is required for normal meiotic crossover formation. *Plant J.* 43, 321–334. doi: 10.1111/j.1365-313X.2005.02461.x
35. Chikashige, Y., Haraguchi, T., and Hiraoka, Y. (2007). Another way to move chromosomes. *Chromosoma* 116, 497–505. doi: 10.1007/s00412-007-0114-8
36. Cigliano, R. A., Sanseverino, W., Cremona, G., Consiglio, F. M., and Conicella, C. (2011). Evolution of parallel spindles like genes in plants and highlight of unique domain architecture. *BMC Evol. Biol.* 11:78. doi: 10.1186/1471-2148-11-78
37. Clift, D., and Marston, A. L. (2011). The role of shugoshin in meiotic chromosome segregation. *Cytogenet. Genome Res.* 133, 234–242. doi: 10.1159/000323793
38. Clissold, P. M., and Ponting, C. P. (2000). PIN domains in nonsense-mediated mRNA decay and RNAi. *Curr. Biol.* 10, 888–890. doi: 10.1016/S0960-9822(00)00858-7
39. Comai, L. (2005). The advantages and disadvantages of being polyploid. *Nat. Rev. Genet.* 6, 836–846. doi: 10.1038/nrg1711
40. Conicella, C., Capo, A., Cammareri, M., Errico, A., Shamina, N., and Monti, L. M. (2003). Elucidation of meiotic nuclear restitution mechanisms in potato through analysis of microtubular cytoskeleton. *Euphytica*, 107–115. doi: 10.1023/A:1025636321757
41. Cooper, K. F., and Strich, R. (2011). Meiotic control of the APC/C: similarities & differences from mitosis. *Cell Div.* 6:16. doi: 10.1186/1747-1028-6-16
42. Corbett, K. D., and Harrison, S. C. (2012). Molecular Architecture of the Yeast Monopolin Complex. *Cell. Rep.* 1, 583–589. doi: 10.1016/j.celrep.2012.05.012
43. Couteau, F., Belzile, F., Horlow, C., Grandjean, O., Vezon, D., and Doutriaux, M. P. (1999). Random chromosome segregation without meiotic arrest in both male and female meiocytes of a dmc1 mutant of Arabidopsis. *Plant Cell* 11, 1623–1634. doi: 10.1105/tpc.11.9.1623
44. Crismani, W., and Mercier, R. (2012). What limits meiotic crossovers? *Cell Cycle* 11, 3527–3528. doi: 10.4161/cc.21963
45. Cromer, L., Jolivet, S., Horlow, C., Chelysheva, L., Heyman, J., De Jaeger, G., et al. (2013). Centromeric cohesion is protected twice at meiosis, by SHUGOSHINs at anaphase I and by PATRONUS at interkinesis. *Curr. Biol.* 23, 2090–2099. doi: 10.1016/j.cub.2013.08.036
46. Da Ines, O., Gallego, M. E., and White, C. I. (2014). Recombination-independent mechanisms and pairing of homologous chromosomes during meiosis in plants. *Mol. Plant* 7, 492–501. doi: 10.1093/mp/sst172
47. Dawe, R. K. (1998). Meiotic chromosome organization and segregation in plants. *Annu. Rev. Plant Physiol. Plant Mol. Biol.* 49, 371–395. doi: 10.1146/annurev.arplant.49.1.371
48. Dawe, R. K., Reed, L. M., Yu, H.-G., Muszynski, G., and Hiatt, E. N. (1999). A maize homolog of mammalian CENPC is a constitutive component of the inner

kinetochore. *Plant Cell* 11, 1227–1238. doi: 10.1105/tpc.11.7.1227

49. Demidov, D., Van Damme, D., Geelen, D., Blattner, F. R., and Houben, A. (2005). Identification and dynamics of two classes of aurora-like kinases in arabidopsis and other plants. *Plant* 17, 836–848. doi: 10.1105/tpc.104.029710.1
50. De Muyt, A., Pereira, L., Vezon, D., Chelysheva, L., Gendrot, G., Chambon, A., et al. (2009). A high throughput genetic screen identifies new early meiotic recombination functions in Arabidopsis thaliana. *PLoS Genet.* 5:e1000654. doi: 10.1371/journal.pgen.1000654
51. De Oliveira, E. A. G., Romeiro, N. C., Ribeiro, E. D. S., Santa-Catarina, C., Oliveira, A. E. A., Silveira, V., et al. (2012). Structural and functional characterization of the protein kinase Mps1 in Arabidopsis thaliana. *PLoS ONE* 7:e45707. doi: 10.1371/journal.pone.0045707
52. D›Erfurth, I., Jolivet, S., Froger, N., Catrice, O., Novatchkova, M., Simon, M., et al. (2008). Mutations in AtPS1 (Arabidopsis thaliana parallel spindle 1) lead to the production of diploid pollen grains. *PLoS Genet.* 4:e1000274. doi: 10.1371/journal.pgen.1000274
53. De Rop, V., Padeganeh, A., and Maddox, P. S. (2012). CENP-A: the key player behind centromere identity, propagation, and kinetochore assembly. *Chromosoma* 121, 527–538. doi: 10.1007/s00412-012-0386-5
54. De Storme, N., and Geelen, D. (2011). The Arabidopsis mutant jason produces unreduced first division restitution male gametes through a parallel/fused spindle mechanism in meiosis II. *Plant Physiol.* 155, 1403–1415. doi: 10.1104/pp.110.170415
55. De Storme, N., and Geelen, D. (2013a). Cytokinesis in plant male meiosis. *Plant Signal. Behav.* 8:e23394. doi: 10.4161/psb.23394
56. De Storme, N., and Geelen, D. (2013b). Sexual polyploidization in plants—cytological mechanisms and molecular regulation.*New Phytol.* 198, 670–684. doi: 10.1111/nph.12184
57. De Storme, N., and Geelen, D. (2014). The impact of environmental stress on male reproductive development in plants: biological processes and molecular mechanisms. *Plant. Cell Environ.* 37, 1–18. doi: 10.1111/pce.12142
58. Ding, D., Muthuswamy, S., and Meier, I. (2012). Functional interaction between the Arabidopsis orthologs of spindle assembly checkpoint proteins MAD1 and MAD2 and the nucleoporin NUA. *Plant Mol. Biol.* 79, 203–216. doi: 10.1007/s11103-012-9903-4
59. Drouaud, J., Camilleri, C., Bourguignon, P.-Y., Canaguier, A., Bérard, A., Vezon, D., et al. (2006). Variation in crossing-over rates across chromosome 4 of Arabidopsis thaliana reveals the presence of meiotic recombination "hot spots." *Genome Res.* 16, 106–114. doi: 10.1101/gr.4319006
60. Du, Y., and Dawe, R. K. (2007). Maize NDC80 is a constitutive feature of the central kinetochore. *Chromosome Res.* 15, 767–775. doi: 10.1007/s10577-007-1160-z
61. Eckert, C. A., Gravdahl, D. J., and Megee, P. C. (2007). The enhancement of pericentromeric cohesin association by conserved kinetochore components promotes high-fidelity chromosome segregation and is sensitive to microtubule-based tension. *Genes Dev.* 21, 278–291. doi: 10.1101/gad.1498707.somes
62. Edlinger, B., and Schlögelhofer, P. (2011). Have a break: determinants of meiotic DNA double strand break (DSB) formation and processing in plants. *J. Exp. Bot.* 62, 1545–1563. doi: 10.1093/jxb/erq421
63. Eijpe, M., Heyting, C., Gross, B., and Jessberger, R. (2000). Association of mammalian SMC1 and SMC3 proteins with meiotic chromosomes and synaptonemal complexes. *J. Cell Sci.* 113(Pt 4), 673–682.
64. Ekwall, K. (2007). Epigenetic control of centromere behavior. *Annu. Rev. Genet.* 41, 63–81. doi: 10.1146/annurev.genet.41.110306.130127
65. Endow, S. A. (1999). Microtubule motors in spindle and chromosome motility. *Eur. J. Biochem.* 262, 12–18. doi: 10.1046/j.1432-1327.1999.00339.x
66. Erilova, A., Brownfield, L., Exner, V., Rosa, M., Twell, D., Mittelsten Scheid, O., et al. (2009). Imprinting of the polycomb group gene MEDEA serves as a ploidy sensor in Arabidopsis. *PLoS Genet.* 5:e1000663. doi: 10.1371/journal.pgen.1000663
67. Fernius, J., Nerusheva, O. O., Galander, S., Alves, F. D. L., Rappsilber, J., and Marston, A. L. (2013). Cohesin-dependent association of scc2/4 with the centromere initiates pericentromeric cohesion establishment. *Curr. Biol.* 23, 599–606. doi: 10.1016/j.cub.2013.02.022
68. Golubovskaya, I. N., Hamant, O., Timofejeva, L., Wang, C.-J. R., Braun, D., Meeley, R., et al. (2006). Alleles of afd1 dissect REC8 functions during meiotic prophase I. *J. Cell Sci.* 119, 3306–3315. doi: 10.1242/jcs.03054
69. Golubovskaya, I. N., Harper, L. C., Pawlowski, W. P., Schichnes, D., and Cande, W. Z. (2002). The pam1 gene is required for meiotic bouquet formation and efficient. *Genetics* 1993, 1979–1993.
70. Gómez, R., Valdeolmillos, A., Parra, M. T., Viera, A., Carreiro, C., Roncal, F., et al. (2007). Mammalian SGO2 appears at the inner centromere domain and redistributes depending on tension across centromeres during meiosis II and mitosis. *EMBO Rep.* 8, 173–180. doi: 10.1038/sj.embor.7400877
71. Gore, M. A., Chia, J. M., Elshire, R. J., Sun, Q., Ersoz, E. S., Hurwitz, B. L., et al. (2009). A first-generation haplotype map of maize. *Science* 326, 1115–1117. doi: 10.1126/science.1177837
72. Graumann, K., Runions, J., and Evans, D. E. (2010). Characterization of SUN-domain proteins at the higher plant nuclear envelope. *Plant J.* 61, 134–144. doi: 10.1111/j.1365-313X.2009.04038.x
73. Gregson, H. C., Schmiesing, J. A., Kim, J. S., Kobayashi, T., Zhou, S., and Yokomori, K. (2001). A potential role for human cohesin in mitotic spindle aster assembly. *J. Biol. Chem.* 276, 47575–47582. doi: 10.1074/jbc.M103364200
74. Gruber, S., Haering, C. H., and Nasmyth, K. (2003).

Chromosomal cohesin forms a ring. *Cell* 112, 765–777. doi: 10.1016/S0092-8674(03)00162-4

75. Gutiérrez-Caballero, C., Cebollero, L. R., and Pendás, A. M. (2012). Shugoshins: from protectors of cohesion to versatile adaptors at the centromere. *Trends Genet*. 28, 351–360. doi: 10.1016/j.tig.2012.03.003

76. Haering, C. H., Farcas, A.-M., Arumugam, P., Metson, J., and Nasmyth, K. (2008). The cohesin ring concatenates sister DNA molecules. *Nature* 454, 297–301. doi: 10.1038/nature07098

77. Hamant, O., Golubovskaya, I., Meeley, R., Fiume, E., Timofejeva, L., Schleiffer, A., et al. (2005). A REC8-dependent plant Shugoshin is required for maintenance of centromeric cohesion during meiosis and has no mitotic functions. *Curr. Biol*. 15, 948–954. doi: 10.1016/j.cub.2005.04.049

78. Hamant, O., Ma, H., and Cande, W. Z. (2006). Genetics of meiotic prophase I in plants. *Annu. Rev. Plant Biol*. 57, 267–302. doi: 10.1146/annurev.arplant.57.032905.105255

79. Harper, J. W., Burton, J. L., and Solomon, M. J. (2002). The anaphase-promoting complex: it's not just for mitosis any more.*Genes Dev*. 16, 2179–2206. doi: 10.1101/gad.1013102

80. Harper, L., Golubovskaya, I., and Cande, W. Z. (2004). A bouquet of chromosomes. *J. Cell Sci*. 117, 4025–4032. doi: 10.1242/jcs.01363

81. Hayashi, T., Fujita, Y., Iwasaki, O., Adachi, Y., Takahashi, K., and Yanagida, M. (2004). Mis16 and Mis18 are required for CENP-A loading and histone deacetylation at centromeres. *Cell* 118, 715–729. doi: 10.1016/j.cell.2004.09.002

82. Henry, I. M., Dilkes, B. P., Miller, E. S., Burkart-Waco, D., and Comai, L. (2010). Phenotypic consequences of aneuploidy in Arabidopsis thaliana. *Genetics* 186, 1231–1245. doi: 10.1534/genetics.110.121079

83. Henry, I. M., Dilkes, B. P., Young, K., Watson, B., Wu, H., and Comai, L. (2005). Aneuploidy and genetic variation in the Arabidopsis thaliana triploid response. *Genetics* 170, 1979–1988. doi: 10.1534/genetics.104.037788

84. Hewitt, L., Tighe, A., Santaguida, S., White, A. M., Jones, C. D., Musacchio, A., et al. (2010). Sustained Mps1 activity is required in mitosis to recruit O-Mad2 to the Mad1-C-Mad2 core complex. *J. Cell Biol*. 190, 25–34. doi: 10.1083/jcb.201002133

85. Higgins, J. D., Armstrong, S. J., Franklin, F. C. H., and Jones, G. H. (2004). The Arabidopsis MutS homolog AtMSH4 functions at an early step in recombination: evidence for two classes of recombination in Arabidopsis. *Genes Dev*. 18, 2557–2570. doi: 10.1101/gad.317504

86. Higgins, J. D., Sanchez-Moran, E., Armstrong, S. J., Jones, G. H., and Franklin, F. C. H. (2005). The Arabidopsis synaptonemal complex protein ZYP1 is required for chromosome synapsis and normal fidelity of crossing over. *Genes Dev*. 19, 2488–2500. doi: 10.1101/gad.354705

87. Higgins, J. D., Vignard, J., Mercier, R., Pugh, A. G., Franklin, F. C. H., and Jones, G. H. (2008). AtMSH5 partners AtMSH4 in the class I meiotic crossover pathway in Arabidopsis thaliana, but is not required for synapsis. *Plant J*. 55, 28–39. doi: 10.1111/j.1365-313X.2008.03470.x

88. Houben, A., and Schubert, I. (2003). DNA and proteins of plant centromeres. *Curr. Opin. Plant Biol*. 6, 554–560. doi: 10.1016/j.pbi.2003.09.007

89. Huang, H., Feng, J., Famulski, J., Rattner, J. B., Liu, S. T., Kao, G. D., et al. (2007). Tripin/hSgo2 recruits MCAK to the inner centromere to correct defective kinetochore attachments. *J. Cell Biol*. 177, 413–424. doi: 10.1083/jcb.200701122

90. Jaspersen, S. L., and Winey, M. (2004). The budding yeast spindle pole body: structure, duplication, and function. *Annu. Rev. Cell Dev. Biol*. 20, 1–28. doi: 10.1146/annurev.cellbio.20.022003.114106

91. Jensen, S., Segal, M., Clarke, D. J., and Reed, S. I. (2001). A novel role of the budding yeast separin Esp1 in anaphase spindle elongation: evidence that proper spindle association of Esp1 is regulated by Pds1. *J. Cell Biol*. 152, 27–40. doi: 10.1083/jcb.152.1.27

92. Jiang, H., Wang, F.-F., Wu, Y.-T., Zhou, X., Huang, X.-Y., Zhu, J., et al. (2009). MULTIPOLAR SPINDLE 1 (MPS1), a novel coiled-coil protein of Arabidopsis thaliana, is required for meiotic spindle organization. *Plant J*. 59, 1001–1010. doi: 10.1111/j.1365-313X.2009.03929.x

93. Jiang, L., Yuan, L., Xia, M., and Makaroff, C. A. (2010). Proper levels of the Arabidopsis cohesion establishment factor CTF7 are essential for embryo and megagametophyte, but not endosperm, development. *Plant Physiol*. 154, 820–832. doi: 10.1104/pp.110.157560

94. Jones, G. H. (1984). The control of chiasma distribution. *Symp. Soc. Exp. Biol*. 38, 293–320.

95. Juranič, M., Srilunchang, K., Krohn, N. G., Leljak-Levanic, D., Sprunck, S., and Dresselhaus, T. (2012). Germline-specific MATH-BTB substrate adaptor MAB1 regulates spindle length and nuclei identity in maize. *Plant Cell* 24, 4974–4991. doi: 10.1105/tpc.112.107169

96. Kawabe, A., Matsunaga, S., Nakagawa, K., Kurihara, D., Yoneda, A., Hasezawa, S., et al. (2005). Characterization of plant Aurora kinases during mitosis. *Plant Mol. Biol*. 58, 1–13. doi: 10.1007/s11103-005-3454-x

97. Kawashima, S. A., Tsukahara, T., Langegger, M., Hauf, S., Kitajima, T. S., and Watanabe, Y. (2007). Shugoshin enables tension-generating attachment of kinetochores by loading Aurora to centromeres. *Genes Dev*. 21, 420–435. doi: 10.1101/gad.1497307

98. Keeney, S., Giroux, C. N., and Kleckner, N. (1997). Meiosis-specific DNA double-strand breaks are catalyzed by Spo11, a member of a widely conserved protein family. *Cell* 88, 375–384. doi: 10.1016/S0092-8674(00)81876-0

99. Kerrebrock, A. W., Moore, D. P., Wu, J. S., and Orr-Weaver, T. L. (1995). Mei-S332, a Drosophila protein required for sister-chromatid cohesion, can localize

to meiotic centromere regions. *Cell* 83, 247–256. doi: 10.1016/0092-8674(95)90166-3

100. Kimbara, J., Endo, T. R., and Nasuda, S. (2004). Characterization of the genes encoding for MAD2 homologues in wheat.*Chromosome Res.* 12, 703–714. doi: 10.1023/B:CHRO.0000045760.42880.8c

101. Kirioukhova, O., Johnston, A. J., Kleen, D., Kägi, C., Baskar, R., Moore, J. M., et al. (2011). Female gametophytic cell specification and seed development require the function of the putative Arabidopsis INCENP ortholog WYRD. *Development*138, 3409–3420. doi: 10.1242/dev.060384

102. Kitajima, T. S., Kawashima, S. A., and Watanabe, Y. (2004). The conserved kinetochore protein shugoshin protects centromeric cohesion during meiosis. *Nature* 427, 510–517. doi: 10.1038/nature02312

103. Kitajima, T. S., Miyazaki, Y., Yamamoto, M., and Watanabe, Y. (2003). Rec8 cleavage by separase is required for meiotic nuclear divisions in fission yeast. *EMBO J.* 22, 5643–5653. doi: 10.1093/emboj/cdg527

104. Klein, F., Mahr, P., Galova, M., Buonomo, S. B., Michaelis, C., Nairz, K., et al. (1999). A central role for cohesins in sister chromatid cohesion, formation of axial elements, and recombination during yeast meiosis. *Cell* 98, 91–103. doi: 10.1016/S0092-8674(00)80609-1

105. Klutstein, M., and Cooper, J. P. (2014). The chromosomal courtship dance-homolog pairing in early meiosis. *Curr. Opin. Cell Biol.* 26, 123–131. doi: 10.1016/j.ceb.2013.12.004

106. Koszul, R., Kim, K. P., Prentiss, M., Kleckner, N., and Kameoka, S. (2009). Actin-mediated motion of meiotic chromosomes.*NIH* 133, 1188–1201. doi: 10.1016/j.cell.2008.04.050

107. Kudo, N. R., Anger, M., Peters, A. H., Stemmann, O., Theussl, H.-C., Helmhart, W., et al. (2009). Role of cleavage by separase of the Rec8 kleisin subunit of cohesin during mammalian meiosis I. *J. Cell Sci.* 122, 2686–2698. doi: 10.1242/jcs.035287

108. Kurzbauer, M.-T., Uanschou, C., Chen, D., and Schlögelhofer, P. (2012). The recombinases DMC1 and RAD51 are functionally and spatially separated during meiosis in Arabidopsis. *Plant Cell* 24, 2058–2070. doi: 10.1105/tpc.112.098459

109. Lam, W. S., Yang, X., and Makaroff, C. A. (2005). Characterization of Arabidopsis thaliana SMC1 and SMC3: evidence that AtSMC3 may function beyond chromosome cohesion. *J. Cell Sci.* 118, 3037–3048. doi: 10.1242/jcs.02443

110. Lampert, F., and Westermann, S. (2011). A blueprint for kinetochores—new insights into the molecular mechanics of cell division. *Nat. Rev. Mol. Cell. Biol.* 12, 407–412. doi: 10.1038/nrm3133

111. Lee, J., Kitajima, T. S., Tanno, Y., Yoshida, K., Morita, T., Miyano, T., et al. (2008a). Unified mode of centromeric protection by shugoshin in mammalian oocytes and somatic cells. *Nat. Cell Biol.* 10, 42–52. doi: 10.1038/ncb1667

112. Lee, S. H., Sterling, H., Burlingame, A., and McCormick, F. (2008b). Tpr directly binds to Mad1 and Mad2 and is important for the Mad1-Mad2-mediated mitotic spindle checkpoint. *Genes Dev.* 22, 2926–2931. doi: 10.1101/gad.1677208

113. Lermontova, I., Koroleva, O., Rutten, T., Fuchs, J., Schubert, V., Moraes, I., et al. (2011). Knockdown of CENH3 in Arabidopsis reduces mitotic divisions and causes sterility by disturbed meiotic chromosome segregation. *Plant J.* 68, 40–50. doi: 10.1111/j.1365-313X.2011.04664.x

114. Lermontova, I., Kuhlmann, M., Friedel, S., Rutten, T., Heckmann, S., Sandmann, M., et al. (2013). Arabidopsis KINETOCHORE NULL2 is an upstream component for centromeric histone H3 variant cenH3 deposition at centromeres. *Plant Cell* 25, 3389–3404. doi: 10.1105/tpc.113.114736

115. Lermontova, I., Schubert, V., Fuchs, J., Klatte, S., Macas, J., and Schubert, I. (2006). Loading of Arabidopsis centromeric histone CENH3 occurs mainly during G2 and requires the presence of the histone fold domain. *Plant Cell* 18, 2443–2451. doi: 10.1105/tpc.106.043174

116. Lhuissier, F. G. P., Offenberg, H. H., Wittich, P. E., Vischer, N. O. E., and Heyting, C. (2007). The mismatch repair protein MLH1 marks a subset of strongly interfering crossovers in tomato. *Plant Cell* 19, 862–876. doi: 10.1105/tpc.106.049106

117. Li, J., Lee, G. I., Van Doren, S. R., and Walker, J. C. (2000). The FHA domain mediates phosphoprotein interactions. *J. Cell Sci.* 113(Pt 23), 4143–4149.

118. Li, W., Chen, C., Markmann-Mulisch, U., Timofejeva, L., Schmelzer, E., Ma, H., et al. (2004). The Arabidopsis AtRAD51 gene is dispensable for vegetative development but required for meiosis. *Proc. Natl. Acad. Sci. U.S.A.* 101, 10596–10601. doi: 10.1073/pnas.0404110101

119. Li, X., and Dawe, R. K. (2009). Fused sister kinetochores initiate the reductional division in meiosis I. *Nat. Cell Biol.* 11, 1103–1108. doi: 10.1038/ncb1923

120. Li, Y., Shen, Y., Cai, C., Zhong, C., Zhu, L., Yuan, M., et al. (2010). The type II Arabidopsis formin14 interacts with microtubules and microfilaments to regulate cell division. *Plant Cell* 22, 2710–2726. doi: 10.1105/tpc.110.075507

121. Liu, B., Cyr, R. J., and Palevitz, B. A. (1996). A kinesin-like protein, KatAp, in the cells of arabidopsis and other plants. *Plant Cell*8, 119–132. doi: 10.1105/tpc.8.1.119

122. Liu, C. M., McElver, J., Tzafrir, I., Joosen, R., Wittich, P., Patton, D., et al. (2002). Condensin and cohesin knockouts in Arabidopsis exhibit a titan seed phenotype. *Plant J.* 29, 405–415. doi: 10.1046/j.1365-313x.2002.01224.x

123. Liu, Z., and Makaroff, C. A. (2006). Arabidopsis separase AESP is essential for embryo development and the release of cohesin during meiosis. *Plant Cell* 18, 1213–1225. doi: 10.1105/tpc.105.036913.2001

124. Ma, H. (2006). A molecular portrait of Arabidopsis meiosis. *Arabidopsis Book* 4:e0095. doi: 10.1199/tab.0095

125. Madlung, A., and Wendel, J. F. (2013). Genetic and epigenetic aspects of polyploid evolution in plants. *Cytogenet. Genome Res*. 140, 270–285. doi: 10.1159/000351430

126. Malmanche, N., Maia, A., and Sunkel, C. E. (2006). The spindle assembly checkpoint: preventing chromosome mis-segregation during mitosis and meiosis. *FEBS Lett*. 580, 2888–2895. doi: 10.1016/j.febslet.2006.03.081

127. Marcus, A. I., Ambrose, J. C., Blickley, L., Hancock, W. O., and Cyr, R. J. (2002). Arabidopsis thaliana protein, ATK1, is a minus-end directed kinesin that exhibits non-processive movement. *Cell Motil. Cytoskeleton* 52, 144–150. doi: 10.1002/cm.10045

128. Marston, A. L. (2014). Chromosome segregation in budding yeast: sister chromatid cohesion and related Mechanisms. *Genetics*196, 31–63. doi: 10.1534/genetics.112.145144

129. Masson, J. Y., and West, S. C. (2001). The Rad51 and Dmc1 recombinases: a non-identical twin relationship. *Trends Biochem. Sci*. 26, 131–136. doi: 10.1016/S0968-0004(00)01742-4

130. May, K. M., and Hardwick, K. G. (2006). The spindle checkpoint. *J. Cell Sci*. 119, 4139–4142. doi: 10.1242/jcs.03165

131. McGuinness, B. E., Hirota, T., Kudo, N. R., Peters, J.-M., and Nasmyth, K. (2005). Shugoshin prevents dissociation of cohesin from centromeres during mitosis in vertebrate cells. *PLoS Biol*. 3:e86. doi: 10.1371/journal.pbio.0030086

132. Mercier, R., Armstrong, S. J., Horlow, C., Kackson, N. P., Makaroff, C. A., Vezon, D., et al. (2003). The meiotic protein SWI1 is required for axial element formation and recombination initiation in Arabidopsis. *Development* 130, 3309–3318. doi: 10.1242/dev.00550

133. Mercier, R., and Grelon, M. (2008). Meiosis in plants: ten years of gene discovery. *Cytogenet. Genome Res*. 120, 281–290. doi: 10.1159/000121077

134. Mercier, R., Jolivet, S., Vezon, D., Huppe, E., Chelysheva, L., Giovanni, M., et al. (2005). Two meiotic crossover classes cohabit in Arabidopsis: one is dependent on MER3,whereas the other one is not. *Curr. Biol*. 15, 692–701. doi: 10.1016/j.cub.2005.02.056

135. Mercier, R., Vezon, D., Bullier, E., Motamayor, J. C., Sellier, A., Lefèvre, F., et al. (2001). SWITCH1 (SWI1): a novel protein required for the establishment of sister chromatid cohesion and for bivalent formation at meiosis. *Genes Dev*. 15, 1859–1871. doi: 10.1101/gad.203201

136. Miki, F., Kurabayashi, A., Tange, Y., Okazaki, K., Shimanuki, M., and Niwa, O. (2004). Two-hybrid search for proteins that interact with Sad1 and Kms1, two membrane-bound components of the spindle pole body in fission yeast. *Mol. Genet. Genomics*270, 449–471. doi: 10.1007/s00438-003-0938-8

137. Moraes, I. C. R., Lermontova, I., and Schubert, I. (2011). Recognition of *A.* thaliana centromeres by heterologous CENH3 requires high similarity to the endogenous protein. *Plant Mol. Biol*. 75, 253–261. doi: 10.1007/s11103-010-9723-3

138. Moschou, P. N., and Bozhkov, P. V (2012). Separases: biochemistry and function. *Physiol. Plant*. 145, 67–76. doi: 10.1111/j.1399-3054.2011.01550.x

139. Motamayor, J. C., Vezon, D., Bajon, C., Sauvanet, A., Grandjean, O., Marchand, M., et al. (2000). Switch (swi1), an Arabidopsis thaliana mutant affected in the female meiotic switch. *Sex. Plant Reprod*. 12, 209–218. doi: 10.1007/s004970050002

140. Murata, M. (2013). "Arabidopsis centromeres," in *Plant Centromere Biology*, eds J. Jiang and J. A. Birchler (Oxford, UK: Wiley-Blackwell). doi: 10.1002/9781118525715.ch1

141. Murphy, S. P., Simmons, C. R., and Bass, H. W. (2010). Structure and expression of the maize (Zea mays L.) SUN-domain protein gene family: evidence for the existence of two divergent classes of SUN proteins in plants. *BMC Plant Biol*. 10:269. doi: 10.1186/1471-2229-10-269

142. Murray, A. (1994). Cell cycle checkpoints. *Curr. Opin. Cell Biol*. 6, 872–876. doi: 10.1016/0955-0674(94)90059-0

143. Musacchio, A. (2011). Spindle assembly checkpoint: the third decade. *Philos. Trans. R. Soc. Lond. B. Biol. Sci*. 366, 3595–604. doi: 10.1098/rstb.2011.0072

144. Nasmyth, K. (2001). Disseminating the genome: joining, resolving, and separating sister chromatids during mitosis and meiosis.*Annu. Rev. Genet*. 35, 673–745. doi: 10.1146/annurev.genet.35.102401.091334

145. Nasmyth, K., and Haering, C. H. (2005). The structure and function of SMC and kleisin complexes. *Annu. Rev. Biochem*. 74, 595–648. doi: 10.1146/annurev.biochem.74.082803.133219

146. Nicklas, R. B., Ward, S. C., and Gorbsky, G. J. (1995). Kinetochore chemistry is sensitive to tension and may link mitotic forces to a cell cycle checkpoint. *J. Cell Biol*. 130, 929–939. doi: 10.1083/jcb.130.4.929

147. Niu, H., Wan, L., Busygina, V., Kwon, Y., Allen, J. A., Li, X., et al. (2009). Regulation of meiotic recombination via Mek1-mediated Rad54 phosphorylation. *Mol. Cell* 36, 393–404. doi: 10.1016/j.molcel.2009.09.029

148. Nogales, E. (2000). Structural insights into microtubule function. *Annu. Rev. Biochem*. 69, 277–302. doi: 10.1146/annurev.biochem.69.1.277

149. Ocampo-Hafalla, M. T., and Uhlmann, F. (2011). Cohesin loading and sliding. *J. Cell Sci*. 124, 685–691. doi: 10.1242/jcs.073866

150. Ogura, Y., Shibata, F., Sato, H., and Murata, M. (2004). Characterization of a CENP-C homolog in Arabidopsis thaliana. *Genes Genet. Syst*. 79, 139–144. doi: 10.1266/ggs.79.139

151. Onn, I., Heidinger-Pauli, J. M., Guacci, V., Unal, E., and Koshland, D. E. (2008). Sister chromatid cohesion: a simple concept with a complex reality. *Annu. Rev. Cell Dev. Biol*. 24, 105–129. doi: 10.1146/annurev.cellbio.24.110707.175350

152. Osman, K., Higgins, J. D., Sanchez-Moran, E., Armstrong, S. J., and Franklin, F. C. H. (2011). Pathways to meiotic

recombination in Arabidopsis thaliana. *New Phytol.* 190, 523–544. doi: 10.1111/j.1469-8137.2011.03665.x

153. O'Toole, E., Greenan, G., Lange, K. I., Srayko, M., and Müller-Reichert, T. (2012). The role of γ-tubulin in centrosomal microtubule organization. *PLoS ONE* 7:e29795. doi: 10.1371/journal.pone.0029795

154. Otto, S. P. (2007). The evolutionary consequences of polyploidy. *Cell* 131, 452–462. doi: 10.1016/j.cell.2007.10.022

155. Pastuglia, M., Azimzadeh, J., Magali, G., Christine, C., Belcram, K., Evrard, J., et al. (2006). g-Tubulin is essential for microtubule organization and development in Arabidopsis. *Plant Cell* 18, 1412–1425. doi: 10.1105/tpc.105.039644.2

156. Pawlowski, W. P., and Cande, W. Z. (2005). Coordinating the events of the meiotic prophase. *Trends Cell Biol.* 15, 674–681. doi: 10.1016/j.tcb.2005.10.005

157. Pawlowski, W. P., Wang, C.-J. R., Golubovskaya, I. N., Szymaniak, J. M., Shi, L., Hamant, O., et al. (2009). Maize AMEIOTIC1 is essential for multiple early meiotic processes and likely required for the initiation of meiosis. *Proc. Natl. Acad. Sci. U.S.A.* 106, 3603–3608. doi: 10.1073/pnas.0810115106

158. Penkner, A. M., Fridkin, A., Gloggnitzer, J., Baudrimont, A., Machacek, T., Woglar, A., et al. (2009). Meiotic chromosome homology search involves modifications of the nuclear envelope protein Matefin/SUN-1. *Cell* 139, 920–933. doi: 10.1016/j.cell.2009.10.045

159. Pereira, G., and Schiebel, E. (1997). Centrosome-microtubule nucleation. *J. Cell Sci.* 110(Pt 3), 295–300.

160. Peters, J.-M. (2006). The anaphase promoting complex/cyclosome: a machine designed to destroy. *Nat. Rev. Mol. Cell Biol.* 7, 644–656. doi: 10.1038/nrm1988

161. Peters, J.-M., Tedeschi, A., and Schmitz, J. (2008). The cohesin complex and its roles in chromosome biology. *Genes Dev.* 22, 3089–3114. doi: 10.1101/gad.1724308

162. Petroski, M. D., and Deshaies, R. J. (2005). Function and regulation of cullin-RING ubiquitin ligases. *Nat. Rev. Mol. Cell Biol.* 6, 9–20. doi: 10.1038/nrm1547

163. Phillips, C. M., and Dernburg, A. F. (2006). A family of zinc-finger proteins is required for chromosome-specific pairing and synapsis during meiosis in *C. elegans*. *Dev. Cell* 11, 817–829. doi: 10.1016/j.devcel.2006.09.020

164. Pidoux, A. L., and Allshire, R. C. (2005). The role of heterochromatin in centromere function. *Philos. Trans. R. Soc. Lond. B. Biol. Sci.* 360, 569–579. doi: 10.1098/rstb.2004.1611

165. Pradillo, M., López, E., Linacero, R., Romero, C., Cuñado, N., Sánchez-Morán, E., et al. (2012). Together yes, but not coupled: new insights into the roles of RAD51 and DMC1 in plant meiotic recombination. *Plant J.* 69, 921–933. doi: 10.1111/j.1365-313X.2011.04845.x

166. Quan, L., Xiao, R., Li, W., Oh, S.-A., Kong, H., Ambrose, J. C., et al. (2008). Functional divergence of the duplicated AtKIN14a and AtKIN14b genes: critical roles in Arabidopsis meiosis and gametophyte development. *Plant J.* 53, 1013–1026. doi: 10.1111/j.1365-313X.2007.03391.x

167. Ramsey, J., and Schemske, D. W. (1998). Pathways, mechanisms, and rates of polyploid formation in flowering plants. *Annu. Rev. Ecol. Syst.* 29, 467–501.

168. Ramsey, J., and Schemske, D. W. (2002). Neopolyploidy in flowering plants. *Annu. Rev. Ecol. Syst.* 33, 589–639. doi: 10.1146/annurev.ecolsys.33.010802.150437

169. Ravi, M., Kwong, P. N., Menorca, R. M. G., Valencia, J. T., Ramahi, J. S., Stewart, J. L., et al. (2010). The rapidly evolving centromere-specific histone has stringent functional requirements in Arabidopsis thaliana. *Genetics* 186, 461–471. doi: 10.1534/genetics.110.120337

170. Ravi, M., Shibata, F., Ramahi, J. S., Nagaki, K., Chen, C., Murata, M., et al. (2011). Meiosis-specific loading of the centromere-specific histone CENH3 in Arabidopsis thaliana. *PLoS Genet.* 7:e1002121. doi: 10.1371/journal.pgen.1002121

171. Reddy, A. S. N., and Day, I. S. (2001). Kinesins in the Arabidopsis genome?: a comparative analysis among eukaryotes. *BMC Genomics* 2:2. doi: 10.1186/1471-2164-2-2

172. Richards, D. M., Greer, E., Martin, A. C., Moore, G., Shaw, P. J., and Howard, M. (2012). Quantitative dynamics of telomere bouquet formation. *PLoS Comput. Biol.* 8:e1002812. doi: 10.1371/journal.pcbi.1002812

173. Ross, K. J., Fransz, P., Armstrong, S. J., Vizir, I., Mulligan, B., Franklin, F. C., et al. (1997). Cytological characterization of four meiotic mutants of Arabidopsis isolated from T-DNA-transformed lines. *Chromosome Res.* 5, 551–559. doi: 10.1023/A:1018497804129

174. Rowland, B. D., Roig, M. B., Nishino, T., Kurze, A., Uluocak, P., Mishra, A., et al. (2009). Building sister chromatid cohesion: smc3 acetylation counteracts an antiestablishment activity. *Mol. Cell* 33, 763–774. doi: 10.1016/j.molcel.2009.02.028

175. Ruchaud, S., Carmena, M., and Earnshaw, W. C. (2007). Chromosomal passengers: conducting cell division. *Nat. Rev. Mol. Cell Biol.* 8, 798–812. doi: 10.1038/nrm2257

176. Sakuno, T., Tada, K., and Watanabe, Y. (2009). Kinetochore geometry defined by cohesion within the centromere. *Nature* 458, 852–858. doi: 10.1038/nature07876

177. Salic, A., Waters, J. C., and Mitchison, T. J. (2004). Vertebrate shugoshin links sister centromere cohesion and kinetochore microtubule stability in mitosis. *Cell* 118, 567–578. doi: 10.1016/j.cell.2004.08.016

178. Sanchez-Moran, E., Osman, K., Higgins, J. D., Pradillo, M., Cuñado, N., Jones, G. H., et al. (2008). ASY1 coordinates early events in the plant meiotic recombination pathway. *Cytogenet. Genome Res.* 120, 302–312. doi: 10.1159/000121079

179. Santaguida, S., and Musacchio, A. (2009). The life and miracles of kinetochores. *EMBO J.* 28, 2511–2531. doi: 10.1038/emboj.2009.173

180. Sato, A., Isaac, B., Phillips, C. M., Rillo, R., Carlton, P. M., Wynne, D. J., et al. (2009). Cytoskeletal forces span the nuclear envelope to coordinate meiotic chromosome

pairing and synapsis. *Cell* 139, 907–919. doi: 10.1016/j.cell.2009.10.039

181. Sato, H., Shibata, F., and Murata, M. (2005). Characterization of a Mis12 homologue in Arabidopsis thaliana. *Chromosome Res.* 13, 827–834. doi: 10.1007/s10577-005-1016-3

182. Schmitt, J., Benavente, R., Hodzic, D., Hoog, C., Stewart, C. L., and Alsheimer, M. (2007). Transmembrane protein Sun2 is involved in tethering mammalian meiotic telomeres to the nuclear envelope. *Proc. Natl. Acad. Sci. U.S.A.* 104, 7426–7431. doi: 10.1073/pnas.0609198104

183. Schubert, V., Kim, Y.-M., Berr, A., Fuchs, J., Meister, A., Marschner, S., et al. (2007). Random homologous pairing and incomplete sister chromatid alignment are common in angiosperm interphase nuclei. *Mol. Genet. Genomics* 278, 167–176. doi: 10.1007/s00438-007-0242-0

184. Schubert, V., Weissleder, A., Ali, H., Fuchs, J., Lermontova, I., Meister, A., et al. (2009a). Cohesin gene defects may impair sister chromatid alignment and genome stability in Arabidopsis thaliana. *Chromosoma* 118, 591–605. doi: 10.1007/s00412-009-0220-x

185. Schubert, V., Weissleder, A., Ali, H., Fuchs, J., Lermontova, I., Meister, A., et al. (2009b). Cohesin gene defects may impair sister chromatid alignment and genome stability in Arabidopsis thaliana. *Chromosoma* 118, 591–605. doi: 10.1007/s00412-009-0220-x

186. Sebastian, J., Ravi, M., Andreuzza, S., Panoli, A. P., Marimuthu, M. P. A., and Siddiqi, I. (2009). The plant adherin AtSCC2 is required for embryogenesis and sister-chromatid cohesion during meiosis in Arabidopsis. *Plant J.* 59, 1–13. doi: 10.1111/j.1365-313X.2009.03845.x

187. Shao, T., Tang, D., Wang, K., Wang, M., Che, L., Qin, B., et al. (2011). OsREC8 is essential for chromatid cohesion and metaphase I monopolar orientation in rice meiosis. *Plant Physiol.* 156, 1386–1396. doi: 10.1104/pp.111.177428

188. Sharp, D. J., Rogers, G. C., and Scholey, J. M. (2000). Microtubule motors in mitosis. *Nature* 407, 41–47. doi: 10.1038/35024000

189. Sharp, D. J., Yu, K. R., Sisson, J. C., Sullivan, W., and Scholey, J. M. (1999). Antagonistic microtubule-sliding motors position mitotic centrosomes in *Drosophila* early embryos. *Nat. Cell Biol.* 1, 51–54. doi: 10.1038/9025

190. Sheehan, M. J., and Pawlowski, W. P. (2009). Live imaging of rapid chromosome movements in meiotic prophase I in maize.*Proc. Natl. Acad. Sci. U.S.A.* 106, 20989–20994. doi: 10.1073/pnas.0906498106

191. Shimamura, M., Brown, R. C., Lemmon, B. E., Akashi, T., Mizuno, K., Nishihara, N., et al. (2004). γ-Tubulin in basal land plants?: characterization, localization, and implication in the evolution of acentriolar microtubule. *Plant Cell* 16, 45–59. doi: 10.1105/tpc.016501.bules

192. Siddiqi, I., Ganesh, G., Grossniklaus, U., and Subbiah, V. (2000). The dyad gene is required for progression through female meiosis in Arabidopsis. *Development* 127, 197–207.

193. Siderakis, M., and Tarsounas, M. (2007). Telomere regulation and function during meiosis. *Chromosome Res.* 15, 667–679. doi: 10.1007/s10577-007-1149-7

194. Singh, D. K., Andreuzza, S., Panoli, A. P., and Siddiqi, I. (2013). AtCTF7 is required for establishment of sister chromatid cohesion and association of cohesin with chromatin during meiosis in Arabidopsis. *BMC Plant Biol.* 13:117. doi: 10.1186/1471-2229-13-117

195. Staiger, C. J., and Cande, W. Z. (1991). Microfilament distribution in maize meiotic mutants correlates with microtubule organization. *Plant Cell* 3, 637–644. doi: 10.1105/tpc.3.6.637

196. Stoop-Myer, C., and Amon, A. (1999). Meiosis: rec8 is the reason for cohesion. *Nat. Cell Biol.* 1, E125–E127. doi: 10.1038/12956

197. Sudakin, V., Chan, G. K., and Yen, T. J. (2001). Checkpoint inhibition of the APC/C in HeLa cells is mediated by a complex of BUBR1, BUB3, CDC20, and MAD2. *J. Cell Biol.* 154, 925–936. doi: 10.1083/jcb.200102093

198. Sumara, I., Vorlaufer, E., Stukenberg, P. T., Kelm, O., Redemann, N., Nigg, E. A., et al. (2002). The dissociation of cohesin from chromosomes in prophase is regulated by Polo-like kinase. *Mol. Cell* 9, 515–525. doi: 10.1016/S1097-2765(02)00473-2

199. Sun, S.-C., and Kim, N.-H. (2012). Spindle assembly checkpoint and its regulators in meiosis. *Hum. Reprod. Update* 18, 60–72. doi: 10.1093/humupd/dmr044

200. Talbert, P. B. (2002). Centromeric localization and adaptive evolution of an arabidopsis histone H3 variant. *Plant Cell Online*14, 1053–1066. doi: 10.1105/tpc.010425

201. Tiang, C.-L., He, Y., and Pawlowski, W. P. (2012). Chromosome organization and dynamics during interphase, mitosis, and meiosis in plants. *Plant Physiol.* 158, 26–34. doi: 10.1104/pp.111.187161

202. Torras-Llort, M., Moreno-Moreno, O., and Azorín, F. (2009). Focus on the centre: the role of chromatin on the regulation of centromere identity and function. *EMBO J.* 28, 2337–2348. doi: 10.1038/emboj.2009.174

203. Tsukahara, T., Tanno, Y., and Watanabe, Y. (2010). Phosphorylation of the CPC by Cdk1 promotes chromosome bi-orientation.*Nature* 467, 719–723. doi: 10.1038/nature09390

204. Uhlmann, F. (2001). Chromosome cohesion and segregation in mitosis and meiosis. *Curr. Opin. Cell Biol.* 13, 754–761. doi: 10.1016/S0955-0674(00)00279-9

205. Uhlmann, F. (2009). A matter of choice: the establishment of sister chromatid cohesion. *EMBO Rep.* 10, 1095–1102. doi: 10.1038/embor.2009.207

206. Vader, G., Maia, A. F., and Lens, S. M. (2008). The chromosomal passenger complex and the spindle assembly checkpoint: kinetochore-microtubule error correction and beyond. *Cell Div.* 3:10. doi: 10.1186/1747-1028-3-10

207. Vanoosthuyse, V., Prykhozhij, S., and Hardwick, K. G. (2007). Shugoshin 2 regulates localization of the chromosomal passenger proteins in fission yeast

mitosis. *Mol. Biol. Cell* 18, 1657–1669. doi: 10.1091/mbc.E06

208. Walczak, C. E., Vernos, I., Mitchison, T. J., Karsenti, E., and Heald, R. (1998). A model for the proposed roles of different microtubule-based motor proteins in establishing spindle bipolarity. *Curr. Biol.* 8, 903–913. doi: 10.1016/S0960-9822(07)00370-3

209. Wang, G., Zhang, X., and Jin, W. (2009). An overview of plant centromeres. *J. Genet. Genomics* 36, 529–537. doi: 10.1016/S1673-8527(08)60144-7

210. Wang, M., Tang, D., Wang, K., Shen, Y., Qin, B., Miao, C., et al. (2011). OsSGO1 maintains synaptonemal complex stabilization in addition to protecting centromeric cohesion during rice meiosis. *Plant J.* 67, 583–594. doi: 10.1111/j.1365-313X.2011.04615.x

211. Wang, Y., Wu, H., Liang, G., and Yang, M. (2004). Defects in nucleolar migration and synapsis in male prophase I in the ask1-1 mutant of Arabidopsis. *Sex. Plant Reprod.* 16, 273–282. doi: 10.1007/s00497-004-0206-z

212. Wasteneys, G. O. (2002). Microtubule organization in the green kingdom: chaos or self-order? *J. Cell Sci.* 115, 1345–1354.

213. Watanabe, Y. (2012). Geometry and force behind kinetochore orientation: lessons from meiosis. *Nat. Rev. Mol. Cell Biol.* 13, 370–382. doi: 10.1038/nrm3349

214. Watanabe, Y., and Nurse, P. (1999). Cohesin Rec8 is required for reductional chromosome segregation at meiosis. *Nature* 400, 461–464. doi: 10.1038/22774

215. Weber, H., and Hellmann, H. (2009). Arabidopsis thaliana BTB/ POZ-MATH proteins interact with members of the ERF/AP2 transcription factor family. *FEBS J.* 276, 6624–6635. doi: 10.1111/j.1742-4658.2009.07373.x

216. Wijnker, E., and Schnittger, A. (2013). Control of the meiotic cell division program in plants. *Plant Reprod.* 26, 143–158. doi: 10.1007/s00497-013-0223-x

217. Wittmann, T., Hyman, A., and Desai, A. (2001). The spindle: a dynamic assembly of microtubules and motors. *Nat. Cell Biol.* 3, E28–E34. doi: 10.1038/35050669

218. Woehlke, G., and Schliwa, M. (2000). Walking on two heads: the many talents of kinesin. *Nat. Rev. Mol. Cell Biol.* 1, 50–58. doi: 10.1038/35036069

219. Woglar, A., and Jantsch, V. (2013). Chromosome movement in meiosis I prophase of Caenorhabditis elegans. *Chromosoma* 123, 15–24. doi: 10.1007/s00412-013-0436-7

220. Wu, S., Scheible, W.-R., Schindelasch, D., Van Den Daele, H., De Veylder, L., and Baskin, T. I. (2010). A conditional mutation in Arabidopsis thaliana separase induces chromosome non-disjunction, aberrant morphogenesis and cyclin B1;1 stability.*Development* 137, 953–961. doi: 10.1242/dev.041939

221. Xu, H., Beasley, M. D., Warren, W. D., van der Horst, G. T. J., and McKay, M. J. (2005). Absence of mouse REC8 cohesin promotes synapsis of sister chromatids in meiosis. *Dev. Cell* 8, 949–961. doi: 10.1016/j.devcel.2005.03.018

222. Xu, Z., Cetin, B., Anger, M., Cho, U. S., Helmhart, W., Nasmyth, K., et al. (2009). Structure and function of the PP2A-shugoshin interaction. *Mol. Cell* 35, 426–441. doi: 10.1016/j.molcel.2009.06.031

223. Yamagishi, Y., Sakuno, T., Shimura, M., and Watanabe, Y. (2008). Heterochromatin links to centromeric protection by recruiting shugoshin. *Nature* 455, 251–255. doi: 10.1038/nature07217

224. Yamamoto, A., Kitamura, K., Hihara, D., Hirose, Y., Katsuyama, S., and Hiraoka, Y. (2008). Spindle checkpoint activation at meiosis I advances anaphase II onset via meiosis-specific APC/C regulation. *J. Cell Biol.* 182, 277–288. doi: 10.1083/jcb.200802053

225. Yang, M., Hu, Y., Lodhi, M., McCombie, W. R., and Ma, H. (1999). The Arabidopsis SKP1-LIKE1 gene is essential for male meiosis and may control homologue separation. *Proc. Natl. Acad. Sci. U.S.A.* 96, 11416–11421. doi: 10.1073/pnas.96.20.11416

226. Yang, X., Boateng, K. A., Strittmatter, L., Burgess, R., and Makaroff, C. A. (2009). Arabidopsis separase functions beyond the removal of sister chromatid cohesion during meiosis. *Plant Physiol.* 151, 323–333. doi: 10.1104/pp.109.140699

227. Yang, X., Boateng, K. A., Yuan, L., Wu, S., Baskin, T. I., and Makaroff, C. A. (2011). The radially swollen 4 separase mutation of Arabidopsis thaliana blocks chromosome disjunction and disrupts the radial microtubule system in meiocytes. *PLoS ONE*6:e19459. doi: 10.1371/journal.pone.0019459

228. Yang, X., Timofejeva, L., Ma, H., and Makaroff, C. A. (2006). The Arabidopsis SKP1 homolog ASK1 controls meiotic chromosome remodeling and release of chromatin from the nuclear membrane and nucleolus. *J. Cell Sci.* 119, 3754–3763. doi: 10.1242/jcs.03155

229. Yao, Y., and Dai, W. (2012). Shugoshins function as a guardian for chromosomal stability in nuclear division. *Cell Cycle* 11, 2631–2642. doi: 10.4161/cc.20633

230. Yokobayashi, S., Yamamoto, M., and Watanabe, Y. (2003). Cohesins determine the attachment manner of kinetochores to spindle microtubules at meiosis I in fission yeast. *Mol. Cell Biol.* 23, 3965–3973. doi: 10.1128/MCB.23.11.3965-3973.2003

231. Yu, H.-G., and Koshland, D. (2005). Chromosome morphogenesis: condensin-dependent cohesin removal during meiosis. *Cell*123, 397–407. doi: 10.1016/j.cell.2005.09.014

232. Yu, H., Muszynski, M. G., and Dawe, R. K. (1999). The maize homologue of the cell cycle checkpoint protein MAD2 localization Patterns. *J. Cell Biol.* 145, 425–435.

233. Zamariola, L., De Storme, N., Tiang, C. L., Armstrong, S. J., Franklin, F. C. H., and Geelen, D. (2013). SGO1 but not SGO2 is required for maintenance of centromere cohesion in Arabidopsis thaliana meiosis. *Plant Reprod.* 26, 197–208. doi: 10.1007/s00497-013-0231-x

234. Zamariola, L., De Storme, N., Vannerum, K., Vandepoele, K., Armstrong, S. J., Franklin, F. C., et al. (2014). SHUGOSHINs and PATRONUS protect meiotic centromere cohesion in Arabidopsis thaliana. *Plant J.*

77, 782–794. doi: 10.1111/tpj.12432

235. Zhang, L., Tao, J., Wang, S., Chong, K., and Wang, T. (2006). The rice OsRad21-4, an orthologue of yeast Rec8 protein, is required for efficient meiosis. *Plant Mol. Biol.* 60, 533–554. doi: 10.1007/s11103-005-4922-z

236. Zhao, D., Han, T., Risseeuw, E., Crosby, W. L., and Ma, H. (2003a). Conservation and divergence of ASK1 and ASK2 gene functions during male meiosis in Arabidopsis thaliana. *Plant Mol. Biol.* 53, 163–173. doi: 10.1023/B:PLAN.0000009273.81702.b5

237. Zhao, D., Ni, W., Feng, B., Han, T., Petrasek, M. G., and Ma, H. (2003b). Members of the arabidopsis-SKP1-like gene family exhibit a variety of expression patterns and may play diverse roles in arabidopsis 1. 133, 203–217. doi: 10.1104/pp.103.024703.to

238. Zhao, D., Yang, X., Quan, L., Timofejeva, L., Rigel, N. W., Ma, H., et al. (2006). ASK1, a SKP1 homolog, is required for nuclear reorganization, presynaptic homolog juxtaposition and the proper distribution of cohesin during meiosis in Arabidopsis. *Plant Mol. Biol.* 62, 99–110. doi: 10.1007/s11103-006-9006-1

239. Zhao, D., Yu, Q., Chen, M., and Ma, H. (2001). The ASK1 gene regulates B function gene expression in cooperation with UFO and LEAFY in Arabidopsis. *Development* 128, 2735–2746.

240. Zhong, C. X., Marshall, J. B., Topp, C., Mroczek, R., Kato, A., Nagaki, K., et al. (2002). Centromeric retroelements and satellites interact with maize kinetochore Protein CENH3. *Plant Cell* 14, 2825–2836. doi: 10.1105/tpc.006106

241. Zhou, S., Wang, Y., Li, W., Zhao, Z., Ren, Y., Wang, Y., et al. (2011). Pollen semi-sterility1 encodes a kinesin-1-like protein important for male meiosis, anther dehiscence, and fertility in rice. *Plant Cell* 23, 111–129. doi: 10.1105/tpc.109.073692

242. Zhou, X., Graumann, K., Evans, D. E., and Meier, I. (2012). Novel plant SUN-KASH bridges are involved in RanGAP anchoring and nuclear shape determination. *J. Cell Biol.* 196, 203–211. doi: 10.1083/jcb.201108098

CHAPTER 2

MEIOTIC TRANSMISSION OF AN IN VITRO–ASSEMBLED AUTONOMOUS MAIZE MINICHROMOSOME

Shawn R. Carlson, Gary W. Rudgers, Helge Zieler, Jennifer M. Mach, Song Luo, Eric Grunden, Cheryl Krol,Gregory P. Copenhaver, Daphne Preuss

INTRODUCTION

Agricultural crops have the potential to meet escalating global demands for affordable and sustainable production of food, fuels, therapeutics, and biomaterials [1]. While standard integrative plant transformation can often meet these needs by safely introducing novel genes into plant chromosomes, they are limited in efficiency. Typically, biological delivery of DNA carried on an *Agrobacterium* T-DNA plasmid, or biolistic delivery of small DNA-coated particles is employed to transfer and integrate desired genes into a host plant chromosome [2]. Integration at random sites results in unpredictable transgene expression due to position effect variegation, variable copy number from tandem integrations, and frequent loss of gene integrity as a result of unpredictable breakage and end joining [2,3]. For highly characterized crops such as maize, transgene integration can also result in genetic linkage of the introduced genes to portions of the genome known to encode loci that confer undesired phenotypes, adding complexity when the transgenic locus is introgressed into other varieties [4,5]. Recent advances in gene integration technologies have aimed to surmount some of these difficulties. For example, zinc finger–mediated homologous recombination or site-specific recombination could eliminate the unpredictable expression that results from random insertion into the plant genome [6,7]. In addition, combining binary T-DNA elements with bacterial artificial chromosome (BAC) technology to produce BiBACs has the potential to introduce larger DNA fragments into the host genome [8,9]. In contrast to these systems, the maize minichromosomes described here remain separate from the host chromosomes, and thus provide an alternative approach with important benefits. Indeed, although precise integration into host chromosomes has long been a routine technique in Saccharomyces cerevisiae, the facile properties of autonomous vectors often make them a preferred choice for numerous applications, including commercial-scale protein production.

The first eukaryotic minichromosomes employed a simple centromere (CEN) sequence from the budding yeast S. cerevisiae, incorporated into versatile circular CEN and linear yeast artificial chromosome (YAC) vectors [10,11]. These yeast vectors were used to define a 125-bp DNA fragment sufficient for mitotic and meiotic centromere function [12]. While circular CEN vectors are most useful for carrying smaller DNA fragments, YAC vectors can carry megabase quantities of DNA and are convenient for manipulating large fragments of DNA [13]. Similarly, with carrying capacities of hundreds of kb, human artificial chromosomes (HACs) provide advantages over other in vitro–assembled vectors used in human cell transfection [14]. HACs containing tandem repeats of a 171-bp alpha satellite sequence can be maintained either as circular or linear, telomere-containing, episomes [15–19].

DNA sequences that can form stable minichromosomes are able to recapitulate centromere functions de novo by recruiting essential DNA binding proteins and epigenetic modifications. In human cells, different satellite arrays vary in their ability to efficiently form HACs, based on their satellite monomer sequence, chromosomal origin, array length, higher-order structure, and even vector composition [20–23]. These DNA sequences recruit centromere binding protein A (CENP-A), which substitutes for histone H3 to form centromeric nucleosomes; this protein marks active centromeres in S. cerevisiae (Cse4p), Schizosaccharomyces pombe (Cnp1),Drosophila melanogaster (Cid), Arabidopsis thaliana (*HTR12*), Zea mays (CENH3), and Homo sapiens (CENP-A) [24–29]. CENP-A complexes are

Citation: Carlson SR, Rudgers GW, Zieler H, Mach JM, Luo S, Grunden E, et al. (2007) Meiotic Transmission of an In Vitro–Assembled Autonomous Maize Minichromosome. PLoS Genet 3(10): e179. doi:10.1371/journal.pgen.0030179.

maintained through mitosis and meiosis [30], resulting in an epigenetic mark that may be more important in perpetuating centromere activity than the underlying DNA sequence. Evidence for this role in centromere maintenance comes from human neocentromeres [31], where, at a very low frequency, ectopic centromeres are nucleated in regions that lack satellite DNA. Once formed, these neocentromeres are efficiently perpetuated. The ability to form centromeres on naked DNA also depends on cell type in mammalian systems; indeed, HAC formation has only been demonstrated in HT1080 fibrosarcoma cells. Yet once established, HACs can be transferred to other mammalian cell types, where they are stably maintained [32].

Maize centromeres contain repetitive sequences that are similar to those found in mammalian centromeres; for example, analogous to the tandem arrays of alpha satellite found in human centromeres, large tandem arrays of the 156-bp maize CentC satellite bind to CENP-A [33,34,28]. These satellite arrays are often interrupted by CRM, a centromere-specific retroelement that also binds CENP-A [28]; the significance of this arrangement for centromere function is unknown. Some maize varieties also have supernumerary B chromosomes with a distinct centromere satellite sequence, ZmBs [35,36]. These B chromosomes lack essential genes, and thus have been particularly useful for discerning the relationship between centromere structure and meiotic transmission [37–39]. A series of deletion derivatives of natural B chromosomes, derived from an A-B translocation event, showed a strong dependence on centromere size—the smallest functional derivative contained a 110-kb centromere and resulted in a meiotic transmission rate of 5%, yet showed a high stability in mitosis [39]. More recently, telomere-mediated chromosomal truncation was used to generate deletion derivatives from both A and B maize chromosomes [40]. Transgenes carried on these derivative chromosomes (or "engineered minichromosomes") were expressed and meiotic inheritance ranged from 12% to 39% [40]. While this telomere-truncation approach can deliver both transgenes and sequences that promote site-directed integration, its utility for commercial applications may be limited—most commercial maize hybrids lack B chromosomes, and the duplications needed to maintain truncated A chromosomes may prove challenging for regulatory approval.

As described below, we developed autonomous minichromosomes that do not rely on alteration of endogenous chromosomes. We constructed plasmids carrying maize centromeric repeats, delivered purified constructs to embryogenic maize tissue, and assessed their ability to promote the formation of MMCs. MMC1 was characterized in detail; this CentC-based construct contained 19 kb of centromeric DNA and conferred efficient mitotic and meiotic inheritance through at least four generations when introduced into plant cells. This approach could be widely used in commercial corn production—a construct with a defined sequence will facilitate regulatory review, while MMC independence from the host genome reduces the risk of alternations that impair host fitness.

RESULTS

Screen for Functional Maize Minichromosomes

We probed a maize genomic BAC library with repetitive sequences, including those typically found in maize centromeres (Materials and Methods; Table 1). Clones enriched in satellite sequences, centromeric retroelements, and other repetitive sequences were chosen to assess whether they can form MMCs when delivered to plant cells. While our study did not explore the interactions between MMC DNA inserts and kinetochore or spindle proteins, we hereafter refer to these fragments as "centromeric," based on the typical genomic location of the sequences they contain. In vitro Cre-lox recombination was used to fuse selected BAC clones to a circular vector containing a plant selectable marker (*nptII*) and a cell-autonomous reporter gene (nuclear-expressed *DsRed*), forming circular constructs. While circular and linear HACs containing repetitive centromeric DNA are mitotically transmitted in human cell lines, circular HACs can confer higher levels of meiotic transmission in transgenic mice [3,20,32]. Consequently, we focused our initial efforts on circular constructs, generating plants by bombarding embryogenic maize tissue with purified candidate MMC DNA, selecting transformed cells expressing the *nptII* marker and resistant to antibiotics, and propagating regenerated plants in the absence of selection (see Materials and Methods). Of the 102 constructs bombarded, 66 gave rise to regenerated plants; 52 of these constructs were randomly chosen and characterized as described below.

Table 1: Classification of Maize BAC Clones Containing Repetitive DNA http://dx.doi.org/10.1371/journal.pgen.0030179.t001

Class[a]	CentA[b]	Cent4[b]	CentC[b]	CRM[b]	MZEHETRO[b]	TR-1[b]	MMCs[c]	Autonomous[d]
I	<7	-	≥7	<7	-	-	13	10
II	≥6	-	≥7	<7	-	-	6	5
III	<6	-	≥7	≥6	-	-	10	10
IV	>6	-	≥7	≥6	-	-	8	8
V	>0	-	≥7	>0	-	-	6	5
VI	>0	>5	>0	>0	-	-	1	1
VII	>0	-	>0	-	<6	>6	2	2
VIII	>0	-	>0	-	>7	<5	3	3
IX	>0	-	>0	-	>6	>6	3	3
Total							52	47

[a]BAC clones were sorted into hybridization classes I through IX;
[b]Phosphorimager signal intensity to the indicated repetitive probes, normalized to a scale of 1–10.
[c]MMCs characterized.
[d]Autonomous MMCs characterized, by class (see Table 2). MMC1 was derived from Class I.
doi:10.1371/journal.pgen.0030179.t001

Table 2: MMC Transformation Events and FISH Analysis http://dx.doi.org/10.1371/journal.pgen.0030179.t002

Construct[a]	Explants[b]	Transformation Events	Analyzed by FISH	Autonomous	Integrated	Autonomous and Integrated
1	264	10	9	5	—	4
2	87	2	2	2	—	—
3	87	1	1	1	—	—
4	150	13	1	1	—	—
5	25	1	1	—	1	—
6	150	8	3	3	—	—
7	204	22	10	4	—	6
8	134	6	1	1	—	—
9	54	4	1	—	—	1
10	54	2	1	—	1	—
11	54	1	1	—	1	—
12	54	2	1	1	—	—
13	50	3	1	1	—	—
14	50	3	1	—	—	1
15	50	3	2	1	—	1
16	168	7	1	1	—	—
17	50	4	2	—	—	2
18	50	6	4	3	—	1
19	50	2	1	1	—	—
20	200	5	5	4	—	1
21	50	1	1	1	—	—
22	50	1	1	1	—	—
23	50	1	1	1	—	—
24	40	1	1	1	—	—
25	60	1	1	1	—	—
26	60	1	1	1	—	—
27	60	2	1	1	—	—
28	126	6	1	1	—	—
29	65	3	1	1	—	—
30	126	2	1	1	—	—
31	126	4	2	1	—	1
32	168	10	5	4	—	1
33	294	15	2	2	—	—
34	168	1	1	1	—	—
35	126	12	5	3	—	2
36	126	7	2	1	—	1
37	110	6	2	1	—	1
38	126	6	1	1	—	—
39	126	8	2	2	—	—
40	126	1	1	1	—	—
41	168	6	1	1	—	—
42	126	4	2	1	—	1
43	206	2	1	—	1	—
44	126	2	1	1	—	—
45	126	6	5	3	1	1
46	126	7	2	1	—	1
47	126	2	2	2	—	—
48	126	10	3	1	—	2
49	206	1	1	—	1	—
50	126	3	1	1	—	—
51	126	6	1	—	—	1
52	126	2	2	2	—	—
Total	5,882	245	104	69	6	29

[a]Candidate MMCs; construct 1 is MMC1.
[b]Number of embryogenic tissues bombarded.
doi:10.1371/journal.pgen.0030179.t002

To evaluate whether the introduced constructs were maintained autonomously or instead had integrated into the genome, we preformed fluorescence in situ hybridization (FISH). We arrested root tip cells in mitosis, and stained chromosome spreads [41] with rhodamine and fluorescein-labeled probes corresponding to centromeric repeats and to MMC-encoded genes, respectively (Figure 1A–1I). FISH labeling of integrated control constructs resulted in adjacent pairs of metaphase FISH signals corresponding to replicated sister chromatids (Figure 1J). While some MMC constructs integrated (see below), we considered MMCs autonomous when (i) ≥70% of the cells examined ($n \geq 15$) contained signals that were clearly distinct from the DAPI-stained host chromosomes, (ii) integrated signals were not detected, and iii) the fluorescent probe corresponding to the MMC-encoded genes colocalized with the probe to repetitive centromeric DNA, suggesting an intact construct and making it unlikely that the signal was due to noise. In many cases, the detection of a DAPI signal that colocalized with the FISH probes provided further evidence of MMC autonomy (Figure 1B–1D and 1F–1H).

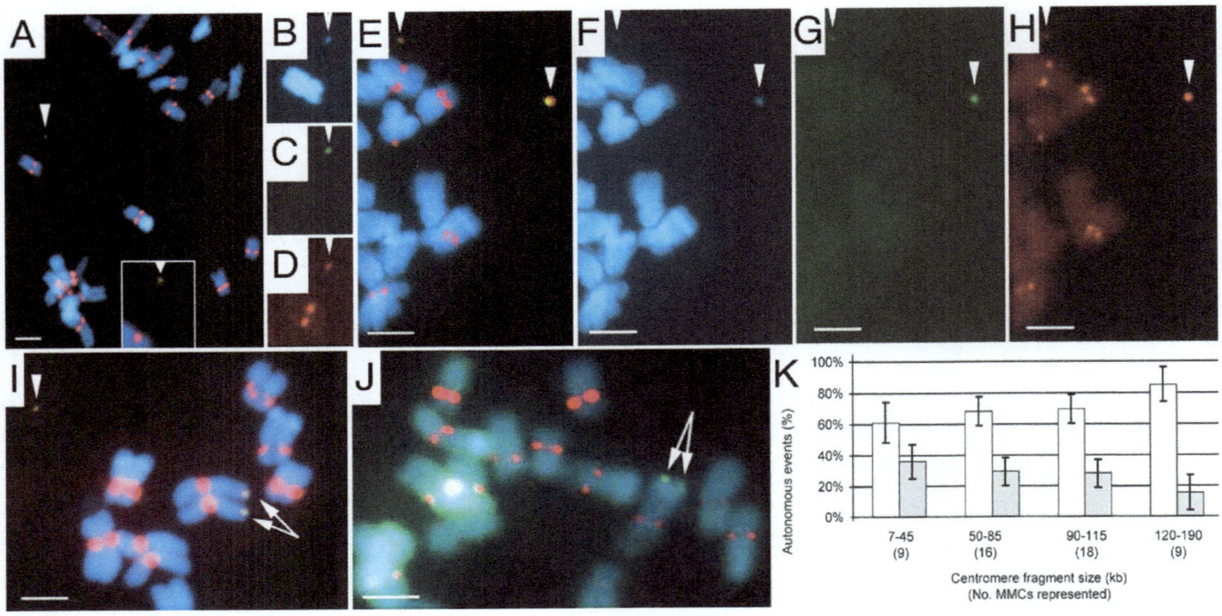

Figure 1: Generation of Autonomous Minichromosomes (A–H) Metaphase chromosome spreads from MMC1 event V-1: (A–D) T1 plant; (B–D) correspond to the region denoted by the arrowhead in (A); (E–H) T2 plant. DNA is stained with DAPI ([B F], blue) and labeled with FISH probes specific for the *DsRed* and*nptII* gene cassette ([C, G], green); or centromere sequences ([D, H], red). (I, J) Event V-4 with autonomous and integrated copies of MMC1 (I); pCHR758 (noncentromeric control) (J). Autonomous minichromosomes (arrowheads); integrated constructs appear as pairs of FISH signals (arrows); size bar, 5 μm. (K) Centromere fragments across a wide size range enable autonomous MMC inheritance. For each size category, the percentage of transformation events (total = 52) that yielded only an autonomous MMC (white bars) or both an autonomous and integrated MMC in the same cell (grey bars) is shown; the number of MMCs in each category is noted parenthetically; error bars indicate standard error. http://dx.doi.org/10.1371/journal.pgen.0030179.g001

Based on these criteria, 47/52 (90%) of the constructs we evaluated with FISH were able to form an autonomous MMC, and 43/52 (with centromeric inserts ranging in size from 7 to 190 kb) gave rise to plants that contained only an autonomous MMC (Table 2). This unexpectedly high rate of recovering autonomous MMCs suggests that embryogenic maize tissue readily establishes MMCs from purified DNA and that the BAC clones that yielded transformed plants contained sequences that efficiently promote MMC formation. The efficiency of forming an autonomous MMC increased slightly, although not significantly (*t*-test), as the size of the genomic DNA insert increased (Figure 1K). A similar analysis of human centromeric fragments showed that as little as 35 kb could generate a HAC, while larger fragments (70–220 kb) were required for efficient HAC formation [42]. As described below, MMCs were often efficiently inherited; nonetheless, MMC integration was detected only during the initial transformation event, and not in subsequent generations (T1 through T4, 0/312 metaphase spreads, 33 plants). Below, we report on the composition and behavior of one of the MMC constructs (MMC1) in detail.

Meiotic Inheritance of MMC1

Control transformations performed with a *DsRed/nptII* plasmid lacking a centromere-derived insert (pCHR758) contained a construct that integrated, as expected, into a native chromosome (7/7 events, Figure 1J). In contrast, for MMC1, 5/9 independent transformation events yielded solely an autonomous chromosome (Figure 1A–1H) and 4/9 generated both integrated and autonomous copies (Figure 1I). We tested the ability of these MMCs to confer inheritance by crossing T0 transformants to wild type, growing the progeny without selection, and monitoring nuclear-localized *DsRed* fluorescence (Figure 2A). Because we typically observed only one MMC per cell (monosomic), we expected these T0 plants to behave as hemizygotes; if the MMC obeyed Mendelian inheritance, then such crosses would yield *DsRed* progeny in a 1:1 ratio. Ten T0 plants (derived from three events) carrying solely an autonomous MMC1 copy were crossed to wild-type pollen. Two of the MMC1 events (V-1 and Q-1) indeed transmitted *DsRed* to T1 offspring in ratios that did not differ significantly from Mendelian predictions (Table 3). However, for a third MMC1 event (Q-2), we saw a significant reduction in $DsRed^{+}$ progeny compared to expectations (52%, Table 3), suggesting genetic instability. PCR analysis of the progeny from this cross confirmed that the plants lacking*DsRed* expression also lacked *DsRed* sequences, indicating that the deviation from Mendelian assortment was not due to silencing of gene expression. Instead, the elevated MMC loss rate in this event could result from in planta modifications of the centromeric insert or from epigenetic effects that led to less robust segregation [43]. As expected, performing a similar analysis of six events carrying an integrated pCHR758 backbone yielded Mendelian inheritance ratios (118:119 $DsRed^{+}$:$DsRed^{-}$; $p > 0.05$).

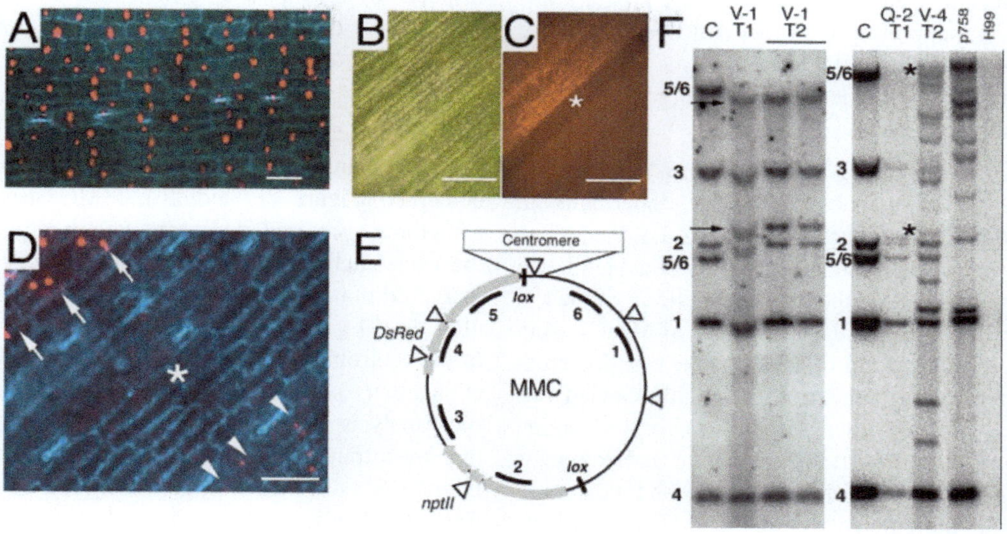

Figure 2: MMC Gene Expression and Structure (A) Fluorescent detection of nuclear-localized *DsRed* in MMC1 maize leaf; size bar, 50 μm. (B, C) Detection of *DsRed* sectors in a T2 plant leaf from event V-1 under (B) bright-field and (C) fluorescence microscopy; size bars, 0.5 mm. (D) high magnification view of image shown in (C) with the corresponding sector, comprising all cell layers, indicated by an asterisk; the edge of a sector that comprises only the adaxial cell layer is indicated by arrowheads, cells with typical *DsRed*expression are indicated by arrows. Size bar, 50 μm. (E) MMC consisting of a pCHR758 backbone and a centromere-derived insert, gene expression cassettes (grey), centromeric inserts (box), BglII restriction sites (arrowheads), and probes used for FISH and Southern blot analyses are indicated. (F) Southern blot of DNA digested with BglII and hybridized to probes 1–6 (E); Bands 1–4 measure 2,067, 3,167, 5,227 and 790 bp, respectively; those hybridizing to probes 5 and 6 vary in size, depending on the location of BglII sites within the centromeric DNA insert. MMC1 Control (c, lanes 1 and 5) DNA was purified from E. coli and hybridization patterns were compared to DNA from plant cell extracts derived from MMC1 events V-1 (lanes 2–4), Q-2 (lane 6), and V-4 (lane 7), as well as from plants transformed with pCHR758 (lane 8) and untransformed wild type (H99, lane 9). For events V-1 and Q-2, bands differing from bacterial grown controls are indicated (arrows and asterisk, respectively). http://dx.doi.org/10.1371/journal.pgen.0030179.g002

Table 3: Meiotic Inheritance of MMC1 http://dx.doi.org/10.1371/journal.pgen.0030179.t003

Construct	Event	Generation	Female[a]	Male[a]	Expected Progeny Ratio	Observed Progeny Ratio (*p*-Value)[b]	% Loss[c]
MMC1	V-1	T0	MMC1 (M)	WT	1:1	78:94 (0.75)	9.3
		T1	MMC1 (M)	WT	1:1	118:149 (0.06)	12
		T1	WT	MMC1 (M)	1:1	31:48 (0.06)	21.5
		T1[d]	MMC1 (M)	WT	1:1	0:48 (*)	100
		T1[d]	WT	MMC1 (M)	1:1	0:35 (*)	100
		T1	MMC1 (M)	MMC1 (M)	3:1	61:86 (*)	44.7
		T2	MMC1 (M)	WT	1:1	67:65 (0.86)	0
		T2	MMC1 (M)	MMC1 (M)	3:1	82:35 (0.22)	6.6
		T2	WT	MMC1 (D)	1:0	184:18 (NA)	8.9
		T2	MMC1 (D)	MMC1 (D)	1:0	48:0 (NA)	0
		T3	MMC1 (M)	WT	1:1	38:37 (0.91)	0
		T3	MMC1 (M)	MMC1 (M)	3:1	80:33 (0.3)	5.6
MMC1	Q-1	T0	MMC1 (M)	WT	1:1	17:19 (0.74)	5.6
MMC1	Q-2	T0	MMC1 (M)	WT	1:1	33:105 (*)	52
pCHR758	1	T0	pCHR758 (M)	WT	1:1	118:119 (0.95)	0.4

[a]M, monosomic for MMC1; D, disomic for MMC1; WT, wild-type maize.
[b]*p*-Value calculations based on chi-square distributions with 1 degree of freedom. *p*-Values significantly different from expectations (χ^2 $p < 0.05$) are indicated with an asterisk. *p*-Values were not calculated for expectations of 1:0 and are noted as NA.
[c]Loss rates calculated as the difference between the expected and observed numbers of *DsRed* positive progeny, expressed as percent of the expected (assuming Mendelian assortment).
[d]Crosses derived from a single V-1 plant that demonstrated sectoring in the T1 generation; loss was confirmed by PCR.
doi:10.1371/journal.pgen.0030179.t003

FISH analysis showed that T1 plants from event V-1 retained an autonomous MMC: a *DsRed*-containing episome was present in 80% of root metaphase cells ($n = 44$), a detection level consistent with previous artificial chromosome studies [44]. Because we consistently observed*DsRed* expression in nearly every cell from these plants (see below), we conclude that the absence of an MMC FISH signal in 20% of root cells likely represents the challenges of retaining and detecting every MMC throughout the FISH protocol. To monitor MMC1 inheritance in subsequent generations and through both male and female gametes, we performed a series of crosses with T1, T2, and T3 plants derived from event V-1 and monitored *DsRed* transmission. When male or female monosomic MMC1 plants were crossed to wild type, *DsRed* segregation was not significantly different from Mendelian inheritance ratios (1:1, Table 3). For one exceptional T1 plant, however, such crosses yielded no progeny containing MMC1 (female: 0:48, male: 0:35; Table

3); the absence of *DsRed*-encoding DNA in these progeny was confirmed by PCR, supporting the view that this MMC was indeed autonomous. Interestingly, the leaf tissue of this plant had prominent mitotic *DsRed* leaf sectors, suggesting a high rate of MMC instability.

When we self-pollinated T2 and T3 hemizygous plants derived from event V-1, we observed *DsRed*$^+$ inheritance in a ratio that did not significantly differ from a 3:1 Mendelian pattern. However, in a second case of non-Mendelian assortment, a self-cross in the T1 generation yielded a 1:1 *DsRed*$^+$ inheritance ratio, suggesting loss of MMC1 from either the male or female floral tissue. Nonetheless, this cross was useful for generating plants that potentially carried two copies of MMC1 (homozygous disomes). Crossing pollen from a candidate T2 disome onto five different maize inbreds yielded 184 *DsRed*$^+$:18 *DsRed*$^-$ offspring ($p > 0.05$ for disomy). Similarly, self-pollinating potentially disomic T2 or T3 plants produced 48:0 and 24:0 *DsRed*$^+$:*DsRed*$^-$ offspring, respectively. Quantitative PCR (qPCR) analysis of the potentially disomic T2 plants confirmed 2.00 and 1.90 (standard error = 0.08) *DsRed* copies per cell, respectively (see Materials and Methods).

MMC1 Stability: Gene Expression and Structure

For most plants carrying an autonomous MMC, nuclear *DsRed* expression was observed in nearly every leaf cell, indicating stability through mitosis. In some cases, however, sectors that lacked *DsRed* expression were found (Figure 2B–2D); these were generally limited to a few cell files. In reproductive tissues, such sectors could be responsible for the aberrant meiotic MMC segregation described above. In total, mitotic sectors of *DsRed* expression from MMC1 were detected in 3.6% of T0 plants ($n = 56$), 3.0% of T1 plants ($n = 404$), 1.9% of T2 plants ($n = 837$), and no T3 ($n = 738$) or T4 ($n = 250$) plants. The reduced sectoring frequency as plants advanced through generations suggests a gradual increase in MMC stability due to changes in DNA composition, epigenetic modifications, or MMC copy number in mitotic cells. A similar stabilization through generations was observed in an oat-maize addition line [45]. We also found that 60 d of crowding and drought stress did not appreciably alter MMC1 stability; *DsRed* expression was found in every T2 and T3 plant from event V-1 grown under stress (151 and 159 plants, respectively). Moreover, pollen from stressed hemizygous T2 plants demonstrated Mendelian *DsRed* segregation (281:238 *DsRed*$^+$:*DsRed*$^-$; $p > 0.05$).

To assess the structure of MMC1 through generations, we performed Southern blot analysis, probing to detect all of the unique sequence bands contained in the MMC construct (Figure 2E and 2F). MMC structural alterations sometimes occurred during transformation, often involving the centromeric insert, rather than the gene cassette (Figure 2F). Additional rearrangements were typically not detected after the T1 generation ($n = 5$), although the repetitive nature of the centromeric fragment made it impossible to thoroughly evaluate its structure on these blots. In addition, Southern blot analysis showed centromeric alterations in event V-1 that were transmitted from the T1 parent to the T2 progeny. Event Q-2 suffered a larger alteration of the centromeric fragments (indicated by an asterisk in Figure 2F), potentially explaining its reduced meiotic stability. In contrast, an event carrying both integrated and autonomous MMC1 copies (V-4) showed a more complicated pattern, as did plants carrying integrated pCHR758. As expected for independently assorting loci, when plants from event V-4 were crossed to wild type, the autonomous and integrated copies segregated: FISH evaluation of *DsRed*-expressing T2 plants yielded a 1:4:2 ratio (autonomous:autonomous and integrated:integrated).

MMC1 Composition

MMC1 was originally identified by its strong hybridization to a CentC probe, suggesting it contained a high percentage of this satellite repeat (Table 1). Sequence analysis confirmed the presence of CentC repeats arranged in an uninterrupted tandem array (GenBank accession number in Supporting Information; Figure 3A and 3B). The repetitive nature of CentC made a precise assembly of this array challenging; we used rare DNA polymorphisms within the repeats to aid in sequence assembly, and confirmed the overall length of the array (approximately 9 kb) with restriction enzyme digestion and gel electrophoresis. Based on these measurements and quantitative dot blot hybridization (see Materials and Methods) the CentC array contains between 59 and 64 (61.4 ± 2.3) copies. CentC repeat alignments showed that each base is conserved at an average frequency of 96.1% (Figure 3C and 3D), a level consistent with previously reported plant satellite conservation [46]. Clustering algorithms failed to detect higher order repeat patterns in MMC1 (unpublished data).

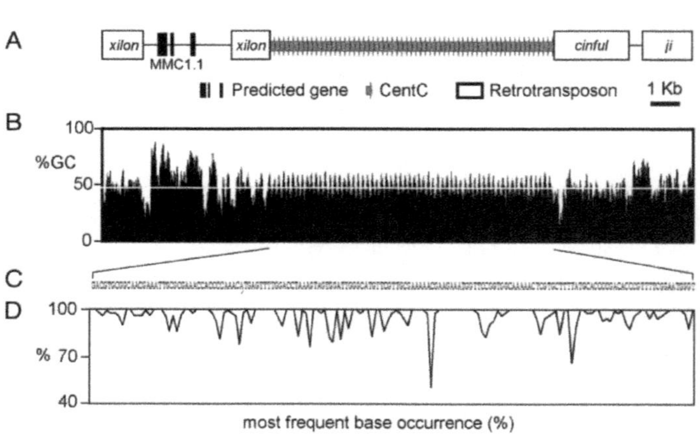

Figure 3: Sequence Analysis of the MMC1 Centromeric Fragment (A) Assembled sequence. (B) Fifty base-pair sliding window analysis of MMC1 GC content; white line, genomic average = 49.5%. (C) Consensus sequence of ~60 MMC1 CentC repeats. (D) The occurrence of the

most frequent base (%) for each nucleotide in the CentC consensus. http://dx.doi.org/10.1371/journal.pgen.0030179.g003

While the maize genome has an average GC content of 49.5%, the 5.6- and 4.8-kb regions flanking the CentC array of MMC1 reach 88% and 70% GC, respectively (Figure 3B). Overall, the GC content of the MMC1 centromeric insert is 48%; by comparison, published sequences from two maize centromeric BACs had 43% and 47% GC content [34] while *Arabidopsis* and rice centromere DNA averages 35%–40% and 39%–48%, respectively [47,48]. MMC1 encodes four regions with similarity to retrotransposons *xilon*, *cinful*, or *ji* [49], as well as a 453-bp open reading frame (MMC1.1) that potentially encodes a novel protein of unknown function, complete with a promoter and poly-A signal (Figure 3A). BLAST searches of GenBank revealed no evidence for MMC1.1 expression, but transcripts >95% identical to CentC and to the MMC1 retrotransposons were abundant. Transcription of centromeric repeats is important for centromere function in S. pombe [50], and *Arabidopsis* satellites are also transcribed [51]. The centromere-specific histone CENH3 binds to transcripts corresponding to CentC and to the retrotransposon CRM, suggesting a role for these RNAs in centromere function [52]; it is not clear if *xilon*, *cinful*, or *ji* transcripts play a similar role. Retrotransposons also can nucleate the formation of heterochromatin that can spread to nearby regions [53], although MMC1-encoded*DsRed* and *nptII* were readily expressed, despite their separation of 3.3 and 6.2 kb, respectively, from retrotransposons.

DISCUSSION

Taken together, the experiments described above strongly support the conclusion that MMC1 can be maintained as an autonomous chromosome: it remains distinct from host chromosomes, its gene cassette is structurally stable through at least four generations, the genes it carries are expressed and transmitted through meiosis and mitosis, and, in some cases, it can be lost from the genome at a frequency higher than that of a native chromosome. Interestingly, classical studies of plant trisomics typically reveal far greater defects in meiotic inheritance [54], while inheritance levels similar to those we observed with MMCs have been reported in other artificial chromosome systems. For example, a monosomic mouse artificial chromosome that showed <1% mitotic loss when carried in human, bovine, or mouse cell lines [55], suffered only 4% meiotic loss through the mouse germline [44]. Furthermore, while classically studied ring chromosomes are often unstable [56], circular MMC inheritance through four generations was reminiscent of that observed for circular chromosomes from yeast [57], mammals [32], and maize [58]. These data suggest that this MMC could be maintained indefinitely.

MMC centromere sequences, like those that make up endogenous centromeres, could rely on the kinetochore and spindle machinery for faithful segregation, or could be inherited through alternative mechanisms. For example, in plants, dense heterochromatic domains known as knobs or neocentromeres migrate to daughter cells by moving along the sides of the spindle, rather than by kinetochore-mediated association with the ends of microtubules. This process results in preferential assortment to gametes, and consequently greater than expected inheritance ratios (termed meiotic drive) [59]. Heterochromatin-based mechanisms of assortment have also been characterized in *Drosophila*, where chromosomes that lack evidence of meiotic exchange (chiasmata) are nonetheless inherited at Mendelian ratios [60]. Further, in S. cerevisiae, which lacks appreciable heterochromatin, the 2-μm circle plasmid is partitioned at an efficiency that rivals that of yeast chromosomes; this assortment relies on microtubule-mediated attachment of cohesin to 2 μm of DNA [61]. The possibility that MMC segregation might rely on alternative mechanisms is intriguing; indeed, the relatively small MMCs may differ from mammalian artificial chromosomes in which large alpha satellite arrays bind essential centromere proteins.

Epigenetic factors have been postulated to play a principal role in establishing higher eukaryotic centromeres [43], with studies of human neocentromeres [62] and *Drosophila*strains overexpressing CENH3 [63] suggesting a lack of dependence on specific DNA sequences. On the other hand, HACs are able to efficiently nucleate centromere activity in a sequence-dependent manner, and HAC sequences tend to expand in vivo [20], suggesting a selection for a preferred size and/or composition. The MMC1 DNA that we delivered to plant cells was purified from E. coli and thus lacked eukaryotic epigenetic marks, yet it formed autonomous chromosomes. This MMC construct contained only a 19-kb centromeric insert and is thus substantially smaller than the centromeric regions that were previously known to provide mitotic and meiotic inheritance. For example, the fully sequenced centromere of rice Chromosome 8 contains a satellite array measuring 69 kb [64], and a deletion derivative of the maize B chromosome that measures 110 kb is sufficient to confer meiotic inheritance, albeit inefficiently [39]. While HACs routinely expand to a larger size in vivo, we did not detect major rearrangements or expansions of MMC1 DNA through two plant generations, suggesting that its composition was adequate to establish a minichromosome. Nonetheless, our analysis was unable to fully assess the structure of the repetitive centromeric DNA, and it remains possible that these regions could expand, contract, or rearrange in some other manner.

While the total size of MMC1 is quite small (35 kb), other MMCs, some measuring over 200 kb, were successfully delivered to plants and transmitted through meiosis (unpublished data). This suggests that MMC1 has the capacity to serve as a platform to carry a large number of genes. As this MMC is optimized to commercial performance levels, it will provide an unprecedented opportunity to deliver gene combinations ("stacks") that confer valuable traits to corn varieties. Long breeding programs are often required to introgress an integrated transgene into desired germplasm, while eliminating undesirable linked loci. Because an MMC forms an independent linkage group, these programs could be accelerated, allowing products to appear in the marketplace sooner. Moreover, the performance and expression of transgenic traits will likely become more predictable and reliable as MMC design rules are understood. Extensions of this minichromosome technology beyond traditional agriculture may enable the construction of multigene pathways to produce pharmaceuticals and other industrial products in plants.

MATERIALS AND METHODS

Construction of Candidate MMCs.

A BAC library was created in pBeloBAC11 using MboI-digested DNA from the maize inbred B73. This library was arrayed on nitrocellulose filters and probed separately with repetitive sequences from maize that are often found in centromeres or neocentromeres: CentA, Cent4, CentC, CRM, MZEHETRO, and TR-1; ^{32}P-labeled probes were hybridized for 14 h at 65 °C and washed with 0.5× SSC, 1% SDS three times at 65 °C. To identify clones carrying centromere DNA, phosphorimager scans of each hybridization experiment were digitally assembled into a MySQL database. BAC clones with strong hybridization signals to one or more of the repetitive sequences were selected for minichromosome construction (Table 1). First, a high copy number plasmid (pCHR758) carrying the *Arabidopsis* UBQ10 promoter to *DsRed* (Clonetech) and the yeast YAT1 promoter fused to *nptII* was constructed. An 8.5-kb fragment encoding the*DsRed* and *nptII* expression constructs (and lacking a bacterial origin) was liberated from pCHR758 with I-PpoI, purified from an agarose gel (QIAquick Gel Extraction Kit, Qiagen), and circularized by Cre-mediated exchange (New England Biolabs) at two *loxP* sites that flanked the gene expression cassette. BAC clones carrying putative centromere DNA insertions were recombined with this vector via the *loxP* site in pBeloBAC11, generating circular candidate MMC constructs (Figure 2E). These constructs were maintained in E. coli DH10B (Invitrogen).

Delivery and propagation of candidate MMCs in plants.

MMC constructs grown in E. coli were purified using alkaline lysis or cesium chloride protocols and delivered to embryogenic H99 maize tissues by biolistic bombardment of DNA-coated gold particles as described [65]. Transformed events were identified by selection on Chu's N6 medium containing G418 sulfate (PhytoTechnology Laboratories) or paromomycin (Sigma) and regenerated. Transformed plants were subsequently grown without selection in a soilless mix (Sunshine LC1) in a greenhouse (16-h d, 26–28 °C). Seedlings were grown in 48-well flats (2 sq ft) with one plant per well to the V3 developmental stage and then transplanted into 1.6-gallon pots containing 1:1:1 soil:peat:perlite and grown to maturity. Plants subjected to stress conditions were maintained in 48-well flats for 60 d with watering limited to once per day. MMC containing plants have been advanced through four generations by backcrossing to H99, outcrossing to public maize inbreds, and by self pollination or sibling mating.

Fluorescence Assays.

For *DsRed* expression, leaf 3 (V2 stage of development) was sampled across its entire width (minimally 2,500 cells per sample) and fluorescence was detected using a Zeiss SV-11 dissecting microscope equipped with a rhodamine filter cube (excitation: D540/25; dichroic 565LP; emission: D605/55). Background autofluorescence was detected with a GFP filter cube (excitation: BP 470/40; beamsplitter: FT495; emission: BP 525/50); bona fide *DsRed*fluorescence was not detectable at this excitation wavelength. *DsRed* expression in pollen was determined after fixing florets in 95% ethanol; aceto-carmine staining was subsequently used to assess pollen viability. For FISH, root tips were collected approximately 10 d after transplanting regenerated T0 plants to soil or after germination (T1 through T4 plants). Sampled roots (3–6 per plant) were moistened and exposed to nitrous oxide at 150 psi for 2.5 h to arrest chromosomes in metaphase [66]. Roots were fixed in 90% acetic acid and spread onto poly-lysine coated glass slides by squashing thin cross sections. FISH was performed essentially as described [41] using probes labeled with Alexa488 (pCHR758, Molecular Probes) and Alexa568 (CentC, Roche). Following hybridization, slides were counterstained with DAPI (0.04 mg/ml) and ≥15 metaphase cells were evaluated per plant using a Zeiss Axio-Imager equipped with rhodamine, FITC, and DAPI filter sets (excitation BP 550/24, emission BP 605/70; excitation BP 470/40, emission: BP525/50; and excitation G 365, emission BP 445/50, respectively). Extrachromosomal signals were only considered to indicate autonomous MMCs if ≥70% of the images ($n \geq 15$ cells analyzed) showed colocalization of the Alexa488 and Alexa568 signals within one nuclear diameter of the endogenous metaphase maize chromosomes. Grayscale images were captured in each panel, merged, and adjusted with pseudo-color using Zeiss AxioVision (Version 4.5) software; fluorescent signals from doubly labeled MMCs were detected in both the red and green channels.

PCR and Southern blot analysis.

PCRs were carried out on genomic DNA isolated from young plants; qPCRs were performed in triplicate using a BioRad Chromo4 machine with TaqMan primers and probes (Sigma-Genosys). Amplification was achieved by incubating at 95 °C for 3 min, and 39 cycles of 95 °C for 15 s and 59 °C for 48 s, with a 1 s reduction per cycle. Copy number determinations were made by comparing qPCR signals from a control plasmid containing one copy of the maize*Adh1* gene and *DsRed* to the signals obtained from MMC-containing plants. For Southern blots, genomic DNA was isolated from young leaf tissue using a Nucleobond Plant Genomic DNA extraction kit (Clontech). Ten micrograms of DNA was digested with BglII (New England Biolabs), separated on a 0.7% agarose gel, vacuum transferred to a nylon membrane (Amersham BioSciences), and probed with a mixture of nonoverlapping pCHR758 fragments labeled with ^{32}P (Rediprime II, Amersham BioSciences). Hybridization was performed overnight at 65 °C and blots were washed three times (15 min each) with 0.25× SSC, 0.1% SDS at 65 °C; signals were detected with a Storm phosphorimager.

Sequencing and sequence analyses.

MMC1 was sequenced to an average of 30× coverage by shotgun sequencing (Lark Technologies) and 454 Technology (454 Life Sciences) and assembled with Phred/Phrap; a small gap was closed by primer walking, using direct dye-terminator cycling sequencing of MMC1. Quantitative dot blotting was used to calculate the total size of the CentC array. Briefly, two sets of blots, each containing samples in triplicate, were hybridized with CentC (CC) and vector specific (V) probes separately. Signals for each spot were captured with a Storm phosphorimager and CC/V ratios were calculated. Plasmids with the vector sequence and one, three, and eight copies of a cloned CentC repeat were used as standards. MMC1 assembly was verified by restriction mapping with panels of enzymes (BamHI, BmgBI, EcoRI and HindIII); this data was consistent with the calculated size of the CentC array. BLASTN (http://www.ncbi.nlm.nih.gov/BLAST/Blast.cgi) was used to assess sequence similarity, GENSCAN (http://genes.mit.edu/GENSCAN.html) to predict promoters and open reading frames, and repeat finder (http://tandem.bu.edu/trf/trf.basic.submit.html) to analyze CentC satellites.

Statistical Analyses.

The significance of MMC1 inheritance data was determined with a chi-square goodness-of-fit test. Differences from Mendelian segregation (based on a 1:1 segregation ratio in crosses to wild type or a 3:1 segregation ratio in self crosses or crosses to sibling plants) were considered significant at $p < 0.05$ (or a chi-square value greater than 3.84).

ACKNOWLEDGMENTS

We thank T. Antigegn, J. Cuenca, K. Faulhaber, T. Mentzel, E. Leach, K. Moore, J. Nykaza, M. Pantelic, R. Schurr, J. Shahan, and M. Shenaut for expert technical assistance; J. Jin for database management; and J. Jurek for help with sequence analysis. We also acknowledge the United States Department of Agriculture, the Agricultural Research Services, the North Central Regional Plant Introduction Station, and S. Moose for generously providing maize seed.

REFERENCES

1. Herrera S (2004) Industrial biotechnology–a chance at redemption. Nat Biotech 22: 671–675.
2. Lorence A, Verpoorte R (2004) Gene transfer and expression in plants. Methods Mol Biol 267: 329–350.
3. Birch RG (1997) Plant transformation: problems and strategies for practical application. Annu Rev Plant Physiol Plant Mol Biol 48: 297–326.
4. Walker D, Boerma HR, All J, Parrott W (2002) Combining *cry1Ac* with QTL alleles from PI 229358 to improve soybean resistance to lepidopteran pests. Molec Breeding 9: 43–51.
5. Yin Z, Plader W, Malepszy S (2004) Transgene inheritance in plants. J Appl Genet 45: 127–144.
6. Kumar S, Allen GC, Thompson WF (2006) Gene targeting in plants: fingers on the move. Trends Plant Sci 11: 159–161.
7. Gilbertson L (2003) Cre-lox recombination: Cre-ative tools for plant biotechnology. Trends Biotech 21: 550–555.
8. Hamilton CM, Frary A, Lewis C, Tanksley SD (1996) Stable transfer of intact high molecular weight DNA into plant chromosomes. Proc Natl Acad Sci U S A 93: 9975–9979.
9. He RF, Wang Y, Shi Z, Ren X, Zhu L, et al. (2003) Construction of a genomic library of wild rice and *Agrobacterium*-mediated transformation of large insert DNA linked to BPH resistance locus. Gene 321: 113–121.
10. Burke DT, Carle GF, Olson MV (1987) Cloning of large segments of exogenous DNA into yeast by means of artificial chromosome vectors. Science 236: 806–812.
11. Clarke L, Carbon J (1980) Isolation of a yeast centromere and construction of functional small circular chromosomes. Nature 287: 504–509.
12. Cottarel G, Shero JH, Hieter P, Hegemann JH (1989) A 125-base-pair CEN6 DNA fragment is sufficient for complete meiotic and mitotic centromere functions inSaccharomyces cerevisiae. Mol Cell Biol 9: 3342–3349.
13. Larin Z, Monaco AP, Lehrach H (1991) Yeast artificial chromosome libraries containing large inserts from mouse and human DNA. Proc Natl Acad Sci U S A 88: 4123–4127.
14. Kuroiwa Y, Tomizuka K, Shinohara T, Kazuki Y, Yoshida H, Ohguma A, et al. (2000) Manipulation of human minichromosomes to carry greater than megabase-sized chromosome inserts. Nat Biotech 18: 1086–1090.
15. Ebersole TA, Ross A, Clark E, McGill N, Schindelhauer D, et al. (2000) Mammalian artificial chromosome formation from circular alphoid input DNA does not require telomere repeats. Human Mol Genet 9: 1623–1631.
16. Schueler MG, Higgins AW, Rudd MK, Gustashaw K, Willard HF (2001) Genomic and genetic definitions of a functional human centromere. Science 294: 109–115.
17. Harrington JJ, Van Bokkelen G, Mays RW, Gustashaw K, Willard HF (1997) Formation of de novo centromeres and construction of first-generation human artificial microchromosomes. Nat Genet 15: 345–355.
18. Ikeno M, Grimes B, Okazaki T, Nakano M, Saitoh K, et al. (1998) Construction of YAC-based mammalian artificial chromosomes. Nat Biotech 16: 431–439.
19. Tsuduki T, Nakano M, Yasuoka N, Yamazaki S, Okada T, et al. (2006) An artificially constructed de novo human chromosome behaves almost identically to its natural counterpart during metaphase and anaphase in living cells. Mol Cell Biol 26: 7682–7695.
20. Grimes BR, Rhoades AA, Willard HF (2002) α-satellite DNA and vector composition influence rates of human artificial chromosome formation. Mol Ther 5: 798–805.
21. Mejia JE, Alazami A, Willmott A, Marschall P, Levy E, et al. (2002) Efficiency of *de Novo* centromere formation in human artificial chromosomes. Genomics 79: 297–304.
22. Ohzeki J, Nakano M, Okada T, Masumoto H (2002) CENP-B box is required for de novo centromere chromatin assembly on human alphoid DNA. J Cell Biol 159: 765–775.
23. Okamoto Y, Nakano M, Ohzeki J, Larionov V, Masumoto H (2007) A minimal CENP-A core is required for nucleation and maintenance of a functional human centromere. EMBO J 26: 1279–1291.
24. Meluh PB, Yang P, Glowczewski L, Koshland D, Smith MM (1998) Cse4p is a component of the core centromere of Saccharomyces cerevisiae. Cell 94: 607–613.
25. Takahashi K, Chen ES, Yanagida M (2000) Requirement of Mis6 centromere connector for localizing a CENP-

A-like protein in fission yeast. Science 288: 2215–2219.

26. Malik HS, Henikoff S (2001) Adaptive evolution of Cid, a centromere-specific histone in Drosophila. Genetics 157: 1293–1298.

27. Talbert PB, Masuelli R, Tyagi AP, Comai L, Henikoff S (2002) Centromeric localization and adaptive evolution of an Arabidopsis histone H3 variant. Plant Cell 14: 1053–1066.

28. Zhong CX, Marshall JB, Topp C, Mroczek R, Kato A, et al. (2002) Centromeric retroelements and satellites interact with maize kinetochore protein CENH3. Plant Cell 14: 2825–2836.

29. Palmer DK, O›Day K, Wener MH, Andrews BS, Margolis RL (1987) A 17-kD centromere protein (CENP-A) copurifies with nucleosome core particles and with histones. J Cell Biol 104: 805–815.

30. Schatten G, Simerly C, Palmer DK, Margolis RL, Maul G, et al. (1988) Kinetochore appearance during meiosis, fertilization and mitosis in mouse oocytes and zygotes. Chromosoma 96: 341–352.

31. Lo AW, Magliano DJ, Sibson MC, Kalitsis P, Craig JM, et al. (2001) A novel chromatin immunoprecipitation and array (CIA) analysis identifies a 460-kb CENP-A-binding neocentromere DNA. Genome Res 11: 448–457.

32. Suzuki N, Nishii K, Okazaki T, Ikeno M (2006) Human artificial chromosomes constructed using the bottom-up strategy are stably maintained in mitosis and efficiently transmissible to progeny mice. J Biol Chem 281: 26615–26623.

33. Ananiev EV, Phillips RL, Rines HW (1998) Chromosome-specific molecular organization of maize (Zea mays L.) centromeric regions. Proc Natl Acad Sci U S A 95: 13073–13078.

34. Nagaki K, Song J, Stupar R, Parokonny A, Yuan Q, et al. (2003) Molecular and cytological analyses of large tracts of centromeric DNA reveal the structure and evolutionary dynamics of maize centromeres. Genetics 163: 759–770.

35. Alfenito MR, Birchler JA (1993) Molecular characterization of a maize B chromosome centric sequence. Genetics 135: 589–597.

36. Jin W, Lamb JC, Vega JM, Dawe RK, Birchler JA, et al. (2005) Molecular and functional dissection of the maize B chromosome centromere. Plant Cell 17: 1412–1423.

37. Kaszas E, Kato A, Birchler JA (2002) Cytological and molecular analysis of centromere misdivision in maize. Genome 45: 759–768.

38. Kato A, Zheng YZ, Auger DL, Phelps-Durr T, Bauer MJ, et al. (2005) Minichromosomes derived from the B chromosome of maize. Cytogenet Genome Res 109: 156–165.

39. Phelps-Durr TL, Birchler JA (2004) An asymptotic determination of minimum centromere size for the maize B chromosome. Cytogenet Genome Res 106: 309–313.

40. Yu W, Han F, Gao Z, Vega JM, Birchler JA (2007) Construction and behavior of engineered minichromosomes in maize. PNAS 104: 8924–8929.

41. Kato A, Lamb JC, Birchler JA (2004) Chromosome painting using repetitive DNA sequences as probes for somatic chromosome identification in maize. Proc Natl Acad Sci U S A 101: 13554–13559.

42. Grimes BR, Warburton PE, Farr CJ (2002) Chromosome engineering: prospects for gene therapy. Gene Therapy 9: 713–718.

43. Dawe RK, Henikoff S (2006) Centromeres put epigenetics in the driver›s seat. Trends Biochem Sci 31: 662–669.

44. Co DO, Borowski AH, Leung JD, van der Kaa J, Hengst S, et al. (2000) Generation of transgenic mice and germline transmission of a mammalian artificial chromosome introduced into embryos by pronuclear microinjection. Chrom Res 8: 183–191.

45. Ananiev EV, Riera-Lizarazu O, Rines HW, Phillips RL (1997) Oat-maize chromosome addition lines: A new system for mapping the maize genome. Proc Natl Acad Sci U S A 94: 3524–3529.

46. Hall SE, Luo S, Hall AE, Preuss D (2005) Differential rates of local and global homogenization in centromere satellites from Arabidopsis relatives. Genetics 170: 1913–1927.

47. Copenhaver GP, Nickel K, Kuromori T, Benito M-I, Kaul S, et al. (1999) Genetic definition and sequence analysis of *Arabidopsis* centromeres. Science 286: 2468–2474.

48. Dong F, Miller JT, Jackson SA, Wang G-L, Ronald PC, et al. (1998) Rice (Oryza sativa) centromeric regions consist of complex DNA. Proc Natl Acad Sci U S A 95: 8135–8140.

49. Kumar A, Bennetzen JL (1999) Plant retrotransposons. Annu Rev Genet 33: 479–532.

50. Volpe T, Schramke V, Hamilton GL, White SA, Teng G, et al. (2003) RNA interference is required for normal centromere function in fission yeast. Chrom Res 11: 137–146.

51. May BP, Lippman ZB, Fang Y, Spector DL, Martienssen RA (2005) Differential regulation of strand-specific transcripts from *Arabidopsis* centromere satellite repeats. PLoS Genet 1: e79.. doi:10.1371/journal.pgen.0010079.

52. Topp CN, Zhong CX, Dawe RK (2004) Centromere-encoded RNAs are integral components of the maize kinetochore. Proc Natl Acad Sci U S A 101: 15986–15991.

53. Talbert PB, Henikoff S (2006) Spreading of silent chromatin: inaction at a distance. Nat Rev Genet 7: 793–803.

54. Einset J (1943) Chromosome length in relation to transmission frequency of maize trisomes. Genetics 28: 349–364.

55. Telenius H, Szeles A, Kereso J, Csonka E, Praznovszky T, et al. (1999) Stability of a functional murine satellite DNA-based artificial chromosome across mammalian species. Chrom Res 7: 3–7.

56. McClintock B (1938) The production of homozygous deficient tissues with mutant characteristics by means of the aberrant mitotic behavior of ring-shaped chromosomes. Genetics 23: 315–376.

57. Runge KW, Zakian VA (1993) Saccharomyces cerevisiae linear chromosome stability (*lcs*) mutants increase the loss rate of artificial and natural linear chromosomes. Chromosoma 102: 207–217.

58. Kaszas E, Birchler JA (1998) Meiotic transmission rates correlate with physical features of rearranged centromeres in maize. Genetics 150: 1683–1692.

59. Dawe RK, Hiatt EN (2004) Plant neocentromeres: fast, focused, and driven. Chrom Res 12: 655–669.

60. Hawley RS, Theurkauf WE (1993) Requiem for distributive segregation: achiasmate segregation in *Drosophila* females. Trends Genet 9: 310–317.

61. Mehta S, Yang X-M, Jayaram M, Velmurugan S (2005) A novel role for the mitotic spindle during DNA segregation in yeast: promoting 2 μm plasmid-cohesin association. Molec Cell Biol 25: 4283–4298.

62. Amor DJ, Choo KHA (2002) Neocentromeres: role in human disease, evolution, and centromere study. Am J Hum Genet 71: 695–714.

63. Heun P, Erhardt S, Blower MD, Weiss S, Skora AD, et al. (2006) Mislocalization of the*Drosophila* centromere-specific histone CID promotes formation of functional ectopic kinetochores. Dev Cell 10: 303–315.

64. Wu J, Yamagata H, Hayashi-Tsugane M, Hijishita S, Fujisawa M, et al. (2004) Composition and structure of the centromeric region of rice Chromosome 8. Plant Cell 16: 967–976.

65. Frame BR, Zhang H, Cocciolone SM, Sidorenko LV, Dietrich CR, et al. (2000) Production of transgenic maize from bombarded type II callus: effect of gold particle size and callus morphology on transformation efficiency. In Vitro Cell Dev Biol Plant 36: 21–29.

66. Kato A (1999) Air drying method using nitrous oxide for chromosome counting in maize. Biotech Histochem 74: 160–166.

CHAPTER 3

GENOME EDITING IN PLANTS VIA DESIGNED ZINC FINGER NUCLEASES

Joseph F. Petolino

Keywords: Designed nucleases, DNA repair, Gene targeting, Site-specific transgene integration, Targeted mutagenesis

INTRODUCTION

All plant traits result from complex arrays of biochemical, physiological, and developmental processes culminating in phenotypes. These processes are largely dictated by the sequences of nucleotide bases in the nuclear, plastid, and mitochondrial genomes that provide both compositional and regulatory instructions to the living cell and, consequently, the growing organism. Indeed, it is the modifications within, and recombinations among, these DNA sequences that are at the foundation of the phenotypic variability observed among organisms. By taking advantage of naturally occurring and/or induced sequence modifications and recombinations, plant breeders have made, and continue to make, substantial progress with respect to enhancing the quality and performance of crops for agricultural and industrial applications.

The last several years have seen remarkable progress in plant cell and molecular biology. Entire genome sequences have been elucidated and annotated (Michael and Jackson 2013), while databases with information on a multitude of genes and their expression have been assembled (Wingender *et al.*2000). In addition to a deeper understanding of the dynamics of genome structure and function, modern biotechnology has provided tools that allow for the controlled alteration of DNA sequences within plant genomes. One such tool, the designed zinc finger nuclease (ZFN), is the subject of this review. ZFNs, consisting of DNA binding and nuclease domains, can be designed to recognize specific DNA sequences and thereby enable targeted cleavage (Urnov *et al.*2010). Alternative nucleases, such as meganucleases (Stoddard 2011), TALENs (Bogdanove and Voytas 2011), and CRISPR/Cas (Shan *et al.*2013), have also been used to generate targeted DNA breaks. The ability to cleave-specific DNA sequences and promote different mechanisms of DNA repair enable various types of genomic modifications ranging from single-nucleotide mutations to large sequence deletions, rearrangements and/or integrations (Curtin *et al.*2012).

Although the creation of targeted DNA modifications is a powerful capability, genome editing requires other supporting technologies—not the least of which is the ability to generate, isolate, and propagate modified genomes. It is here that *in vitro* biology plays a particularly critical role. Genome editing reagents, such as designed nucleases and donor template DNA, need to be effectively delivered to target cells, and rare genetic variants must be isolated and captured as 'events' for propagation and maintenance. Typically, these operations take advantage of cell and tissue culture methods whereby plant transformation, *in vitro* selection, and regeneration are employed.

The potential value of genome editing technology to basic and applied plant science cannot be overestimated. The ability to modify genetic information in a precise and specific manner and recover modified plants enables not only studies of gene function and biological mechanisms, but also, potentially, the creation of novel phenotypes. As the demand for agricultural output grows with an expanding human population, the need to engineer more complex traits in crops, such as enhanced yield and stress tolerance, will require more sophisticated approaches. Genome editing *via* designed nucleases represents one of the critical enabling capabilities for future crop improvement.

DESIGNED ZFNS

ZFNs consist of zinc finger protein domains, capable of sequence-specific DNA binding, fused to a nuclease domain for DNA cleavage (Fig. 1). DNA binding is the result of a tethered array of 4–6 zinc finger protein domains that each recognize approximately 3 bp of DNA. Although more-or-less modular

Citation: Joseph F. Petolino, Genome editing in plants via designed zinc finger nucleases, DOI: 10.1007/s11627-015-9663-3.

with respect to binding specificity, there are considerable context effects, i.e., interactions with neighboring domains, which make their binding more specific but their design somewhat challenging (Urnov *et al.*2010). They are most effectively assembled from an archive of two-finger modules that each recognize specific 6-bp DNA sequences whereby domain junctions within each module are optimized for sequence recognition (Moore *et al.*2001). For DNA cleavage, the catalytic domain of the type II restriction enzyme *Fok*I has been used (Kim *et al.*1996). A critical property of the catalytic domain is that it must dimerize to cleave DNA, so two adjacent ZFN pairs must orient themselves with appropriate spacing at the target site (Fig. 1). Although a somewhat larger gene product needs to be expressed, the longer recognition sequences (24–36 bp) required for binding result in a higher level of specificity. In addition,*Fok*I variants requiring heterodimerization have been developed, thereby further enhancing sequence specificity and reducing off-site cleavage (Miller *et al.*2007).

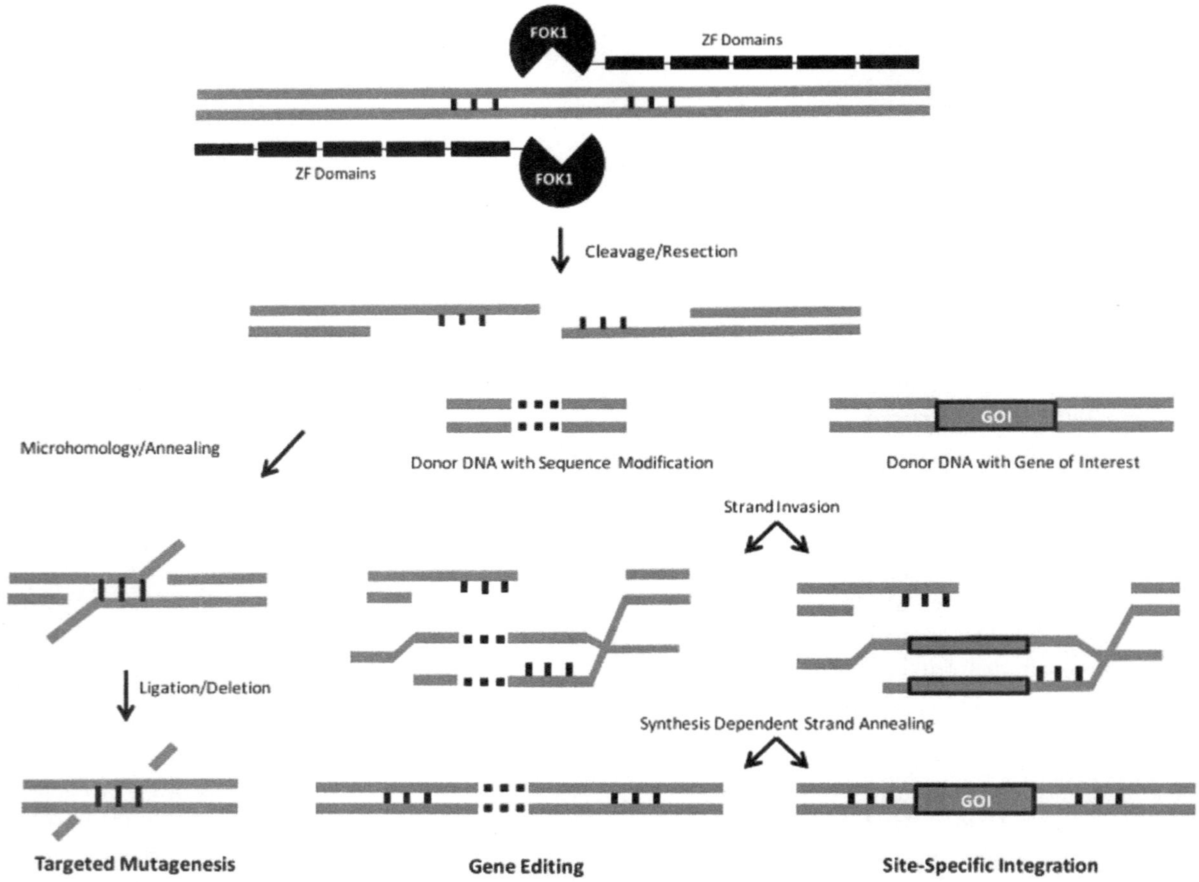

Figure 1: Targeted genome modification via double-strand break (DSB) repair. Zinc finger nucleases (ZFNs) bind to a target sequence, thereby dimerizing *Fok*I nuclease. The DSB generated by ZFN cleavage induces DNA repair processes. In the absence of donor template DNA, error-prone non-homologous end joining (NHEJ) can result in 'targeted mutagenesis' (*left*). In the presence of homologous sequences, homology-directed repair (HDR) can result in 'gene editing' (*center*) or 'site-specific integration' (*right*).

ZFNs can be designed to bind and cleave virtually any stretch of DNA sequence, thereby allowing for the creation of DNA double-strand breaks (DSBs) at specified loci. Expression of genes encoding ZFNs and concomitant cleavage at endogenous genomic loci has been demonstrated in a broad spectrum of organisms, including human (Urnov *et al.*2005; Lombardo *et al.*2007; Moehle *et al.*2007; Perez *et al.*2008; Sebastiano *et al.*2011; Wilen *et al.*2011; Provasi *et al.*2012), hamster (Santiago *et al.*2008), mouse (Osiak *et al.*2011), pig (Hauschild *et al.*2011), frog (Young *et al.*2011), zebra fish (Doyon *et al.*2008), insect (Beumer *et al.*2006; Bibikova *et al.*2002), roundworm (Morton *et al.*2006), and plasmodium (Straimer *et al.*2012). The present review focuses on the use of ZFNs for cleaving genomic loci in plants.

REPAIRING DSBS

A cell's capacity to create and repair DSBs is central to the facilitation of recombination between DNA sequences—so essential to both the maintenance of genomic integrity and the generation of genetic variability. DSB repair can occur using homologous sequences as templates for synthesis (Moynahan and Jasin 2010). Such homology-directed repair (HDR) uses sister chromatids, homologous chromosomes, or other related DNA. Alternative repair pathways involve non-homologous end joining (NHEJ) of broken ends (Lieber 2010). Targeted sequence modification (i.e., genome editing) is predicated on the ability to create sequence-specific DSBs and exploit the cell's DNA repair machinery to generate desired genetic outcomes.

The genetic outcome of the resolution of a DSB generated by ZFN cleavage is a function of the process used to repair it

(Fig. 1). In plants, the most common DNA repair mechanism appears to involve NHEJ, where broken ends are simply religated (Puchta 2005). This repair process can be error-prone, resulting in small insertions, deletions, and/or rearrangements (Gorbunova and Levy 1999). If imprecise DSB repair occurs in a gene sequence, mutations can be introduced which may affect function. Indeed, this represents a general method of targeted mutagenesis whereby ZFNs designed to bind and cleave a specific genomic locus can be introduced and expressed, resulting in NHEJ-mediated sequence alteration (Fig. 1).

In the presence of DNA sequences homologous to those flanking the DSB, HDR can use such sequences as templates for synthesis-dependent strand annealing (Haber 2000). If the homologous sequences are investigator-designed donor DNA, the resolution of a DSB can be dictated by the composition of the repair template. In other words, sequence modifications can be copied into the targeted cleavage site and can range from single- to few-base-pair modifications (i.e., gene editing) to the integration of complete transgene expression cassettes (i.e., site-specific integration; Fig. 1).

ZFN-MEDIATED TARGETED MUTAGENESIS

Mutation breeding dates back to the 1920s with the discovery of the mutagenic effect of X-rays on plant genes (Stadler 1928). Since then, a wide range of characters including plant architecture, flowering, and maturity have been modified through mutation breeding, and over 3000 novel cultivars have been developed and deployed in agriculture (Mba 2013). Historically, 'forward genetic' approaches have been used whereby direct screening for specific phenotypes has resulted in germplasm with enhanced performance for use in breeding. These approaches are limited by the ability to generate, identify, and isolate rare individuals and the fact that, since most mutations are recessive, gene redundancies tend to mask any phenotypic effects accompanying a given sequence modification. More recently, 'reverse genetic' approaches have become possible whereby sequence modifications of specific genes are identified using high-throughput genome analysis (Chen *et al.*2014). Difficulties with these approaches include the lack of understanding of which genes to mutate to generate useful phenotypes and the random nature of the sequence modifications induced by physical or chemical means. The use of designed nucleases to target specific modifications of selected genome sequences adds a level of control to this process and represents a significant improvement of the 'reverse genetic' approach.

Sequence modification of preintegrated reporter constructs.

Early proof-of-concept studies of ZFN-mediated targeted mutagenesis in plants involved modification of preintegrated sequences comprising ZFN cleavage sites. One such study involved the expression of a ZFN under the control of a heat-shock protein promoter whereby an*Eco*RI restriction site could be lost upon mutation (Lloyd *et al.*2005). Mutation frequencies as high as 19.6% were observed in heat-treated Arabidopsis seedlings, and sequencing revealed that most mutations were simple deletions of 1–52 bp. In a similar study, constitutive expression of a ZFN in stably transformed Arabidopsis resulted in a 2% mutation frequency and deletions ranging from 1 to 80 bp (de Pater *et al.*2009). A preintegrated reporter gene containing a ZFN cleavage site has also been used to study targeted mutagenesis. A gene encoding the enzyme β-glucuronidase (*GUS*) with a stop codon within a ZFN cleavage site was stably integrated into Arabidopsis, and subsequent co-cultivation with an *Agrobacterium* strain harboring a corresponding ZFN expression cassette resulted in sectors of GUS staining (Tovkach *et al.*2009). Sequence analysis of the target site revealed several single-nucleotide deletions and substitutions in the stop codon, resulting in a *GUS* open reading frame. Similar results were obtained when the ZFN was expressed using a viral vector (Vainstein *et al.*2011).

Sequence modification of endogenous genes.

Endogenous genomic loci have also been mutated following ZFN expression. Genes encoding ZFNs designed to cleave within the *ABA-INSENSITIVE-4 (ABI4)* gene, driven by a heat-shock protein promoter, were stably integrated into Arabidopsis (Osakabe *et al.*2010). After heat induction, mutations in *ABI4* were observed in somatic cells at frequencies up to 3%, and homozygous mutant T_3 progeny displayed the expected loss-of-function phenotype for this gene, i.e., ABA and glucose insensitivity. In a similar study, genes encoding ZFNs that recognize the Arabidopsis *ALCOHOL DEHYDROGENASE-1 (ADH1)* and *TRANSPARENT TESTA-4 (TT4)*genes were expressed in Arabidopsis under the control of an estrogen-inducible promoter (Zhang *et al.*2010). Induced T_1 plants exhibited somatic mutation frequencies of 7 and 16% for*ADH1* and *TT4*, respectively, and the mutations ranged from insertions of 1–2 bp to deletions of 3–142 bp. No evidence of ZFN-induced 'off-target' cleavage was observed following PCR amplification and sequencing of ectopic Arabidopsis genomic sequences most similar to *ADH1*and *TT4*. Mutations were transmitted to progeny at frequencies of 69 and 33% for *ADH1* and*TT4*, respectively, and 20% appeared to be homozygous, suggesting biallelic mutation. In soybean, a gene encoding a ZFN targeting two paralogous *DICER-LIKE (DCL4b)* genes, *DCL4a*and *DCL4b*, under the control of an estrogen-inducible promoter, was delivered using*Agrobacterium* in the presence of estrogen (Curtin *et al.*2011). Three T_0 plants were recovered, and sequence analysis of PCR-amplified products revealed that one of the plants had an adenine-base insertion at the *DCL4a* locus and another had a two-base thymine and adenine insertion into *DCL4b*. Both plants appeared to be heterozygous for the mutation. The plant with the *dcl4a* mutation exhibited phenotypic abnormalities, including aborted seed. The *dcl4b* plant appeared normal and produced T_1 progeny in which the *dcl4b* mutation segregated 1:2:1 as expected. These results provide clear evidence that expression of genes encoding ZFNs can generate heritable mutations at targeted endogenous loci.

Gene deletion

In addition to small, NHEJ-induced mutations, ZFN-mediated cleavage can result in larger DNA sequence deletions (Fig. 2). A reporter construct containing a tandem repeat of 540 bp of partial, i.e., 3′/5′, *GREEN FLUORESCENT PROTEIN (GFP)* gene fragments with 2.8 kb of intervening heterologous DNA sequence containing a ZFN cleavage site was stably integrated into tobacco (Cai *et al.*2009). Subsequent expression of the corresponding ZFN gene resulted in targeted DSB formation, recombination between the *GFP* gene fragments, and deletion of the intervening 2.8-kb sequence. Deletion of a 4.3-kb integrated *GUS* gene sequence flanked by ZFN cleavage sites was observed in 35% of the F_1 progenies when crossed to a ZFN-expressing plant (Petolino*et al.*2010). Even larger deletions (e.g., 55 kb) were observed following nuclease cleavage within tandem gene clusters (Voytas 2013). These results provide proof of concept for a potentially powerful means of creating targeted genome sequence deletions.

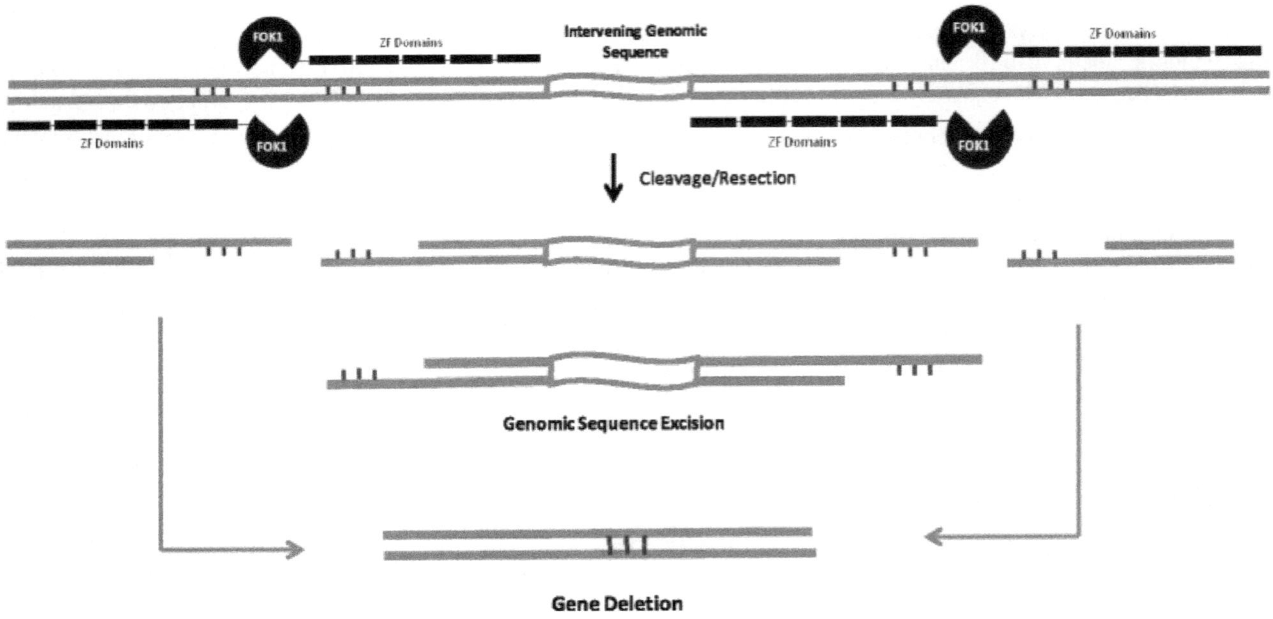

Figure 2: Gene deletion. Two concurrent zinc finger nuclease (ZFN)-mediated double-strand breaks (DSBs) can result in the loss of intervening sequences.

Gene editing using HDR.

A more precise means of generating targeted mutations at specific genomic locations involves the use of donor template DNA sequences to facilitate HDR of DSBs. Typically, donor template DNA for HDR contains approximately 750- to 1000-kb stretches of sequence homologous to that flanking the genomic cleavage site and the desired sequence modification. Unfortunately, the frequency of HDR in somatic plant cells appears to be extremely low, so the identification and isolation of such modifications can be a challenge. Nonetheless, selectable phenotypes have been used to demonstrate gene editing *via* HDR. Specific mutations to the *SULFONYLUREA RECEPTOR* genes *SuRA* and *SuRB* in tobacco confer resistance to certain acetolactate synthase (ALS)-inhibiting herbicides. Herbicide-resistant mutants were isolated following co-introduction of genes encoding designed ZFNs and donor template DNA into tobacco protoplasts (Townsend*et al.*2009). Mutation frequencies exceeding 2% were observed and mutations as far as 1.3 kb from the ZFN cleavage site were obtained. The Arabidopsis *PROTOPORPHYRINOGEN OXIDASE (PPOX)* gene has also been edited using HDR of a ZFN-mediated DSB (de Pater *et al.*2013). In this case, two mutations resulting in resistance to the herbicide butafenacil were delivered on a donor template DNA, and an editing frequency of 3.1×10^{-3} was observed and transmitted to the next generation. One of the future challenges for gene editing *via* HDR will involve the targeted modification of endogenous genes without selectable phenotypes. This may necessitate the use of high-throughput analyses to screen for the desired outcome.

TARGETED TRANSGENE INTEGRATION

Integrating DNA sequences into predetermined genomic locations would greatly enhance the precision of plant transformation and predictability of transgenic plant performance. Random DNA integration carries not only the risk of potential unwanted insertional mutations, but also the unpredictable consequences of position effect on transgene behavior. In addition, transgene stacking *via* sequential random transformation requires resource-intensive introgression *via*conventional breeding. Targeting DNA sequences to specified genomic loci allows for sequential transgene stacking into a single locus, thereby simplifying downstream breeding (Ainley *et al.*2013).

Targeted integration into preintegrated loci.

The initial demonstrations of ZFN-mediated, site-specific transgene integration involved correcting preintegrated selectable marker genes (Wright *et al.*2005; Cai *et al.*2009). In one study using tobacco protoplasts, a defective fusion between genes encoding the GUS reporter and neomycin phosphotransferase (NPTII) selectable marker proteins was corrected *via* co-delivery of a gene encoding a ZFN, which cleaved in the junction between the two genes, and a donor template DNA comprising sequences homologous to those flanking the ZFN cleavage site and the missing *GUS/NPTII* gene sequences (Wright *et al.*2005). HDR occurred in more than 10% of the protoplasts as estimated by positive *GUS* expression. Targeting in this system was enhanced more than two orders of magnitude when the ZFN and donor template DNA were expressed from a geminivirus (Baltes *et al.*2014). In a similar study, a preintegrated construct containing a 3′ partial *PHOSPHINOTHRICIN ACETYL TRANSFERASE (PAT)* gene fragment flanked by a ZFN binding site and an Arabidopsis *4-COUMARYL SYNTHASE (4-CoAS)* intron allowed for *in vitro* selection on Herbiace®-containing medium following ZFN-mediated cleavage and targeted integration of a complementary 5′ *PAT* gene fragment with a homologous*4-CoAS* intron delivered *via Agrobacterium* (Cai *et al.*2009). Preintegrated loci have also been successfully targeted using NHEJ (Tzfira *et al.*2012). In this system, identical ZFN cleavage sites were present in the genome and on the donor template DNA. Co-delivery of a ZFN expression construct and donor template DNA resulted in targeted DSB formation at the preintegrated locus and linearization and subsequent targeted integration of the donor template DNA *via* NHEJ (Weinthal *et al.*2013).

Targeted integration into endogenous loci.

ZFNs designed to cleave endogenous genomic sequences have been used to facilitate site-specific transgene integration into native genes. A gene encoding a ZFN designed to cleave a tobacco *ENDOCHITINASE-50 (CHN50)* gene sequence was co-delivered *via Agrobacterium*along with a *PAT* herbicide-resistance gene flanked on each side by 750 bp of *CHN50* gene sequence (Cai *et al.*2009). The majority of transgenic events were the result of random integration; however, 5–10% of the events appeared to have targeted the *CHN50* locus. In a similar study, genes encoding ZFNs designed to cleave in exon 2 of the maize *INOSITOL-1,3,4,5,6-PENTAKISPHOSPHATE KINASE 1 (IPK1)* gene were co-delivered with donor template DNA containing a promoterless *PAT* gene with a 2A 'stutter' sequence flanked by 815 bp of sequence homologous to *IPK1* (Shukla *et al.*2009). Targeted cleavage at the *IPK1* locus and precise 'trapping' of the *IPK1* promoter resulted in site-specific integration and herbicide resistance.

TARGETING IN DNA REPAIR MUTANTS

The genetic outcome of the resolution of a DSB is largely a function of the pathway used for DNA repair, i.e., HDR versus NHEJ. In an attempt to direct the resolution of repair and enhance mutation and/or targeting frequencies, genes encoding ZFNs have been delivered into various genetic backgrounds in which specific genes involved in DNA repair have been mutated. The frequency of ZFN-mediated deletions larger than 4 bp in the Arabidopsis *ABI4* gene was 2.6-fold greater in a genetic background deficient in KU80, an enzyme involved in NHEJ DNA repair (Osakabe *et al.*2010). Similarly, although the frequency of ZFN-mediated mutagenesis at the endogenous *ADH1* locus in Arabidopsis was not increased in genetic backgrounds with mutations in genes encoding KU70 or LIG4, two other DNA repair proteins, DNA repair appeared to be shifted to microhomology-dependent NHEJ, resulting in larger deletions than in a wild-type background (Qi *et al.*2013). In addition, *ADH1* mutagenesis increased up to eightfold in a genetic background with a mutated gene encoding SMC6B, a protein involved in DSB repair using sister chromatids. Targeted sequence integration into the *ADH1* locus *via* HDR increased 16-fold and 4-fold, respectively, in genetic backgrounds with mutations in genes encoding KU70 and LIG4 (Qi *et al.*2013). Using a different approach to DNA repair pathway manipulation, coexpression of genes encoding a ZFN and RAD54, an enzyme involved in HDR in yeast, both driven by an egg cell-specific promoter, resulted in a tenfold increase in gene targeting in Arabidopsis (Even-Faitelson *et al.*2011). These results suggest that, although it adds a level of complexity to the overall genome editing process, manipulating DNA repair pathways provides a level of control over the types of sequence modifications resulting from ZFN-mediated cleavage and a potential path to increased mutagenesis and gene targeting frequencies.

IN VITRO BIOLOGY AND GENOME EDITING

The first prerequisite to ZFN-mediated genome editing is the ability to effectively deliver macromolecules such as DNA, RNA, and/or protein to cells to facilitate DNA sequence modification. Standard plant transformation methods such as polyethylene glycol delivery into protoplasts (Wright *et al.*2005; Townsend *et al.*2009), microparticle bombardment (Ainley *et al.*2013), WHISKERS™ (Shukla *et al.*2009), and *Agrobacterium* (Cai *et al.*2009; de Pater *et al.*2009, 2013) have been successfully used to deliver such reagents for genome editing. Most recently, genome editing reagents have been successfully delivered to plant cells using non-integrating viruses (Vainstein *et al.*2011; Baltes *et al.*2014). The goal of any delivery system is to assure that active ZFNs and, if necessary, donor template DNA is present at the appropriate stage of the cell cycle to enable the targeted cleavage and subsequent DNA repair.

In addition to making specific DNA sequence modifications, it is critical to capture the genetic change in a transmissible form, e.g., in meristems and germlines. Since genetic modification is fundamentally a cellular phenomenon, being able to grow complex organisms such as higher plants as cultured cells and tissues allows for the isolation of clonal variants. Indeed, as with conventional transgenic production, the development and deployment of *in vitro* methods for plant species is equally important to genome editing. In addition, the ability to induce tissue proliferation (e.g., callus) and regeneration of modified plants enables the performance of various compositional and functional analyses critical for both genome editing and capture of genetic variants for potential breeding applications. It is no coincidence, therefore, that the efficacy of genome editing technology for a given species is directly proportional to the ability to perform the full array of somatic cell genetics techniques.

In an attempt to circumvent the need for robust plant cell and tissue culture systems, *in planta*gene targeting has been attempted using the meganuclease I-*Sce*I (Fauser *et al.*2012; Ayar *et al.*2013). This involved the preintegration of target and donor template DNA and the expression of a genes encoding I-*Sce*I to facilitate cleavage and intrachromosomal DNA repair. In Arabidopsis, the expression of a gene encoding I-*Sce*I resulted in the concurrent cleavage of preintegrated sequences flanking donor template DNA comprising a promoterless *GUS* reporter gene and a genomic target sequence with a corresponding nuclease cleavage site (Fauser *et al.*2012). The preintegrated donor template was excised and used to repair the DSB at the target site, resulting in *GUS* expression. Roughly 1% of the progenies from *GUS*-expressing plants were shown to be targeted. The authors concluded that this method could avoid the need for high-frequency plant transformation and be particularly valuable in more recalcitrant species. However, a similar study in maize required *in vitro* selection and plant regeneration to isolate targeted progeny (Ayar *et al.*2013). In this case, a gene encoding an NPTII selectable marker was functionalized and kanamycin selection was used to identify and capture targeted events.

Genome editing for the development of complex traits is expected to involve the need to deliver many experimental constructs into cells and tissues and to analyze numerous genetic variants. Automated cell and tissue handling opens up the possibility for such high-throughput manipulations (Knoll *et al.*2004) but may require growing protoplasts or cell suspension cultures in small volumes in a multi-well format. In addition, phenotyping complex traits such as stress tolerance or yield will require the recovery of modified plants. Should genome editing move in this direction, the need for robust *in vitro* cell and tissue culture systems will be even more evident.

CONCLUSIONS AND FUTURE PROSPECTS

The use of ZFNs and other designed nucleases to make targeted DSBs has opened up the possibility of precision genome editing. It is anticipated that genome editing will have a major impact on at least two broad areas of plant biology. The first is functional genomics, whereby the ability to make precise DNA sequence modifications to targeted endogenous loci will advance basic understanding of genome structure and function as well as the mechanisms that give rise to phenotypes. Such enhanced understanding should lead to more effective hypothesis-driven gene discovery efforts for novel trait development. The second area where genome editing technology will have a major impact is in applied crop improvement and commercial product development. 'Reverse genetic' approaches to mutation breeding will no doubt be revolutionized by virtue of being able to make targeted DNA sequence modifications rather than random changes. Products developed from such non-transgenic enhancements are likely to require an altered path to deregulation, i.e., similar to conventionally bred genotypes, and thus be more cost-effective to bring to market. In addition, current transgenic production technology and the random insertion of DNA sequences involve the need to generate and screen numerous unproductive events. Targeting transgenes to predetermined genomic loci will not only reduce undesired side-effects and increase the predictability of transgene performance but will also simplify event characterization and allow for sequential transgene stacking, thereby reducing product development time and cost.

Clearly, genome editing shows great promise for both basic and applied plant science. For this promise to be met, supporting technologies such as high resolution molecular methods for genome analysis and bioinformatics for trait characterization are required. However, no supporting technology is more critical than the ability to create and isolate genetic variants. As such, *in vitro* manipulation of plant cells and tissues will continue to play a central role in the further development of genome editing technology.

REFERENCES

1. Ainley WM, Sastry-Dent L, Welter ME, Murray MG, Zeitler B, Amora R, Corbin DR, Miles RR, Arnold NL, Strange TL, Simpson MA, Cao Z, Carroll C, Pawelczak KS, Blue R, West K, Rowland LM, Perkins D, Samuel P, Dewes CM, Shen L, Sriram S, Evans SL, Rebar EJ, Zhang L, Gregory PD, Urnov FD, Webb SR, Petolino JF (2013) Trait stacking via targeted genome editing. Plant Biotechnol J 11:1126–113
2. Ayar A, Wehrkamp-Richter S, Laffaire JB, Le Goff S, Levy J, Chaignon S, Salmi H, Lepicard A, Sallaud C, Gallego ME, White CI, Paul W (2013) Gene targeting in maize by somatic ectopic recombination. Plant Biotechnol J 11:305–314
3. Baltes NJ, Gil-Humanes J, Cermak T, Atkins PA, Voytas DF (2014) DNA replicons for plant genome engineering. Plant Cell 26:151–163
4. Beumer K, Bhattacharyya G, Bibikova M, Trautman JK, Carroll D (2006) Efficient gene targeting in drosophila with zinc-finger nucleases. Genetics 172:2391–2403
5. Bibikova M, Golic M, Golic KG, Carroll D (2002) Targeted chromosomal cleavage and mutagenesis in drosophila using zinc-finger nucleases. Genetics 161:1169–1175
6. Bogdanove AJ, Voytas DF (2011) TAL effectors: Customizable proteins for DNA targeting. Science 333:1843–1846
7. Cai CQ, Doyon Y, Ainley WM, Miller JC, Dekelver RC, Moehle EA, Rock JM, Lee YL, Garrison R, Schulenberg L, Blue R, Worden A, Baker L, Faraji F, Zhang L, Holmes MC, Rebar EJ, Collingwood TN, Rubin-Wilson B, Gregory PD, Urnov FD, Petolino JF (2009) Targeted transgene integration in plant cells using designed zinc finger nucleases. Plant Mol Biol 69:699–709
8. Chen L, Hao L, Parry MAJ, Phillips AL, Hu Y-G (2014) Progress in TILLING as a tool for functional genomics and improvement of crops. J Integr Plant Biol 56:425–443
9. Curtin SJ, Voytas DF, Stupar RM (2012) Genome engineering of crops with designer nucleases. Plant Genome 5:42–50
10. Curtin SJ, Zhang F, Sander JD, Haun WJ, Starker C, Baltes NJ, Reyon D, Dahlborg EJ, Goodwin MJ, Coffman AP, Dobbs D, Joung JK, Voytas DF, Stupar RM (2011) Targeted mutagenesis of duplicated genes in soybean with zinc-finger nucleases. Plant Physiol 156:466–473
11. de Pater S, Neuteboom LW, Pinas JE, Hooykaas PJ, van der Zaal BJ (2009) ZFN-induced mutagenesis and gene-targeting in Arabidopsis through *Agrobacterium*-mediated floral dip transformation. Plant Biotechnol J 7:821–835
12. de Pater S, Pinas JE, Hooykaas PJ, van der Zaal BJ (2013) ZFN-mediated gene targeting of the Arabidopsis protoporphyrinogen oxidase gene through *Agrobacterium*-mediated floral dip transformation. Plant Biotechnol J 11:510–515
13. Doyon Y, McCammon JM, Miller JC, Faraji F, Ngo C, Katibah GE, Amora R, Hocking TD, Zhang L, Rebar EJ, Gregory PD, Urnov FD, Amacher SL (2008) Heritable targeted gene disruption in zebrafish using designed zinc-finger nucleases. Nat Biotechnol 26:702–708
14. Even-Faitelson L, Samach A, Melamed-Bessudo C,

Avivi-Ragolsky N, Levy AA (2011) Localized egg-cell expression of effector proteins for targeted modification of the Arabidopsis genome. Plant J 68:929–937

15. Fauser F, Roth N, Pacher M, Ilg G, Sanchez-Fernandez R, Biesgen C, Puchta H (2012) *In planta*gene targeting. Proc Natl Acad Sci U S A 109:7535–7540
16. Gorbunova V, Levy AA (1999) How plants make ends meet: DNA double-strand break repair. Trends Plant Sci 4:263–269
17. Haber JE (2000) Partners and pathways: repairing a double-strand break. Trends Genet 16:259–264
18. Hauschild J, Petersen B, Santiago Y, Queisser A-L, Carnwath JW, Lucas-Hahn A, Zhang L, Meng X, Gregory PD, Schwinzer R, Cost GJ, Niemann H (2011) Efficient generation of a biallelic knockout in pigs using zinc-finger nucleases. Proc Natl Acad Sci U S A 108:12013–12017
19. Kim YG, Cha J, Chandrasegaran S (1996) Hybrid restriction enzymes: Zinc finger fusions to *Fok*I cleavage domain. Proc Natl Acad Sci U S A 93:1156–1160
20. Knoll A, Scherer T, Poggendorf I, Lütkemeyer D, Lehmann J (2004) Flexible automation of cell culture and tissue engineering tasks. Biotechnol Prog 20:1825–1835
21. Lieber MR (2010) The mechanism of double-strand DNA break repair by the nonhomologous DNA end-joining pathway. Annu Rev Biochem 79:181–211
22. Lloyd A, Plaisier CL, Carroll D, Drews GN (2005) Targeted mutagenesis using zinc-finger nucleases in Arabidopsis. Proc Natl Acad Sci U S A 102:2232–2237
23. Lombardo A, Genovese P, Beausejour CM, Colleoni S, Lee YL, Kim KA, Ando D, Urnov FD, Galli C, Gregory PD, Holmes MC, Naldini L (2007) Gene editing in human stem cells using zinc finger nucleases and integrase-defective lentiviral vector delivery. Nat Biotechnol 25:1298–1306
24. Mba C (2013) Induced mutations unleash the potentials of plant genetic resources for food and agriculture. J Agron 3:200–231
25. Michael TP, Jackson S (2013) The first 50 plant genomes. Plant Genome 6. doi:10.3835/plantgenome2013.03.0001in
26. Miller JC, Holmes MC, Wang J, Guschin DY, Lee YL, Rupniewski I, Beausejour CM, Waite AJ, Wang NS, Kim KA, Gregory PD, Pabo CO, Rebar EJ (2007) An improved zinc-finger nuclease architecture for highly specific genome editing. Nat Biotechnol 25:778–785
27. Moehle EA, Rock JM, Lee YL, Jouvenot Y, DeKelver RC, Gregory PD, Urnov FD, Holmes MC (2007) Targeted gene addition into a specified location in the human genome using designed zinc finger nucleases. Proc Natl Acad Sci U S A 104:3055–3060
28. Moore M, Klug A, Choo Y (2001) Improved DNA binding specificity from polyzinc finger peptides by using strings of two-finger units. Proc Natl Acad Sci U S A 98:1437–1441
29. Morton J, Davis MW, Jorgensen EM, Carroll D (2006) Induction and repair of zinc-finger nuclease-targeted double-strand breaks in *Caenorhabditis elegans* somatic cells. Proc Natl Acad Sci U S A 103:16370–16375
30. Moynahan ME, Jasin M (2010) Mitotic homologous recombination maintains genomic stability and suppresses tumorigenesis. Nat Rev Mol Cell Biol 11:196–207
31. Osakabe K, Osakabe Y, Toki S (2010) Site-directed mutagenesis in Arabidopsis using custom-designed zinc finger nucleases. Proc Natl Acad Sci U S A 107:12034–12039
32. Osiak A, Radecke F, Guhl E, Radecke S, Dannemann N, Lütge F, Glage S, Rudolph C, Cantz T, Schwarz K, Heilbronn R, Cathomen T (2011) Selection-independent generation of gene knockout mouse embryonic stem cells using zinc-finger nucleases. PLoS One 6:e28911
33. Perez EE, Wang J, Miller JC, Jouvenot Y, Kim KA, Liu O, Wang N, Lee G, Bartsevich VV, Lee YL, Guschin DY, Rupniewski I, Waite AJ, Carpenito C, Carroll RG, Orange JS, Urnov FD, Rebar EJ, Ando D, Gregory PD, Riley JL, Holmes MC, June CH (2008) Establishment of HIV-1 resistance in $CD4^+$ T cells by genome editing using zinc-finger nucleases. Nat Biotechnol 26:808–816
34. Petolino JF, Worden A, Curlee K, Connell J, Strange Moynahan TL, Larsen C, Russell S (2010) Zinc finger nuclease-mediated transgene deletion. Plant Mol Biol 73:617–628
35. Provasi E, Genovese P, Lombardo A, Magnani Z, Liu P-Q, Reik A, Chu V, Paschon DE, Zhang L, Kuball J, Camisa B, Bondanza A, Casorati G, Ponzoni M, Ciceri F, Bordignon C, Greenberg PD, Holmes MC, Gregory PD, Naldini L, Bonini C (2012) Editing T cell specificity towards leukemia by zinc finger nucleases and lentiviral gene transfer. Nat Med 18:807–815
36. Puchta H (2005) The repair of double-strand breaks in plants: mechanisms and consequences for genome evolution. J Exp Bot 56:1–14
37. Qi Y, Zhang Y, Zhang F, Baller JA, Cleland SC, Ryu Y, Starker CG, Voytas DF (2013) Increasing frequencies of site-specific mutagenesis and gene targeting in Arabidopsis by manipulating DNA repair pathways. Genome Res 23:547–554
38. Santiago Y, Chan E, Liu PQ, Orlando S, Zhang L, Urnov FD, Holmes MC, Guschin D, Waite A, Miller JC, Rebar EJ, Gregory PD, Klug A, Collingwood TN (2008) Targeted gene knockout in mammalian cells by using engineered zinc-finger nucleases. Proc Natl Acad Sci U S A 105:5809–5814
39. Sebastiano V, Maeder ML, Angstman JF, Haddad B, Khayter C, Yeo DT, Goodwin MJ, Hawkins JS, Ramirez CL, Batista LFZ, Artandi SE, Wernig M, Joung JK (2011) *In situ* genetic correction of the sickle cell anemia mutation in human induced pluripotent stem cells using engineered zinc finger nucleases. Stem Cells 29:1717–1726
40. Shan Q, Wang Y, Li J, Zhang Y, Chen K, Liang Z, Zhang K, Liu J, Xi JJ, Qiu JL, Gao C (2013) Targeted genome modification of crop plants using a CRISPR-Cas system.

Nat Biotechnol 31:686–688

41. Shukla VK, Doyon Y, Miller JC, DeKelver RC, Moehle EA, Worden SE, Mitchell JC, Arnold NL, Gopalan S, Meng X, Choi VM, Rock JM, Wu YY, Katibah GE, Zhifang G, McCaskill D, Simpson MA, Blakeslee B, Greenwalt SA, Butler HJ, Hinkley SJ, Zhang L, Rebar EJ, Gregory PD, Urnov FD (2009) Precise genome modification in the crop species *Zea mays* using zinc-finger nucleases. Nature 459:437–441

42. Stadler LJ (1928) Genetic effects of X-rays in maize. Proc Natl Acad Sci U S A 14:69–75

43. Stoddard BL (2011) Homing endonucleases: from microbial genetic invaders to reagents for targeted DNA modification. Structure 19:7–15

44. Straimer J, Lee MCS, Lee AH, Zeitler B, Williams AE, Pearl JR, Zhang L, Rebar EJ, Gregory PD, Llinas M, Urnov FD, Fidock DA (2012) Site-specific genome editing in *Plasmodium falciparum*using engineered zinc-finger nucleases. Nat Methods 9:993–998

45. Tovkach A, Zeevi V, Tzfira T (2009) A toolbox and procedural notes for characterizing novel zinc finger nucleases for genome editing in plant cells. Plant J 57:747–757

46. Townsend JA, Wright DA, Winfrey RJ, Fu F, Maeder ML, Joung JK, Voytas DF (2009) High-frequency modification of plant genes using engineered zinc-finger nucleases. Nature 459:442–445

47. Tzfira T, Weinthal D, Marton I, Zeevi V, Zuker A, Vainstein A (2012) Genome modifications in plant cells by custom-made restriction enzymes. Plant Biotechnol J 10:373–389

48. Urnov FD, Miller JC, Lee YL, Beausejour CM, Rock JM, Augustus S, Jamieson AC, Porteus MH, Gregory PD, Holmes MC (2005) Highly efficient endogenous human gene correction using designed zinc-finger nucleases. Nature 435:646–651

49. Urnov FD, Rebar EJ, Holmes MC, Zhang HS, Gregory PD (2010) Genome editing with engineered zinc finger nucleases. Nat Rev Genet 11:636–646

50. Vainstein A, Marton I, Zuker A, Danziger M, Tzfira T (2011) Permanent genome modifications in plant cells by transient viral vectors. Trends Biotechnol 29:363–369

51. Voytas DF (2013) Plant genome engineering with sequence-specific nucleases. Annu Rev Plant Biol 64:327–350

52. Weinthal DM, Taylor RA, Tzfira T (2013) Nonhomologous end joining-mediated gene replacement in plant cells. Plant Physiol 162:390–400

53. Wilen CB, Wang J, Tilton JC, Miller JC, Kim KA, Rebar EJ, Sherrill-Mix SA, Patro SC, Secreto AJ, Jordan APO, Lee G, Kahn J, Aye PP, Bunnell BA, Lackner AA, Hoxie JA, Danet-Desnoyers GA, Bushman FD, Riley JL, Gregory PD, June CH, Holmes MC, Doms RW (2011) Engineering HIV-resistant human CD4+ T cells with CXCR4-specific zinc-finger nucleases. PLoS Pathog 7:e1002020

54. Wingender E, Chen X, Hehl R, Karas H, Liebich I, Matys V, Meinhardt T, Pruss M, Reuter I, Schacherer F (2000) TRANSFAC: an integrated system for gene expression regulation. Nucleic Acids Res 28:316–319

55. Wright DA, Townsend JA, Winfrey RJ Jr, Irwin PA, Rajagopal J, Lonosky PM, Hall BD, Jondle MD, Voytas DF (2005) High-frequency homologous recombination in plants mediated by zinc-finger nucleases. Plant J 44:693–705

56. Young JJ, Cherone JM, Doyon Y, Ankoudinova I, Faraji FM, Lee AH, Ngo C, Guschin DY, Paschon DE, Miller JC, Zhang L, Rebar EJ, Gregory PD, Urnov FD, Harland RM, Zeitler B (2011) Efficient targeted gene disruption in the soma and germ line of the frog *Xenopus tropicalis* using engineered zinc-finger nucleases. Proc Natl Acad Sci U S A 108:7052–7057

57. Zhang F, Maeder ML, Unger-Wallace E, Hoshaw JP, Reyon D, Christian M, Li X, Pierick CJ, Dobbs D, Peterson T, Joung JK, Voytas DF (2010) High frequency targeted mutagenesis in*Arabidopsis thaliana* using zinc finger nucleases. Proc Natl Acad Sci U S A 107:12028–12033

CHAPTER 4

CRISPR-BASED TOOLS AND THEIR IMPROVEMENTS: NEW OPPORTUNITIES IN UNDERSTANDING PLANT–PATHOGEN INTERACTIONS FOR BETTER CROP PROTECTION

Abdellah Barakate and Jennifer Stephens

Keywords: CRISPR/Cas9, DNA double-stranded break, gene editing, homologous recombination, non-homologous end joining, plant–pathogen interactions

INTRODUCTION

Genetic crosses and mutagenesis based breeding are time consuming and laborious. The recent development of next generation sequencing is making available fast and cost effective genomic platforms of an increasing number of species including pests, plant models and crops. Now it is easier than ever to perform genome-wide association studies and determine the genes and pathways involved in any particular aspect of pathogen resistance (Olukolu et al., 2014), and pipelines are now well established for genomics-informed breeding (Varshney et al., 2015). It is also quicker and cheaper to obtain the transcriptome of any crop under pathogen attack and determine the virulence and defense pathways and genes that are deployed by both (Kawahara et al., 2012; O'Connell et al., 2012). Proteomics is also starting to make a dent in understanding plant–pathogen interactions (Lodha et al., 2013). A complex network of nuclear processes regulating gene expression and function is emerging from this gene discovery phase but association of a particular pathogen effector with the corresponding host target(s) is still poorly understood (Motion et al., 2015).

Omics technologies generate a huge amount of data and require powerful computational tools to integrate these high-throughput platforms in order to fully understand the multi-layered networks of biomolecules underpinning plant–pathogen interactions. Gene silencing has been extensively used to validate the function of candidate host resistance genes (Duan et al., 2012) and pathogen virulence factors (Yin et al., 2015). However, pathogens have evolved effective suppressors against host RNA silencing, making the system unsuitable for engineering strong and durable resistance in crops (Pumplin and Voinnet, 2013). A more attractive option is gene targeting (GT) which allows any endogenous gene to be disrupted or replaced with a copy that has been manipulated *in vitro*. In GT experiments, double-stranded break (DSB) at the target gene is repaired by one of the two main competing DNA repair pathways: the more frequent non-homologous end joining (NHEJ) pathway or the rare but precise homologous recombination (HR) (Chapman et al., 2012). GT could have a huge impact as a 'clean transgenesis' technology for precise gene manipulation or transfer of novel traits into crops. Despite huge efforts, this powerful tool has been elusive in plant science for a long time as it relied on extremely rare spontaneous DSBs (Puchta and Fauser, 2013). However, this

Citation: Abdellah Barakate and Jennifer Stephens, CRISPR-Based Tools and Their Improvements: New Opportunities in Understanding Plant–Pathogen Interactions for Better Crop Protection, doi.org/10.3389/fpls.2016.00765.

barrier has been recently overcome by the development of novel endonucleases that break DNA specifically at chosen genomic targets. Unfortunately, gene replacement by HR is still inefficient in plants and will need further improvement.

Initially, two endonucleases were engineered by fusing a programmable DNA-binding domain to the cleavage domain of the bacterial restriction enzyme *Fok*I. The first endonuclease was generated by linking the DNA-binding domain of a zinc-finger transcription factor to make the first truly flexible chimeric nuclease zinc-finger nuclease (ZFN) (Smith et al., 2000). Similarly, the DNA-binding domain of a transcription activator-like effector of the plant pathogen *Xanthomonas* was used to make the second, and relatively easier to design, nuclease transcription activator-like effector nuclease (TALEN) (Christian et al., 2010). These two big breakthroughs were superseded by an even simpler system based on the clustered, regularly interspaced, short palindromic repeats (CRISPR) and CRISPR-associated genes (Cas) used by some bacteria and Archaea to destroy invading genetic material (Jinek et al., 2012). Our knowledge of CRISPR/Cas is rapidly evolving and the findings are extensively reported and reviewed. Here we will briefly describe the natural and engineered CRISPR–Cas systems followed by the latest and future applications in plant–pathogen interactions.

THE NATIVE CRISPR–CAS SYSTEM

The CRISPR–Cas system was discovered in bacterial genomes as early as 1987 (Ishino et al., 1987) but its biological role was determined only in 2007 (Barrangou et al., 2007). These evolving adaptive immune systems against invading phages and plasmids are now re-classified into five types I–V (Makarova et al., 2015). During the first invasion, the hosts capture short DNA sequences of about 20 nucleotides, known as spacers, from the foreign genetic material and integrate them between two repeats at the CRISPR locus (Nuñez et al., 2015). Upon subsequent encounters, CRISPR arrays with the acquired spacers are transcribed and processed into small CRISPR RNAs containing the spacer (crRNAs). This chimeric molecule interacts with another auxiliary *trans*-activating CRISPR RNA (tracrRNA), forming a duplex RNA or guide RNA (gRNA) that guides the Cas nuclease to the homologous target (protospacer), resulting in an R-loop structure. The tracrRNA activates Cas nuclease activity, cleaving both strands of the target DNA upstream of a conserved protospacer-adjacent motif (PAM). Cas nuclease has two domains, RuvC and HNH, that cut the PAM-containing strand and its complementary strand, respectively, thus producing a single DSB (Heler et al., 2015). The spacer and PAM requirements depend on CRISPR–Cas type (Xue et al., 2015). In the case of the widely used type II CRISPR–Cas9 system, the last 12 ribonucleotides at the 3′-end of the RNA spacer, known as the seed sequence, dictates the specificity of the complementary target. Mismatches at its 5′-end were thought to be tolerated during gRNA–Cas9 binding to the target. However, the interaction of this region and PAM-distal sequences turned out to be necessary for the activation of Cas9 endonuclease activity (Cencic et al., 2014). PAM sequences are 2–5 bp motifs essential for spacer acquisition and target cleavage (Shah et al., 2013).

REPURPOSING OF THE CRISPR–CAS9 SYSTEM FOR GENE EDITING IN EUKARYOTES

The knowledge of the biological function and mechanism of CRISPR–Cas inspired its reprogramming to target any chosen DNA sequence. CRISPR–Cas9 of *Streptococcus pyogenes* was engineered by simply replacing the first 20 nucleotides of crRNA with the intended target sequence and fusing both crRNA and tracrRNA molecules to make a single guide RNA (sgRNA) (Jinek et al., 2012). This newly programmable system was first adopted to target eukaryotic genes in animals, followed by several successful applications in plants including crops (Bortesi and Fischer, 2015;Butler et al., 2015; Lawrenson et al., 2015). The ease of implementation of CRISPR–Cas9 by anyone with basic molecular biology skills has made it the tool of choice for gene editing in any species of interest. Upon generating a DSB at the desired site by the Cas9–gRNA complex, the host cell repairs the DNA lesion by NHEJ pathway resulting in short insertions or deletions, leading to gene knockout. The flexibility of the CRISPR–Cas9 system allows targeting of adjacent sites in a single gene for specific removal of a region, which will be extremely useful for the studies of gene and mRNA *cis*-elements and protein domains (Brooks et al., 2014). CRISPR–Cas9 can also be used in plants to knockout all or single multigene family members (Endo et al., 2015) and even multiple unrelated genes (Lowder et al., 2015).

The DSB lesion can also be repaired by the HR mechanism in the presence of a donor template, leading to precise gene replacement (knock-in). HR-based gene replacement is still inefficient and has been demonstrated in only a few plant species (Bortesi and Fischer, 2015). The efficiency of homology directed repair (HDR) of CRISPR–Cas9 induced DSBs was recently increased by inhibiting the NHEJ pathway in mammalian cells (Chu et al., 2015; Maruyama et al., 2015). Cas9 was recently found to dissociate slowly from DSB by releasing first the 3′end of the cleaved DNA strand that is not complementary to the sgRNA. Consequently HDR was increased to 60% in human cells by using rationally designed single-stranded DNA donor template of the optimal length complementary to the strand that is released first (Richardson et al., 2016). Maize was the first crop where CRISPR–Cas9 was successfully used to generate plants with precise modifications (Svitashev et al., 2015). Precise gene modifications have been achieved at high frequency in tomato by combining the CRISPR–Cas9 nuclease with a geminivirus-based vector for donor DNA template delivery (Čermák et al., 2015). The combination of some or all of the incremental improvements in different animal and plant species could enhance gene replacement efficiency for all crops.

In pathogens, GT without DSB induction was only improved by inhibiting the NHEJ pathway, as in the ku70 mutant of *Verticillium dahliae* (Qi et al., 2015). CRISPR–Cas9 has now made gene editing possible in fungi (Matsu-ura et al., 2015; Nødvig et al., 2015). The effector Avr4/6 of the soybean pathogen *Phytophthora sojae* was efficiently knocked out or even precisely replaced by the selectable marker *npt*II, uncovering additional roles for the corresponding R gene loci *RPS4* and*RPS6* (Fang and Tyler, 2016). The establishment of gene editing tools in *P. sojae* will speed up studies for crop protection in other oomycetes.

Resistance to geminiviruses has been long sought after and was achieved recently in three independent studies using CRISPR–Cas9 in *Nicotiana benthamiana* (Ali et al., 2015; Baltes et al., 2015; Ji et al., 2015). In these works, CRISPR–Cas9 was shown to mutate the viral genome, resulting in reduced viral replication and attenuated infection symptoms. A single gRNA targeting a conserved sequence in the replication origin resulted in efficient inhibition of multiple monopartite and bipartite geminiviruses in the same host. However, further studies will be required to monitor the evolution of this resistance over generations and in more challenging environments (Chaparro-Garcia et al., 2015).

Viral vectors can also be targeted by CRISPR–Cas9 technology to abolish pathogen transmission or even reduce insect population by the so-called mutagenic chain reaction (Gantz and Bier, 2015). This system is initiated when both Cas9 and gRNA transgenes are inserted by homology directed repair at the intended target in males. The transgenes are then copied into the homologous chromosome by HR in the germ-line cells. During fertilization, the males transfer the CRISPR–Cas9 cassette into the next generation and the chain continues. This gene drive system has been demonstrated to be very efficient in manipulating two species of mosquito which are vectors for malaria (Esvelt et al., 2014; Gantz et al., 2015; Hammond et al., 2016). Though attractive, gene drive will not work in self-fertilizing weeds and non-native invasive plant species but it could potentially be used against flies that are vectors of plant pathogens provided that they are amenable to transgenesis. However, safeguarding against the unintended ecological impact of manipulated insect populations is of great importance and biosafety concerns are starting to be addressed by developing antidote systems to reverse gene drive effects (DiCarlo et al., 2015).

IMPROVEMENTS TO THE CRISPR–CAS9 SYSTEM EFFICIENCY AND SPECIFICITY

Since the conception of the CRISPR–Cas9 gene editing system, its components Cas9 and gRNA have been continuously optimized to improve the efficiency and accuracy of GT (Bolukbasi et al., 2015; Graham and Root, 2015). The repurposing of the CRISPR–Cas9 system to alter eukaryotic genes necessitated targeting the bacterial Cas9 to the nucleus by adding a nuclear localization signal at one or both termini of the protein. To improve translation efficiency, the Cas9 gene was initially codon optimized for human cells and was quickly followed by several plant versions, for both dicots and monocots (Bortesi and Fischer, 2015). The endonuclease Cas9 can easily be converted into a DNA nickase by a single amino acid change in either of its two domains (D10A in RuvC and H840A in HNH; Cong et al., 2013) to cut only one strand. A DSB can still be introduced at the target by these nickases in the presence of two gRNAs that target opposing strands at neighboring sites. This feature has been exploited to improve the specificity of CRISPR–Cas9 and reducing potential off-targets (Ran et al., 2013), a major concern with engineered endonucleases in animals (Hendel et al., 2015) and in plants (Bortesi and Fischer, 2015). Several assays for quantifying on- and off-targets have been developed and inspired strategies for minimizing off-target effects (Hendel et al., 2015; Zhang et al., 2015). In plants, the use of whole genome sequencing as the most accurate method is limited to *Arabidopsis* and rice with good genome reference (Bortesi and Fischer, 2015). Unlike in human gene therapy, off-targets are less problematic in plants where one could eliminate such events by backcrosses. The determination of Cas9 structure (Nishimasu et al., 2014, 2015) has also inspired rational engineering of new Cas9 variants with altered PAM recognition (Kleinstiver et al., 2015) and greater specificity (Slaymaker et al., 2016). Orthologs of commonly used Cas9 from *S. pyogenes* (SpCas9) have been reported to have different features and requirements. The *S. aureus* Cas9 (SaCas9) gene is 1 kbp shorter than SpCas9, improving its stability in viral vectors (Ran et al., 2015). In the screening effort for SpCas9 orthologs, another protein, Cpf1 (CRISPR from *Prevotella* and *Francisella* 1) of type V CRISPR–Cas systems, has been reported to function in a completely different way to Cas9. Cpf1 does not need a tracrRNA but requires a T-rich PAM motif upstream of the target site and generates a DSB with 5′ overhangs (Zetsche et al., 2015).

The design of guide RNAs for efficient and specific gene editing has also been the focus of many studies combining experimental and computing analyses. Several user-friendly algorithms have been developed and freely shared online with the scientific community[1]. Most of these bioinformatics tools are designed to score the efficiency of all potential targets with a PAM motif in the input gene sequence (Wiles et al., 2015). The chance of off-target effects elsewhere in the genome can also be accounted for where the genome sequence is available. These bioinformatics tools are continuously being refined with the availability of new experimental data (Malina et al., 2015; Wong et al., 2015). The structure of the artificial single guide RNA has been revisited recently and improved by lengthening the duplex crRNA/tracrRNA and improving gRNA transcription by shortening its thymine repeat (Dang et al., 2015).

Several systems for delivering Cas9 and gRNA molecules into the cell are available, depending on the species of interest. Plasmid constructs are often used to express Cas9 from RNA polymerase II-driven promoters and gRNAs with polymerase III-mediated transcription. A new strategy based on endogenous tRNA maturation has been developed for expressing multiple gRNAs from a single pol III promoter (Lowder et al., 2015; Xie et al., 2015). While a pol II promoter can be chosen to drive tissue-specific expression of Cas9, snoRNA U3, and U6 pol III promoters are constitutive. However, newly reconstructed sgRNAs can now be expressed from pol II promoters (Wang et al., 2015). Inducible promoters can also be used to induce gene editing *in vivo*, yet reduce potential off-target effects and Cas9-associated toxicity (Dow et al., 2015). Even better, Cas9 and the gRNA can be simultaneously expressed from a single promoter allowing for more spatio-temporal control of each component (Yoshioka et al., 2015). These conditional gene editing methods present new opportunities in crop research but have not yet been tested in plants.

OTHER FACETS OF THE CRISPR–CAS9 SYSTEM

The CRISPR–Cas9 system has become the tool of choice for gene manipulation owing to its simplicity and the willingness of researchers to share the necessary plasmids and methods, including the various algorithms for designing gRNAs. Most of these ingredients are now deposited with the nonprofit plasmid repository Addgene[2] (Harrison et al., 2014). Although most studies focus on knocking out a single gene or a combination of a few targets (multiplexing), the CRISPR–Cas9 system is so powerful that it has been successfully used for genome-wide mutagenesis in mammalian (Malina et al., 2014; Peng et al., 2015) and *Drosophila* (Bassett et al., 2015) cells. The CRISPR–Cas9 based genetic screen uses thousands of unique gRNAs covering the genome of interest and relies on efficient delivery of Cas9/gRNA cargo. This type of forward genetic screen will be very useful in studies of plant–pathogen interactions, but the transformation of plant or pathogen cells must be optimized. This goal can be achieved with at least some pathogens and plant models like *Arabidopsis* and tobacco.

When both RuvC and HNH nuclease domains are mutated, Cas9 becomes an inactive or dead endonuclease (dCas9). Qi et al. (2013) were the first to demonstrate that dCas9 can specifically repress gene expression in *Escherichia coli* in the presence of the gene specific gRNA. This work was quickly followed by another report where dCas9 was fused to transcriptional effectors to silence or activate gene expression in eukaryotes, thereby reversibly manipulating gene expression (Gilbert et al., 2013). Similarly, the epigenome can be manipulated at a specific site by fusing dCas9 with various DNA effectors or histone methylases and acetylases (Hilton et al., 2015; Laufer and Singh, 2015). dCas9-based gene regulation platforms can be used for both genome-wide loss-of-function and gain-of-function screens and the system is amenable to controlled induction (Dominguez et al., 2016). When tagged with fluorescent proteins, dCas9 can be used instead of fluorescence *in situ*hybridization (FISH) to detect chromosomal loci in living (Chen et al., 2013) and fixed (Deng et al., 2015) cells. In this application, dCas9-fluorescent protein fusions can be targeted by a gRNA to a specific locus in the genome for cytological detection. The simultaneous detection of multiple loci in the same cell is feasible by simply fusing different dCas9 orthologs with different fluorescent proteins. Most of the dCas9-based tools will be very useful in deciphering plant–pathogen interactions. Inducible activation or inhibition of master regulators could have huge practical agronomical applications but the down-side is that the dCas9/gRNA transgenes must be kept permanently in the plant.

CONCLUDING REMARKS AND OUTLOOK

Different omics platforms have opened the flood gate of potential disease resistance genes that need a more efficient validation pipeline than earlier gene manipulation tools like gene silencing. Plant–pathogen omics data could be improved even further by reducing the background noise in the biological samples. This can now be achieved, for example, by performing cell-type specific RNA or chromatin profiling with novel tools like INTACT (Deal and Henikoff, 2010). Cell-type enrichment will help monitor the dynamics of post-translational modifications during plant–pathogen interactions (Park and Yun, 2013; Motion et al., 2015). CRISPR–Cas9 technology has revolutionized gene manipulation capabilities in many species including crops. The multitude of functions that can be performed with CRISPR–Cas9 and its many derivatives (Sander and Joung, 2014) make it a molecular tool that will open new opportunities in the complicated world of plant–pathogen interactions and help design durable crop resistance to pathogens. Only the gene editing function of CRISPR–Cas9 has so far been used in plants and pathogens. However, the future use of dCas9-based tools will also help to unmask the master regulators of disease resistance (Seo and Choi, 2015). GT tools will help integrate omics data in order to fully understand and improve crop defense mechanisms. The complexity of the plant microbiome with good and bad microbes is beginning to be unraveled (Bai et al., 2015). CRISPR–Cas9 tools will help future studies of plant–pathogen interactions to transcend individual genes or pathogens and become more holistic in approaches to elucidate plant microbiome systems.

FUNDING

We acknowledge funding for this research from the Scottish Government Rural and Environment Science and Analytical Services division (RESAS) and the European Research Council (SHUFFLE_692631).

CONFLICT OF INTEREST STATEMENT

The authors declare that the research was conducted in the absence of any commercial or financial relationships that could be construed as a potential conflict of interest.

REFERENCES

1. Ali, Z., Abulfaraj, A., Idris, A., Ali, S., Tashkandi, M., and Mahfouz, M. M. (2015). CRISPR/Cas9-mediated viral interference in plants. *Genome Biol.* 16:238. doi: 10.1186/s13059-015-0799-6
2. Bai, Y., Müller, D. B., Srinivas, G., Garrido-Oter, R., Potthoff, E., Rott, M., et al. (2015). Functional overlap of the *Arabidopsis*leaf and root microbiota. *Nature* 528, 364–369. doi: 10.1038/nature16192
3. Baltes, N. J., Hummel, A. W., Konecna, E., Cegan, R., Bruns, A. N., Bisaro, D. M., et al. (2015). Conferring resistance to geminiviruses with the CRISPR-Cas prokaryotic immune system. *Nat. Plants* 1:15145. doi: 10.1038/nplants.2015.145
4. Barrangou, R., Fremaux, C., Deveau, H., Richards, M., Boyaval, P., Moineau, S., et al. (2007). CRISPR provides acquired resistance against viruses in prokaryotes. *Science* 315, 1709–1712. doi: 10.1126/science.1138140
5. Bassett, A. R., Kong, L., and Liu, J.-L. (2015). A genome-wide CRISPR library for high-throughput genetic screening in*Drosophila* cells. *J. Genet. Genomics* 42, 301–309. doi: 10.1016/j.jgg.2015.03.011
6. Bolukbasi, M. F., Gupta, A., and Wolfe, S. A. (2015). Creating and evaluating accurate CRISPR-Cas9 scalpels for genomic surgery. *Nat. Methods* 13, 41–50. doi: 10.1038/nmeth.3684
7. Bortesi, L., and Fischer, R. (2015). The CRISPR/Cas9 system for plant genome editing and beyond. *Biotechnol. Adv.* 33, 41–52. doi: 10.1016/j.biotechadv.2014.12.006
8. Brooks, C., Nekrasov, V., Lippman, Z. B., and Van Eck, J. (2014). Efficient gene editing in tomato in the first generation using the clustered regularly interspaced short palindromic repeats/CRISPR-associated9 system. *Plant Physiol.* 166, 1292–1297. doi: 10.1104/pp.114.247577

9. Butler, N. M., Atkins, P. A., Voytas, D. F., and Douches, D. S. (2015). Generation and inheritance of targeted mutations in potato (*Solanum tuberosum* L.) using the CRISPR/Cas system. *PLoS ONE* 10:e0144591. doi: 10.1371/journal.pone.0144591

10. Cencic, R., Miura, H., Malina, A., Robert, F., Ethier, S., Schmeing, T. M., et al. (2014). Protospacer adjacent motif (PAM)-distal sequences engage CRISPR Cas9 DNA target cleavage. *PLoS ONE* 9:e109213. doi: 10.1371/journal.pone.0109213

11. Čermák, T., Baltes, N. J., Čegan, R., Zhang, Y., and Voytas, D. F. (2015). High-frequency, precise modification of the tomato genome. *Genome Biol.* 16:232. doi: 10.1186/s13059-015-0796-9

12. Chaparro-Garcia, A., Kamoun, S., and Nekrasov, V. (2015). Boosting plant immunity with CRISPR/Cas. *Genome Biol.* 16:254. doi: 10.1186/s13059-015-0829-4

13. Chapman, J. R., Taylor, M. R., and Boulton, S. J. (2012). Playing the end game: DNA double-strand break repair pathway choice.*Mol. Cell* 47, 497–510. doi: 10.1016/j.molcel.2012.07.029

14. Chen, B., Gilbert, L. A., Cimini, B. A., Schnitzbauer, J., Zhang, W., Li, G.-W., et al. (2013). Dynamic imaging of genomic loci in living human cells by an optimized CRISPR/Cas system. *Cell* 155, 1479–1491. doi: 10.1016/j.cell.2013.12.001

15. Chu, V. T., Weber, T., Wefers, B., Wurst, W., Sander, S., Rajewsky, K., et al. (2015). Increasing the efficiency of homology-directed repair for CRISPR-Cas9-induced precise gene editing in mammalian cells. *Nat. Biotechnol.* 33, 543–548. doi: 10.1038/nbt.3198

16. Christian, M., Cermak, T., Doyle, E. L., Schmidt, C., Zhang, F., et al. (2010). Targeting DNA double-strand breaks with TAL effector nucleases. *Genetics* 186, 757–761. doi: 10.1534/genetics.110.120717

17. Cong, L., Ran, F. A., Cox, D., Lin, S., Barretto, R., Habib, N., et al. (2013). Multiplex genome engineering using CRISPR/Cas systems. *Science* 339, 819–823. doi: 10.1126/science.1231143

18. Dang, Y., Jia, G., Choi, J., Ma, H., Anaya, E., Ye, C., et al. (2015). Optimizing sgRNA structure to improve CRISPR-Cas9 knockout efficiency. *Genome Biol.* 16:280. doi: 10.1186/s13059-015-0846-3

19. Deal, R. B., and Henikoff, S. (2010). A simple method for gene expression and chromatin profiling of individual cell types within a tissue. *Dev. Cell* 18, 1030–1040. doi: 10.1016/j.devcel.2010.05.013

20. Deng, W., Shi, X., Tjian, R., Lionnet, T., and Singer, R. H. (2015). CASFISH: CRISPR/Cas9-mediated in situ labeling of genomic loci in fixed cells. *Proc. Natl. Acad. Sci. U.S.A.* 112, 11870–11875. doi: 10.1073/pnas.1515692112

21. DiCarlo, J. E., Chavez, A., Dietz, S. L., Esvelt, K. M., and Church, G. M. (2015). Safeguarding CRISPR-Cas9 gene drives in yeast.*Nat. Biotechnol.* 33, 1250–1255. doi: 10.1038/nbt.3412

22. Dominguez, A. A., Lim, W. A., and Qi, L. S. (2016). Beyond editing: repurposing CRISPR–Cas9 for precision genome regulation and interrogation. *Nat. Rev. Mol. Cell Biol.* 17, 5–15. doi: 10.1038/nrm.2015.2

23. Dow, L. E., Fisher, J., O'Rourke, K. P., Muley, A., Kastenhuber, E. R., Livshits, G., et al. (2015). Inducible in vivo genome editing with CRISPR-Cas9. *Nat. Biotechnol.* 33, 390–394. doi: 10.1038/nbt.3155

24. Duan, C.-G., Wang, C.-H., and Guo, H.-S. (2012). Application of RNA silencing to plant disease resistance. *Silence* 3:5. doi: 10.1186/1758-907X-3-5

25. Endo, M., Mikami, M., and Toki, S. (2015). Multigene knockout utilizing off-target mutations of the CRISPR/Cas9 system in rice. *Plant Cell Physiol.* 56, 41–47. doi: 10.1093/pcp/pcu154

26. Esvelt, K. M., Smidler, A. L., Catteruccia, F., and Church, G. M. (2014). Emerging technology: concerning RNA-guided gene drives for the alteration of wild populations. *eLife* 3:e03401. doi: 10.7554/eLife.03401

27. Fang, Y., and Tyler, B. M. (2016). Efficient disruption and replacement of an effector gene in the oomycete *Phytophthora sojae*using CRISPR/Cas9. *Mol. Plant Pathol.* 17, 127–139. doi: 10.1111/mpp.12318

28. Gantz, V. M., and Bier, E. (2015). The mutagenic chain reaction: a method for converting heterozygous to homozygous mutations. *Science* 348, 442–444. doi: 10.1126/science.aaa5945

29. Gantz, V. M., Jasinskiene, N., Tatarenkova, O., Fazekas, A., Macias, V. M., Bier, E., et al. (2015). Highly efficient Cas9-mediated gene drive for population modification of the malaria vector mosquito *Anopheles stephensi*. *Proc. Natl. Acad. Sci. U.S.A.*112:e6736-43. doi: 10.1073/pnas.1521077112

30. Gilbert, L. A., Larson, M. H., Morsut, L., Liu, Z., Brar, G. A., Torres, S. E., et al. (2013). CRISPR-mediated modular RNA-guided regulation of transcription in eukaryotes. *Cell* 154, 442–451. doi: 10.1016/j.cell.2013.06.044

31. Graham, D. B., and Root, D. E. (2015). Resources for the design of CRISPR gene editing experiments. *Genome Biol.* 16, 260. doi: 10.1186/s13059-015-0823-x

32. Hammond, A., Galizi, R., Kyrou, K., Simoni, A., Siniscalchi, C., Katsanos, D., et al. (2016). A CRISPR-Cas9 gene drive system targeting female reproduction in the malaria mosquito vector *Anopheles gambiae*. *Nat. Biotechnol.* 34, 78–83. doi: 10.1038/nbt.3439

33. Harrison, M. M., Jenkins, B. V., O'Connor-Giles, K. M., and Wildonger, J. (2014). A CRISPR view of development. *Gene Dev.* 28, 1859–1872. doi: 10.1101/gad.248252.114

34. Heler, R., Samai, P., Modell, J. W., Weiner, C., Goldberg, G. W., Bikard, D., et al. (2015). Cas9 specifies functional viral targets during CRISPR-Cas adaptation. *Nature* 519, 199–202. doi: 10.1038/nature14245

35. Hendel, A., Fine, E. J., Bao, G., and Porteus, M. H. (2015). Quantifying on- and off-target genome editing. *Trends Biotechnol.*33, 132–140. doi: 10.1016/j.tibtech.2014.12.001

36. Hilton, I. B., D'Ippolito, A. M., Vockley, C. M., Thakore, P. I., Crawford, G. E., Reddy, T. E., et al. (2015). Epigenome editing by a CRISPR-Cas9-based acetyltransferase activates genes from promoters and enhancers. *Nat. Biotechnol.* 33, 510–517. doi: 10.1038/nbt.3199

37. Ishino, Y., Shinagawa, H., Makino, K., Amemura, M., and Nakata, A. (1987). Nucleotide sequence of the iap gene, responsible for alkaline phosphatase isozyme conversion in *Escherichia coli*, and identification of the gene product. *J. Bacteriol.* 169, 5429–5433.

38. Ji, X., Zhang, H., Zhang, Y., Wang, Y., and Gao, C. (2015). Establishing a CRISPR-Cas-like immune system conferring DNA virus resistance in plants. *Nat. Plants* 1:15144. doi: 10.1038/nplants.2015.144

39. Jinek, M., Chylinski, K., Fonfara, I., Hauer, M., Doudna, J. A., and Charpentier, E. (2012). A programmable dual-RNA-guided DNA endonuclease in adaptive bacterial immunity. *Science* 337, 816–821. doi: 10.1126/science.1225829

40. Kawahara, Y., Oono, Y., Kanamori, H., Matsumoto, T., Itoh, T., and Minami, E. (2012). Simultaneous RNA-Seq analysis of a mixed transcriptome of rice and blast fungus interaction. *PLoS ONE* 7:e49423. doi: 10.1371/journal.pone.0049423

41. Kleinstiver, B. P., Prew, M. S., Tsai, S. Q., Topkar, V. V., Nguyen, N. T., Zheng, Z., et al. (2015). Engineered CRISPR-Cas9 nucleases with altered PAM specificities. *Nature* 523, 481–485. doi: 10.1038/nature14592

42. Laufer, B. I., and Singh, S. M. (2015). Strategies for precision modulation of gene expression by epigenome editing: an overview.*Epigenetics Chromatin* 8:34. doi: 10.1186/s13072-015-0023-7

43. Lawrenson, T., Shorinola, O., Stacey, N., Li, C., Østergaard, L., Patron, N., et al. (2015). Induction of targeted, heritable mutations in barley and *Brassica oleracea* using RNA-guided Cas9 nuclease. *Genome Biol.* 16:258. doi: 10.1186/s13059-015-0826-7

44. Lodha, T. D., Hembram, P., Tep, N., and Basak, J. (2013). Proteomics: a successful approach to understand the molecular mechanism of plant-pathogen interaction. *Am. J. Plant Sci.* 4, 1212–1226. doi: 10.4236/ajps.2013.46149

45. Lowder, L. G., Zhang, D., Baltes, N. J., Paul, J. W. III, Tang, X., Zheng, X., et al. (2015). A CRISPR/Cas9 toolbox for multiplexed plant genome editing and transcriptional regulation. *Plant Physiol.* 169, 971–985. doi: 10.1104/pp.15.00636

46. Makarova, K. S., Wolf, Y. I., Alkhnbashi, O. S., Costa, F., Shah, S. A., Saunders, S. J., et al. (2015). An updated evolutionary classification of CRISPR-Cas systems. *Nat. Rev. Microbiol.* 13, 722–736. doi: 10.1038/nrmicro3569

47. Malina, A., Cameron, C. J. F., Robert, F., Blanchette, M., Dostie, J., and Pelletier, J. (2015). PAM multiplicity marks genomic target sites as inhibitory to CRISPR-Cas9 editing. *Nat. Commun.* 6:10124. doi: 10.1038/ncomms10124

48. Malina, A., Katigbak, A., Cencic, R., Maïga, R. I., Robert, F., Miura, H., et al. (2014). Adapting CRISPR/Cas9 for functional genomics screens. *Methods Enzymol.* 546, 193–213. doi: 10.1016/B978-0-12-801185-0.00010-6

49. Maruyama, T., Dougan, S. K., Truttmann, M. C., Bilate, A. M., Ingram, J. R., and Ploegh, H. L. (2015). Increasing the efficiency of precise genome editing with CRISPR-Cas9 by inhibition of nonhomologous end joining. *Nat. Biotechnol.* 33, 538–542. doi: 10.1038/nbt.3190

50. Matsu-ura, T., Baek, M., Kwon, J., and Hong C. (2015). Efficient gene editing in Neurospora crassa with CRISPR technology.*Fungal Biol. Biotechnol.* 2, 1–7. doi: 10.1186/s40694-015-0015-1

51. Motion, G. B., Amaro, T. M., Kulagina, N., and Huitema, E. (2015). Nuclear processes associated with plant immunity and pathogen susceptibility. *Brief. Funct. Genomics* 14, 243–252. doi: 10.1093/bfgp/elv013

52. Nishimasu, H., Cong, L., Yan, W. X., Ran, F. A., Zetsche, B., Li, Y., et al. (2015). Crystal structure of *Staphylococcus aureus* Cas9.*Cell* 162, 1113–1126. doi: 10.1016/j.cell.2015.08.007

53. Nishimasu, H., Ran, F. A., Hsu, P. D., Konermann, S., Shehata, S. I., Dohmae, N., et al. (2014). Crystal structure of Cas9 in complex with guide RNA and target DNA. *Cell* 156, 935–949. doi: 10.1016/j.cell.2014.02.001

54. Nødvig, C. S., Nielsen, J. B., Kogle, M. E., and Mortensen, U. H. (2015). A CRISPR-Cas9 system for genetic engineering of filamentous fungi. *PLoS ONE* 10:e0133085. doi: 10.1371/journal.pone.0133085

55. Nuñez, J. K., Lee, A. S. Y., Engelman, A., and Doudna, J. A. (2015). Integrase-mediated spacer acquisition during CRISPR-Cas adaptive immunity. *Nature* 519, 193–198. doi: 10.1038/nature14237

56. O'Connell, R. J., Thon, M. R., Hacquard, S., Amyotte, S. G., Kleemann, J., Torres, M. F., et al. (2012). Lifestyle transitions in plant pathogenic *Colletotrichum fungi* deciphered by genome and transcriptome analyses. *Nat. Genet.* 44, 1060–1065. doi: 10.1038/ng.2372

57. Olukolu, B. A., Wang, G.-F., Vontimitta, V., Venkata, B. P., Marla, S., Ji, J., et al. (2014). A genome-wide association study of the maize hypersensitive defense response identifies genes that cluster in related pathways. *PLoS Genet.* 10:e1004562. doi: 10.1371/journal.pgen.1004562

58. Park, H. J., and Yun, D. J. (2013). New insights into the role of the small ubiquitin-like modifier (SUMO) in plants. *Int. Rev. Cell. Mol. Biol.* 300, 161–209. doi: 10.1016/B978-0-12-405210-9.00005-9

59. Peng, J., Zhou, Y., Zhu, S., and Wei, W. (2015). High-throughput screens in mammalian cells using the CRISPR-Cas9 system.*FEBS J.* 282, 2089–2096. doi: 10.1111/febs.13251

60. Puchta, H., and Fauser, F. (2013). Gene targeting in plants: 25 years later. *Int. J. Dev. Biol.* 57, 629–637. doi: 10.1387/ijdb.130194hp

61. Pumplin, N., and Voinnet, O. (2013). RNA silencing suppression by plant pathogens: defence, counter-defence and counter-counter-defence. *Nat. Rev. Microbiol.* 11, 745–760. doi: 10.1038/nrmicro3120

62. Qi, L. S., Larson, M. H., Gilbert, L. A., Doudna, J. A., Weissman, J. S., Arkin, A. P., et al. (2013). Repurposing CRISPR as an RNA-guided platform for sequence-specific control of gene expression. *Cell* 152, 1173–1183. doi: 10.1016/j.cell.2013.02.022

63. Qi, X., Su, X., Guo, H., Qi, J., and Cheng, H. (2015). A ku70 null mutant improves gene targeting frequency in the fungal pathogen *Verticillium dahliae*. *World J. Microbiol. Biotechnol.* 31, 1889–1897. doi: 10.1007/s11274-015-1907-1

64. Ran, F. A., Cong, L., Yan, W. X., Scott, D. A., Gootenberg, J. S., Kriz, A. J., et al. (2015). In vivo genome editing using*Staphylococcus aureus* Cas9. *Nature* 520, 186–191. doi: 10.1038/nature14299

65. Ran, F. A., Hsu, P. D., Lin, C. Y., Gootenberg, J. S., Konermann, S., Trevino, A. E., et al. (2013). Double nicking by RNA-guided CRISPR Cas9 for enhanced genome editing specificity. *Cell* 154, 1380–1389. doi: 10.1016/j.cell.2013.08.021

66. Richardson, C. D., Ray, G. J., DeWitt, M. A., Curie, G. L., and Corn, J. E. (2016). Enhancing homology-directed genome editing by catalytically active and inactive CRISPR-Cas9 using asymmetric donor DNA. *Nat. Biotechnol.* 34, 339–344. doi: 10.1038/nbt.3481

67. Sander, J. D., and Joung, J. K. (2014). CRISPR-Cas systems for editing, regulating and targeting genomes. *Nat. Biotechnol.* 32, 347–355. doi: 10.1038/nbt.2842

68. Seo, E., and Choi, D. (2015). Functional studies of transcription factors involved in plant defences in the genomics era. *Brief. Funct. Genomics* 14, 260–267. doi: 10.1093/bfgp/elv011

69. Shah, S. A., Erdmann, S., Mojica, F. J., and Garrett, R. A. (2013). Protospacer recognition motifs: mixed identities and functional diversity. *RNA Biol.* 10, 891–899. doi: 10.4161/rna.23764

70. Slaymaker, I. M., Gao, L., Zetsche, B., Scott, D. A., Yan, W. X., and Zhang, F. (2016). Rationally engineered Cas9 nucleases with improved specificity. *Science* 351, 84–88. doi: 10.1126/science.aad5227

71. Smith, J., Bibikova, M., Whitby, F. G., Reddy, A. R., Chandrasegaran, S., and Carroll, D. (2000). Requirements for double-strand cleavage by chimeric restriction enzymes with zinc finger DNA-recognition domains. *Nucleic Acids Res.* 28, 3361–3369. doi: 10.1093/nar/28.17.3361

72. Svitashev, S., Young, J. K., Schwartz, C., Gao, H., Falco, S. C., and Cigan, A. M. (2015). Targeted mutagenesis, precise gene editing, and site-specific gene insertion in maize using Cas9 and guide RNA. *Plant Physiol.* 169, 931–945. doi: 10.1104/pp.15.00793

73. Varshney, R. K., Singh, V. K., Hickey, J. M., Xun, X., Marshall, D. F., Wang, J., et al. (2015). Analytical and decision support tools for genomics-assisted breeding. *Trends Plant Sci.* doi: 10.1016/j.tplants.2015.10.018 [Epub ahead of print].

74. Wang, J., Li, X., Zhao, Y., Li, J., Zhou, Q., and Liu, Z. (2015). Generation of cell-type-specific gene mutations by expressing the sgRNA of the CRISPR system from the RNA polymerase II promoters. *Protein Cell* 6, 689–692. doi: 10.1007/s13238-015-0169-x

75. Wiles, M. V., Qin, W., Cheng, A. W., and Wang, H. (2015). CRISPR-Cas9-mediated genome editing and guide RNA design.*Mamm. Genome* 26, 501–510. doi: 10.1007/s00335-015-9565-z

76. Wong, N., Liu, W., and Wang, X. (2015). WU-CRISPR: characteristics of functional guide RNAs for the CRISPR/Cas9 system.*Genome Biol.* 16:218. doi: 10.1186/s13059-015-0784-0

77. Xie, K., Minkenberg, B., and Yang Y. (2015). Boosting CRISPR/Cas9 multiplex editing capability with the endogenous tRNA-processing system. *Proc. Natl. Acad. Sci. U.S.A.* 112, 3570–3575. doi: 10.1073/pnas.1420294112

78. Xue, C., Seetharam, A. S., Musharova, O., Severinov, K., Brouns, S. J., Severin, A. J., et al. (2015). CRISPR interference and priming varies with individual spacer sequences. *Nucleic Acids Res.* 43, 10831–10847. doi: 10.1093/nar/gkv1259

79. Yin, C., Downey, S. I., Klages-Mundt, N. L., Ramachandran, S., Chen, X., Szabo, L. J., et al. (2015). Identification of promising host-induced silencing targets among genes preferentially transcribed in haustoria of *Puccinia*. *BMC Genomics* 16:579. doi: 10.1186/s12864-015-1791-y

80. Yoshioka, S., Fujii, W., Ogawa, T., Sugiura, K., and Naito K. (2015). Development of a mono-promoter-driven CRISPR/Cas9 system in mammalian cells. *Sci. Rep.* 5:18341. doi: 10.1038/srep18341

81. Zetsche, B., Gootenberg, J. S., Abudayyeh, O. O., Slaymaker, I. M., Makarova, K. S., Essletzbichler, P., et al. (2015). Cpf1 is a single RNA-guided endonuclease of a class 2 CRISPR-Cas system. *Cell* 163, 759–771. doi: 10.1016/j.cell.2015.09.038

82. Zhang, X. H., Tee, L. Y., Wang, X. G., Huang, Q. S., and Yang, S. H. (2015). Off-target effects in CRISPR/Cas9-mediated genome engineering. *Mol. Ther. Nucleic Acids* 4:e264. doi: 10.1038/mtna.2015.37

CHAPTER 5

MOLECULAR TOOLS FOR EXPLORING POLYPLOID GENOMES IN PLANTS

Riccardo Aversano, Maria Raffaella Ercolano, Immacolata Caruso, Carlo Fasano, Daniele Roselliniand Domenico Carputo

Keywords: Genome doubling, molecular markers, microarrays, gene expression, next generation sequencing, transcriptomics

POLYPLOIDY IN THE PLANT KINGDOM: OCCURRENCE AND SIGNIFICANCE

The genome sizes of eukaryotes can differ 10,000-fold and part of these differences may be attributed to changes in the ploidy level. Polyploids are organisms having more than two complete sets of chromosomes in their cells. They are common in angiosperms, where at least 70% of the species experienced one or more events of genome doubling during their evolutionary history [1,2]. Many crop species are polyploids (Table 1), and it was stated that "life on earth is predominantly a polyploid phenomenon and civilization depends mainly on use of polyploid tissues—noteworthy is the endosperm of cereals" [3]. Polyploidization is considered a major evolutionary force in plants. It is a definitive cause of sympatric speciation due to the immediate reproductive isolation between newly formed polyploids and their parents [4]. Polyploidization also makes it possible to overcome hybrid sterility and produce viable offspring following interspecific hybridization. However, there are still several open questions related to polyploidy and polyploidization. For example, Soltis*et al.* [5] reported that polyploidy frequency in angiosperms is high, even if the number of lineages that really derived from genome-wide duplication (WGD) events is still largely unknown. Similarly, it is not clear if polyploidy causes a change in the interaction with herbivores and pollinators.

Citation: Riccardo Aversano, Maria Raffaella Ercolano , Immacolata Caruso, Carlo Fasano, Daniele Rosellini and Domenico Carputo, Molecular Tools for Exploring Polyploid Genomes in Plants, doi:10.3390/ijms130810316.

Table 1: Examples of polyploid crops. The somatic chromosome number is reported in brackets.

Crop	Species
Cereals	*Triticum aestivum* (6× = 42); *T. durum* (4× = 28); *Avena sativa* (6× = 42); *A. nuda* (6× = 42)
Forage grasses	*Dactylis glomerata* (4× = 28); *Festuca arundinacea* (4× = 28); *Agropyrum repens* (4× = 28); *Paspalum dilatatum* (4× = 40)
Legumes	*Medicago sativa* (4× = 32); *Lupinus alba* (4× = 40); *Trifolium repens* (4× = 32); *Arachis hypogaea* (4× = 40); *Lotus corniculatus* (4× = 32); *Glycine max* (4× = 40)
Industrial plants	*Nicotiana tabacum* (4× = 48); *Coffea* spp. (4× = 44 fino a 8×); *Brassica napus* (4× = 38); *Saccharum officinalis* (8× = 80); *Gossypium hirsutum* (4× = 52)
Tuber plants	*Solanum tuberosum* (4× = 48); *Ipomoea batatas* (6× = 96); *Dioscorea sativa* (6× = 60)
Fruit trees	*Prunus domestica* (6× = 48); *Musa* spp. (3× = 33; 4× = 44); *Citrus aurantifolia* (3× = 27); *Actinidia deliciosa* (4× = 116); *P. cerasus* (4× = 32)

Based on their origin, polyploids are classified into two major groups, autopolyploids and allopolyploids [6]. The former are the result of doubling homologous genomes (e.g., autotetraploid AAAA) from a single, or closely related, species. The latter are the result of hybridization between different species. Therefore, they combine two or more different genomes (e.g., allotetraploid AABB). As a consequence, autopolyploids may form multivalents at meiosis, and have polysomic inheritance. By contrast allopolyploids show bivalent pairing and have disomic inheritance [7]. There are several mechanisms that may lead to an increase in ploidy level in plants. However, there is strong circumstantial evidence that sexual polyploidization through gametes with unreduced chromosome number (2*n*gametes) represents the main route for polyploidization [8,9]. 2*n*gametes generally result from the expression of mutations affecting micro- and megasporogenesis. Such mutations have been extensively studied in a number of genera, including*Solanum*,*Medicago*,*Manihot*,*Malus*,*Arachis*,*Lolium*, and*Agropyrum*, and have been generally attributed to the action of single recessive genes [10]. The first gene (*AtPS1*) involved in 2*n*gamete production has been identified in*A. thaliana*[11].*AtPS1*mutants display an anomalous (parallel, fused, tripolar) orientation of spindles at metaphase II of male meiosis, leading to the production of 2*n*pollen. Similarly, 2*n*gametes have been observed in the*jason*(*jas*) [12],*switch1(swi1)/dyad*[13,14],*osd1*and*tam*(*CYCAS1;2*) mutants [15,16].

In a given species and habitat, the acquisition of the polyploid condition may bring about several advantages. Following polyploidization, often novel phenotypes appear or variation exceeds the range observed in the diploid parental species. Fawcett*et al.*[17] suggested that polyploids had a better chance to survive the Cretaceous-Tertiary extinction event. Phenotypic advantages may include, among the others, changes in morphology, physiology and secondary metabolism that confer an increased fitness. Some of these traits, such as increased drought tolerance, pathogen resistance, longer flowering time, larger vegetative and reproductive organs (Figure 1) may represent important plant breeding targets and, therefore, increase the potential use of polyploids in agriculture. From a genetic point of view, the most significant advantages associated with polyploidy are probably heterosis and gene redundancy [18]. Heterosis is due to non-additive inheritance of traits in a newly formed polyploid compared to its parents. Notably, it can be present also at the gametophytic level. The main factors that affect non-additive inheritance are likely novel regulatory interactions and allelic dosage [19]. Gene redundancy promotes neofunctionalization of duplicated genes, in the long term, but also immediately protects against deleterious recessive alleles. In a recent treatment, Mayrose*et al.* [20] shed more light on the evolutionary dynamics and consequences of polyploidy. The authors, computing the net diversification rates of polyploid lineages for ferns, lycophytes, gymnosperms and angiosperms, hypothesized that polyploidy is often an evolutionary dead end. However, the possible longer-term evolutionary success of those polyploids that survive, needs to be tested. Several studies provided evidence that extensive and reproducible genetic and epigenetic changes are possible following polyploidization [9]. They include DNA and histone methylation, DNA elimination, gene losses, gene neo- and subfunctionalization, translocations, amplification and reduction of repetitive sequences and alteration of gene expressions through different mechanisms (methylation-mediated silencing, transposon activation, intergenomic interactions,*etc.*). Given the recent advances in the field of plant molecular biology and biotechnology, through this review we provide an overview of the most suitable technologies and strategies that can allow, at the molecular level, efficient studies on polyploid plants, possibly promoting research in areas that have been ignored or underestimated so far. Reference to some recent significant findings will also be made.

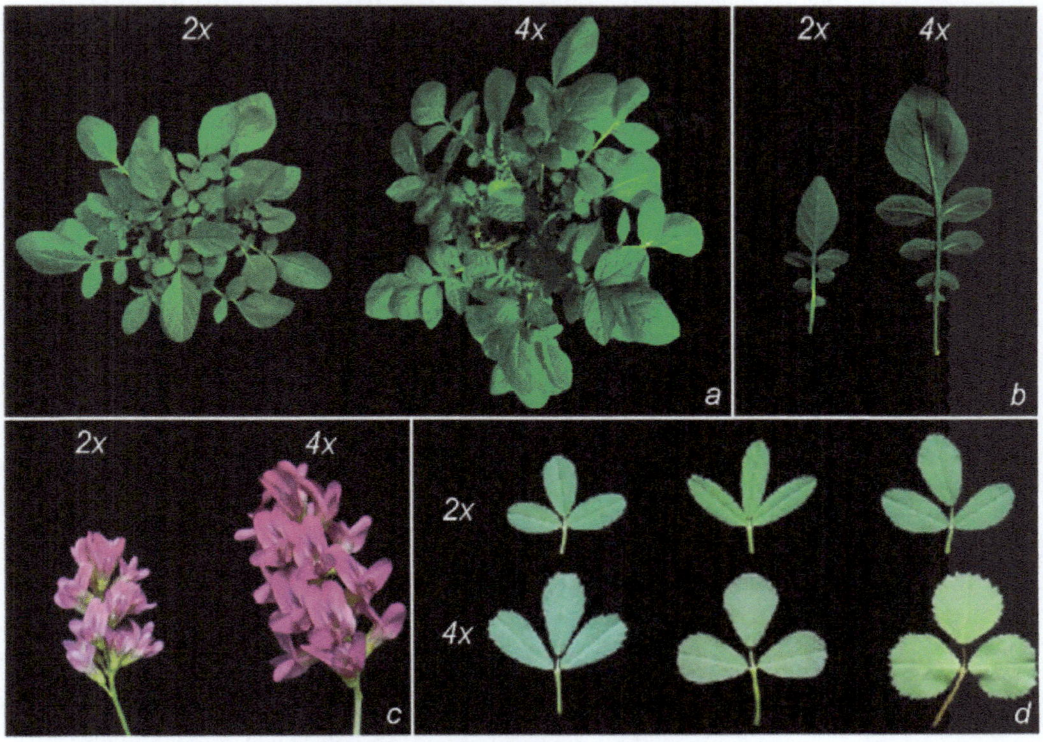

Figure 1: Phenotypic variation between diploids and tetraploids in*Solanum commersonii*(**a**,**b**) and in*Medicago sativa*(**c**,**d**). A diploid ($2n = 2\times = 24$) clone of*S. commersonii*was subjected to oryzaline treatment, an antimitotic drug commonly employed to induce chromosome doubling in plants. The autotetraploid ($2n = 4\times = 48$) genotype displayed larger size at both whole plant (**a**) and leaf (**b**) level. (**c**) Diploid*M. sativa*subsp. *coerulea*($2n = 2\times = 16$, left) and its cultivated tetraploid counterpart,*M. sativa*subsp.*sativa*($2n = 4\times = 32$, right) differ clearly for flower size. (**d**) Leaves of diploid (upper row) and tetraploid (bottom row) plants obtained from crossing two diploid*M. sativa*plants producing both*n*and 2*n*gametes. Leaves are the best qualitative component of forage: tetraploid*M. sativa*has larger leaves and is cultivated.

METHODS FOR GENOME ANALYSIS

A combination of genetic mapping, molecular cytogenetics, sequence and comparative analysis has shed new light and opened perspectives on the nature of ploidy evolution at all timescales, from the base of the plant kingdom, to intra- and interspecific hybridization events associated with plant domestication and breeding. Strong evidences on the mechanisms of genomic modification have come from the use of physical analysis of chromosomes by*in situ*hybridization techniques and from genome-wide molecular marker analyses.

In Situ*Hybridization*

*In situ*hybridization represents the bridge between the chromosomal and molecular level of genome investigation. In recent years it has received a renewed interest for detecting chromosome rearrangements. It is very powerful for reliable identification of chromosomes, allowing the positioning of unique sequences and repetitive DNAs along the chromosome(s). Fluorescent*in situ*hybridization (FISH) is based on fluorescent labels linked to DNA probes and visualized under a fluorescence microscope. Genomic*in situ*hybridization (GISH) involves the use of total genomic DNA of species as a probe on chromosomes, thus leading to whole genome discrimination rather than the localization of specific sequences. There are several examples on the use of these techniques. Studies on the distribution of four tandem repeats in allotetraploids*Tragopogon mirus*,*Tragopogon miscellus*and their diploid parents provided evidence that chromosomal rearrangements did not occur following polyploidization, as suggested by the additive patterns of polyploids [21]. By contrast, in newly synthesized allotetraploid genotypes of*B. napus*, Szadkowski*et al.*[22] demonstrated extensive genome remodeling due to homeologous pairing between the chromosomes of the A and C genomes. Based on high-resolution cytogenetic maps, Wang*et al.*[23] demonstrated that genome size difference between the A and D sub-genomes in allotetraploid cotton was mainly associated with uneven expansion or contraction between different regions of homoeologous chromosomes. Recently, Chester and co-workers [24] combined GISH and FISH analysis to demonstrate that in natural populations of*T. miscellus*extensive chromosomal variation (mainly due to chromosome substitutions and homeologous rearrangements) was present up to the 40th generation following polyploidization.

Molecular Marker-Based Genetic Mapping

Genetic mapping in polyploids presents unique problems with respect to diploid species, since segregations and statistical methods are much more complicated, and large segregating populations are needed to obtain reliable genetic distance estimates [25]. Important results were obtained using classical Restriction Fragment Length Polymorphism markers (RFLP) in*B. napus*[22,26–28],*Draba norvegica*and*T. miscellus*[29], and wheat [30] or employing Amplified Fragment Length Polymorphism (AFLP) in*Arabidopsis*[31],*Brassica*[32], and*Spartina*[33].

Autopolyploids are particularly intractable because segregation depends on chromosome pairing behavior (preferential *vs.* random pairing) and double reduction [34]. A simplification that has been generally adopted for mapping dominant molecular markers is to only utilize the so-called single dose markers from each parent, *i.e.*, those segregating 1:1 in the mapping population (for example, a population obtained from the cross Mmmm × mmmm in a tetraploid species) [35,36]. Statistical models for QTL mapping have been developed for polyploids with bivalent pairing, taking into account preferential *vs.* random chromosome pairing behavior by incorporating a preferential pairing factor [37,38]. Molecular marker segregation allows estimating chromosome pairing behavior. For instance, significant preferential pairing was evidenced in a population obtained by crossing *M. sativa* subsp. *falcata* × *M. sativa* subsp. *sativa* autotetraploid genotypes [39], whereas the data of Julier *et al.* [40] did not support preferential pairing in a mapping study involving cultivated *M. sativa* genotypes. In *Salix*, an allotetraploid origin was inferred based on molecular marker segregation supporting preferential pairing [41]. The recourse to diploid species in search of synteny can facilitate genetic mapping of agronomic traits in complex polyploids [42].

Recently, improvements have been proposed in polysomic inheritance analysis through the determination of the number of allele copies in microsatellite loci [43,44]. In particular, a method named Microsatellite Allele Dose and Configuration Establishment (MADCE) was proposed to establish the exact dose and configuration of microsatellite alleles in allo-octoploid cultivated strawberry (*Fragaria × ananassa*) [44]. This technique enhances the utility of SSR markers in polyploids by overcoming the limitation of single dose markers. A recently introduced type of molecular marker, Diversity Arrays Technology (DArT), was applied to mapping in a tetraploid recombinant inbred line population obtained by crossing wild *Avena magna* ($2n = 4\times = 28$) with cultivated *A. sativa* ($2n = 6\times = 42$) [45].

Methylation Sensitive Molecular Markers

The use of an AFLP-like method using restriction enzymes sharing the same recognition site but having differential sensitivity to DNA methylation (isoschizomeres) was proved to be efficient and reliable for the determination of genome-wide DNA methylation patterns [46]. This technique, termed Methylation-Sensitive Amplified Polymorphism (MSAP), is based on the use of the isoschizomers *Hpa*II and *Msp*I both recognizing the 5′CCGG sequence, but affected by the methylation state of the outer or inner cytosine residues. Using this method, noteworthy results were obtained in newly synthesized polyploids. In *Arabidopsis*, Madlung *et al.* [47] demonstrated that frequent changes occurred in F_4 allotetraploids when compared with the parents. Changes involved both increases and decreases in methylation, but no overall hyper- or hypomethylation. Similarly, alterations in cytosine methylation in wheat occurred in about 13% of the loci, either in F_1 hybrids or in allopolyploids. Notably, alterations in methylation patterns affected both repetitive DNA sequences and low-copy DNA in approximately equal proportions [25]. On the other hand, lack of rapid DNA methylation changes at symmetric CCGG sites was hypothesized in allopolyploid cotton [48]. A similar behavior was observed also in *Brassica* [49] and sugarcane [50].

Comparative Genome Analysis

Comparative genomics research has gained importance as a powerful tool for addressing both fundamental and applied questions in genome evolution [51–53]. The implementation of these methodologies, however, requires consideration of the variable rates at which different aspects of genome evolution occur [52]. Comparing different wheat species, genomic rearrangements originating by illegitimate DNA recombination were identified as a major evolutionary mechanism [54,55]. Innes *et al.* [56], comparing homologous regions in several related legume species, demonstrated that retroelements were the largest contributor to duplicated regions. Comparative analysis of *Brassica oleracea* triplicated segments showed that 35% of the genes were lost. Retained genes were dosage-sensitive and not randomly located. Duplicates of transcription factors and members of signal transduction pathways were significantly over-retained following WGD, whereas these same functional gene categories exhibited lower retention rates following smaller scale duplications [57]. For instance, in four independent polyploid wheat lineages, recurrent deletions of Puroindoline (*Pin*) gene at the grain *Hardness* (Ha) locus were identified [58].

Phylogenetic and taxonomic studies have been conducted in order to pinpoint the exact placement of the ancient polyploidy events within lineages and to determine when novel genes resulting from polyploidy have enabled adaptive processes. Recent genomic investigations not only indicated that polyploidy is ubiquitous among angiosperms, but also suggested several ancient WGD events [59–62], even in basal angiosperm lineages. Phylogenetic reconstruction with completely sequenced genomes suggested that genome doubling led to a dramatic increase in species richness in several angiosperms, including *Poaceae*, *Solanaceae* and *Brassicaceae*, thus contributing to the dominance of seed plants and angiosperms [63]. To date, only a few reports investigated the fate of the genome after polyploid formation [64]. The probability of fixation and maintenance of duplicated genes depends on many variables. Transposable elements may play a key role in fuelling genome reorganization and functional changes following allopolyploidization. A pivotal example of using comparative approaches to investigate the role of retroelements in polyploids is provided by Wawrzynski *et al.* [65]. The authors, investigating the nonautonomous retrotransposon replication in soybean estimated a much greater impact of such transposable elements on genome size than previously appreciated. More recently, in wheat Kraitshtein *et al.* [66] reported a retrotransposition bursts in subsequent generations. By contrast, no evidence for a transposition burst was found in different allopolyploid species [31,67,68]. Comparative approaches in which genetic events are considered both in a phylogenetic and genetics framework should be conceptualized and modeled.

High-Throughput DNA Sequencing and High Resolution Melting (HRM) Analysis

High-throughput DNA sequencing associated with computational analysis provides general solutions for the genetic analysis of polyploids [69]. However, ploidy is a substantial challenge in sequencing and assembly of plant genomes. A number of biological factors influence the feasibility of discrimination, including the degree of gene family complexity, and the reproductive system. Of course, the level of knowledge concerning the progenitor diploid species is also very important. To date, all attempts to sequence polyploids have relied on either a reduction in ploidy or a physical separation of chromosomes. Attempts to sequence a heterozygous diploid potato genome (RH89-039-16) were challenging due to the high degree of heterozygosity [70]. In order to bypass the difficulties of sequencing the polyploid genome of cultivated strawberry (*Fragaria x ananassa*), woodland strawberry (*Fragaria vesca*) was sequenced [71]. In*B. napus*, the polyploidy issue was addressed by sequencing leaf transcriptome across a mapping population and representative ancestors of the parents of the population [72]. The Wheat Genome Initiative (http://www.wheatgenome.org/) has focused on flow cytometry separation of individual or groups of homeologous chromosomes [73]. To better understand the nature and extent of variation in functionally relevant regions of a polyploid genome, a sequence capture assay to compare exonic sequences of allotetraploid wheat accessions was developed [74]. In cultivated wheat gene duplications were predominant, while in wild wheat mainly deletions were identified. Exon capture proved to be a powerful approach for variant discovery in polyploids. This technique has the potential to identify variation that can play a critical role in the origin of new adaptations and important agronomic traits.

A wealth of SNP detection approaches has been applied to study polyploidy in plants. Akhunov*et al.*[75], using the Illumina GoldenGate assay, identified a high number of SNPs in tetraploid and hexaploid wheat. More recently, Allen*et al.* [76] from Illumina GAIIx data identified more than 14,000 putative SNPs in 6225 distinct hexaploid bread wheat reference sequences. In elite inbred maize lines, more than 1 million SNPs have been identified an Illumina sequencing platform [77]. In the heterozygous polyploid sugarcane, a targeted SNP discovery approach based on 454 sequencing technology was developed by Bundock*et al.*[78]. Using a 454 and Illumina expressed sequence tag sequencing of the parental diploid species of the allotetraploid*Tragopogon miscellus*, Buggs*et al.* [79] identified more than 7,700 SNPs differing between the two progenitor genomes the allotetraploid derived from. The Sequenom MassARRAY iPlex platform [80] was used by to validate 92 SNP markers at the genomic level that were diagnostic for the two parental genomes. SNP discovery was also pursed through 454 technology coupled to High Resolution Melting (HRM) curve analysis in tetraploid alfalfa (*Medicago sativa*) [81]. HRM is a technique that can identify mismatches, even for single bases, in amplicons containing heteroduplex molecules [82], and is emerging as a powerful tool for polyploid genetics [83]. It was demonstrated that the 454 system is a cost-effective approach for SNP discovery targeted to genes of interest in polyploid genomes, and that HRM can identify different alleles in polyploids [66,68]. Salmon*et al.*[84] detected homoeologous SNPs in*G. arboreum*(A genome),*G. raimondii*(D genome), and*G. hirsutum*(AD genomes). The authors estimated that the proportion of genome in*G. hirsutum*that has experienced non reciprocal homoeologous exchanges since the origin of polyploid cotton 1–2 Mya was between 1.8% and 1.9%. SNPs have also been discovered in transcriptome sequences of polyploidy*B. napus*[85]. Next-generation sequencing has been used to mine SNPs in elite wheat germplasm [76].

METHODS FOR GENE EXPRESSION AND REGULATION ANALYSIS

Several methods have been developed for quantifying gene transcription and regulation patterns in polyploids. Although studying gene expression changes in allopolyploids is more complicated than in autopolyploids, most studies on ploidy-related gene expression changes were carried out on synthetic allopolyploids. Indeed, genome merger and doubling can determine widespread transcriptome modifications, generating cascades of novel expression patterns, regulatory interactions, and new phenotypic variations that subsequent natural selection may act upon.

Northern Hybridization and cDNA-AFLP

When sequence information was still scanty, comprehensive transcript-profiling for quantitatively measuring gene expression variation was carried out by Northern blot analysis. This technique involves the use of electrophoresis to separate RNA samples and detection with a labeled probe complementary to a specific RNA target sequence. There are only a few examples on the use of this type of analysis in polyploids. It was used by Guo and co-workers [86] to investigate the dosage effects of 18 genes in an autopolyploid maize series (1×, 2×, 3×, and 4×). Expression levels of genes were dependent on chromosomes dosage, although some varied their expression in response to the "odd" or "even" ploidy. By contrast, several examples are available on the use of cDNA-AFLP. It is a PCR-based technique, which relies on digestion of cDNA by two restriction enzymes and ligation of specific adapters. A set of specific primers designed for these adapters allow simultaneous amplification of fragments under stringent conditions. In synthetic allotetraploids between A. thaliana and A. arenos*a*, Comai*et al.*[87] found 20 suppressed genes out of 700 examined. Similarly, Lee and Chen [88], by extending the analysis also to*A. suecica*(a natural allopolyploid likely formed through pollination of*A. arenosa*with 2*n*gametes from*A. thaliana*) were able to identify a set of 10 different genes differentially expressed in*A. suecica*and its progenitors. In synthetic*Triticum aestivum*allohexaploids [89], about 8% of transcripts displayed altered expression, and >95% of them were reduced or absent. Similar gene expression changes have been found in cotton allopolyploids [90]. Another example on the use of this technique is offered by the work carried out by Tate*et al.*[91]. In newly synthesized*Tragopogon*allotetraploids, preferential expression of parental homeologues displayed a correlation with a loss of parental genomic fragments. Notably, such changes were not observed in newly developed*Tragopogon*F_1hybrids, implying that they arose following genome duplication.

Single-Strand Conformational Polymorphism (SSCP) Analysis

This technique detects sequence variations (single-point mutations and other small-scale DNA changes) through electrophoretic mobility differences. DNA that contains a sequence mutation (even a single base pair change) displays a different measurable mobility compared to reference DNA when electrophoresed in non-denaturing, or partially denaturing conditions. Due to these features, SSCP was employed to distinguish between homoeologous cDNA molecules. This approach has been applied to*A. suecica*[88], cotton [90] and wheat [92] leading to the finding that, basically, genes duplicated by polyploidy are rarely expressed at similar levels, and that there is a biased expression or silencing of some homeologous gene pairs.

Microarrays

Technologies to monitor gene expression achieved a breakthrough through the introduction of microarrays [93]. Genome and transcriptome sequencing have speed up probe development, which consequently resulted in the commercial availability of whole-genome microarrays for many model and crop species. This also gave the possibility to design custom arrays at affordable costs. Approaches of comparative expression profiling have mainly focused on synthetic allotetraploids revealing both additive and non-additive gene expression. The former occurs when gene expression level in the tetraploid is either the sum of the parental values or equal to the mid-parent value (MPV). For instance, tanscriptional profiling of resynthesized*A. suecica*lines from newly created autotetraploid*A. thaliana*and the natural tetraploid*A. arenosa*revealed that, albeit most of genes were additively expressed (from 65% to 95%), more than 1,400 genes diverged from the MPV. The combination of diverged parental genomes in a common nucleus during allopolyploidization implies the reunion of previously diverged regulatory hierarchies, which likely entails non-additive gene expression. This hypothesis has been validated by genome-wide expression analyses also in synthetic polyploids of wheat, cotton,*Senecio*,*Brassica*, and*Spartina*[94–99]. These studies demonstrated that allopolyploid plants exhibit considerable transcriptome alterations as compared with their diploid progenitors. Transcriptome analyses of autopolyploids suggested that there are less dramatic alterations of gene expression compared to allopolyploids [100]. Expression profiling analyses of autotetraploid*A. thaliana*of two different accessions revealed that transcriptome alterations caused by autopolyploidy depend on genome or genetic composition [101]. Microarray analysis provided evidence that ~10% of the ~9,000 potato genes tested displayed expression changes (within the twofold level) among a potato autopolyploid series (1×, 2×, and 4×) [102]. A similar twofold level change was detected in a corn ploidy series (1×–4×) [103].

DNA microarray technology has been also used to profile expression of noncoding RNA molecules naturally occurring in the plant genomes, such as micro RNA (miRNA). They are a class of 20–24 nucleotide small RNAs that repress their target genes by mRNA degradation or translational repression. Therefore, identification and quantification of miRNAs is deemed essential to understanding an organism's or tissue's gene regulatory network [104]. MicroRNA expression profiling was performed with custom designed chips in both natural*A. suecica*and resynthesized*Arabidopsis*genotypes. It indicated that many miRNA and trans-acting siRNA (tasiRNA),*i.e.*, endogenous siRNAs that direct the cleavage of non-identical transcripts, are non-additively expressed [105]. Among the differentially expressed miRNAs, miR163 is severely repressed in leaves and flowers of*A. arenosa*and allotetraploids, but is highly expressed in*A. thaliana*. Analysis by Ng*et al.*[106] demonstrated that miR163 expression differences results from cis-acting effects, as well as from trans-acting repressor(s) that are present in*A. arenosa*and allotetraploids but absent in*A. thaliana*.

High Throughput RNA Sequencing

Next-generation sequencing (NGS) technologies are changing the ways in which gene expression is studied. The principle behind these applications of high throughput sequencing technologies, which have been termed RNA-seq, is simple: complex RNA samples are directly sequenced to determine their content. Therefore, unlike hybridization-based data requiring the estimation of RNA amount by image analysis, RNA-seq data consists of absolute numbers of reads from each gene. These data are highly suitable for the analysis of gene expression since, not relying on probes, they are less error-prone than previous methods and allow to determine absolute expression levels [107]. Sequencing-based methods also permit the genome-wide study of small RNA expression. In*T. miscellus*allopolyploids, Buggs*et al.* [79] profiled almost 3000 SNP markers using an Illumina RNA-seq approach to study differential expression of duplicate homologous genes derived from the parental genomes. The authors found expression biases among tissues in the diploid parents (*T. dubius*and*T. pratensis*) in comparison to the natural allopolyploids, as well as uniform expression in $F_1$and first-generation synthetic allopolyploids. To explain the observed "transcriptomic shock", they hypothesized a loosening of gene expression regulation, which may set the stage for gradual evolution of novel patterns of expression in the early generations of polyploidy. Croate and Doyle [108] used quantitative reverse transcriptase-polymerase chain reaction and RNAseq in allopolyploid*Glycine dolichocarpa*and its diploid progenitors. They inferred dosage responses for several thousand genes and showed that most of them had partial dosage compensation. In*G. max*, RNA-seq allowed the identification of the gene family likely contributing to differences in photosynthetic rate between the allotetraploid and its progenitors [109]. The authors also provided evidences that the tetraploid appeared to use the "redundant" gene copies in novel ways. In*A. thaliana*allopolyploids, transcriptome profiling was carried out by Ha and colleagues [105] through high-throughput cDNA pyrosequencing. For the first time, these authors gained insight into small RNA expression diversity and evolution in closely related species as well as in interspecific hybrids. The data suggested a role for small RNAs in buffering against genomic shock in*Arabidopsis*interspecific hybrids and allopolyploids. In particular, they seem to have a central role in maintaining genome and chromatin stability as well as in modulating non additive gene expression. In addition, Ha*et al.* [105] found that repeat- and transposon-associated siRNAs (rasiRNA and TE-siRNA, respectively) were highly divergent between*A. thaliana*and*A. arenosa*and their non additive gene expression in allopolyploids were not correlated. By contrast, miRNA and tasiRNA sequences were conserved between species, but their expression patterns were highly variable between the allotetraploids and their progenitors.

METHODS FOR PROTEIN ANALYSIS

Compared with genomic and gene expression variations, changes in proteins and gene products in polyploids and their progenitors were rarely examined. An early study on genome-wide protein profiling was performed in maize lines of different ploidies by sodium dodecyl sulfate–polyacrylamide gel electrophoresis (SDS–PAGE) [110]. This is a technique for the separation of proteins according to their molecular weight, in the presence of a reducing agent (2-mercaptoethanol). Data obtained showed that expression per genome for most maize proteins did not change with ploidy, even though ploidy-modulated expression changes were detected for a few proteins. A more powerful method to analyze complex protein mixtures is the protein two-dimensional electrophoresis (2-DE) analysis, in which proteins are separated according to their isoelectric point and mass. In diploid, tetraploid and hexaploid wheat 2-DE experiments showed that the expression of homeologous proteins in hexaploid wheat depended on interactions among the parental A, B and D genomes [111]. Other recent studies using 2-DE indicated numerous and unbiased variations of proteins in newly synthesized*B. napus*[112,113] and wheat hybrids [114]. Using protein 2-DE coupled with mass spectrometry (MS) assays, a recent study in maize showed a positive correlation of differentially expressed proteins with ploidy levels [115]. The highest correlations were found in diploid–hexaploid and tetraploid–hexaploid comparisons. Recently, Ng and co-workers [116] were able to study quantitative changes in the proteome of*Arabidopsis*autopolyploids and allotetraploids and their progenitors using the isobaric tags for relative and absolute quantification (iTRAQ) technique, coupled with mass spectrometry. The levels of protein divergence ranged from ~18% between*A. thaliana*and*A. arenosa*to ~7% between an*A. thaliana*diploid and autotetraploid. In F_1- and F_8-resynthesized allotetraploids the proteomic divergence relative to MPV was intermediate (~8%). These data suggest that, during polyploidization, rapid changes occurring in post-transcriptional regulation and translational modifications of proteins can lead to high protein discrepancy between species.

CONCLUSIONS AND PERSPECTIVES

With the speed of technology improvement and the application of genomic tools, polyploidy research is undergoing a renaissance. It can be expected that comprehensive studies using multidisciplinary approaches will push the boundaries of current methodologies to translate the knowledge gained into practical applications. Particularly significant will be the high-throughput genome-wide approaches to unraveling the genetic and epigenetic consequences of polyploidization and the availability of phenotyping platforms. They are all reaching an unprecedented level of resolution at relatively affordable costs to the point that genotyping-by-sequencing [117] and targeted sequence capture [118] are now feasible also for high diversity, large genome species. NGS not only will extend the possibilities of gene and marker discovery, but will enable genome-wide quantification of gene expression. It will also allow direct genome-scale investigation of chromatin and DNA methylation cross-talk, by ChIP-Seq, bisulfite sequencing,*etc*. Characterizing transcripts through sequencing is advantageous to circumvent problems posed by highly redundant and extremely large genomes. It should be pointed out that the rapid pace at which new sequencing technologies are emerging is generating a growing disparity between the rate of data generation and its full and biologically meaningful analysis. However, there are outstanding examples addressing successful strategies for dealing with these challenges [109,118–120].

Genetic mapping can exploit robust statistical models, and will be crucial for identifying the genes underlying the polyploidization process in the bulk of the fast growing genome sequence information. Merging results from genetic, genomics and proteomics investigations will help to understand to what extent polyploid genome flexibility is associated with amplified responses to selection. We have recently hypothesized that defense response plasticity of potato could be correlated to gene number and category and cluster organization [121]. Understanding polyploid evolution requires knowledge to be integrated at the population level, and will have not only to rely on suitable experimental designs, but also on surveys of variation at multiple levels. Recent and forthcoming sequencing technologies are providing a wealth of genomic data to be released soon, also for wild species that can be employed for evolutionary studies. Until recently, sequencing complex genomes was considered very challenging due, for example, to the difficulties in discriminating among paralogous, hortologous, and homoeologous sequences. However, the availability of the genome sequence of the ancestors, the reduction in the ploidy level or the physical separation of chromosomes offered the possibility to circumvent these challenges and examples of polyploid genomes fully sequenced have become available [122,123]. The accumulation of knowledge on polyploid formation, maintenance, and divergence at the whole-genome and subgenome levels will not only help plant biologists to understand how plants have evolved and diversified, but also assist plant breeders in designing new strategies for crop improvement.

ACKNOWLEDGMENTS

This research was funded in part by the Ministry of University and Science, Project PRIN 2008: "Investigations of the genetic and epigenetic changes in induced polysomic polyploids of crop plants", PI Fabio Veronesi. This paper received the contribution number 272 by the Department of Soil, Plant, Environmental and Animal Production Sciences.

REFERENCES

1. Masterson, J. Stomatal size in fossil plants: Evidence for polyploidy in majority of angiosperms.*Science*1994,*264*, 421–424.
2. Wendel, J.F. Genome evolution in polyploids.*Plant. Mol. Biol*2000,*42*, 225–249.
3. Bennett, M.D. Perspectives on polyploidy in plants—Ancient and neo.*Biol. J. Linn. Soc*2004,*82*, 411–423.
4. Hendry, A.P.; Bolnick, D.I.; Berner, D.; Peichel, C.L. Along the speciation continuum in sticklebacks.*J. Fish. Biol*2009,*75*, 2000–2036.
5. Soltis, D.E.; Buggs, R.J.A.; Doyle, J.J.; Soltis, P.S. What we still don't know about polyploidy.*Taxon*2010,*59*, 1387–1403.

6. Ramsey, J.; Schemske, D.W. Pathways, mechanisms, and rates of polyploid formation in flowering plants. *Annu. Rev. Ecol. Syst*1998,*29*, 467–501.

7. Ramsey, J.; Schemske, D.W. Neopolyploidy in flowering plants.*Annu. Rev. Ecol. Syst*2002,*33*, 589–639.

8. Carputo, D.; Frusciante, L.; Peloquin, S.J. The role of 2*n*gametes and endosperm balance number in the origin and evolution of polyploids in the tuber-bearing Solanums.*Genetics*2003,*163*, 287–294.

9. Chen, Z.J. Genetic and epigenetic mechanisms for gene expression and phenotypic variation in plant polyploids. *Annu. Rev. Plant. Biol*2007,*58*, 377–406.

10. Bretagnolle, F.; Thompson, J.D. Tansley review no. 78. Gametes with the stomatic chromosome number: Mechanisms of their formation and role in the evolution of autopolypoid plants.*New Phytol*1995,*129*, 1–22.

11. D'Erfurth, I.; Jolivet, S.; Froger, N.; Catrice, O.; Novatchkova, M.; Simon, M.; Jenczewski, E.; Mercier, R. Mutations in ATPS1 (*Arabidopsis Thaliana Parallel Spindle 1*) lead to the production of diploid pollen grains. *PLoS Genet*2008,*4*, e1000274.

12. Erilova, A.; Brownfield, L.; Exner, V.; Rosa, M.; Twell, D.; Scheid, O.M.; Hennig, L.; Köhler, C. Imprinting of the polycomb group gene*MEDEA*serves as a ploidy sensor in*Arabidopsis*.*PLoS Genet*2009,*5*, e1000663.

13. Mercier, R.; Vezon, D.; Bullier, E.; Motamayor, J.C.; Sellier, A.; Lefèvre, F.; Pelletier, G.; Horlow, C. SWITCH1 (SWI1): A novel protein required for the establishment of sister chromatid cohesion and for bivalent formation at meiosis.*Genes Dev*2001,*15*, 1859–1871.

14. Agashe, B.; Prasad, C.K.; Siddiqi, I. Identification and analysis of*DYAD*: A gene required for meiotic chromosome organisation and female meiotic progression in*Arabidopsis*.*Development*2002,*129*, 3935–3943.

15. D'Erfurth, I.; Jolivet, S.; Froger, N.; Catrice, O.; Novatchkova, M.; Mercier, R. Turning meiosis into mitosis.*PLoS Biol*2009,*7*, e1000124.

16. D'Erfurth, I.; Cromer, L.; Jolivet, S.; Girard, C.; Horlow, C.; Sun, Y.; To, J.P.C.; Berchowitz, L.E.; Copenhaver, G.P.; Mercier, R. The CYCLIN-A CYCA1;2/TAM is required for the meiosis I to meiosis II transition and cooperates with OSD1 for the prophase to first meiotic division transition.*PLoS Genet*2010,*6*, e1000989.

17. Fawcett, J.A.; Maere, S.; van de Peer, Y. Plants with double genomes might have had a better chance to survive the cretaceous-tertiary extinction event.*Proc. Natl. Acad. Sci. USA*2009,*106*, 5737–5742.

18. Comai, L. The advantages and disadvantages of being polyploid.*Nat. Rev. Genet*2005,*6*, 836–846.

19. Osborn, T.C.; Pires, J.C.; Birchler, J.A.; Auger, D.L.; Chen, Z.J.; Lee, H.-S.; Comai, L.; Madlung, A.; Doerge, R.W.; Colot, V.;*et al.* Understanding mechanisms of novel gene expression in polyploids.*Trends Genet*2003,*19*, 141–147.

20. Mayrose, I.; Zhan, S.H.; Rothfels, C.J.; Magnuson-Ford, K.; Barker, M.S.; Rieseberg, L.H.; Otto, S.P. Recently formed polyploid plants diversify at lower rates. *Science*2011,*333*.

21. Pires, J.C.; Lim, K.Y.; Kovarík, A.; Matyásek, R.; Boyd, A.; Leitch, A.R.; Leitch, I.J.; Bennett, M.D.; Soltis, P.S.; Soltis, D.E. Molecular cytogenetic analysis of recently evolved*Tragopogon*(Asteraceae) allopolyploids reveal a karyotype that is additive of the diploid progenitors.*Am. J. Bot*2004,*91*, 1022–1035.

22. Szadkowski, E.; Eber, F.; Huteau, V.; Lode, M.; Huneau, C.; Belcram, H.; Coriton, O.; Manzanares-Dauleux, M.J.; Delourme, R.; King, G.J.;*et al.* The first meiosis of resynthesized*Brassica napus*, a genome blender.*New Phytol*2010,*186*, 102–112.

23. Wang, K.; Guo, W.; Yang, Z.; Hu, Y.; Zhang, W.; Zhou, B.; Stelly, D.; Chen, Z.; Zhang, T. Structure and size variations between 12A and 12D homoeologous chromosomes based on high-resolution cytogenetic map in allotetraploid cotton.*Chromosoma*2010,*119*, 255–266.

24. Chester, M.; Gallagher, J.P.; Symonds, V.V.; Cruz da Silva, A.V.; Mavrodiev, E.V.; Leitch, A.R.; Soltis, P.S.; Soltis, D.E. Extensive chromosomal variation in a recently formed natural allopolyploid species,*Tragopogon miscellus*(Asteraceae).*Proc. Natl. Acad. Sci. USA*2012,*109*, 1176–1181.

25. Porceddu, A.; Albertini, E.; Barcaccia, G.; Falistocco, E.; Falcinelli, M. Linkage mapping in apomictic and sexual Kentucky bluegrass (*Poa pratensis*L.) genotypes using a two way pseudo-testcross strategy based on AFLP and SAMPl markers.*Theor. Appl. Genet*2002,*104*, 273–280.

26. Gaeta, R.T.; Pires, J.C.; Iniguez-Luy, F.; Leon, E.; Osborn, T.C. Genomic changes in resynthesized*Brassica napus*and their effect on gene expression and phenotype. *Plant Cell*2007,*19*, 3403–3417.

27. Nicolas, S.D.; Leflon, M.; Monod, H.; Eber, F.; Coriton, O.; Huteau, V.; Chèvre, A.-M.; Jenczewski, E. Genetic regulation of meiotic cross-overs between related genomes in*Brassica napus*haploids and hybrids.*Plant Cell*2009,*21*, 373–385.

28. Cifuentes, M.; Eber, F.; Lucas, M.-O.; Lode, M.; Chèvre, A.-M.; Jenczewski, E. Repeated polyploidy drove different levels of crossover suppression between homoeologous chromosomes in*Brassica napus*allohaploids.*Plant Cell*2010,*22*, 2265–2276.

29. Leitch, I.J.; Bennett, M.D. Polyploidy in angiosperms. *Trends Plant Sci*1997,*2*, 470–476.

30. Liu, B.; Vega, J.M.; Feldman, M. Rapid genomic changes in newly synthesized amphiploids of*Triticum*and*Aegilops*. II. Changes in low-copy coding DNA sequences.*Genome*1998,*41*, 535–542.

31. Madlung, A.; Tyagi, A.P.; Watson, B.; Jiang, H.; Kagochi, T.; Doerge, R.W.; Martienssen, R.; Comai, L. Genomic changes in synthetic*Arabidopsis*polyploids. *Plant J*2005,*41*, 221–230.

32. Song, K.; Lu, P.; Tang, K.; Osborn, T.C. Rapid genome change in synthetic polyploids of*Brassica*and its implications for polyploid evolution.*Proc. Natl. Acad. Sci. USA*1995,*92*, 7719–7723.

33. Salmon, A.; Ainouche, M.L.; Wendel, J.F. Genetic and epigenetic consequences of recent hybridization and

polyploidy in *Spartina* (Poaceae). *Mol. Ecol* 2005, *14*, 1163–1175.

34. Wu, K.K.; Burnquist, W.; Sorrells, M.E.; Tew, T.L.; Moore, P.H.; Tanksley, S.D. The detection and estimation of linkage in polyploids using single-dose restriction fragments. *Theor. Appl. Genet* 1992, *83*, 294–300.

35. Hackett, C.A.; Bradshaw, J.E.; Meyer, R.C.; McNicol, J.W.; Milbourne, D.; Waugh, R. Linkage analysis in tetraploid species: A simulation study. *Genet. Res* 1998, *71*, 143–154.

36. Ripol, M.I.; Churchill, G.A.; da Silva, J.A.G.; Sorrells, M. Statistical aspects of genetic mapping in autopolyploids. *Gene* 1999, *235*, 31–41.

37. Wu, R.; Ma, C.-X.; Casella, G. A bivalent polyploid model for mapping quantitative trait loci in outcrossing tetraploids. *Genetics* 2004, *166*, 581–595.

38. Li, J.; Das, K.; Liu, J.; Fu, G.; Li, Y.; Tobias, C.; Wu, R. Statistical models for genetic mapping in polyploids: Challenges and opportunities. *Methods Mol. Biol* 2012, *871*, 245–261.

39. Ma, J.F.; Shen, R.; Zhao, Z.; Wissuwa, M.; Takeuchi, Y.; Ebitani, T.; Yano, M. Response of rice to Al stress and identification of quantitative trait loci for Al tolerance. *Plant Cell Physiol* 2002, *43*, 652–659.

40. Julier, B.; Flajoulot, S.; Barre, P.; Cardinet, G.; Santoni, S.; Huguet, T.; Huyghe, C. Construction of two genetic linkage maps in cultivated tetraploid alfalfa (*Medicago sativa*) using microsatellite and AFLP markers. *BMC Plant Biol* 2003, *3*.

41. Barcaccia, G.; Meneghetti, S.; Albertini, E.; Triest, L.; Lucchin, M. Linkage mapping in tetraploid willows: Segregation of molecular markers and estimation of linkage phases support an allotetraploid structure for *Salix alba*×*Salix fragilis* interspecific hybrids. *Heredity* 2003, *90*, 169–180.

42. Le Cunff, L.; Garsmeur, O.; Raboin, L.M.; Pauquet, J.; Telismart, H.; Selvi, A.; Grivet, L.; Philippe, R.; Begum, D.; Deu, M.; *et al.* Diploid/polyploid syntenic shuttle mapping and haplotype-specific chromosome walking toward a rust resistance gene (*Bru1*) in highly polyploid sugarcane (2*n*~12x~115). *Genetics* 2008, *180*, 649–660.

43. Esselink, G.D.; Nybom, H.; Vosman, B. Assignment of allelic configuration in polyploids using the MAC-PR (microsatellite DNA allele counting—peak ratios) method. *Theor. Appl. Genet* 2004, *109*, 402–408.

44. Van Dijk, T.; Noordijk, Y.; Dubos, T.; Bink, M.; Meulenbroek, B.; Visser, R.; van de Weg, E. Microsatellite allele dose and configuration establishment (MADCE): An integrated approach for genetic studies in allopolyploids. *BMC Plant Biol* 2012, *12*.

45. Oliver, R.; Jellen, E.; Ladizinsky, G.; Korol, A.; Kilian, A.; Beard, J.; Dumlupinar, Z.; Wisniewski-Morehead, N.; Svedin, E.; Coon, M.; *et al.* New Diversity Arrays Technology (DArT) markers for tetraploid oat (*Avena magna* Murphy et Terrell) provide the first complete oat linkage map and markers linked to domestication genes from hexaploid *A. sativa* L. *Theor. Appl. Genet* 2011, *123*, 1159–1171.

46. Reyna-López, G.E.; Simpson, J.; Ruiz-Herrera, J. Differences in DNA methylation patterns are detectable during the dimorphic transition of fungi by amplification of restriction polymorphisms. *Mol. Gen. Genet* 1997, *253*, 703–710.

47. Madlung, A.; Masuelli, R.W.; Watson, B.; Reynolds, S.H.; Davison, J.; Comai, L. Remodeling of DNA methylation and phenotypic and transcriptional changes in synthetic Arabidopsis allotetraploids. *Plant Physiol* 2002, *129*, 733–746.

48. Liu, B.; Brubaker, C.L.; Mergeai, G.; Cronn, R.C.; Wendel, J.F. Polyploid formation in cotton is not accompanied by rapid genomic changes. *Genome* 2001, *44*, 321–330.

49. Axelsson, T.; Bowman, C.M.; Sharpe, A.G.; Lydiate, D.J.; Lagercrantz, U. Amphidiploid *Brassica juncea* contains conserved progenitor genomes. *Genome* 2000, *43*, 679–688.

50. Jannoo, N.; Grivet, L.; Chantret, N.; Garsmeur, O.; Glaszmann, J.C.; Arruda, P.; D'Hont, A. Orthologous comparison in a gene-rich region among grasses reveals stability in the sugarcane polyploid genome. *Plant J* 2007, *50*, 574–585.

51. Paterson, A.H. Leafing through the genomes of our major crop plants: Strategies for capturing unique information. *Nat. Rev. Genet* 2006, *7*, 174–184.

52. Schranz, M.E.; Song, B.-H.; Windsor, A.J.; Mitchell-Olds, T. Comparative genomics in the Brassicaceae: A family-wide perspective. *Curr. Opin. Plant Biol* 2007, *10*, 168–175.

53. Margulies, E.H.; Birney, E. Approaches to comparative sequence analysis: Towards a functional view of vertebrate genomes. *Nat. Rev. Genet* 2008, *9*, 303–313.

54. Chantret, N.; Salse, J.; Sabot, F.; Rahman, S.; Bellec, A.; Laubin, B.; Dubois, I.; Dossat, C.; Sourdille, P.; Joudrier, P.; *et al.* Molecular basis of evolutionary events that shaped the *hardness* locus in diploid and polyploid wheat species (Triticum and Aegilops). *Plant Cell* 2005, *17*, 1033–1045.

55. Gao, S.; Gu, Y.; Wu, J.; Coleman-Derr, D.; Huo, N.; Crossman, C.; Jia, J.; Zuo, Q.; Ren, Z.; Anderson, O.; *et al.* Rapid evolution and complex structural organization in genomic regions harboring multiple prolamin genes in the polyploid wheat genome. *Plant Mol. Biol* 2007, *65*, 189–203.

56. Innes, R.W.; Ameline-Torregrosa, C.; Ashfield, T.; Cannon, E.; Cannon, S.B.; Chacko, B.; Chen, N.W.G.; Couloux, A.; Dalwani, A.; Denny, R.; *et al.* Differential accumulation of retroelements and diversification of NB-LRR disease resistance genes in duplicated regions following polyploidy in the ancestor of soybean. *Plant Physiol* 2008, *148*, 1740–1759.

57. Town, C.D.; Cheung, F.; Maiti, R.; Crabtree, J.; Haas, B.J.; Wortman, J.R.; Hine, E.E.; Althoff, R.; Arbogast, T.S.; Tallon, L.J.; *et al.* Comparative genomics of *Brassica oleracea* and *Arabidopsis thaliana* reveal gene loss, fragmentation, and dispersal after polyploidy. *Plant Cell* 2006, *18*, 1348–1359.

58. Li, W.; Huang, L.; Gill, B.S. Recurrent deletions of puroindoline genes at the grain *hardness* locus in

four independent lineages of polyploid wheat.*Plant Physiol*2008,*146*, 200–212.

59. De Bodt, S.; Maere, S.; van de Peer, Y. Genome duplication and the origin of angiosperms.*Trends Ecol. Evol*2005,*20*, 591–597.

60. Soltis, D.E.; Bell, C.D.; Kim, S.; Soltis, P.S. Origin and early evolution of angiosperms.*Ann. N. Y. Acad. Sci*2008,*1133*, 3–25.

61. Velasco, R.; Zharkikh, A.; Affourtit, J.; Dhingra, A.; Cestaro, A.; Kalyanaraman, A.; Fontana, P.; Bhatnagar, S.K.; Troggio, M.; Pruss, D.;*et al.* The genome of the domesticated apple (*Malus domestica*Borkh.).*Nat. Genet*2010,*42*, 833–839.

62. The Tomato Genome Consortium. The tomato genome sequence provides insights into fleshy fruit evolution. *Nature*2012,*485*, 635–641.

63. Jiao, Y.; Wickett, N.J.; Ayyampalayam, S.; Chanderbali, A.S.; Landherr, L.; Ralph, P.E.; Tomsho, L.P.; Hu, Y.; Liang, H.; Soltis, P.S.;*et al.* Ancestral polyploidy in seed plants and angiosperms.*Nature*2011,*473*, 97–100.

64. Levasseur, A.; Pontarotti, P. The role of duplications in the evolution of genomes highlights the need for evolutionary-based approaches in comparative genomics.*Biol. Direct*2011,*6*.

65. Wawrzynski, A.; Ashfield, T.; Chen, N.W.G.; Mammadov, J.; Nguyen, A.; Podicheti, R.; Cannon, S.B.; Thareau, V.; Ameline-Torregrosa, C.; Cannon, E.;*et al.* Replication of nonautonomous retroelements in soybean appears to be both recent and common.*Plant Physiol*2008,*148*, 1760–1771.

66. Kraitshtein, Z.; Yaakov, B.; Khasdan, V.; Kashkush, K. Genetic and epigenetic dynamics of a retrotransposon after allopolyploidization of wheat.*Genetics*2010,*186*, 801–812.

67. Ainouche, M.; Fortune, P.; Salmon, A.; Parisod, C.; Grandbastien, M.A.; Fukunaga, K.; Ricou, M.; Misset, M.T. Hybridization, polyploidy and invasion: Lessons from*Spartina*(Poaceae).*Biol. Invasions*2009,*11*, 1159–1173.

68. Beaulieu, J.; Jean, M.; Belzile, F. The allotetraploid*Arabidopsis thaliana–Arabidopsis lyrata*subsp.*petraea*as an alternative model system for the study of polyploidy in plants.*Mol. Genet. Genomics*2009,*281*, 421–435.

69. Kaur, S.; Francki, M.G.; Forster, J.W. Identification, characterization and interpretation of single-nucleotide sequence variation in allopolyploid crop species.*Plant Biotechnol. J*2012,*10*, 125–138.

70. Hamilton, J.P.; Buell, R.C. Advances in plant genome sequencing.*Plant J*2012,*70*, 177–190.

71. Shulaev, V.; Sargent, D.J.; Crowhurst, R.N.; Mockler, T.C.; Folkerts, O.; Delcher, A.L.; Jaiswal, P.; Mockaitis, K.; Liston, A.; Mane, S.P.;*et al.* The genome of woodland strawberry (*Fragaria vesca*).*Nat. Genet*2011,*43*, 109–116.

72. Bancroft, I.; Morgan, C.; Fraser, F.; Higgins, J.; Wells, R.; Clissold, L.; Baker, D.; Long, Y.; Meng, J.; Wang, X.;*et al.* Dissecting the genome of the polyploid crop oilseed rape by transcriptome sequencing.*Nat. Biotech*2011,*29*, 762–766.

73. Paux, E.; Sourdille, P.; Salse, J.; Saintenac, C.; Choulet, F.; Leroy, P.; Korol, A.; Michalak, M.; Kianian, S.; Spielmeyer, W.;*et al.* A physical map of the 1-gigabase bread wheat chromosome 3B.*Science*2008,*322*, 101–104.

74. Saintenac, C.; Jiang, D.; Akhunov, E. Targeted analysis of nucleotide and copy number variation by exon capture in allotetraploid wheat genome.*Genome Biol*2011,*12*.

75. Akhunov, E.; Nicolet, C.; Dvorak, J. Single nucleotide polymorphism genotyping in polyploid wheat with the Illumina GoldenGate assay.*Theor. Appl. Genet*2009,*119*, 507–517.

76. Allen, A.; Islamovic, E.; Kaur, J.; Gold, S.; Shah, D.; Smith, T.J. Transgenic maize plants expressing the Totivirus antifungal protein, KP4, are highly resistant to corn smut.*Plant Biotechnol. J*2011,*9*, 857–864.

77. Lai, J.; Li, R.; Xu, X.; Jin, W.; Xu, M.; Zhao, H.; Xiang, Z.; Song, W.; Ying, K.; Zhang, M.;*et al.* Genome-wide patterns of genetic variation among elite maize inbred lines.*Nat. Genet*2010,*42*, 1027–1030.

78. Bundock, P.C.; Eliott, F.G.; Ablett, G.; Benson, A.D.; Casu, R.E.; Aitken, K.S.; Henry, R.J. Targeted single nucleotide polymorphism (SNP) discovery in a highly polyploid plant species using 454 sequencing.*Plant Biotechnol. J*2009,*7*, 347–354.

79. Buggs, R.J.; Chamala, S.; Wu, W.; Gao, L.; May, G.D.; Schnable, P.S.; Soltis, D.E.; Soltis, P.S.; Barbazuk, W.B. Characterization of duplicate gene evolution in the recent natural allopolyploid*Tragopogon miscellus*by next-generation sequencing and Sequenom iPLEX MassARRAY genotyping.*Mol. Ecol*2010,*19*, S132–S146.

80. Gabriel, S.; Ziaugra, L.; Tabbaa, D.*Current Protocols in Human Genetics*; John Wiley & Sons, Inc: 2009; Volume Chapter 2.12, pp. 1–16.

81. Han, Y.; Kang, Y.; Torres-Jerez, I.; Cheung, F.; Town, C.; Zhao, P.; Udvardi, M.; Monteros, M. Genome-wide SNP discovery in tetraploid alfalfa using 454 sequencing and high resolution melting analysis.*BMC Genomics*2011,*12*.

82. Cho, M.H.; Ciulla, D.; Klanderman, B.J.; Raby, B.A.; Silverman, E.K. High-resolution melting curve analysis of genomic and whole-genome amplified DNA.*Clin. Chem*2008,*54*, 2055–2058.

83. Han, Y.; Khu, D.M.; Monteros, M.J. High-resolution melting analysis for SNP genotyping and mapping in tetraploid alfalfa (*Medicago sativa*L.).*Mol. Breed*2012,*29*, 489–501.

84. Salmon, A.; Flagel, L.; Ying, B.; Udall, J.A.; Wendel, J.F. Homoeologous nonreciprocal recombination in polyploid cotton.*New Phytol*2010,*186*, 123–134.

85. Trick, M.; Long, Y.; Meng, J.; Bancroft, I. Single nucleotide polymorphism (SNP) discovery in the polyploid*Brassica napus*using Solexa transcriptome sequencing.*Plant Biotechnol. J*2009,*7*, 334–346.

86. Guo, M.; Davis, D.; Birchler, J.A. Dosage effects on gene expression in a maize ploidy series.*Genetics*1996,*142*,

1349–1355.

87. Comai, L.; Tyagi, A.P.; Winter, K.; Holmes-Davis, R.; Reynolds, S.H.; Stevens, Y.; Byers, B. Phenotypic instability and rapid gene silencing in newly formed*Arabidopsis*allotetraploids.*Plant Cell*2000,*12*, 1551–1567.

88. Lee, H.-S.; Chen, Z.J. Protein-coding genes are epigenetically regulated in*Arabidopsis*polyploids.*Proc. Natl. Acad. Sci. USA*2001,*98*, 6753–6758.

89. He, P.; Friebe, B.R.; Gill, B.S.; Zhou, J.-M. Allopolyploidy alters gene expression in the highly stable hexaploid wheat.*Plant Mol. Biol*2003,*52*, 401–414.

90. Adams, K.L.; Cronn, R.; Percifield, R.; Wendel, J.F. Genes duplicated by polyploidy show unequal contributions to the transcriptome and organ-specific reciprocal silencing.*Proc. Natl. Acad. Sci. USA*2003,*100*, 4649–4654.

91. Tate, J.A.; Ni, Z.; Scheen, A.-C.; Koh, J.; Gilbert, C.A.; Lefkowitz, D.; Chen, Z.J.; Soltis, P.S.; Soltis, D.E. Evolution and expression of homeologous loci in*Tragopogon miscellus*(Asteraceae), a recent and reciprocally formed allopolyploid.*Genetics*2006,*173*, 1599–1611.

92. Bottley, A.; Xia, G.M.; Koebner, R.M.D. Homoeologous gene silencing in hexaploid wheat.*Plant J*2006,*47*, 897–906.

93. Schena, M.; Shalon, D.; Davis, R.W.; Brown, P.O. Quantitative monitoring of gene expression patterns with a complementary DNA microarray.*Science*1995,*270*, 467–470.

94. Chagué, V.; Just, J.; Mestiri, I.; Balzergue, S.; Tanguy, A.-M.; Huneau, C.; Huteau, V.; Belcram, H.; Coriton, O.; Jahier, J.;*et al*. Genome-wide gene expression changes in genetically stable synthetic and natural wheat allohexaploids.*New Phytol*2010,*187*, 1181–1194.

95. Pumphrey, M.; Bai, J.; Laudencia-Chingcuanco, D.; Anderson, O.; Gill, B.S. Nonadditive expression of homoeologous genes is established upon polyploidization in hexaploid wheat.*Genetics*2009,*181*, 1147–1157.

96. Chaudhary, B.; Hovav, R.; Flagel, L.; Mittler, R.; Wendel, J. Parallel expression evolution of oxidative stress-related genes in fiber from wild and domesticated diploid and polyploid cotton (*Gossypium*).*BMC Genomics*2009,*10*.

97. Marmagne, A.; Brabant, P.; Thiellement, H.; Alix, K. Analysis of gene expression in resynthesized B*rassica napus*allotetraploids: Transcriptional changes do not explain differential protein regulation.*New Phytol*2010,*186*, 216–227.

98. Chelaifa, H.; Monnier, A.; Ainouche, M. Transcriptomic changes following recent natural hybridization and allopolyploidy in the salt marsh species*Spartina×townse ndii*and*Spartina anglica*(Poaceae).*New Phytol*2010,*186*, 161–174.

99. Buggs, R.J.A.; Doust, A.N.; Tate, J.A.; Koh, J.; Soltis, K.; Feltus, F.A.; Paterson, A.H.; Soltis, P.S.; Soltis, D.E. Gene loss and silencing in*Tragopogon miscellus*(Asteraceae): Comparison of natural and synthetic allotetraploids. *Heredity*2009,*103*, 73–81.

100. Doyle, J.J.; Flagel, L.E.; Paterson, A.H.; Rapp, R.A.; Soltis, D.E.; Soltis, P.S.; Wendel, J.F. Evolutionary genetics of genome merger and doubling in plants.*Annu. Rev. Genet*2008,*42*, 443–461.

101. Yu, Z.; Haberer, G.; Matthes, M.; Rattei, T.; Mayer, K.F.X.; Gierl, A.; Torres-Ruiz, R.A. Impact of natural genetic variation on the transcriptome of autotetraploid*Arabidopsis thaliana*.*Proc. Natl. Acad. Sci. USA*2010,*107*, 17809–17814.

102. Stupar, R.M.; Bhaskar, P.B.; Yandell, B.S.; Rensink, W.A.; Hart, A.L.; Ouyang, S.; Veilleux, R.E.; Busse, J.S.; Erhardt, R.J.; Buell, C.R.;*et al*. Phenotypic and transcriptomic changes associated with potato autopolyploidization.*Genetics*2007,*176*, 2055–2067.

103. Riddle, N.; Jiang, H.; An, L.; Doerge, R.; Birchler, J. Gene expression analysis at the intersection of ploidy and hybridity in maize.*Theor. Appl. Genet*2010,*120*, 341–353.

104. Havecker, E.R. Detection of small RNAs and microRNAs using deep sequencing technology.*Methods Mol. Biol*2011,*732*, 55–68.

105. Ha, M.; Lu, J.; Tian, L.; Ramachandran, V.; Kasschau, K.D.; Chapman, E.J.; Carrington, J.C.; Chen, X.; Wang, X.-J.; Chen, Z.J. Small RNAs serve as a genetic buffer against genomic shock in*Arabidopsis*interspecific hybrids and allopolyploids.*Proc. Natl. Acad. Sci. USA*2009,*106*, 17835–17840.

106. Ng, D.W.-K.; Zhang, C.; Miller, M.; Palmer, G.; Whiteley, M.; Tholl, D.; Chen, Z.J.*Cis*- and*trans*-regulation of miR163 and target genes confers natural variation of secondary metabolites in two*Arabidopsis*species and their allopolyploids.*Plant Cell*2011,*23*, 1729–1740.

107. Wang, Z.; Gerstein, M.; Snyder, M. RNA-Seq: A revolutionary tool for transcriptomics.*Nat. Rev. Genet*2009,*10*, 57–63.

108. Croate, E.C.; Doyle, J.J. Quantifying whole transcriptome size, a prerequisite for understanding transcriptome evolution across species: An example from a plant allopolyploid.*Genome Biol. Evol*2010,*2*, 534–546.

109. Ilut, D.C.; Coate, J.E.; Luciano, A.K.; Owens, T.G.; May, G.D.; Farmer, A.; Doyle, J.J. A comparative transcriptomic study of an allotetraploid and its diploid progenitors illustrates the unique advantages and challenges of RNA-seq in plant species.*Am. J. Bot*2012,*99*, 383–396.

110. Birchler, J.A.; Newton, K.J. Modulation of protein levels in chromosomal dosage series of maize: The biochemical basis of aneuploid syndromes.*Genetics*1981,*99*, 247–266.

111. Islam, N.; Tsujimoto, H.; Hirano, H. Proteome analysis of diploid, tetraploid and hexaploid wheat: Towards understanding genome interaction in protein expression. *Proteomics*2003,*3*, 549–557.

112. Albertin, W.; Balliau, T.; Brabant, P.; Chèvre, A.-M.; Eber, F.; Malosse, C.; Thiellement, H. Numerous and rapid nonstochastic modifications of gene products in newly synthesized*Brassica napus*allotetraploids. *Genetics*2006,*173*, 1101–1113.

113. Albertin, W.; Langella, O.; Joets, J.; Négroni, L.; Zivy, M.; Damerval, C.; Thiellement, H. Comparative proteomics of leaf, stem, and root tissues of synthetic *Brassica napus*. *Proteomics* 2009, *9*, 793–799.

114. Song, X.; Ni, Z.; Yao, Y.; Xie, C.; Li, Z.; Wu, H.; Zhang, Y.; Sun, Q. Wheat (*Triticum aestivum* L.) root proteome and differentially expressed root proteins between hybrid and parents. *Proteomics* 2007, *7*, 3538–3557.

115. Yao, H.; Kato, A.; Mooney, B.; Birchler, J. Phenotypic and gene expression analyses of a ploidy series of maize inbred Oh43. *Plant Mol. Biol* 2011, *75*, 237–251.

116. Ng, D.W.K.; Zhang, C.; Miller, M.; Shen, Z.; Briggs, S.P.; Chen, Z.J. Proteomic divergence in Arabidopsis autopolyploids and allopolyploids and their progenitors. *Heredity* 2012, *108*, 419–430.

117. Elshire, R.J.; Glaubitz, J.C.; Sun, Q.; Poland, J.A.; Kawamoto, K.; Buckler, E.S.; Mitchell, S.E. A robust, simple genotyping-by-sequencing (GBS) approach for high diversity species. *PLoS One* 2011, *6*, e19379.

118. Grover, C.E.; Salmon, A.; Wendel, J.F. Targeted sequence capture as a powerful tool for evolutionary analysis. *Am. J. Bot* 2012, *99*, 312–319.

119. Cronn, R.; Knaus, B.J.; Liston, A.; Maughan, P.J.; Parks, M.; Syring, J.V.; Udall, J. Targeted enrichment strategies for next-generation plant biology. *Am. J. Bot* 2012, *99*, 291–311.

120. Kvam, V.M.; Liu, P.; Si, Y. A comparison of statistical methods for detecting differentially expressed genes from RNA-seq data. *Am. J. Bot* 2012, *99*, 248–256.

121. Ercolano, M.R. University of Naples Federico II: Portici (NA), Italy, Unpublished work; 2012.

122. Schmutz, J.; Cannon, S.B.; Schlueter, J.; Ma, J.; Mitros, T.; Nelson, W.; Hyten, D.L.; Song, Q.; Thelen, J.J.; Cheng, J.; *et al.* Genome sequence of the palaeopolyploid soybean. *Nature* 2010, *463*, 178–183.

123. Potato Genome Sequencing Consortium. Genome sequence and analysis of the tuber crop potato. *Nature* 2012, *475*, 189–195.

CHAPTER 6

ADVANCES IN BAC-BASED PHYSICAL MAPPING AND MAP INTEGRATION STRATEGIES IN PLANTS

Ruvini Ariyadasa and Nils Stein

INTRODUCTION

Access to a high-quality reference genome sequence is the ultimate tool for advanced genomic research both in plants and biomedical sciences [1]. A genome sequence fundamentally can be obtained from two different approaches [2]. The widespread whole-genome shotgun sequencing (WGS) strategy randomly shears the genomic DNA into small pieces and subsequently clones into plasmids prior to sequencing [3]. In contrast, the conservative hierarchical/clone-by-clone method relies on a physical map that is used to define a minimum tilling path (MTP) representing the least redundant overlapping clones [4]. Due to the revolutionary advances in next-generation sequencing (NGS) platforms, the whole-genome shotgun (WGS) strategy soon became the method of choice for many sequencing projects as it enabled sequencing any given organism irrespective of the genome size within a short time and for a fraction of the cost it used to be [5, 6]. These sequencing projects indeed have provided rapid access to new gene models from many more organisms than previously anticipated [2, 7]. In spite of the constant improvements on the assemblers, read lengths, and mate pair technologies [8–10], to date the resulted assemblies still remain highly fragmented with incomplete genomic representation and prone to misassemblies [2, 7]. Although the quality of WGS assemblies over the hierarchical method has been questioned ever since its invention, this was not critically assessed, mainly due to the lack of resources such as complete cDNA libraries or a finished genome sequence [7]. The first attempt to address this issue was performed by Aach et al. [11] by comparing the two draft sequences of the human genome generated by human genome project [4] (HGP) and Celera genomics [3]. The "Celera assembly" remained far from being independent as it used the sequences from HGP in a "shredded" manner that retained nearly all the information from the original sequences [12]. Recently the availability of two completely de novo assembled human draft genomes [13] enabled a systematic comparison to the HGP reference sequence [3, 4]. The results indicated that the most adversely affected areas of de novo NGS alone drafts are the segmental duplications and larger common repeats [7, 14]. The depiction became actuate when the sequence identity exceeded 85% resulting in a loss of 16% of the genome due to sequence collapse while 99% of the true pairwise segmental duplications were absent [7, 14]. Further analysis on impact of genomic reduction revealed that only 57% of the genes had sufficient representation in the assembly (≥95% of the gene) while over 2300 exons were completely missing [7]. In contrast, as the hierarchical strategy primarily relies on an ordered, densely anchored physical map, the resulted sequences mainly remain local, thereby reducing the complexity [15–17] and facilitating efficient handling of repeats and segmentally duplicated regions. Furthermore, as the hierarchical method can deliver nearly the complete genome sequence, it facilitates the accuracy of sequence assembly as well as gene annotation processes [2, 7]. As the shortcomings of draft sequences generated by NGS platforms alone and its adverse effects on downstream applications (such as copy number variations (CNVs), accurate gene prediction, functional annotations, and evolutionary and comparative studies) are becoming increasingly apparent, hybrid sequencing approaches combining the conventional

Citation: Ruvini Ariyadasa and Nils Stein, "Advances in BAC-Based Physical Mapping and Map Integration Strategies in Plants," Journal of Biomedicine and Biotechnology, vol. 2012, Article ID 184854, 11 pages, 2012. doi:10.1155/2012/184854.

hierarchical strategy with multiplatform NGS sequencing including the third-generation long read technologies are gaining more recognition [2, 7, 9], especially when dealing with large and repeat-rich genomes. While physical maps are identified as an important intermediate layer in modern sequencing, considerable amount of technical advances have been established in contig map construction. This paper outlines the recent technical advances of BAC-based physical mapping and its integration with high-density genetic maps.

DIVERSE BAC LIBRARIES ENSURE HIGH GENOMIC REPRESENTATION

A physical map is a model of a genome of interest reconstructed on the basis of overlapping fragments of genomic DNA. A prerequisite for building a physical map either being genomewide, chromosome specific, or for a region of interest is access to BAC libraries that represent several folds of haploid genome coverage [18–24]. It is best to use a combination of BAC libraries with various origins such as different restriction enzymes and/or constructed by mechanical shearing of genomic DNA [24, 25]. If only a single BAC library originated from partial digest with a single enzyme is solely used for physical map construction, the resulted map is likely to contain gaps due to cloning bias and segments that are noncloneable or unstable in the E. colihost [26, 27]. To overcome the cloning bias, BAC libraries originated from different restriction enzymes are employed [24, 28, 29]. The rationale behind constructing independent BAC libraries by partial digestion with different restriction endonucleases is that the frequency of occurrence of a specific restriction site in the DNA sequence is a function of the base pair composition of a species genome and recognition site [29]. For instance a restriction site composed of 4 nucleotides should theoretically generate fragments once in 256 bp ($(1/4)^4 = 1/256$) while a six-base cutter will generate fragments in every 4096 ($(1/4)^6 = 1/4096$) nucleotides in a completely random sequence. However, as for most DNA sequences the nucleotide distribution is not random, a bias in restriction site distribution is observed. Thus, selecting multiple enzymes with different recognition sequences is a measure to limit the risk of underrepresentation of specific regions in the resulted BAC map [29]. Including a mechanically sheared library is an alternative to further overcome cloning bias [24, 30, 31].

FOSMIDS AS A TOOL FOR PHYSICAL MAPPING

Fosmids represent an alternative cloning system derived from low copy number cosmid vectors. Similar to BACs, the Fosmid inserts remain under the control of E. coli F factor replicon enhancing its stable maintenance and propagation. However, in contrast to BACs, Fosmids carry much narrower insert sizes with an average of 40 Kb. As the fragmentation of DNA is performed via mechanical shearing, no cloning bias is introduced to the resulted libraries [32]. As opposed to the high technical skills and costs required to develop large-insert deep-coverage BAC libraries, Fosmid library construction remains quite trivial [33]. The narrower insert range, stable maintenance, and easy production have made Fosmids suitable for a wide range of applications [30, 34]. For instance, Fosmid libraries per se are useful for whole-genome physical mapping [35–37] and refining the BAC/PAC-based physical maps by effective gap filling [26, 30]. Furthermore, paired Fosmid end sequences, due to the narrower insert size range of Fosmid clones, are useful for sequence scaffolding in large-scale sequencing projects and studies related to structural variation and organization of the genomes [34, 38–41].

METHODS OF PHYSICAL MAP CONSTRUCTION—STATUS QUO AND NEW APPROACHES

Physical maps are generated based on technologies to detect clone overlaps. Two distinct approaches, namely (1) screening BACs to assess presence of mapped markers and (2) restriction map-based fingerprinting, are used to identify the overlap among clones. Each of these approaches is represented by numerous techniques [42]. The underlying principle of these techniques, advantages, and limitations have been extensively discussed in previous reviews [42, 43] and thus will only be outlined below. Briefly the first method relies on PCR amplification of short fragments (sequence tagged sited STS) [44, 45] and/or hybridization of labeled cDNA clones/gene-specific oligonucleotides [46]. The method became less attractive as it was laborious and required extremely large number of markers that were limited for many species [42]. The second approach is based on DNA fingerprinting, essentially performed by restriction mapping [42, 44, 47]. Soon this approach became the method of choice in physical map construction as it is suited better for unexploited genomes and was more amenable for automation [42]. A genomewide restriction fragment map was first produced based on acrylamide [47] and agrose gel electrophoresis [44]. As these methods largely remained of low throughput while delivering relatively low amount of information and imprecise fragment sizing, a set of new fingerprinting methods collectively known as high information content fingerprinting (HICF) has been developed [48–50]. All HICF methods are based on capillary electrophoresis leading to substantial increase in both throughput and sensitivity of the resulted fingerprints. After a comparative evaluation of the proposed HICF fingerprinting methods it was concluded that the method developed by Luo et al. [50] was the most effective and thus has been used in many recent fingerprinting projects thereafter [16, 24, 42].

WHOLE-GENOME PROFILING (WGP)

Whole-genome profiling (WGP) is a novel next-generation, sequence-based physical mapping strategy. The method makes use of short sequence tags adjacent to restriction enzyme sites generated either by Illumina GA or SOLiD platforms [51]. A key difference between WGP and the current gold standard HICF method [50] is that WGP uses short sequence tags (20 bp or larger with recent read length improvements of the short read sequencing technologies) generated across BACs (every 2-3 Kb) to identify the overlapping BACs uniquely and for the subsequent contigs building process [52], while the latter relies on fragment sizes obtained after capillary electrophoresis [50]. As WGP orders BACs into contigs by identifying the clone overlaps through sequence data, it simultaneously generates a partial genome sequence. To trade off the costs involved in BAC DNA preparation and Illumina GA/SOLiD sequencing, an efficient BAC pooling strategy is a must for WGP. The complexity of the pools and the redundancy of the WGP tags per BAC (i.e., coverage) are related to the complexity and size of the genome under investigation. It was demonstrated that WGP on small genomes requires less complex pools, whereas large and complex genomes require complex pools with more BACs per pool [51]. The pooled DNA is then digested with two enzymes MseI and EcoRI prior to ligating with universalMseI P7 and a sample-specific tagged EcoRI P5 adaptors, respectively. The subsets of fragments that contain both adaptors are selected by PCR amplification and sequenced on Illumina GA or SOLiD platforms. In contrast to HICF, WGP allows the identification and filtration of repeat originated sequences reducing the chances of chimeric contig formation [51]. The method was tested by developing a WGP-based physical map ofArabidopsis thaliana Col-1 by using ~6100 BAC clones (6X genome coverage). The comparison of the resulted sequence-based physical contigs against the "Columbia" reference sequence revealed ~98% of the assembled contigs (i.e., 80% of the map length) matched the reference sequence. To substantiate the robustness of the method, four additional genomes of melon (450 Mb), tomato (950 Mb), rape seed (1.2 Gb), and lettuce (2.6 Gb) were subjected to WGP. Furthermore, in silico analysis of the maize genome indicated the reliability of the method when applied to repeat-rich genomes [51].

Interestingly, the comparative studies on WGP and HICF have revealed that WGP results in significantly less number of contigs while reducing the chimeric and misassembled contigs by 3% and 4%, respectively [53]. Though WGP offers attractive and immediate advantages, the method still remains expensive. Nevertheless, the dramatic decrease in sequencing costs and employing smarter BAC pools will help to trade off the associated costs.

GENETIC ANCHORING OF PHYSICAL MAPS—INCREASING THE EFFICIENCY

To achieve the full benefits of a physical map, bacterial artificial chromosome (BAC) contigs are required to be integrated with genetic maps at high density and resolution [54–56] (Figure 1). Screening BAC libraries is an indispensable step for physical and genetic map integration, by which BAC clones and contigs can be placed and ordered along a genetic map [16, 54–56]. Until few years ago anchoring per se was a rate-limiting step mainly due to technical limitations in screening procedures and partly due to the lack of high-quality dense genetic maps [56–58]. The steady technological advancements in screening procedures resulted in cost-effective and highly parallelized multiplexing platforms.

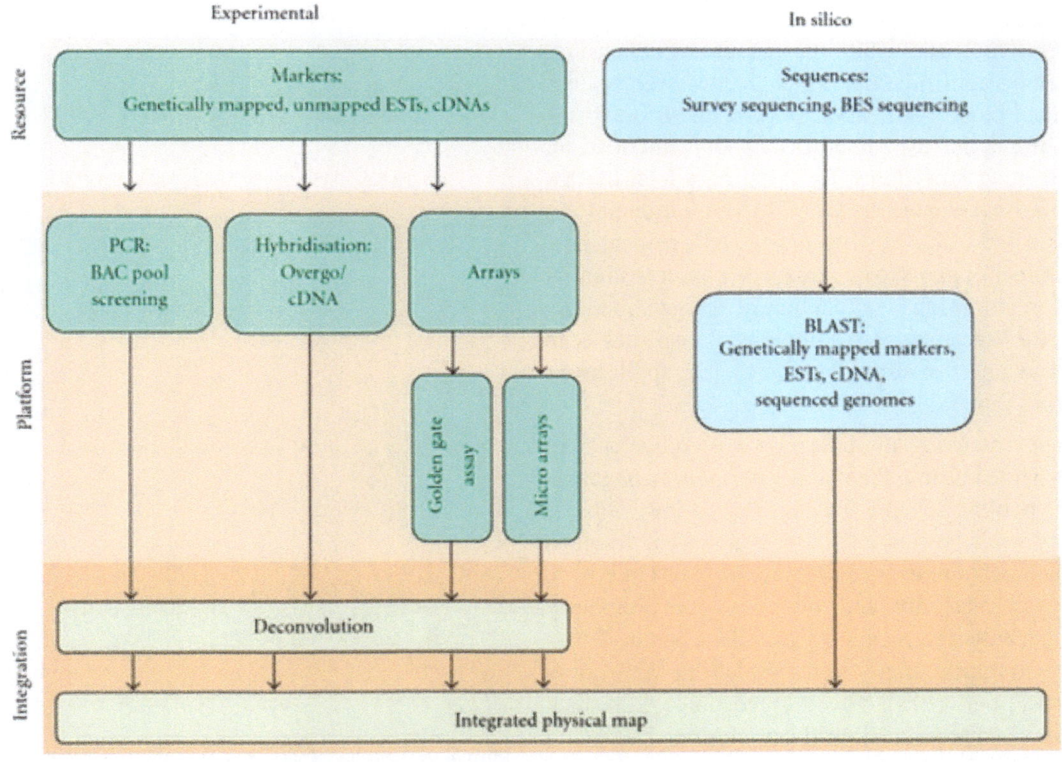

Figure 1: Approaches to BAC library screening. Genetically mapped markers and unmapped sequence information such as cDNA and ETSs can be used to screen multidimensional BAC pools via PCR-based methods, hybridization of BAC filters, or highly parallel array platforms. The resulted data are deconvoluted to obtain marker BAC relationships prior to integrating to the contig map. Alternatively the BAC sequence information obtained through survey sequencing (NGS/Sanger sequencing) or BES could be searched against genetically mapped markers, ESTs, cDNA, and so forth of the same organism or different species.

ADVANCES IN MULTIDIMENSIONAL BAC POOLING

At the end of constructing a BAC library the recombinant colonies are robotically picked and arrayed into microtiter plates generating gridded bacterial stock cultures that are used for storage, routine work as well as shipping [18, 24, 59, 60]. The gridded cultures can be used for generating multidimensional BAC pools, a prerequisite to maximize the effectiveness of BAC library screening [61, 62]. A BAC pool is composed by combining a set of predefined BACs within or between microtiter plates [62]. Several pooling strategies have been proposed, all of which rely on concentrating the BACs into pools composed of overlapping groups of clones [56, 58, 61]. One of the earliest and frequently used pooling schemes is the three-dimensional (3D) plate, row, and column pools (PP, CP, and RP) [61, 62]. The 3D pools are generated by subdividing the library into several groups, each containing equal number of microtiter plates representing a known amount of genome equivalents. Prior to pooling, the DNA is isolated separately and pooled in an equimolar fashion to generate plate, row, and column pools within the predefined subset of microtiter plates [61, 62]. Although 3D pools reduce the number of assays required to identify a marker-BAC relationship, it is prone to generate false-positives [56, 58, 63]. Therefore, to identify the positive BAC clones for a given marker reliably while performing minimum number of PCR assays, pooling additional dimensions was tested [61]. For instance the 6D pooling strategy includes three-dimensional pools called plate, face, and side pools (3D stack), together with additional three types of pools, namely, row, column, and diagonal pools [63]. Only the 3D stack is screened to identify the positive clones while the remaining dimensions are used for clone confirmation, which reduces the tedious task of individual clone verification [63]. A major drawback of the 6D scheme is the large number of pools generated for subsequent screening procedures, for instance, a total of 184, 208, and 288 pools were generated for 730 Mbp sorghum [64], 1.1 Gbp soybean [65], and 2.5 Gbp maize [63], respectively. Because the number of pools to be screened in order to identify a marker-BAC relationship is a function of the genome size, the workload for screening BAC libraries for species with larger genomes is unacceptably high, limiting its applications on large genomes [56, 58].

To facilitate screening multiple genome equivalents of large genomes while keeping the number of pools manageably low, a new 5D pooling scheme was introduced [58]. Similar to other pooling strategies the microtiter plates were arranged into groups, each containing equal number of plates that were then pooled to form plate, row, and column pools as previously described. To generate the remaining two dimensions, BACs with a single microtiter plate were pooled to form low-complexity plate pools (PPs). The resulted pools were then arranged in a two-dimensional grid and further pooled to form row super pools (RSPs) and column super pools (CSPs), respectively [58].

Recently, conventional 3D BAC pools were further modified into convenient two-step screening, namely, super pool and matrix pool, a proprietary method invented by a commercial supplier [66–68]. The total number of super pools in a kit is determined by the number of BACs in the library [67]. For instance, the barley BAC library HVVMRXALLeA with 147 and 456 BACs (i.e., 3.7x genome equivalent) contained 55 super pools. A single super pool is composed of seven 384 microtiter plates (i.e., 2,688 BACs per super pool) (Figure 2). The BACs included in a single super pool are further pooled on a predefined matrix to form plate, row, and column pools (matrix pools) [67, 68]. An important feature of this two-step pooling scheme is that it does not only reduce the number of PCR assays, but also decreases the false-positives as the matrix pool screening results demand for two signals to identify a given coordinate, that is, a total of six signals is required to identify a positive BAC (Figure 3) [67]. This is because within a given matrix pool each coordinate (plate, row, or column) is represented twice; thus both coordinates are required to identify the corresponding BAC address (Figure 3).

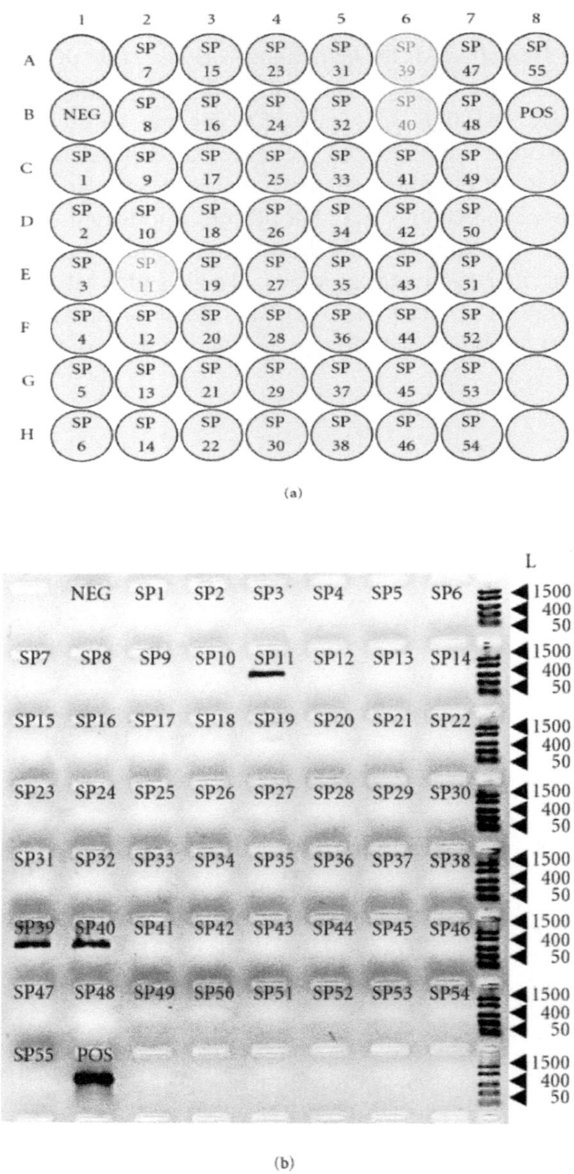

Figure 2: Results of super pool (SP) screening for the barley BAC library HVVMRXALLeA. (a) Schematic representation of 55 barley super pools originated from the BAC library HVVMRXALLeA [24]. SP1 to SP55 represent the super pools 1 to 55. Each super pool is composed of 7 × 3 8 4 microtiter plates, that is, 2688 BACs per super pool. NEG: negative control. POS: positive control, in this case genomic DNA of the barley cultivar Morex where the BAC library was derived. (b) Ethidium-bromide-stained agarose gel (1.5%) image of the super pool screening. Only the positive super pools (highlighted

with a green border in (a)) amplified a single fragment that is of identical size to the product in Morex. Arrowheads indicate in each panel the position of the 1500, 400, and 50 bp fragments of the FastRuler Low Range marker (Fermentas). L (size standard in bp).

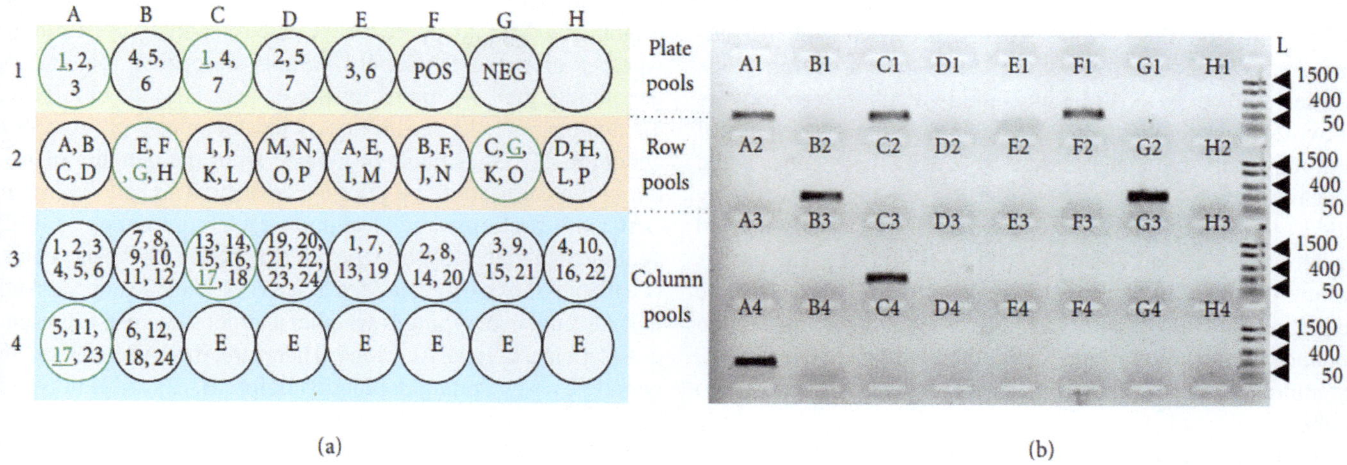

Figure 3: Results of matrix pool screening for the barley BAC library HVVMRXALLeA. (a) Schematic representation of a matrix pool composed of 23 pools for the BAC library HVVMRXALLeA [24]. The colored panels in A indicate the plate, row, and column matrix pools. Each circle represents a plate/row/column matrix pool where its composition is indicated within the circle. A matrix pool is composed of clones from seven 384 well BAC library plates. The clones within the seven plates are pooled based on a predefined matrix to form plate, row, and column matrix pools. A1—H4 indicate the position of each pool in a 96-well plate. NEG: negative control. POS: positive control, in this case genomic DNA of barley cultivar "Morex" where the BAC library was derived from. E: empty well. (b) Ethidium-bromide-stained agarose gel (1.5%) image of matrix pool screening. A1—H4 indicate the well position of each pool in the PCR plate. F1 is the positive control where G1 represents the negative control. Two signals are required to identify a given coordinate of the BAC address. Positive pools are highlighted by a green border and the shared coordinates between positive pools are underlined in (a). Arrowheads indicate in each panel the position of the 1500, 400, and 50 bp fragments of the FastRuler Low Range marker. L (size standard in bp).

PCR-BASED METHODS FOR BAC LIBRARY SCREENING

Screening BAC libraries by PCR assays is one of the earliest methods of library screening and has been extensively used in early physical mapping projects [15, 69]. The key advantage of PCR screening over hybridizations is the low rate of false-positives compared to conventional hybridization techniques [56–58, 63]. Additionally, the appropriate design of primers to amplify locus-specific products would allow mapping of genes that belong to a gene family [63]. On the other hand it enables sequencing the resulted amplicons to identify paralogous genes that often complicate the assignment between FPC contigs and genetic markers [65]. Recently, PCR-based techniques have also benefited from the increased automation and throughput [55, 70]. For example the advancements in thermocyclers have focused on scaling sample volume down, increasing the number of samples run simultaneously and decreasing overall cycling time [57, 70]. Additionally BAC library screening has also been performed on quantitative PCR machines with melt curve analysis capabilities [55, 70]. In principle the technique utilizes a standard PCR reaction performed in presence of a saturated/unsaturated intercalating dye such as Eva or SYBR green. The method accelerates the sample analysis and automated scoring by eliminating post-PCR agarose gel electrophoresis. The specificity of the amplicons generated in a given sample is verified by comparing the melting profiles of the samples to a known positive control that is essentially the genomic DNA of the cultivar/accession the BAC library was developed from [55, 70].

To enhance the throughput of BAC pool screening via PCR assays, new in silico analysis algorithms have been developed [55]. These algorithms facilitated the reduction of cost and labor involved in screening procedures, especially when dealing with large genomes. For example "electronic physical map anchoring tool" (ELEPHANT) combines the contig information generated from standard physical mapping tool "FPC" [52] along with the screening data originated from single-dimension plate pools. The programme is capable of identifying marker-BAC relationships unambiguously, simultaneously reducing the number of PCR assays. Nevertheless, this tool can only be applied if the library represents a genome coverage between 6- and 29-fold. Being below or above this range decreases the confidence of the identified contigs-BAC relationships [55]. In spite of these advancements, the PCR-based methods still remain cumbersome and of rather low throughput when large numbers of markers are to be screened [46, 54, 58, 68, 71].

BAC FILTERS FOR LIBRARY SCREENING

Until a couple of years ago, hybridization of single DNA probes or combinatorial pools of probes to BAC filters was a common approach for BAC library screening [15, 54, 71–73]. The BAC filters are composed of immobilized and lysed BAC colonies, which were robotically grided onto Nylon membranes [28, 57, 71]. Several types of probes can be used for screening BAC filters. The most commonly used are subcloned DNA fragments, PCR amplicons, cDNA inserts [74], or DNA oligonucleotides [64]. These probes can be used either individually or as a group [54, 71]. Major disadvantages of the hybridization method are the tendency of single probes hybridizing to multiple clones belonging to independent loci or of multiple probes with contradicting genetic information hybridizing to the same single BAC [58, 71]. Thus, the unequivocal assignment of contigs to loci on a genetic map may become troublesome [58]. Such contradictions could arise if the probes have originated from duplicated genes, conserved motifs, and repeat content, or due to contaminations that may have occurred during library construction and/or handling steps [71]. To overcome the drawbacks largely associated with hybridizing PCR/cDNA or EST inserts directly onto BAC filters, Ross et al. [46] introduced the concept of "overgo probes." An overgo probe is often a 40 bp dsDNA oligonucleotide. Each overgo consists of two 24 mer oligonucleotides, which share eight base pairs of complimentary sequence at their 3′ ends. After synthesis and annealing, they create a sixteen-base overhang subsequently filled in using "Klenow fragment" and radio-labeled nucleotides ([$^{\infty}P^{32}$]dCTP and [$^{\infty}P^{32}$]dATP) [46, 71]. The probes can be designed after consulting a complete database where sequences are already repeat masked [75]. The precaution enhances the unequivocal assignment of contigs to genetic loci by improving probe specificity [71, 75]. Furthermore, overgo probes facilitated probe pooling/multiplexing as they could be designed to have uniform thermodynamic properties [71]. The drawback of the original Ross et al. [46] protocol is that if extrapolated to large number of probes (10,000 or more) and applied to larger genomes (e.g.,: barley, 5.1 Gbp), the method demanded a problematic level of human exposure to radio isotopes [71]. Therefore, Madishetty et al. [71] modified and optimized the protocol for its application in large-scale multiplex screening. In contrast to the original method, the modified protocol utilizes all four radio isotopes in a hot pulse followed by a cold nucleotide chase and extends the exposure time to compensate for reduced specific activity of the probes [71]. The rationale behind using four radio nucleotides was that this would help to equalize the intensity of all probes by reducing the influence of nucleotide frequency. The modified method was successfully applied to identify a set of gene-bearing barley BACs (~65,000) from a BAC library composed of 313,344 clones (6.3-fold genome coverage) by using 12,600 overgo probes [71]. Although the multiplex hybridization screening is both cost- and time-efficient, it generates large amount of data that must be analyzed, assimilated, and disambiguated efficiently. Thus, to facilitate the data handling and rapid deconvolution, computational solutions are in place and have been successfully applied in large-scale physical mapping projects [76].

ARRAY-BASED METHODS FOR BAC LIBRARY SCREENING

Illumina Golden Gate Assay

To increase the efficiency of BAC library screening, highly parallel array-based platforms have been tested on multidimensional BAC pools [58, 68]. Luo et al. [58] employed the highly parallel SNP genotyping platform, the Illumina golden gate assay, to screen a BAC library [77, 78]. Principally the golden gate assay targets a specific SNP locus and utilizes two allele-specific oligonucleotides to discriminate between SNP alleles while the third locus-specific oligonucleotide contains an address sequence downstream of the SNP. The two allele-specific primers each containing an allele discriminating nucleotide at its 3′ end are labeled either with Cy3 or Cy5 fluorochromes, respectively. After annealing one of the two allele-specific oligonucleotides to the genomic DNA template, it is extended by DNA polymerase and ligated to the locus-specific oligonucleotide downstream forming a contiguous PCR template. The ratio of Cy3 and Cy5 fluorescence is used to determine the genotype at a SNP locus [77, 78]. Luo et al. [58] demonstrated that annealing of allele- and locus-specific oligonucleotides to pooled BAC DNA enables determining whether or not a BAC pool harbors a specific locus, thus genotyping the BAC pools for presence or absence of a locus. To demonstrate the feasibility of the technique on complex genomes, 5D BAC pools originated from the genome of Ae. tauschii were used in the pilot experiment. The genotyping of BACs was performed on plate pools each representing ~1-fold genome equivalents. While the query asked in SNP genotyping is which of the two nucleotides is at a SNP site, the query asked in BAC pool genotyping is whether a target sequence is present or absent in a given BAC clone. Therefore, it is a prerequisite to genotype a segregating population before BACs are genotyped as the former provides the reference for defining call regions/clusters to identify the positive BACs. Furthermore, the BAC library used for genotyping should originate from one of the parents of the segregating population. Though this method primarily utilizes the golden gate assay, it still requires PCR screening of row super pools (RSPs) and column super pools (CSP) to verify the positive plates while clone verification required PCR screening of row pools (RPs) and column pools (CPs) [58] (see Section 7 for pooling details). In order to enhance the efficiency by eliminating the necessity of PCR screening, You et al. proposed a new implementation of the 5D pooling where four of the five dimensions (RSP, CSP, RP, and CP) could be screened directly on the golden gate platform [56]. The fifth dimension is the information about overlapping BAC clones in the contig map that is used to detect and verify true-positives among putative positive clones [56]. The main advantage of the golden gate screening as opposed to general PCR and hybridization approaches is a higher tendency of revealing one-to-one relationships between BAC contigs and genetic markers [56, 58].

Microarray Platforms

Although originally designed for studying multiparallel gene expressions, the feasibility of microarrays for high-throughput BAC library screening was recently explored [68, 79]. Pilot studies explored the possibility of using barley transcript-derived markers as a surrogate to anchor and order physical contigs of wheat chromosome 3B [79]. The BAC DNA of the three-dimensional MTP pools of wheat 3B chromosome was directly hybridized to a custom microarray of barley. The wheat-barley cross hybridizations allowed to identify the gene space across the wheat 3B chromosome while overcoming the limited availability of transcript-derived markers originated from wheat. Prior to hybridization, the pooled DNA was mixed with a reference barley mRNA template that facilitated the grid alignment and spot identification. The reference mRNA and BAC DNA was tagged with Cy5 (red) and Cy3 (green), respectively. After signal quantification and normalization, hybridization data was evaluated with four complementary scoring methods to reliably locate as many barley gene homologs as possible on wheat BACs. A complementary scoring scheme was applied to identify all positive signals, as missing coordinates in one approach could be recovered from the remaining methods. The method allowed mapping of 738 barley orthologous genes on wheat chromosome 3B. The reliability of the identified genes was validated by homology searches against the sequenced contigs of 3B [79].

This approach was further extended to anchor the HICF-derived barley BAC contigs to the unigenes deposited at HarvEST database [68, 80]. For this study a custom microarray (SCRI_Hv35_44k_v1) that contained 42,300 barley expressed sequence tags (ESTs) was used together with previously described matrix pools originated from barley BAC library HVVMRXALLeA (see Section 7). The hybridizations were performed by using two-channel processing of the microarray. Two matrix pools each representing 1/16th of the whole barley genome were uniquely labeled with either Cy3 or Cy5 and directly hybridized to the barley array. The signals were extracted, normalized, and subjected to downstream data processing through a customized semiautomated pipeline. The data processing scripts facilitated the rapid deconvolution of BAC addresses. The resulted marker BAC relations were independently validated and confirmed by four different approaches namely (1) shotgun sequencing of individual BACs, (2) plate/super pool DNA, (3) gene-specific qPCR assays, and (4) BLAST analysis against Brachypodium genes [68]. Microarray-based screening comes at the advantage of high-throughput and accurate gene-to-BAC addressing. Furthermore, the costs are defined by the overall number of arrays required to screen sufficient pool depth. For the golden gate assay-based screening cost degression can be achieved by using the same assay design for many samples, which, however, is not necessarily a limitation since the most efficient golden gate assay-based BAC library screening will be achieved in conjunction with genotyping a segregating population. Both multiplex screening approaches offer efficient improvements over previously applied marker-BAC address screening methods.

In Silico Anchoring

The term in silico anchoring refers to the computational assignment of the marker/gene sequences to a particular BAC unambiguously based on sequence similarities. Sequencing paired ends of BAC clones (BAC end sequencing (BES)), ideally of fingerprinted BAC clones, has become a standard practice in physical mapping projects [16, 81–83]. Paired BAC end sequencing refers to the bidirectional end sequencing of the genomic DNA insert with the help of universal priming sites in the cloning vector [84]. Access to end sequence information of BAC clones can facilitate in silico anchoring of the contig map to the genetic map and has proven particularly useful in small, gene-dense genomes [84, 85]. Their utility diminishes in repeat-rich genomes as only a limited percentage of BES will account for coding/low copy sequences, which can be reliably used for in silico anchoring [86]. Usage of BES information for in silico anchoring is two-fold. First they could be directly searched for gene-based/genomic markers within the same species, a standard practice from the early genome sequencing projects [15, 69, 87]. For instance during the rice genome sequencing project [87] all BESs originated from BAC libraries used to construct the rice physical map were preprocessed (by cleaning E. coli contamination and vector trimming) and repeat masked to avoid repetitive DNA sequences [87]. The preprocessed sequences were then searched for over 2000 rice markers via rapid sequence comparison programmes. This approach delivered over 400 putative marker BAC relations on average with 98.5% sequence identity over 216 bases. The verification experiments indicated that in silico anchored BACs grouped into the same fingerprint contig thus confirming the stringency level of alignment parameters helping to avoid the paralogous gene detection. The assignments were further confirmed by amplifying the corresponding markers from the overlapping BACs during wet lab experiments [87].

Secondly cross-referencing BES against the sequenced genomes in order to assign the BAC contigs to genetic bins became a method of choice due to the availability of large number of sequenced genomes and extensive knowledge on conserved syntenic relationships [69, 88]. This approach facilitated assigning physical contigs into genetic bins even in absence of dense genetic maps for the species of interest. For instance in the physical mapping project of Brachypodium distachyon BAC end sequences were aligned against the wheat ESTs and rice genome sequences, which immediately facilitated anchoring a large number of Brachypodium BAC contigs to the rice genome sequence and wheat deletion maps [89]. The indirect anchoring of the Brachypodium BAC contigs bypassed the lack of a dense genetic map of Brachypodium.

Additionally BES provides the raw material for "end sequence profiling" (ESP), a standard practice in human genomics to identify the genomic architecture of tumor/disease genomes [90, 91]. The ESP is essentially based on the concept of sequence-tagged connectors developed to facilitate de novo genome sequencing [92]. The technique is capable of providing high-resolution copy number (CNV) and structural aberrations/variation (SV) maps on selected tumor/disease tissues. To this end BAC libraries are required to be generated from the rearranged genomes followed by subsequent generation of BES. Mapping these BES to a reference genome facilitates the detection of CNV/SV. For instance the aberrations in distance and/or orientation between end sequence read pairs reveals the presence of larger-scale SV while the use of comparative genomic hybridization (cGH) arrays allows mainly insights into CVs. The ESP method has since been adapted to characterize CNV/SV in human genomes using paired-end sequencing on NGS platforms [93–95].

SUMMARY AND OUTLOOK

The next-generation sequencing (NGS) platforms have offered the potential to generate whole-genome sequences (WGS) for any organism irrespective of its genome size for a fraction of cost and time it used to be [1, 2, 7]. Though the resulted assemblies remain highly fragmented with lack of context and encompassing limited information on gene location and order, they provide immediate access to large number of gene models than ever before. Nevertheless, often this type of output remains far from even a true draft genome sequence, not to mention a gold standard reference sequence, which is the primary goal of most research communities [2,7]. To this end it is noteworthy that the plant genomes are of average representative genome size of 6 Gbp and composed of a high proportion of repetitive DNA making WGS-based sequencing alone a daunting task [2,96]. Although the access to genome sequences of different quality and status of finishing can be very enabling and utilized in many aspects of genome analysis and crop improvement, only high-quality drafts or finished genome sequences will allow advanced and accurate gene prediction including the determination of all regulatory sequences of a genome. The understanding of the extent structural/copy number variations and their impact on crop performance is only in its infancy and studied only in a limited number of plant genomes like maize and rice [97–99]. To gain further and deeper insights towards all aspects of the structure of economically important crop plant genomes and their impact on crop performance will require access to high-quality physical maps. This is not only true for different crop species but applies furthermore to genotypes representing key haplotypes of the available crop germplasm. Thus, the possibilities provided by new sequencing technologies eventually lead to a reinforced need for physical maps rather than making those obsolete.

REFERENCES

1. R. K. Varshney, S. N. Nayak, G. D. May, and S. A. Jackson, "Next-generation sequencing technologies and their implications for crop genetics and breeding," Trends in Biotechnology, vol. 27, no. 9, pp. 522–530, 2009.
2. C. Feuillet, J. E. Leach, J. Rogers, P. S. Schnable, and K. Eversole, "Crop genome sequencing: lessons and rationales," Trends in Plant Science, vol. 16, no. 2, pp. 77–88, 2011.
3. J. Craig Venter, M. D. Adams, E. W. Myers et al., "The sequence of the human genome," Science, vol. 291, no. 5507, pp. 1304–1351, 2001.
4. E. S. Lander, L. M. Linton, B. Birren et al., "Initial sequencing and analysis of the human genome,"Nature, vol. 409, no. 6822, pp. 860–921, 2001.
5. S. Huang, R. Li, Z. Zhang et al., "The genome of the cucumber, Cucumis sativus L," Nature Genetics, vol. 41, no. 12, pp. 1275–1281, 2009.
6. R. Li, W. Fan, G. Tian et al., "The sequence and de novo assembly of the giant panda genome," Nature, vol. 463, no. 7279, pp. 311–317, 2010.
7. C. Alkan, S. Sajjadian, and E. E. Eichler, "Limitations of next-generation genome sequence assembly,"Nature Methods, vol. 8, no. 1, pp. 61–65, 2011.
8. D. G. Hert, C. P. Fredlake, and A. E. Barron, "Advantages and limitations of next-generation sequencing technologies: a comparison of electrophoresis and non-electrophoresis methods," Electrophoresis, vol. 29, no. 23, pp. 4618–4626, 2008.
9. M. C. Schatz, A. L. Delcher, and S. L. Salzberg, "Assembly of large genomes using second-generation sequencing," Genome Research, vol. 20, no. 9, pp. 1165–1173, 2010.
10. J. Shendure and H. Ji, "Next-generation DNA sequencing," Nature Biotechnology, vol. 26, no. 10, pp. 1135–1145, 2008.
11. J. Aach, M. L. Bulyk, G. M. Church, J. Comander, A. Derti, and J. Shendure, "Computational comparison of two draft sequences of the human genome," Nature, vol. 409, no. 6822, pp. 856–859, 2001.
12. P. Green, "Whole-genome disassembly," Proceedings of the National Academy of Sciences of the United States of America, vol. 99, no. 7, pp. 4143–4144, 2002.
13. R. Li, H. Zhu, J. Ruan et al., "De novo assembly of human genomes with massively parallel short read sequencing," Genome Research, vol. 20, no. 2, pp. 265–272, 2010.
14. C. Alkan, J. M. Kidd, T. Marques-Bonet et al., "Personalized copy number and segmental duplication maps using next-generation sequencing," Nature Genetics, vol. 41, no. 10, pp. 1061–1067, 2009.
15. J. D. McPherson, M. Marra, L. Hillier et al., "A physical map of the human genome," Nature, vol. 409, no. 6822, pp. 934–941, 2001.
16. E. Paux, P. Sourdille, J. Salse et al., "A physical map of the 1-gigabase bread wheat chromosome 3B,"Science, vol. 322, no. 5898, pp. 101–104, 2008.
17. F. Wei, J. Zhang, S. Zhou et al., "The physical and genetic framework of the maize B73 genome," PLoS Genetics, vol. 5, no. 11, Article ID e1000715, 2009.
18. F. Adam-Blondon, A. Bernole, G. Faes et al., "Construction and characterization of BAC libraries from major grapevine cultivars," Theoretical and Applied Genetics, vol. 110, no. 8, pp. 1363–1371, 2005.
19. Y. S. Yim, G. L. Davis, N. A. Duru et al., "Characterization of three maize bacterial artificial chromosome libraries toward anchoring of the physical map to the genetic map using high-density bacterial artificial chromosome filter hybridization," Plant Physiology, vol. 130, no. 4, pp. 1686–1696, 2002.
20. B. Yüksel and A. H. Paterson, "Construction and characterization of a peanut HindIII BAC library,"Theoretical and Applied Genetics, vol. 111, no. 4, pp. 630–639, 2005.
21. R. Ming, P. H. Moore, F. Zee, C. A. Abbey, H. Ma, and A. H. Paterson, "Construction and characterization of a papaya BAC library as a foundation for molecular dissection of a tree-fruit genome,"Theoretical and Applied Genetics, vol. 102, no. 6-7, pp. 892–899, 2001.
22. U. J. Kim, B. W. Birren, T. Slepak et al., "Construction and characterization of a human bacterial artificial chromosome library," Genomics, vol. 34, no. 2, pp. 213–218, 1996.
23. S. S. Woo, J. Jiang, B. S. Gill, A. H. Paterson, and R. A. Wing, "Construction and characterization of a bacterial artificial chromosome library of Sorghum bicolor," Nucleic Acids Research, vol. 22, no. 23, pp. 4922–4931, 1994.
24. D. Schulte, R. Ariyadasa, B. Shi et al., "BAC library resources for map-based cloning and physical map construction in barley (Hordeum vulgare L.)," BMC Genomics, vol. 12, article 247, 2011.
25. C. C. Wu, P. Nimmakayala, F. A. Santos et al., "Construction and characterization of a soybean bacterial artificial chromosome library and use of multiple complementary libraries for genome physical mapping," Theoretical and Applied Genetics, vol. 109, no. 5, pp. 1041–1050, 2004.
26. J. S. S. Ammiraju, Y. Yu, M. Luo et al., "Random sheared fosmid library as a new genomic tool to accelerate complete finishing of rice (Oryza sativa spp. Nipponbare) genome sequence: sequencing of gap-specific fosmid

clones uncovers new euchromatic portions of the genome," Theoretical and Applied Genetics, vol. 111, no. 8, pp. 1596–1607, 2005.

27. T. J. Yang, Y. Yu, G. Nah et al., "Construction and utility of 10-kb libraries for efficient clone-gap closure for rice genome sequencing," Theoretical and Applied Genetics, vol. 107, no. 4, pp. 652–660, 2003

28. E. Coe, K. Cone, M. McMullen et al., "Access to the maize genome: an integrated physical and genetic map," Plant Physiology, vol. 128, no. 1, pp. 9–12, 2002.

29. C. Wu, S. Sun, P. Nimmakayala et al., "A BAC- and BIBAC-based physical map of the soybean genome,"Genome Research, vol. 14, no. 2, pp. 319–326, 2004.

30. J. S. S. Ammiraju, M. Luo, J. L. Goicoechea et al., "The Oryza bacterial artificial chromosome library resource: construction and analysis of 12 deep-coverage large-insert BAC libraries that represent the 10 genome types of the genus Oryza," Genome Research, vol. 16, no. 1, pp. 140–147, 2006.

31. R. Godiska, D. A. Mead, V. Dhodda, et al., Bias-Free Cloning of "Unclonable" DNA for Simplified Genomic Finishing, Jones and Bartlett Publishers, Sudbury, Mass, USA, 2008.

32. U. J. Kim, H. Shizuya, P. J. De Jong, B. Birren, and M. I. Simon, "Stable propagation of cosmid sized human DNA inserts in an F factor based vector," Nucleic Acids Research, vol. 20, no. 5, pp. 1083–1085, 1992.

33. W. De Tomaso and I. L. Weissman, "Construction and characterization of large-insert genomic libraries (BAC and fosmid) from the ascidian Botryllus schlosseri and initial physical mapping of a histocompatibility locus," Marine Biotechnology, vol. 5, no. 2, pp. 103–115, 2003.

34. D. Hao, L. Yang, and P. Xiao, "The first insight into the Taxus genome via fosmid library construction and end sequencing," Molecular Genetics and Genomics, vol. 285, no. 3, pp. 197–205, 2011.

35. J. C. Gingrich, D. M. Boehrer, J. A. Garnes et al., "Construction and characterization of human chromosome 2-specific cosmid, fosmid, and PAC clone libraries," Genomics, vol. 32, no. 1, pp. 65–74, 1996.

36. C. G. Kim, A. Fujiyama, and N. Saitou, "Construction of a gorilla fosmid library and its PCR screening system," Genomics, vol. 82, no. 5, pp. 571–574, 2003.

37. U. J. Kim, H. Shizuya, J. Sainz et al., "Construction and utility of a human chromosome 22-specific Fosmid library," Genetic Analysis—Biomolecular Engineering, vol. 12, no. 2, pp. 81–84, 1995.

38. C. Lange, D. Holtgräwe, B. Schulz, B. Weisshaar, and H. Himmelbauer, "Construction and characterization of a sugar beet (Beta vulgaris) fosmid library," Genome, vol. 51, no. 11, pp. 948–951, 2008.

39. T. L. Newman, E. Tuzun, V. A. Morrison et al., "A genome-wide survey of structural variation between human and chimpanzee," Genome Research, vol. 15, no. 10, pp. 1344–1356, 2005.

40. E. Tuzun, A. J. Sharp, J. A. Bailey et al., "Fine-scale structural variation of the human genome," Nature Genetics, vol. 37, no. 7, pp. 727–732, 2005.

41. J. D. F. Meyer, W. Deleu, J. Garcia-Mas, and M. J. Havey, "Construction of a fosmid library of cucumber (Cucumis sativus) and comparative analyses of the eIF4E and eIF(iso)4E regions from cucumber and melon (Cucumis melo)," Molecular Genetics and Genomics, vol. 279, no. 5, pp. 473–480, 2008.

42. B. C. Meyers, S. Scalabrin, and M. Morgante, "Mapping and sequencing complex genomes: let's get physical!," Nature Reviews Genetics, vol. 5, no. 8, pp. 578–588, 2004.

43. W. M. Nelson, A. K. Bharti, E. Butler et al., "Whole-genome validation of high-information-content fingerprinting," Plant Physiology, vol. 139, no. 1, pp. 27–38, 2005.

44. M. Olson, L. Hood, C. Cantor, and D. Botstein, "A common language for physical mapping of the human genome," Science, vol. 245, no. 4925, pp. 1434–1435, 1989.

45. E. D. Green and P. Green, "Sequence-tagged site (STS) content mapping of human chromosomes: theoretical considerations and early experiences," PCR Methods and Applications, vol. 1, no. 2, pp. 77–90, 1991.

46. M. Ross, T. LaBrie, S. McPherson, and V. P. Stanton, "Screening large-insert libraries by hybridization," inurrent Protocols in Human Genetics, pp. 5.6.1–5.6.32, John Wiley & Sons, New York, NY, USA, 1999.

47. Coulson, J. Sulston, S. Brenner, and J. Karn, "Toward a physical map of the genome of the nematode Caenorhabditis elegans," Proceedings of the National Academy of Sciences of the United States of America, vol. 83, no. 20, pp. 7821–7825, 1986.

48. Y. Ding, M. D. Johnson, W. Q. Chen et al., "Five-color-based high-information-content fingerprinting of bacterial artificial chromosome clones using type IIS restriction endonucleases," Genomics, vol. 74, no. 2, pp. 142–154, 2001.

49. Y. Ding, M. D. Johnson, R. Colayco et al., "Contig assembly of bacterial artificial chromosome clones through multiplexed fluorescence-labeled fingerprinting," Genomics, vol. 56, no. 3, pp. 237–246, 1999.

50. M. C. Luo, C. Thomas, F. M. You et al., "High-throughput fingerprinting of bacterial artificial chromosomes using the SNaPshot labeling kit and sizing of restriction fragments by capillary electrophoresis," Genomics, vol. 82, no. 3, pp. 378–389, 2003.

51. J. Van Oeveren, M. De Ruiter, T. Jesse et al., "Sequence-based physical mapping of complex genomes by whole genome profiling," Genome Research, vol. 21, no. 4, pp. 618–625, 2011.

52. C. Soderlund, I. Longden, and R. Mott, "FPC: a system for building contigs from restriction fingerprinted clones," Computer Applications in the Biosciences, vol. 13, no. 5, pp. 523–535, 1997.

53. "20th International Triticeae Mapping Initiative (ITMI) programme and abstract web page,"

54. J. Gardiner, S. Schroeder, M. L. Polacco et al.,

"Anchoring 9,371 maize expressed sequence tagged unigenes to the bacterial artificial chromosome contig map by two-dimensional overgo hybridization,"Plant Physiology, vol. 134, no. 4, pp. 1317–1326, 2004.

55. E. Paux, F. Legeai, N. Guilhot et al., "Physical mapping in large genomes: accelerating anchoring of BAC contigs to genetic maps through in silico analysis," Functional and Integrative Genomics, vol. 8, no. 1, pp. 29–32, 2008.

56. F. M. You, M. C. Luo, K. Xu, K. R. Deal, O. D. Anderson, and J. Dvorak, "A new implementation of high-throughput five-dimensional clone pooling strategy for BAC library screening," BMC Genomics, vol. 11, no. 1, article 692, 2010.

57. T. N. Campbell and F. Y. M. Choy, "Approaches to library screening," Journal of Molecular Microbiology and Biotechnology, vol. 4, no. 6, pp. 551–554, 2002.

58. M. C. Luo, K. Xu, Y. Ma, K. R. Deal, C. M. Nicolet, and J. Dvorak, "A high-throughput strategy for screening of bacterial artificial chromosome libraries and anchoring of clones on a genetic map constructed with single nucleotide polymorphisms," BMC Genomics, vol. 10, article 28, 2009.

59. K. Farrar and I. S. Donnison, "Construction and screening of BAC libraries made from Brachypodium genomic DNA," Nature Protocols, vol. 2, no. 7, pp. 1661–1674, 2007.

60. J. Safar, J. Bartoš, J. Janda et al., "Dissecting large and complex genomes: flow sorting and BAC cloning of individual chromosomes from bread wheat," Plant Journal, vol. 39, no. 6, pp. 960–968, 2004

61. E. Barillot, B. Lacroix, and D. Cohen, "Theoretical analysis of library screening using a N-dimensional pooling strategy," Nucleic Acids Research, vol. 19, no. 22, pp. 6241–6247, 1991.

62. W. J. Bruno, E. Knill, D. J. Balding et al., "Efficient pooling designs for library screening," Genomics, vol. 26, no. 1, pp. 21–30, 1995.

63. Y. S. Yim, P. Moak, H. Sanchez-Villeda et al., "A BAC pooling strategy combined with PCR-based screenings in a large, highly repetitive genome enables integration of the maize genetic and physical maps," BMC Genomics, vol. 8, article 47, 2007.

64. P. E. Klein, R. R. Klein, S. W. Cartinhour et al., "A high-throughput AFLP-based method for constructing integrated genetic and physical maps: progress toward a sorghum genome map," Genome Research, vol. 10, no. 6, pp. 789–807, 2000.

65. X. Wu, G. Zhong, S. D. Findley, P. Cregan, G. Stacey, and H. T. Nguyen, "Genetic marker anchoring by six-dimensional pools for development of a soybean physical map," BMC Genomics, vol. 9, article 28, 2008.

66. M. F. Bouzidi, J. Franchel, Q. Tao et al., "A sunflower BAC library suitable for PCR screening and physical mapping of targeted genomic regions," Theoretical and Applied Genetics, vol. 113, no. 1, pp. 81–89, 2006.

67. "Users manual for BAC pools produced by Amplon Express," http://www.genomex.com/aexDownloads.php.

68. H. Liu, J. McNicol, M. Bayer et al., "Highly parallel gene-to-BAC addressing using microarrays,"BioTechniques, vol. 50, no. 3, pp. 165–172, 2011.

69. S. G. Gregory, M. Sekhon, J. Schein et al., "A physical map of the mouse genome," Nature, vol. 418, no. 6899, pp. 743–750, 2002.

70. G. T. H. Vu, P. D. S. Caligari, and M. J. Wilkinson, "A simple, high throughput method to locate single copy sequences from Bacterial Artificial Chromosome (BAC) libraries using High Resolution Melt analysis," BMC Genomics, vol. 11, no. 1, article 301, 2010.

71. K. Madishetty, P. Condamine, J. T. Svensson, E. Rodriguez, and T. J. Close, "An improved method to identify BAC clones using pooled overgos," Nucleic Acids Research, vol. 35, no. 1, article e5, 2007.

72. M. Chen, G. Presting, W. B. Barbazuk et al., "An integrated physical and genetic map of the rice genome,"Plant Cell, vol. 14, no. 3, pp. 537–545, 2002.

73. M. N. Romanov, J. A. Price, and J. B. Dodgson, "Integration of animal linkage and BAC contig maps using overgo hybridization," Cytogenetic and Genome Research, vol. 102, no. 1–4, pp. 277–281, 2003.

74. J. Xu, D. Yang, J. Domingo, J. Ni, and N. Huang, "Screening for overlapping bacterial artificial chromosome clones by PCR analysis with an arbitrary primer," Proceedings of the National Academy of Sciences of the United States of America, vol. 95, no. 10, pp. 5661–5666, 1998.

75. J. Zheng, J. T. Svensson, K. Madishetty, T. J. Close, T. Jiang, and S. Lonardi, "OligoSpawn: a software tool for the design of overgo probes from large unigene datasets," BMC Bioinformatics, vol. 7, article 7, 2006

76. D. C. Jamison, J. W. Thomas, and E. D. Green, "ComboScreen facilitates the multiplex hybridization-based screening of high-density clone arrays," Bioinformatics, vol. 16, no. 8, pp. 678–684, 2000.

77. "Illumina golden gate assay genotyping web page," http://www.illumina.com/technology/goldengate_genotyping_assay.ilmn.

78. J. B. Fan, A. Oliphant, R. Shen et al., "Highly parallel SNP genotyping," Cold Spring Harbor Symposia on Quantitative Biology, vol. 68, pp. 69–78, 2003.

79. C. Rustenholz, P. E. Hedley, J. Morris et al., "Specific patterns of gene space organisation revealed in wheat by using the combination of barley and wheat genomic resources," BMC Genomics, vol. 11, article 714, 2010.

80. "HarvEST web page," http://www.harvest.ucr.edu/.

81. F. Cheung and C. D. Town, "A BAC end view of the Musa acuminata genome," BMC Plant Biology, vol. 7, article 29, 2007.

82. Y. Han and S. S. Korban, "An overview of the apple genome through BAC end sequence analysis," Plant Molecular Biology, vol. 67, no. 6, pp. 581–588, 2008.

83. L. Mao, T. C. Wood, Y. Yu et al., "Rice transposable elements: a survey of 73,000 sequence-tagged-connectors," Genome Research, vol. 10, no. 7, pp. 982–990, 2000.

84. R. Ragupathy, R. Rathinavelu, and S. Cloutier, "Physical mapping and BAC-end sequence analysis provide initial insights into the flax (Linum usitatissimum L.) genome," BMC Genomics, vol. 12, article 217, 2011.

85. M. Febrer, J. L. Goicoechea, J. Wright et al., "An integrated physical, genetic and cytogenetic map of Brachypodium distachyon, a model system for grass research," PLoS ONE, vol. 5, no. 10, Article ID e13461, 2010.

86. Y. Q. Gu, Y. Ma, N. Huo et al., "A BAC-based physical map of Brachypodium distachyon and its comparative analysis with rice and wheat," BMC Genomics, vol. 10, article 1471, p. 496, 2009.

87. Q. Yuan, F. Liang, J. Hsiao et al., "Anchoring of rice BAC clones to the rice genetic map in silico," Nucleic Acids Research, vol. 28, no. 18, pp. 3636–3641, 2000.

88. J. E. Bowers, M. A. Arias, R. Asher et al., "Comparative physical mapping links conservation of microsynteny to chromosome structure and recombination in grasses," Proceedings of the National Academy of Sciences of the United States of America, vol. 102, no. 37, pp. 13206–13211, 2005.

89. N. Huo, G. R. Lazo, J. P. Vogel et al., "The nuclear genome of Brachypodium distachyon: analysis of BAC end sequences," Functional and Integrative Genomics, vol. 8, no. 2, pp. 135–147, 2008.

90. S. Volik, S. Zhao, K. Chin et al., "End-sequence profiling: sequence-based analysis of aberrant genomes,"Proceedings of the National Academy of Sciences of the United States of America, vol. 100, no. 13, pp. 7696–7701, 2003.

91. B. J. Raphael, S. Volik, C. Collins, and P. A. Pevzner, "Reconstructing tumor genome architectures,"Bioinformatics, vol. 19, no. 2, pp. 162–171, 2003.

92. G. G. Mahairas, J. C. Wallace, K. Smith et al., "Sequence-tagged connectors: a sequence approach to mapping and scanning the human genome," Proceedings of the National Academy of Sciences of the United States of America, vol. 96, no. 17, pp. 9739–9744, 1999.

93. J. O. Korbel, A. E. Urban, J. P. Affourtit et al., "Paired-end mapping reveals extensive structural variation in the human genome," Science, vol. 318, no. 5849, pp. 420–426, 2007.

94. P. J. Campbell, P. J. Stephens, E. D. Pleasance et al., "Identification of somatically acquired rearrangements in cancer using genome-wide massively parallel paired-end sequencing," Nature Genetics, vol. 40, no. 6, pp. 722–729, 2008.

95. K. J. McKernan, H. E. Peckham, G. L. Costa et al., "Sequence and structural variation in a human genome uncovered by short-read, massively parallel ligation sequencing using two-base encoding,"Genome Research, vol. 19, no. 9, pp. 1527–1541, 2009.

96. T. R. Gregory, J. A. Nicol, H. Tamm et al., "Eukaryotic genome size databases," Nucleic Acids Research, vol. 35, no. 1, pp. D332–D338, 2007.

97. H. R. Kim, B. Hurwitz, Y. Yu et al., "Construction, alignment and analysis of twelve framework physical maps that represent the ten genome types of the genus Oryza," Genome Biology, vol. 9, no. 2, article R45, 2008.

98. N. M. Springer, K. Ying, Y. Fu et al., "Maize inbreds exhibit high levels of copy number variation (CNV) and presence/absence variation (PAV) in genome content," PLoS Genetics, vol. 5, no. 11, Article ID e1000734, 2009.

99. R. A. Swanson-Wagner, S. R. Eichten, S. Kumari et al., "Pervasive gene content variation and copy number variation in maize and its undomesticated progenitor," Genome Research, vol. 20, no. 12, pp. 1689–1699, 2010.

CHAPTER 7 CHROMOSOMES OF BRACHYPODIUM

Dominika Idziak Alexander Betekhtin, Elzbieta Wolny, Karolina Lesniewska, Jonathan Wright, Melanie Febrer, Michael W. Bevan, Glyn Jenkins Robert Hasterok

INTRODUCTION

Chromosome painting (CP), also known as chromosomal in situ suppression, is one of the most powerful and spectacular tools of molecular cytogenetics. It refers to the method of visualisation of specific chromosome regions or entire chromosomes using fluorescence in situ hybridisation (FISH) with chromosome-specific DNA probes (Pinkel et al. 1988; Ried et al. 1998). For many years, this technique was successfully applied only to studies of animal chromosomes, giving unprecedented insight into the structural and functional compartmentalisation of the nucleus (Cremer et al. 2008; Cremer et al. 1993; Kurz et al. 1996). It has also been widely used for molecular cytotaxonomy, for example of primates (Muller et al. 1999; Wienberg et al. 1990), mutagenicity testing (Marshall and Obe 1998) and radiation biology (Gray et al. 1992). High resolution CP-based analyses have become especially important to pre-, neo- and postnatal clinical diagnostics of various chromosomal aberrations associated with human diseases and malignancies (Gray and Pinkel 1992; Popp et al. 1991; Tkachuk et al. 1991; Weise et al. 2003; Xu et al. 2010).

For more than a decade, efforts to apply CP in plants failed to yield satisfactory results, which was attributed to the presence of vast amounts of ubiquitous repetitive DNA in most plant genomes (Schubert et al. 2001). The first true painting of plant chromosomes was achieved by Lysak et al. (2001) in the dicotyledonous model organism, *Arabidopsis thaliana*, whose nuclear genome had just been sequenced (Arabidopsis Genome Initiative 2000). The availability of genome sequence and ordered libraries of genomic DNA in BAC (Bacterial Artificial Chromosome) vectors, along with some features of *A. thaliana* genome, such as its compactness, low amount of repetitive DNA and low number of chromosomes, were key prerequisites for successful painting of its chromosomes. Initially restricted to the chromosomes of *A. thaliana* (Pecinka et al. 2004), CP in later studies was also successfully applied to other closely related members of the Brassicaceae. This comparative chromosome painting (CCP) allowed unprecedented analyses of the evolution of the genomes of this group of plants at the chromosomal level (Lysak et al. 2006; Lysak et al. 2010; Mandakova and Lysak 2008).

In contrast to the dicots, chromosome painting has not been applied so far to any monocotyledonous plant. Despite the availability of its long-published genomic sequence (Goff et al. 2002), no chromosome painting has been performed in rice. Another species of Poaceae,*Brachypodium distachyon*, emerged several years ago as a model system for temperate cereals and grasses. Like *A. thaliana*, *B. distachyon* possesses numerous 'model' attributes, including inter alia a small (~300 Mb) genome with a low ($x=5$) basic chromosome number, small physical stature, self-fertility, a short life cycle and undemanding growth requirements (Draper et al. 2001; Garvin et al. 2008). The genome of *B. distachyon* has been extensively studied at the chromosomal level using fluorescence in situ hybridisation (FISH) with a wide range of DNA probes, which enabled unambiguous identification of all the chromosomes and their arms of the complement (Hasterok et al. 2004; Hasterok et al. 2006).

The well-established cytogenetic platform, combined with resources such as BAC DNA libraries and bioinformatic data from the recently completed genome sequencing project (Febrer et al.2010; Gu et al. 2009; International Brachypodium Initiative 2010), opened up the possibility of painting for the first time the chromosomes of a monocot species. In this

Citation: Dominika Idziak, Alexander Betekhtin, Elzbieta Wolny, Karolina Lesniewska, Jonathan Wright, Melanie Febrer, Michael W. Bevan, Glyn Jenkins, Robert Hasterok, chromosomes of Brachypodium, DOI 10.1007/s00412-011-0326-9.

paper, we present the current status of CP in *B. distachyon*, which allows reliable recognition and tracking of individual chromosome regions during different stages of meiosis. Additionally, we show that CCP with BAC pools representing individual *B. distachyon* chromosomes can be used as an effective tool for studying karyotype evolution of its close relatives in the genus *Brachypodium*. Possible future prospects are also discussed.

MATERIALS AND METHODS

Plant material and chromosome preparations

Three cytotypes of *B. distachyon* were used in this study. Detailed information on the plant material is provided in Table 1. Seeds were sown at high density in compost. All plants were grown as described in Jenkins and Hasterok (2007). The plants of cytotypes ABR114 and ABR113 did not require vernalisation and usually reached the generative stage of their life cycle within 1 month after sowing.

Table 1: The original identities, sources, origins and chromosome numbers of the *B. distachyon*material used in this study

Species	**Accession no.**	**Origin**	**Source**	**2*n***
B. distachyon	Bd21	Iraq	*a*	10
*'B. distachyon'**	ABR114	Spain	*b*	20
*'B. distachyon'**	ABR113	Portugal	*b*	30

a US Department of Agriculture–National Plant Germplasm System, USA, *b* Institute of Biological, Environmental and Rural Sciences, Aberystwyth University, UK

*According to our previous cytomolecular studies (Hasterok et al. 2004; Hasterok et al. 2006) ABR114 and ABR113 represent, respectively, a different unnamed diploid and an allotetraploid species within the *Brachypodium* genus. As their proposed taxonomical status is not yet officially accepted, we refer in this paper to ABR114 and ABR113 as the cytotypes of '*B. distachyon*'. The cytotype with $2n = 10$ chromosomes is referred to as *B. distachyon*without apostrophes

Mitotic chromosome preparations were made using the methodology described by Hasterok et al. (2006). Preparation of anther tissue for meiotic chromosome squashes was adopted from Jenkins and Hasterok (2007) with minor modifications. In brief, immature inflorescences were collected, fixed immediately in fresh 3:1 absolute methanol:glacial acetic acid mixture for 3 × 24 h at room temperature (RT) and stored at −20°C until required. Individual anthers were isolated and washed in 10 mM citric acid–sodium citrate buffer and digested enzymatically for 2 h at 37°C in a mixture comprising 10% (*v/v*) pectinase (Sigma), 0.65% (*w/v*) cellulase Onozuka R-10 (Serva), 0.5% (*w/v*) cellulase (Calbiochem), 0.15% (*w/v*) cytohelicase and 0.15 % (*w/v*) pectolyase in 10 mM citrate buffer. Several anthers per slide were squashed in drops of 45% acetic acid, frozen, postfixed in 4% formaldehyde for 10 min, washed briefly in distilled water and air dried.

Selection of BAC clones for chromosome painting

BAC clones used for CP were selected from the five assemblies of FPCs (FingerPrinted Contigs) previously aligned to the *B. distachyon* karyotype (Febrer et al. 2010; International Brachypodium Initiative 2010). Clones from centromeric and pericentromeric regions were excluded from the painting experiments as they were likely to contain large amounts of highly repetitive and dispersed DNA sequences which might cause unspecific hybridisation signals. For the same reason, the BACs constituting the assemblies were examined for repetitive DNA content. Low repeat BAC clones were identified using the same method as described by Febrer et al. (2010). With a few exceptions, clones chosen for painting experiments contained less than 30% of repeats. The number of BACs selected for each arm of every chromosome in the complement is given in Table 2. The total length of BAC clones spanning a chromosome arm ranged from 9.7% of total arm length for chromosome 5 short (top) arm (Bd5S) to nearly 34% for the long (bottom) arm of *B. distachyon* chromosome 3 (Bd3L; Table 2).

Table 2: Characteristics of the BAC pools used for painting *B. distachyon* chromosome arms

Chromosome	**Bd1**		**Bd2**		**Bd3**		**Bd4**		**Bd5**	
Arm	**S**	**L**	**S**	**L**	**S**	**L**	**S**	**L**	**S**	**L**
Number of clones selected	70	72	26	29	42	54	20	39	4	19
Average distance between neighbouring clones (kbp)	347	358	874	707	385	377	594	406	381	543
Percentage of arm length covered by BAC inserts	28.1	28.7	13.1	12.3	24.6	33.9	12.8	19.5	9.7	13.5

Selected BACs were divided into pools of eight to ten clones each. Instead of isolating DNA from each clone separately, liquid cultures of bacteria carrying clones belonging to the same pool were mixed and DNA was isolated. Individual pools were labelled with either digoxigenin- or tetramethyl-rhodamine-dUTP and in order to verify their position on *B. distachyon*chromosomes, hybridised independently to pachytene or zygotene nuclei (data not shown). In most cases, the hybridisation sites of the selected clones corresponded to their predicted positions on the FPC-derived physical map. The pool assigned to the interstitial region of the Bd4S yielded dispersed signals in the entire chromosome set, probably due to the presence of highly repetitive and ubiquitous DNA in some of the clones. BACs constituting this pool were then examined separately using FISH, and the clones responsible for cross-hybridisation were identified and removed from the pool before subsequent experiments.

Probe labelling and fluorescence in situ hybridisation

BAC DNA was isolated using a standard alkaline method and labelled by nick translation with digoxigenin-11-dUTP (Roche) for short chromosome arms and with tetramethyl-rhodamine-5-dUTP (Roche) for long chromosome arms as described in Hasterok et al. (2002) with the exception of the experiment shown on Figs. 1f and 3. The list of BACs comprising the pools for individual chromosomes may be requested from the authors.

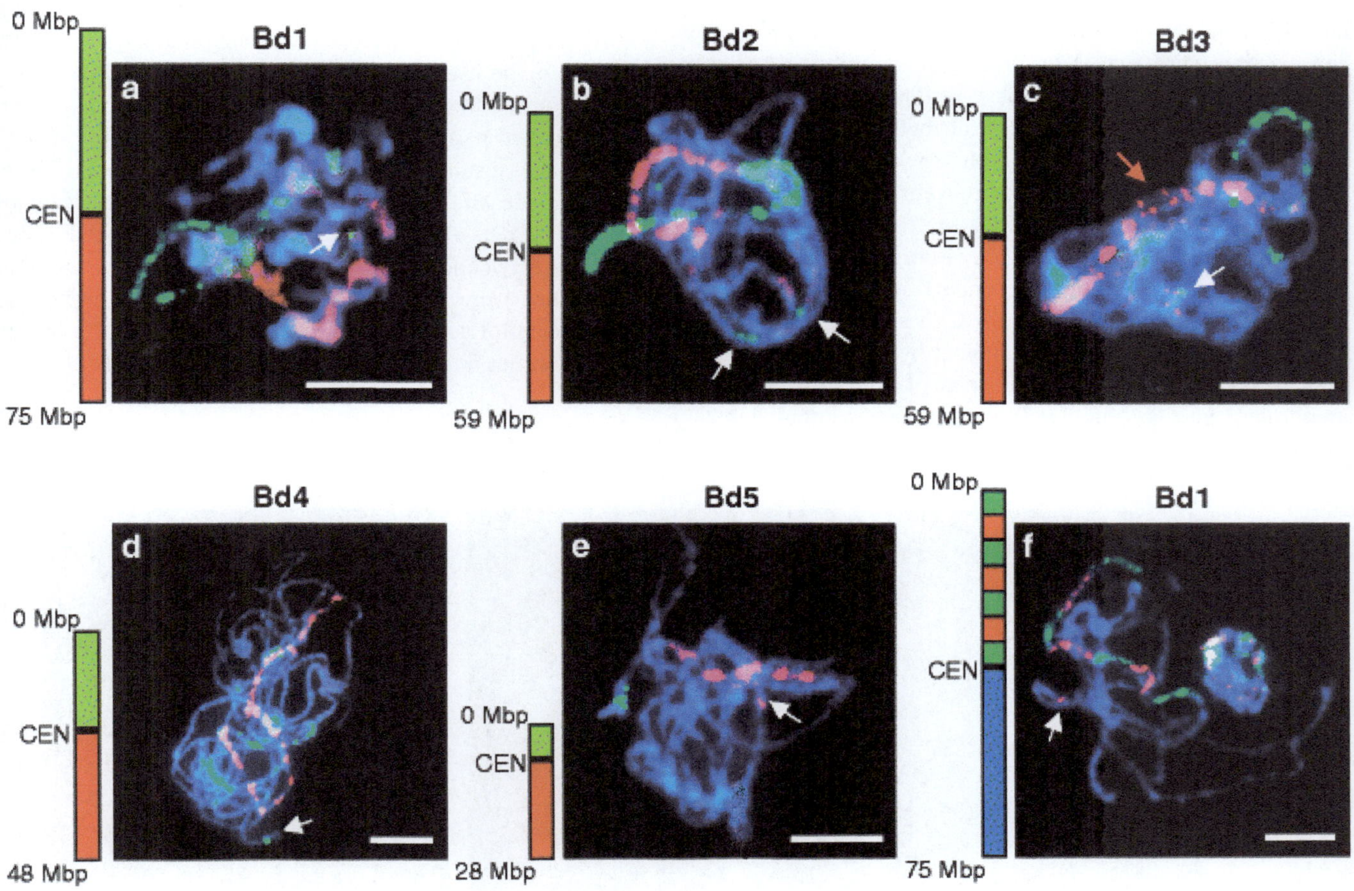

Figure 1: Chromosome painting in PMCs of *B. distachyon* at late zygotene/pachytene. **a–e** Painting using BAC pools spanning short (*green*) and long (*red*) arms of a given chromosome. The ideograms next to the photographs indicate the approximate physical positions of the pools used. *CEN*=centromere. **a** chromosome Bd1; **b** chromosome Bd2; **c** chromosome Bd3; **d**chromosome Bd4; **e** chromosome Bd5; **f** painting of the short arm of chromosome Bd1 with 70 BAC clones arranged into seven alternating pools. *White arrows* indicate additional unexpected signals. The *red arrow* in **c** indicates the position of an unpaired chromosome region on Bd3. All chromosomes were counterstained with DAPI. *Bars*=5 µm

The FISH procedure was carried out according to the protocol of Jenkins and Hasterok (2007) with minor modifications. The BAC DNAs were pooled, precipitated and dissolved in hybridisation mixture consisting of 50% deionised formamide and 10% dextran sulphate in 2× saline sodium citrate (SSC). Chromosome preparations and pre-denatured (10 min at 75°C) hybridisation mixtures were denatured together for 4.5 min at 73°C and allowed to hybridise in a humid chamber for about 20 h at 37°C. Post-hybridisation washes were performed in 10% formamide in 0.1× SSC at 42°C, which is equivalent to 79% stringency. The hybridisation signals were detected by anti-digoxigenin-fluorescein conjugated antibodies (Roche) or—in the case of tetramethyl-rhodamine-5-dUTP—directly visualised. For comparative chromosome painting, the hybridisation time was occasionally extended up to 48–55 h. The chromosomes were mounted and counterstained in Vectashield (Vector Laboratories) containing 2.5 mg/ml of 4′,6-diamidino-2-phenylindole (DAPI; Serva).

All microphotographs were acquired using a monochromatic CCD camera attached to a wide-field epifluorescence microscope and digitally processed and superimposed using Wasabi (Hamamatsu Photonics), Photoshop (Adobe) and Picture Publisher (Micrografx/Corel) software.

RESULTS

Painting of *B. distachyon* chromosomes

To paint a chromosome along its entire length, pools of BAC DNA were combined into arm-specific probes. Discrimination between the arms of a chromosome was enabled by differential labelling of the probes specific for its short and long arm. Labelled DNA was mapped to pachytene or late zygotene spreads which provide the best resolution for physical mapping by fluorescence in situ hybridisation. Every painting probe gave a strong and distinct hybridisation signal allowing unambiguous identification and tracking of homologous chromosome pairs (Fig. 1). In some cases, unspecific signals on other chromosomes were also observed. Most of such signals were faint and presumably could be attributed to the presence of repetitive DNA in the painting probe. However, signals of high fluorescence intensity also appeared in chromosomal positions in addition to those expected (Fig. 1a–f; *white arrows*). This suggests misalignment of several clones in the FPC contig map. Due to exclusion from the experiments of high repeat clones from the pools, the BACs used for CP rarely overlapped, but usually were separated from each other by tens or hundreds of kilobases. These breaks in the tiling path of clones are visualised on chromosomes as blue, DAPI-stained gaps without signals. The exclusion of centromeric and pericentromeric BACs, as well as clones containing 45S rDNA repeats, was particularly noticeable as large unlabelled chromosomal segments (Fig. 1c, e). The discrimination between different chromosome regions was achieved by differential labelling of BAC pools. Simultaneous hybridisation of 70 BAC clones arranged into seven alternating pools resulted in a banding pattern of red and green segments along the short arm of chromosome Bd1 (Fig. 1f). Similar results were obtained after differential labelling of BAC pools assigned to other chromosomes of the karyotype (data not shown). The size of a particular block was always consistent with the length of the BAC assembly covered by the respective pool, and the sequential order of the pools along the chromosome was easily discernible.

A pool of 40 clones corresponding to the distal part of the short arm of chromosome Bd1 was hybridised to chromosomes at different stages of meiosis (Fig. 2). The low condensation and frequent entanglement of the chromatin at zygotene (Fig. 2a) often made it difficult to track the painted region among other chromosomes. At pachytene (Fig. 2b), the arrangement of the labelled part of chromosome Bd1 was much easier to follow. Gaps resulting from excluded BACs were visible at both zygotene and pachytene nuclei. During later stages of meiosis (Fig. 2c–e), the gaps were no longer detectable due to increasing condensation of the chromosomes. Similar observations were made when the same painting probe was applied to mitotic metaphase chromosomes (Fig. 2f), which are 30–40 times as condensed as pachytene bivalents.

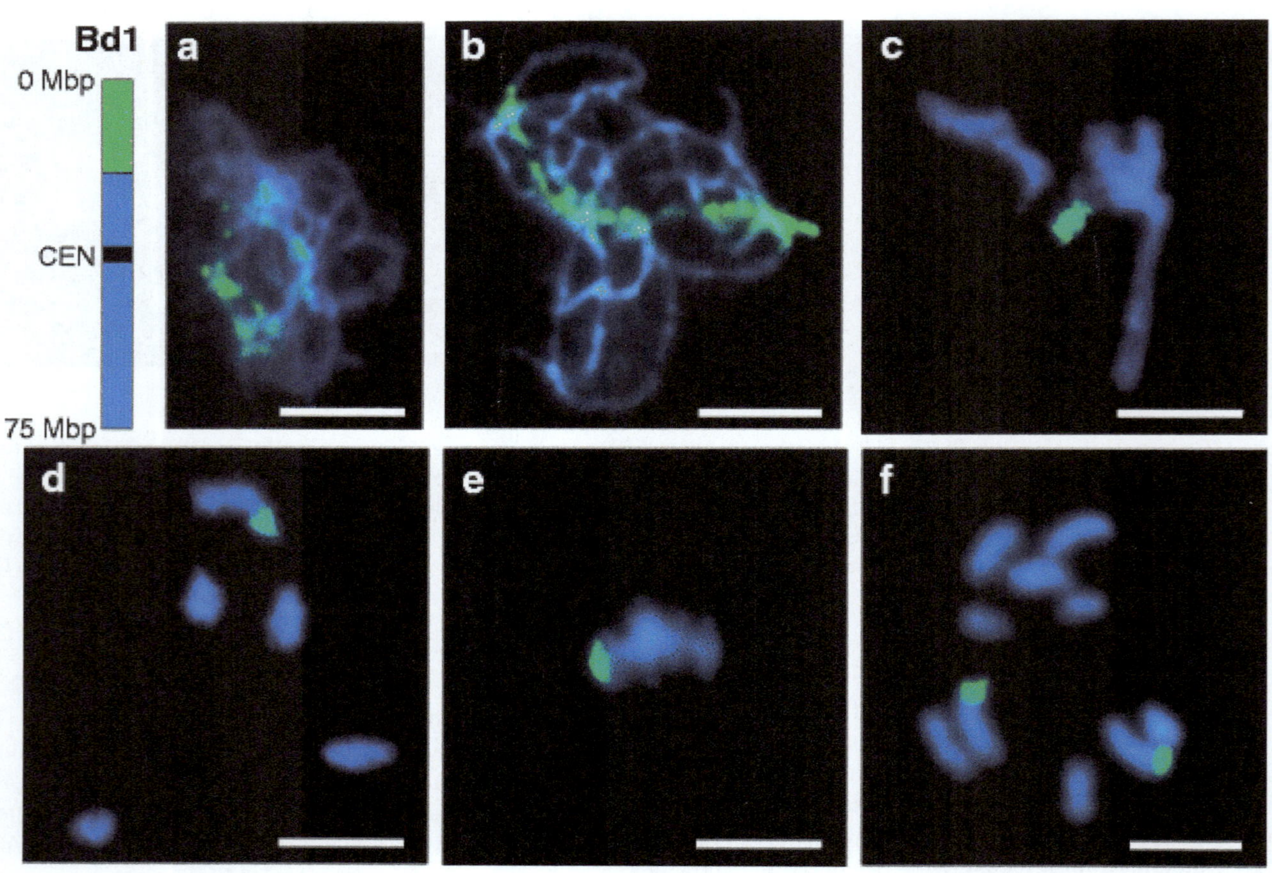

Figure 2: Tracking a distal region of Bd1 short arm (*green*) at different stages of meiosis and mitotic metaphase using chromosome painting with a BAC pool containing 40 clones spanning 19.5 Mbp. The ideogram next to the photographs indicates the approximate physical position of the pool on the chromosome. *CEN* = centromere. **a** zygotene; **b** pachytene; **c**diplotene; **d** diakinesis; **e** meiotic metaphase I; **f** mitotic metaphase. All chromosomes were counterstained with DAPI. *Bars* = 5 μm

In interphase nuclei of anther tissues, chromosome painting identified the territories occupied by different chromosomes. Simultaneous hybridisation of differentially labelled chromosomes Bd1 and Bd2 revealed frequent association of homologues (Fig. 3a), while heterologous chromosomes were usually separated (Fig. 3a–b). Analyses of the respective positions of the short and long arms of Bd1 indicated that association of both (Fig. 3c) or only one chromosome arm (Fig. 3d) is common in interphase nuclei, although two distinct nuclear domains corresponding to each of the Bd1 homologues were also occasionally observed (Fig. 3e). When seven alternating BAC pools specific for different regions of the Bd1 short arm were used as a hybridisation probe, the result was visible as a mixture of green and red signals that could not be attributed to particular pools (Fig. 3f).

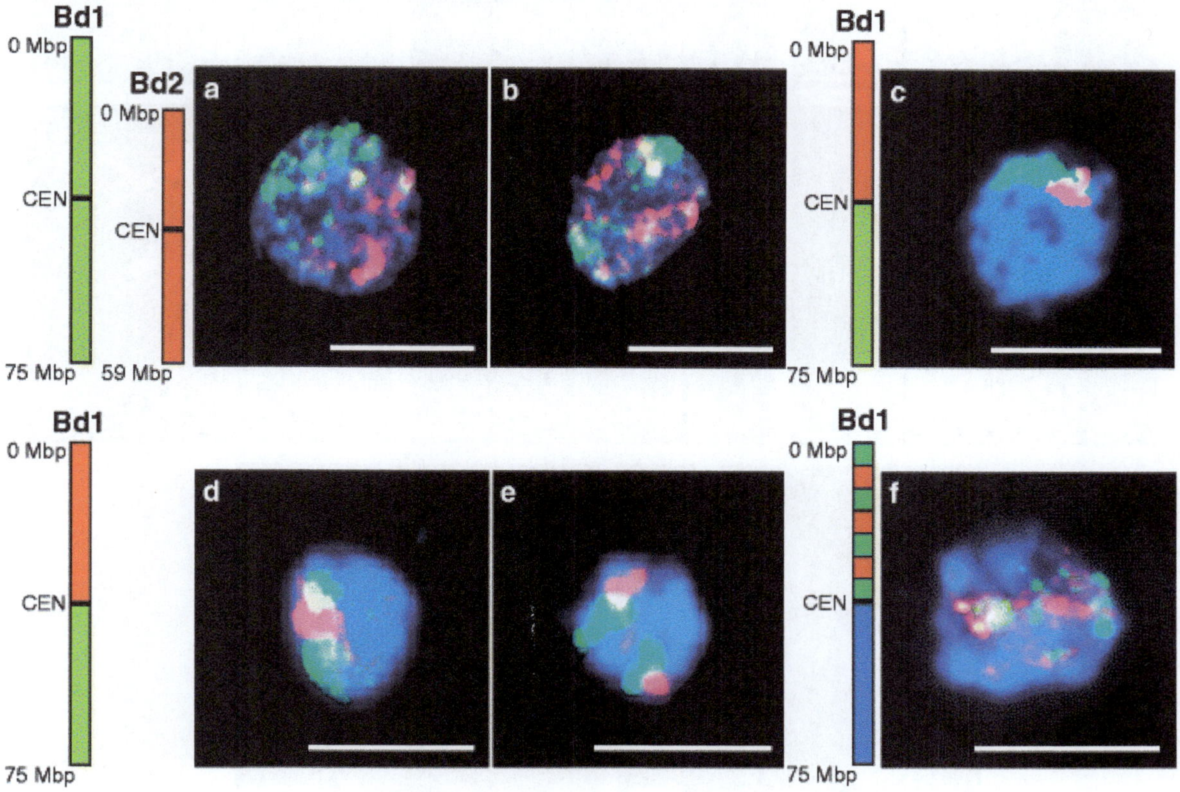

Figure 3: Identification of chromosome territories in interphase nuclei of *B. distachyon* anther tissue using chromosome painting. **a–b** Simultaneous painting of chromosomes Bd1 (*green*) and Bd2 (*red*). **a** Interphase nucleus showing association of homologous chromosomes while territories occupied by heterologues remain separated; **b** interphase nucleus showing separation of all four chromosome territories; **c–e** painting of the short (*red*) and long (*green*) arm of chromosome Bd1; **c** association of territories of both arms;**d** association of territories of the short arm; **e** separation of territories occupied by chromosome Bd1; **f** chromosome painting of interphase nuclei with seven alternating BAC pools specific to different regions of Bd1 short arm: the result is visible as a mixture of red and green signals and the territories occupied by particular pools are not discernible. The drawings next to the photographs indicate approximate positions of the BAC pools used for painting. *CEN* = centromere. All chromosomes were counterstained with DAPI. *Bars* = 5 µm

Comparative Chromosome Painting

The feasibility of using the BAC pools for comparative physical mapping of other*Brachypodium* species was tested by cross-hybridising pools that represented each chromosome of Bd21 to pachytene bivalents of two '*B. distachyon*' cytotypes: diploid ABR114 ($2n = 20$) and allotetraploid ABR113 ($2n = 30$). The probe comprising BAC clones specific for chromosome Bd1 identified two bivalents in ABR114 PMCs at pachytene (Fig. 4a1). Both bivalents were painted green and red, corresponding to the short and long arms of Bd1, respectively. In both bivalents, terminal regions of the bivalents proximal to the green segments were unlabelled. Hybridisation of the same BAC pools to ABR113 highlighted three bivalents (Fig. 4b1). The painted bivalents differ in size, one (designated I) was at least twice as long as the other two (II and III). The hybridisation pattern on the three homoeologues was similar, with the green signal corresponding to the short arm of Bd1 abutting the red signal representing the long arm of Bd1. However, whilst the bivalents numbered I and II were painted along their entire length, bivalent number III was only partially decorated by the probe for Bd1. The border between labelled and unlabelled regions of bivalent II is clearly visible (Fig. 4b1, red arrow), but the length of the non-painted chromosome segment was difficult to discern due to the entanglement of the pachytene chromosomes.

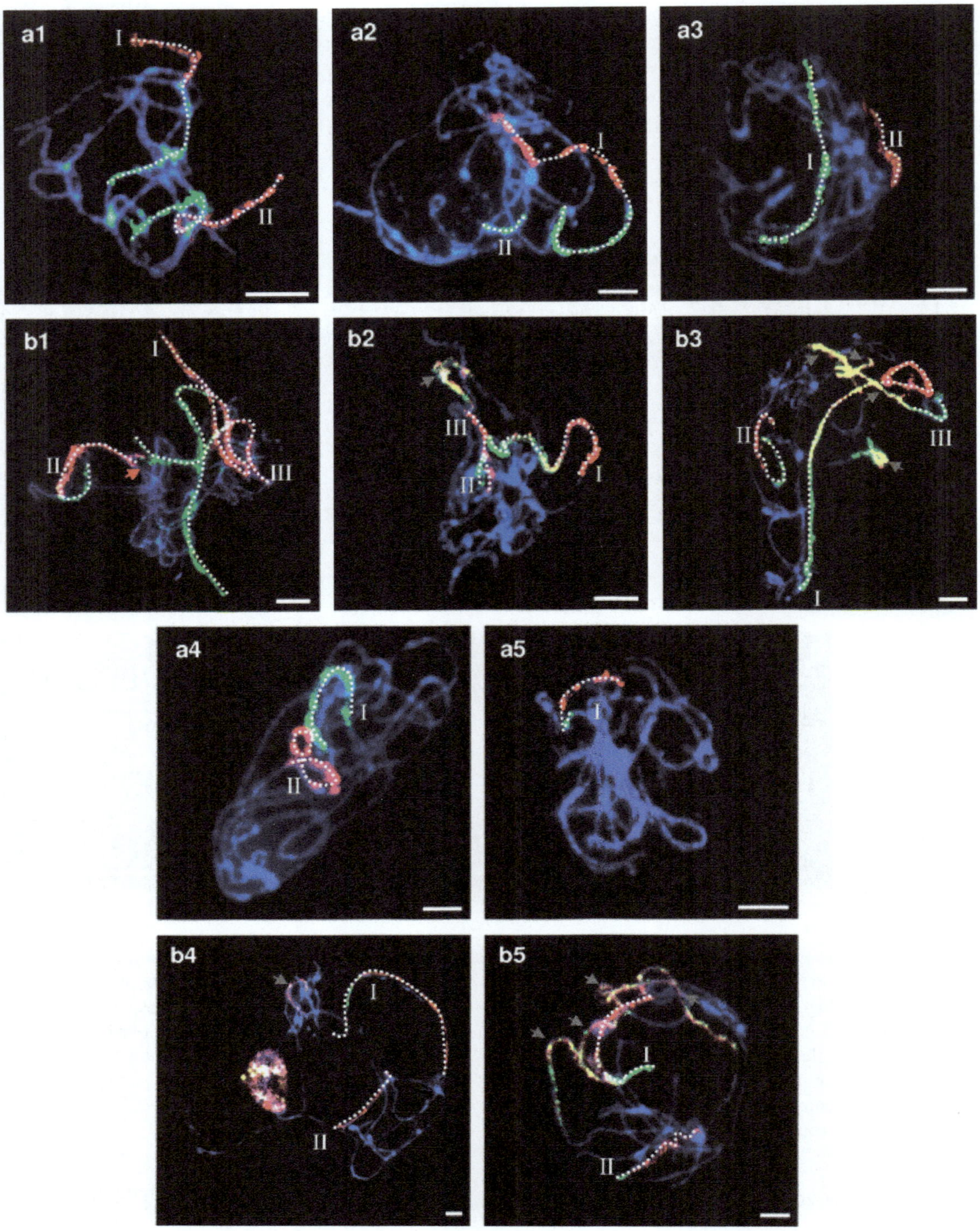

Figure 4: Comparative chromosome painting of *'B. distachyon'* cytotypes ABR114 and ABR113 using BAC pools spanning the short (*green*) and long (*red*) arms of the five *B. distachyon*chromosomes. **a1–a5** chromosome painting of ABR114 pachytene chromosomes; **b1–b5**chromosome painting of ABR113 pachytene. *Roman numerals* indicate homoeologues of a given chromosome in the cytotypes. The *red arrow* in **b1** indicates the border between labelled and unlabelled regions of the homoeologue III in ABR113. The labelled bivalent axes have been highlighted with dots to emphasise continuity. *Grey arrows* indicate non-specific hybridisation. All chromosomes were counterstained with DAPI. *Bars* = 5 µm

Painting with probes specific for the long and short arms of Bd2 revealed two bivalents in ABR114 (Fig. 4a2). One of them (designated I) has long tracts of green and red signal, representing the short and long arms of Bd2, respectively. A short fragment of another bivalent (II) was painted green only. Analysis of the Bd2 hybridisation pattern to ABR113 (Fig. 4b2) showed contiguous painting of the entire arms of one bivalent (I). Additional short green (II) and short red (III) segments were found to belong to the same bivalent. An additional pair of bivalents is consistently labelled as a mosaic of green, red and colocalising yellow signals.

The painting probes comprising BAC clones specific for short and long arm of chromosome Bd3 identified two bivalents in ABR114 PMCs at pachytene (Fig. 4a3). One of them (I) hybridised only with the pool representing the short arm of Bd3, whilst the other (II) only with the pool for the long arm. Bd3-specific probes painted three pachytene bivalents of ABR113 (Fig. 4b3). Bivalents I and II differed significantly in terms of signal length, but both had green and red segments corresponding to the short and long arms of Bd3. The former also has an interstitial yellow tract, presumably delimiting sequences in common between the two pools. Bivalent III has a short green segment adjoining a longer yellow tract, presumably representing non-specific repeat sequences common to the two arms. A short green and yellow tract was also observed.

Chromosome painting of Bd4 probes to ABR114 chromosomes highlighted two bivalents (Fig. 4a4), one (I) hybridising with the Bd3 short arm probe, and the other (II) with the long arm probe. The two tracts do not belong to the same bivalent. The probe comprising BAC clones specific for Bd4 identified only two bivalents in ABR113 (Fig. 4b4). One bivalent (I), presumably derived from a *B. distachyon*-like genome, hybridised with both the Bd4-short arm and Bd4-long arm probe, whilst the other (II) showed only red fluorescence.

Bd5-specific paints hybridised with only one bivalent of ABR114 (Fig. 4a5; I), whereas they hybridised to two bivalents in ABR113 (Fig. 4b5; I and II). In addition, some of the other bivalents comprise patchworks of green, red and yellow signals revealing the sites of sequences common to several of the chromosomes of the complement.

A summary of the pattern of hybridisation of the five chromosome-specific paints to the two '*B. distachyon*' cytotypes is shown in Fig. 5. Since the continuity of unlabelled bivalents cannot be tracked with certainty in most cases, the diagrammatic representations of such bivalents have been left open ended. The relative lengths of bivalents in the three cytotypes is subject to stage-specific variation.

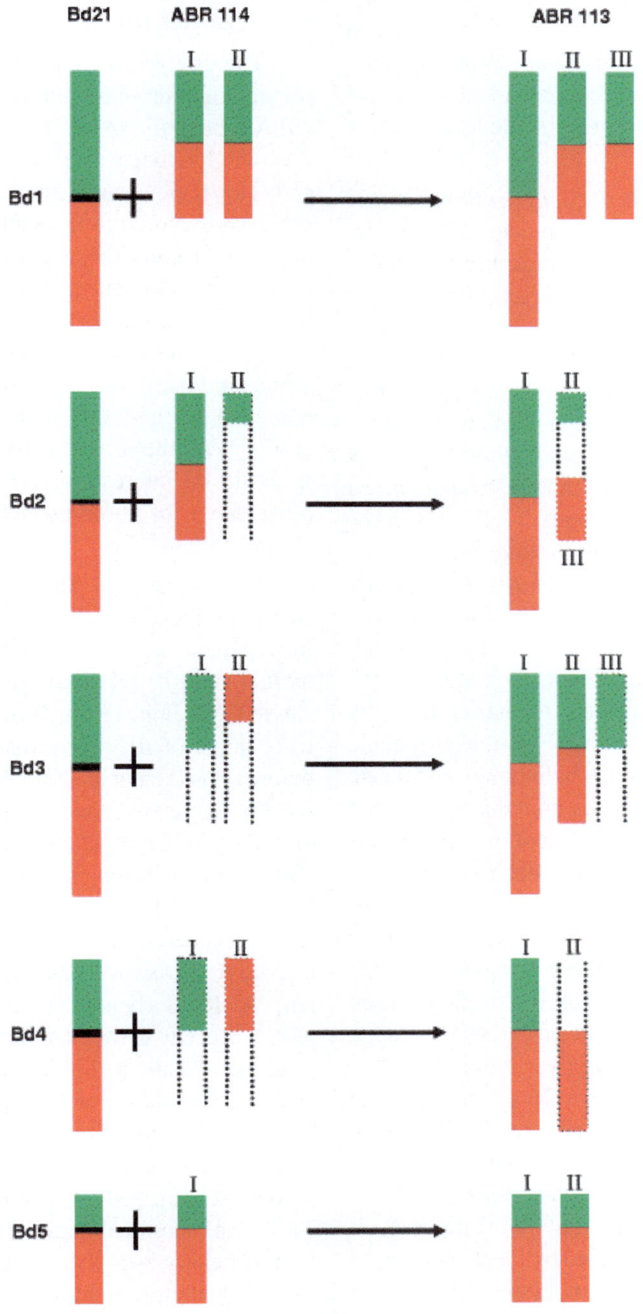

Figure 5: Diagrammatic summary of the pattern of chromosome painting in pachytene bivalents of Bd21, ABR114 and ABR113. Hybridisation with short-arm probes is shaded *green*, and long arm probes *red*. Centromeres are marked with *horizontal black bars*. Bivalents that cannot be traced from end to end have *dotted open-ended lines*. Relative lengths are only approximate. *Roman numerals* correspond to those in Fig. 4

DISCUSSION

CP in plants has not been demonstrated since the first painting of *A. thaliana* chromosome 4 by Lysak et al. (2001) and subsequent painting of the rest of the complement as well as the complements of related species within the Brassicaceae (Lysak and Lexer 2006). The main requirement for successful chromosome painting is the availability of a contiguous tiling path of sequenced large-insert clones integrated with the karyotype of a particular species. Moreover, the effectiveness and specificity of the technique depend upon eliminating highly repetitive DNA sequences from the painting probe. CP is potentially useful, therefore, only in species with small genomes and with repetitive DNA localised in pericentromeric regions. *B. distachyon*, a recently established model grass species, has one of the smallest genomes in the Poaceae and very low repetitive DNA content. Recently developed resources for investigating its genome, including genetic, physical and cytogenetic maps fully integrated with genome sequence assemblies (Febrer et al. 2010; International Brachypodium Initiative 2010), have made this species an ideal candidate for CP, thus enabling the application of this method to study monocots.

A crucial step in making CP of *B. distachyon* chromosomes feasible was the selection of the chromosome-specific, single-locus BAC clones. One of the available approaches to identify clones that contain low amount of interspersed DNA repeats and thus presumably map to single locus in the genome is dot-blot or Southern-blot hybridisation of labelled genomic DNA to all analysed BACs and subsequent selection of clones that have less intensive hybridisation signals (Jenkins and Hasterok 2007; Kim et al. 2005; Pedrosa-Harand et al. 2009; Walling et al. 2006). However, it was demonstrated that the strategy based on in silico sequence scanning can be equally efficient in the evaluation of repeated sequence content in large insert clones (Kim et al.2005). In the case of *A. thaliana,* the suitability of the clones for CP was double-tested using bioinformatic analyses of BAC sequences as well as dot-blot hybridisation (Lysak et al. 2003). In the presented work, BACs from FPC contigs were assayed for interspersed repeat content by comparison of their sequence data with the Pooideae-specific Repeat Element Database (http://mips.helmholtz-muenchen.de/plant/recat/) from MIPS. Only clones containing less than 30% of DNA repeats within their inserts were included in the painting probes. As it turned out, this approach was in most cases sufficient to guarantee the chromosomal specificity of selected BACs. As a result of setting the threshold of permitted repeated DNA amount at 30%, the number of clones selected for painting was much lower than the number of clones used for CP of *A. thaliana* chromosomes (Pecinka 2005), despite the genome size of *B. distachyon* being over twice that of *A. thaliana* (Arabidopsis Genome Initiative 2000; International Brachypodium Initiative 2010). These clones rarely constituted a contiguous tiling path, but rather were distributed along chromosomes at intervals of ~347 to ~874 kbp on average, depending on the chromosome arm (Table 2). This lack of continuity of probe sequence was reflected by the discontinuity of the hybridisation signal but did not hamper the identification and tracking of any painted chromosome. It proves that CP is feasible even where the total length of BAC inserts used for painting is less than one third the length of the painted chromosome if the clones used are spread uniformly along its arms (Table 2).

The painting probes enabled not only discrimination of individual chromosomes of *B. distachyon* but also delimited different chromosome regions. As shown in Fig. 2, any chromosomal segment can be unambiguously identified and observed in PMCs at different developmental stages. In late zygotene cells, CP highlighted paired and unpaired chromosome regions (Figs. 1c and 4b1, red arrows). This means that CP can be used to investigate in detail the pairing behaviour of homologous chromosomes during early prophase I. Coupled with immunolabelling of recombination proteins, there is the potential to correlate the assembly of meiotic bivalents with recombination events, as has been recently demonstrated for barley (Phillips et al. 2010). The development of a similar assay for *B. distachyon* based on immunolocalisation of key meiotic proteins and chromosome painting could allow physical mapping of recombination events to specific chromosomes and their segments with unprecedented resolution and precision.

The analysis of chromosome territories in somatic and premeiotic interphase nuclei is another possible application of CP. Although the spatial distribution of particular chromosome domains has been extensively studied in human and animal cells (for review, see Cremer and Cremer2010), this topic has not been explored to the same extent in plants. Painting of *A. thaliana*chromosomes significantly broadened our knowledge of nuclear architecture in plants, showing that save for NOR-bearing chromosomes there is no association of homologues at interphase (Pecinka et al. 2004) and that the three-dimensional arrangement of nuclear chromosome domains is similar in meristematic and differentiated tissues (Berr and Schubert 2007). With the advent of painting probes for *B. distachyon*, it would be interesting and profitable to compare territorial organisation of chromosomes in the nuclei of a dicot and a monocot plant.

After successful painting of *B. distachyon* chromosomes, we focused on determining whether the same probes had utility for comparative mapping in the genus *Brachypodium* as in *A. thaliana* and its relatives. The effectiveness of cross-mapping single BAC clones from *B. distachyon* genomic libraries to the wide range of *Brachypodium* species has been demonstrated before (Hasterok et al. 2006; Wolny et al. 2011). In order to test whether large BAC pools can be cross-mapped with the same efficiency as single clones, we hybridised the painting probes (Bd1-5) to the chromosomes of two '*B. distachyon*' cytotypes, known as ABR114 ($2n=20$) and ABR113 ($2n=30$). These were considered until recently to be polyploid derivatives of diploid *B. distachyon* ($2n=10$); ABR114 being autotetraploid, and ABR113 being autohexaploid cytotypes (Robertson 1981). Recent cytomolecular analyses suggest that ABR114 is rather a distinct diploid species, while ABR113 is in fact an allotetraploid formed by natural hybridisation between two diploid progenitors, one of which is similar to *B. distachyon* with $2n=10$ chromosomes and the other resembling ABR114 (Hasterok et al. 2004; Hasterok et al.2006; Wolny et al. 2011). As the taxonomical status of the two cytotypes has not yet been officially established, comparative chromosome painting should shed more light on their true phylogeny.

CCP with the Bd1 painting probe showed that *B. distachyon* chromosome 1 is represented in the ABR114 genome by two different chromosomes bearing close structural similarity to one another as well as to their homoeologue in the *B. distachyon* karyotype. Results of the experiment with the Bd4-specific painting probes are in concordance with the data obtained earlier (Hasterok et al. 2006) for single *B. distachyon* BAC clones, as we observed signals specific for the short and for the long arm of Bd4 on two different bivalents in ABR114. A similar pattern was observed for Bd3-specific probes in ABR114. However, Bd5 probes specific for the short and long arms were represented in ABR114 on a single bivalent homoeologous to a *B. distachyon*-like genome. Based on earlier experiments of landing single *B. distachyon* BAC clones (Hasterok et al. 2006), it would be expected that the painting pattern of the Bd2-specific probes in ABR114 should resemble that of Bd3 and Bd4. Surprisingly, in ABR114 signals for the short and for the long arm of Bd2 are present on the same bivalent. Additionally, there is a small fragment corresponding to a part of the short arm on another bivalent. One possible explanation is that the short green fragment on the bivalent designated II resulted from a local translocation event and the BAC clone ABR1-41-E10 used in the previous experiment landed in the translocated region. The results of CCP seem to support our previous conclusion that multiple fusion/fission events involving centromeric regions of the chromosomes were one of the mechanisms shaping the present structure of the karyotypes of ABR114 and *B. distachyon* (Hasterok et al. 2006). The painting patterns of Bd1-, Bd2- and Bd5-specific probes in ABR114 do not contradict the fusion/fission hypothesis because it cannot be ruled out that different chromosomes of the *B. distachyon* karyotype have evolved in different ways and that various chromosomal rearrangements, e.g., translocations or duplications also played an important role in the evolution of the karyotypes of *B. distachyon* and ABR114.

The most convincing evidence for a hybrid origin of ABR113 came from genomic in situ hybridisation (GISH) with total nuclear DNA from both putative progenitors. Hybridisation of genomic DNA from the diploid ($2n=10$) *B. distachyon* highlighted the ten largest chromosomes of the complement, while the genomic DNA from ABR114 distinguished the other 20 smaller chromosomes (Hasterok et al. 2004). Comparative mapping of single locus BAC clones from *B. distachyon* libraries revealed that for the vast majority of clones, the number of loci observed in the putative parental genomes of the allotetraploid is additive (Hasterok et al. 2006). CCP of ABR113 has corroborated the evidence for its parental origin from ABR114 and *B. distachyon*. Similar to the data obtained with mapping single BACs, the number of bivalents in ABR113 painted with the Bd1 probe was equal to the sum of the painted chromosomes in the putative ancestors. Also, the size and structure of homoeologues closely resembled their counterparts in the respective parental genomes. The longest bivalent is presumably derived from the *B. distachyon* genome. The two shorter ones might originate from ABR114. The absence of the unlabelled segment at the terminal position on one of the shorter bivalents in comparison to ABR114 chromosomes might be the result of chromosome restructuring which is a frequent phenomenon in polyploid genomes. It was also the case for CCP with Bd5 probe, where the number of painted chromosomes was equal to the sum of ABR114 and *B. distachyon* genomes. The analysis of orthologous gene pairs in diverse grass genomes, such as rice, wheat and sorghum, demonstrated that chromosome 5 is the most conserved (International Brachypodium Initiative 2010). The CCP analysis of Bd2, Bd3 and Bd4 in ABR113 revealed significant chromosome rearrangements. Probably, the longest chromosomes in all cases were from a *B. distachyon*-like ancestral genome, while different fragments of the short and long arm originated from translocations of ABR114 chromosomes. Multiple chromosome rearrangements during the evolution of genomes has been demonstrated in allopolyploids of *Brassica napus*(Udall et al. 2005) as well as in some species in the *Tragopogon* (Lim et al. 2008) and *Lilium*(Xie et al. 2010) genera. Painting with probes that allow only discrimination between chromosome arms obviously cannot reveal all chromosomal rearrangements or the exact locations of breakpoints. CP with denser probes targeting particular chromosome regions may uncover karyotypical changes with higher resolution and facilitate studies of chromosomal evolutionary processes.

Our data indicate that comparative chromosome painting with BAC pools from *B. distachyon* is a useful tool to investigate phylogenetic relationships between different *Brachypodium* species and the mechanisms which shape their karyotypes. The genus *Brachypodium* comprises relatively few species of different genome sizes and complexities. It includes diploid species with basic chromosome numbers of 5, 7, 8, 9 and 10 (Robertson 1981) as well as allopolyploids (Hasterok et al. 2004; Wolny and Hasterok 2009). Because of this diversity in chromosome number and ploidy, *Brachypodium* could become an interesting model system for studying karyotype evolution and divergence in grasses. It has yet to be determined whether or not *B. distachyon*-specific BAC pools could be used for comparative analysis of grasses outside the genus. Presumably, CCP can be applied successfully to the genomes of species such as *Lolium,Bromus* or *Festuca* that are phylogenetically relatively closely related to *Brachypodium.* It will be much more problematic to paint the large and complex genomes of crop cereals, such as barley and wheat or more distantly related species such as sorghum or rice. Cross hybridisation of *B. distachyon* single-locus BAC clones to the chromosomes of Triticale and rice did not yield hybridisation signals. As these clones were identified as syntenic to rice chromosome 6, it is likely that the homologous rice sequences are interspersed with blocks of species-specific repeats which preclude the effective visualisation of the probe signal (Hasterok et al. 2006). On the other hand, it was demonstrated that BAC clones from sorghum genomic libraries can effectively hybridise to maize chromosomes (Koumbaris and Bass 2003) despite the fact that the genome of maize is three times larger than the genome of sorghum (Paterson et al. 2009). Moreover, successful mapping of two *B. distachyon* BAC clones to barley chromosomes has been recently reported by Ma et al. (2010). These results are promising in the context of comparative chromosome painting in Triticeae species. Undoubtedly, the development of interspecific chromosome painting with *B. distachyon*-specific BAC pools significantly broadens the repertoire of cytogenetic techniques available for the analysis of grass genomes.

ACKNOWLEDGEMENTS

The authors are grateful to Dr. Martin Lysak and Ms. Terezie Mandakova from Masaryk University, Brno, Czech Republic, for their invaluable advice on the chromosome painting procedure as well as to two anonymous reviewers for their helpful comments on the manuscript. This work was supported by the Polish Ministry of Science and Higher Education Grant N N303 570738 to DI, EW, KL and RH as well as by the Biological and Biotechnological Sciences Research Council Grant BB/E004725/1 to MWB and the John Innes Centre Core Strategic Grant. DI and AB acknowledge scholarships from the UPGOW project co-financed by the European Social Fund.

REFERENCES

1. Arabidopsis Genome Initiative (2000) Analysis of the genome sequence of the flowering plant*Arabidopsis thaliana*. Nature 408:796–815
2. Berr A, Schubert I (2007) Interphase chromosome arrangement in *Arabidopsis thaliana* is similar in differentiated and meristematic tissues and shows a transient mirror symmetry after nuclear division. Genetics 176:853–863
3. Cremer T, Cremer M (2010) Chromosome territories. Cold Spring Harbor Press Biol 2:a003889
4. Cremer T, Kurz A, Zirbel R, Dietzel S, Rinke B et al (1993) Role of chromosome territories in the functional compartmentalization of the cell nucleus. Cold Spring Harb Symp Quant Biol 58:777–792
5. Cremer M, Grasser F, Lanctot C, Muller S, Neusser M et al (2008) Multicolor 3D fluorescence in situ hybridization for imaging interphase chromosomes. Methods Mol Biol 463:205–239
6. Draper J, Mur LA, Jenkins G, Ghosh-Biswas GC, Bablak P et al (2001) *Brachypodium distachyon*. A new model system for functional genomics in grasses. Plant Physiol 127:1539–1555
7. Febrer M, Goicoechea JL, Wright J, McKenzie N, Song X et al (2010) An integrated physical, genetic and cytogenetic map of *Brachypodium distachyon*, a model system for grass research. PLoS One 5:e13461
8. Garvin DF, Gu YQ, Hasterok R, Hazen SP, Jenkins G et al (2008) Development of genetic and genomic research resources for *Brachypodium distachyon*, a new model system for grass crop research. Crop Sci Plant Genome 48:S69–S84
9. Goff SA, Ricke D, Lan TH, Presting G, Wang R et al (2002) A draft sequence of the rice genome (*Oryza sativa* L. ssp. *japonica*). Science 296:92–100
10. Gray JW, Pinkel D (1992) Molecular cytogenetics in human cancer diagnosis. Cancer 69:1536–1542
11. Gray JW, Lucas JN, Pinkel D, Awa A (1992) Structural chromosome analysis by whole chromosome painting for assessment of radiation-induced genetic damage. J Radiat Res (Tokyo) 33(Suppl):80–86
12. Gu YQ, Ma Y, Huo N, Vogel JP, You FM et al (2009) A BAC-based physical map of*Brachypodium distachyon* and its comparative analysis with rice and wheat. BMC genomics 10:496
13. Hasterok R, Langdon T, Taylor S, Jenkins G (2002) Combinatorial labelling of DNA probes enables multicolour fluorescence in situ hybridisation in plants. Folia Histochem Cytobiol 40:319–323
14. Hasterok R, Draper J, Jenkins G (2004) Laying the cytotaxonomic foundations of a new model grass, *Brachypodium distachyon* (L.) Beauv. Chromosome Res 12:397–403
15. Hasterok R, Marasek A, Donnison IS, Armstead I, Thomas A et al (2006) Alignment of the genomes of *Brachypodium distachyon* and temperate cereals and grasses using bacterial artificial chromosome landing with fluorescence in situ hybridization. Genetics 173:349–362
16. International Brachypodium Initiative (2010) Genome sequencing and analysis of the model grass *Brachypodium distachyon*. Nature 463:763–768
17. Jenkins G, Hasterok R (2007) BAC 'landing' on chromosomes of *Brachypodium distachyon* for comparative genome alignment. Nat Protoc 2:88–98
18. Kim JS, Klein PE, Klein RR, Price HJ, Mullet JE et al (2005) Molecular cytogenetic maps of sorghum linkage groups 2 and 8. Genetics 169:955–965
19. Koumbaris GL, Bass HW (2003) A new single-locus cytogenetic mapping system for maize (*Zea mays* L.): overcoming FISH detection limits with marker-selected sorghum (*S. propinquum* L.) BAC clones. Plant J 35:647–659
20. Kurz A, Lampel S, Nickolenko JE, Bradl J, Benner A et al (1996) Active and inactive genes localize preferentially in the periphery of chromosome territories. J Cell Biol 135:1195–1205
21. Lim KY, Soltis DE, Soltis PS, Tate J, Matyasek R et al (2008) Rapid chromosome evolution in recently formed polyploids in *Tragopogon* (Asteraceae). PloS one 3:e3353
22. Lysak MA, Lexer C (2006) Towards the era of comparative evolutionary genomics in Brassicaceae. Pl Syst Evol 259:175–198
23. Lysak MA, Fransz PF, Ali HB, Schubert I (2001) Chromosome painting in *Arabidopsis thaliana*. Plant J 28:689–697
24. Lysak MA, Pecinka A, Schubert I (2003) Recent progress in chromosome painting of*Arabidopsis* and related species. Chromosome Res 11:195–204
25. Lysak MA, Berr A, Pecinka A, Schmidt R, McBreen K et al (2006) Mechanisms of chromosome number reduction in *Arabidopsis thaliana* and related Brassicaceae species. Proc Natl Acad Sci U S A 103:5224–5229
26. Lysak MA, Mandakova T, Lacombe E (2010) Reciprocal and multi-species chromosome BAC painting in crucifers (Brassicaceae). Cytogenet Genome Res 129:184–189
27. Ma L, Vu GT, Schubert V, Watanabe K, Stein N et al (2010) Synteny between *Brachypodium distachyon* and *Hordeum vulgare* as revealed by FISH. Chromosome Res 18:841–850
28. Mandakova T, Lysak MA (2008) Chromosomal phylogeny and karyotype evolution in $x=7$ crucifer species (Brassicaceae). Plant Cell 20:2559–2570
29. Marshall R, Obe G (1998) Application of chromosome painting to clastogenicity testing in vitro. Environ Mol Mutagen 32:212–222

30. Muller S, Stanyon R, O'Brien PC, Ferguson-Smith MA, Plesker R et al (1999) Defining the ancestral karyotype of all primates by multidirectional chromosome painting between tree shrews, lemurs and humans. Chromosoma 108:393–400

31. Paterson AH, Bowers JE, Bruggmann R, Dubchak I, Grimwood J et al (2009) The *Sorghum bicolor* genome and the diversification of grasses. Nature 457:551–556

32. Pecinka A (2005) Chromosome painting and arrangement of interphase chromosome territories in *Arabidopsis thaliana* [PhD thesis]. Halle (Saale), Germany: Martin-Luther-Universität Halle-Wittenberg, pp 92

33. Pecinka A, Schubert V, Meister A, Kreth G, Klatte M et al (2004) Chromosome territory arrangement and homologous pairing in nuclei of *Arabidopsis thaliana* are predominantly random except for NOR-bearing chromosomes. Chromosoma 113:258–269

34. Pedrosa-Harand A, Kami J, Gepts P, Geffroy V, Schweizer D (2009) Cytogenetic mapping of common bean chromosomes reveals a less compartmentalized small-genome plant species. Chromosome Res 17:405–417

35. Phillips D, Nibau C, Ramsay L, Waugh R, Jenkins G (2010) Development of a molecular cytogenetic recombination assay for barley. Cytogenet Genome Res 129:154–161

36. Pinkel D, Landegent J, Collins C, Fuscoe J, Segraves R et al (1988) Fluorescence in situ hybridization with human chromosome-specific libraries: detection of trisomy 21 and translocations of chromosome 4. Proc Natl Acad Sci U S A 85:9138–9142

37. Popp S, Jauch A, Qiu JY, Smialek B, Cremer T et al (1991) Translocation (8;21) in acute nonlymphocytic leukemia delineated by chromosomal in situ suppression hybridization. Cancer Genet Cytogenet 57:103–107

38. Ried T, Schrock E, Ning Y, Wienberg J (1998) Chromosome painting: a useful art. Hum Mol Genet 7:1619–1626

39. Robertson IH (1981) Chromosome numbers in *Brachypodium* Beauv. (Gramineae). Genetica 56:55–60

40. Schubert I, Fransz PF, Fuchs J, de Jong JH (2001) Chromosome painting in plants. Methods Cell Sci 23:57–69

41. Tkachuk DC, Pinkel D, Kuo WL, Weier HU, Gray JW (1991) Clinical applications of fluorescence in situ hybridization. Genet Anal Tech Appl 8:67–74

42. Udall JA, Quijada PA, Osborn TC (2005) Detection of chromosomal rearrangements derived from homologous recombination in four mapping populations of *Brassica napus* L. Genetics 169:967–979

43. Walling JG, Shoemaker R, Young N, Mudge J, Jackson S (2006) Chromosome-level homeology in paleopolyploid soybean (*Glycine max*) revealed through integration of genetic and chromosome maps. Genetics 172:1893–1900

44. Weise A, Heller A, Starke H, Mrasek K, Kuechler A et al (2003) Multitude multicolor chromosome banding (mMCB)—a comprehensive one-step multicolor FISH banding method. Cytogenet Genome Res 103:34–39

45. Wienberg J, Jauch A, Stanyon R, Cremer T (1990) Molecular cytotaxonomy of primates by chromosomal in situ suppression hybridization. Genomics 8:347–350

46. Wolny E, Hasterok R (2009) Comparative cytogenetic analysis of the genomes of the model grass *Brachypodium distachyon* and its close relatives. Ann Bot 104:873–881

47. Wolny E, Lesniewska K, Hasterok R, Langdon T (2011) Compact genomes and complex evolution in the genus *Brachypodium*. Chromosoma 120:199–212

48. Xie S, Khan N, Ramanna MS, Niu L, Marasek-Ciolakowska A et al (2010) An assessment of chromosomal rearrangements in neopolyploids of *Lilium* hybrids. Genome 53:439–446

49. Xu W, Li JY, Liu Q, Zhu Y, Pan JL et al (2010) Multiplex fluorescence in situ hybridization in identifying chromosome involvement of complex karyotypes in de novo myelodysplastic syndromes and acute myeloid leukemia. Int J Lab Hematol 32:e86–e95

CHAPTER 8

THE MOLECULAR CLONING AND CLARIFICATION OF A PHOTORESPIRATORY MUTANT, OSCDM1, USING ENHANCER TRAPPING

Jinxia Wu, Zhiguo Zhang, Qian Zhang, Xiao Han, Xiaofeng Gu and Tiegang Lu

Keywords: Hydrogen peroxide, mitochondria, photorespiratory, rice, serine hydroxymethyltransferase

INTRODUCTION

Leaves are the most important organs for photosynthesis in plants; indeed, the photosynthetic efficiency of leaves determines crop productivity. Improving photosynthesis in leaves is a reliable method for breeding cultivars with high photosynthetic efficiency (Govaerts et al., 1996; Zhang et al., 2009). The key to this method is cloning and analyzing the function of genes involved in leaf photosynthesis. Leaf color mutants are ideal genetic materials for studying the molecular mechanisms that regulate leaf photosynthesis and photorespiration in plants. To date, a considerable number of leaf color mutations have been reported in rice (http://www.shigen.nig.ac.jp/rice/oryzabaseV4), and some of them have been cloned successfully. For example, *OsCHLH* was the first gene from rice to be identified from a T-DNA insertion mutant using gene trapping methods. *OsCHLH* encodes the OsChlH subunit of magnesium chelatase, and its mutation produces plants that lack chlorophyll in their thylakoids (Jung et al., 2003; Goh et al., 2004). *YGL1* encodes a chlorophyll synthase; its mutation reduces chlorophyll accumulation and delays chloroplast development (Wu et al., 2007). *Chl1* and *Chl9* encode key enzymes for chlorophyll synthesis and chloroplast development, respectively. Ultrastructural analyses have revealed that the grana of Chl1 and Chl9 mutants are poorly stacked, resulting in underdeveloped chloroplasts (Zhang et al., 2006). *OsPPR1* encodes a pentatricopeptide repeat protein that may play an essential role in chloroplast biogenesis; its mutant exhibits typical phenotypes of chlorophyll-deficient plants (Gothandam et al., 2005). *OsHAP3* regulates normal chloroplast development and chlorophyll biosynthesis, and its mutant has pale green leaves (Miyoshi et al., 2003).

The mutants and genes described above are mostly related with chlorophyll synthesis or degradation in rice. In recent years, several genes from other pathways affecting leaf color have been identified in monocotyledonous plants. For example, several tie-dyed mutants, including tie-dyed1, tie-dyded2, camoflauge1, psychedelic, and sucrose export defective1, have been described that show chlorotic/variegated leaves and carbon hyperaccumulation in their leaf blades (Braun et al., 2006; Baker and Braun, 2007, 2008; Ma et al., 2008; Huang et al., 2009; Slewinski and Braun, 2010; Slewinski et al., 2012; Baker et al., 2013). Tie-dyed1 (Tdy1) is a novel transmembrane protein that is present only in grasses. Tdy1 functions in carbon partitioning by promoting phloem loading; accordingly, it is expressed exclusively in the phloem cells of both source and

Citation: Wu J, Zhang Z, Zhang Q, Han X, Gu X and Lu T (2015) The molecular cloning and clarification of a photorespiratory mutant, oscdm1, using enhancer trapping. Front. Genet. 6:226. doi: 10.3389/fgene.2015.00226.

sink tissues. The *tie-dyed2* (*tdy2*) mutant exhibits variegated green and yellow leaves. Tie-dyed2 is a callose synthase that functions in vein development and affects symplastic trafficking within the phloem of maize leaves. Several other maize leaf color genes, including *Oil Yellow1* (*Oy1*), those in certain lesion mimic mutants, and *Iojap* (*Ij*), have also been cloned. The semi-dominant *Oy1* mutants of maize are deficient in the first committed step of chlorophyll biosynthesis: the conversion of protoporphyrin IX to magnesium protoporphyrin IX. The *Oy1* gene encodes subunit I of magnesium chelatase (Sawers et al., 2006). The lethal leaf spot 1 lesion mimic locus of maize (*ZmLls1*) encodes pheophorbide a oxygenase. This locus corresponds to gene At3g44880 on chromosome 3 of *Arabidopsis thaliana* (Gray et al., 2002; Yang et al., 2004a). *Iojap* (*ij*) is a recessive striped mutant of maize in which plastid development is locally altered in a position-dependent manner on the leaves; *ij*-affected plastids can be transmitted to progeny even when the function of the nuclear gene is restored. The *ij* mutant is characterized by a number of independent transposon insertion mutations created using Robertson's Mutator. The *Ij* gene encodes a 24.8-kDa protein with no significant sequence similarity to proteins in public databases (Han et al., 1992;Byrne and Taylor, 1996; Silhavy and Maliga, 1998). In addition to these leaf color mutants, the barley mutants *tigrina-d* (Lee et al., 2003; Khandal et al., 2009) and *chlorina* have been well characterized (Knoetzel et al., 1998; Krol et al., 2009).

Plant photorespiration is an essential prerequisite for oxygenic photosynthesis. This metabolic repair pathway bestrides four compartments and requires several metabolite transporters for pathway function and the well-studied enzymatic steps of the core photorespiratory cycle (Eisenhut et al., 2014). Photorespiration dissipates excess photochemical energy to provide protection against oxidative damage under stressful conditions in which CO_2 assimilation is reduced. In addition, photorespiration provides metabolites for protection against stress, including glycine for the synthesis of glutathione (Lakshmanan et al., 2013). Mutants with photorespiration pathway deficiencies exhibit chlorotic and lethal phenotypes when grown in ambient CO_2 (Somerville and Ogren, 1981; Jamai et al., 2009). The enzymatic steps involved in the photorespiratory pathway have been well established. For example, mitochondrial serine hydroxymethyltransferase (SHMT) has been reported to be an essential component of the pathway; it plays an important role in one-carbon metabolism. Specifically, the enzyme produces N^5,N^{10}-methylene tetrahydrofolate (THF) glycine (Gly) from Ser and THF, and generates one-carbon units for cellular use (Mouillon et al., 1999). Seven SHMT genes have been found in *Arabidopsis* and other species. The *Arabidopsis*SHMT mutant *shm1* was one of the first photorespiratory mutants described by Somerville and Ogren (1981); the affected gene was later cloned by map-based cloning (Voll et al., 2006). AtSHMT1 was reported to play a critical role in controlling cell damage caused by abiotic stress, whereas its mutant *shm1-1* has a conditional lethal photorespiratory phenotype (Moreno et al., 2005). Thus, SHMT has acquired a new or altered function. A soybean (*Glycine max* L. Merr.) SHMT conferring resistance to nematodes plays an important role in plant resistance mechanisms against the pathogen (Liu et al., 2012).

T-DNA tagging methods, including gene trapping, promoter trapping, and enhancer trapping, have been shown in rice to be more effective and faster than positional cloning. These techniques are also advantageous when dealing with early embryogenic or gametophytic lethality. Several genes have been cloned by gene trapping or enhancer trapping, including *OsCHLH* (Jung et al., 2003),*Wda1* (Jung et al., 2006), *UDT1* (Jung et al., 2005), *Rip1* (Han et al., 2006), and *MIT* (Bashir et al., 2011). We previously reported the establishment of a large T-DNA enhancer trapping population (Yang et al., 2004b; Peng et al., 2005; Wan et al., 2009). The vector used to establish the pool carried a minimal cauliflower mosaic virus (CaMV) 35S promoter containing only the TATA box and transcription start site, which was placed adjacent to the T-DNA border. Only when the insertion site is next to an enhancer can the minimal promoter drive reporter gene expression. The GAL4/VP16-UAS enhancer trap was used in our insertion pool; this system improves the efficiency of enhancer trapping greatly. Insertion of the T-DNA into a gene in the proper orientation enables fusion between the endogenous gene and the GUS reporter gene; thus, T-DNA-tagged lines can be verified by a GUS assay. Screening of our insertion lines revealed that 40% were positive for GUS in the leaves or seeds (Peng et al., 2005).

To understand the molecular mechanism controlling leaf color, we first screened for leaf color mutants in the T_1 generation. Leaf color phenotypes co-segregated with GUS staining in nine independent lines. One line, *oscdm1*, showed co-segregation between its mutant phenotype and the T-DNA; thus, it was selected for further analysis. The *oscdm1* plants had yellow-green chlorotic leaves; this trait was found to be controlled by a single nuclear gene. The *oscdm1* plants displayed a lethal phenotype under natural conditions. Amplification of the T-DNA insertion-flanking sequence by PCR-based genome walking and subsequent sequencing resulted in the identification of an SHMT gene, *OsSHMT1*, which is predicted to encode the largest subunit of SHMT. The mutant phenotype was complemented by transformation with the wild-type gene, and plants in which*OsSHMT1* expression was reduced by RNA interference (RNAi) exhibited the *oscdm1* mutant phenotype. Moreover, the *oscdm1* plants exhibited extensive thylakoid membrane damage compared to wild type, especially at high light intensities. This damage may have resulted from excessive H_2O_2 levels. The identification of *OsSHMT1* using enhancer trapping and the characterization of *oscdm1* will aid future studies of the exact role of OsSHMT1 in photorespiration in rice.

MATERIALS AND METHODS

Plant Materials

The enhancer trapping population used in the present study for the mutant screen was described previously (Peng et al., 2005). The functional regions of the enhancer trap vector pFX-E24.2-15R are shown in Additional File 1. Within the T-DNA, the reporter gene was comprised of −48CaMV, *GAL4*/*Vp*16, 6×UAS, BoGUS, and EGFP; the reporter was located at the right border (RB) of the T-DNA. The minimal promoter −48CaMV cannot drive expression of the reporter gene alone; however, when the reporter gene is inserted into a gene or nearby enhancer it can be activated by the minimal promoter with the help of these enhancer elements.

Leaf Color Mutant Isolation

About 30 T_1 seeds per line were sterilized in 1% sodium hypochlorite for 30 min, washed with running tap water, and then imbibed in tap water for 3 days at 20°C. The seeds were then sown on a plastic net floating on nutrient fluid (Yoshida et al., 1976). The pH was adjusted every 2 days, with replacement of the nutrient fluid every week. The plants were grown for 14–30 days under greenhouse conditions, with natural sunlight and a minimum nighttime temperature of 25°C.

Histochemical Staining for GUS Activity in Plant Tissues

GUS assays were performed essentially as described by Jefferson et al. (1987). Rice tissues were removed, transferred to Eppendorf tubes containing GUS staining solution (50 mM sodium phosphate, pH 7.0, 10 mM EDTA, 0.1% Triton X-100, 1 mg/ml X-gluc, 0.1 mM potassium ferricyanide, and 10% methanol), and incubated at 37°C for 10–12 h. The staining solution was then removed and the tissues were stored in 70% ethanol. The stained tissues were examined under a zoom stereomicroscope (Nikon SMZ1000; Tokyo, Japan) and photos were taken with a digital camera (Nikon DXM1200F).

Amplification and Analysis of the T-DNA Flanking Sequences

Genomic DNA from the enhancer trap mutants was isolated using an improved version of the CTAB method (Murray and Thompson, 1980), and the T-DNA left border (LB) flanking regions were rescued using a modified version of the PCR-based genome walking protocols of Siebert et al. (1995) and Sallaud et al. (2003). The method consisted of three steps: digestion of genomic DNA using a blunt-end restriction enzyme and ligation of an asymmetrical adaptor, amplification by PCR using primers specific for the T-DNA and the adaptor, respectively, and successive PCR using two nested, specific primers. The adaptor (ADAR, 50 mM) was prepared by heating a mixture comprised of equal volumes of the complementary oligonucleotides ADAR1 (100 mM) and ADAR2 (100 mM) to 80°C for 10 min, and then allowing it to cool gradually to room temperature for annealing. The specific primers for the adaptor were APR1 (20 mM) and APR2 (20 mM); the specific primers for the T-DNA LB in pFX-E24.2-15R were LB1 (20 mM) and LB2 (20 mM). The distance between the LB2 binding site and the LB was about 180 bp. Unwanted amplification resulting from the two adaptors was inhibited by the asymmetry of the adaptor and its NH_2-blocked internal 3′ end; the second round of nested PCR also reduced non-specific amplification reactions (Siebert et al., 1995). Each DNA sample (40–100 ng) was digested using *Dra*I, *Eco*RV, or*Pvu*II (2 U; Takara, Dalian, China) and ligated with ADAR (0.1 ml) using T4 DNA ligase (70 U; Takara) at 25°C in a total volume of 10 µl (1 µl of 10× ligase buffer). After digestion/ligation for 10–12 h, aliquots (2 µl each) were used for the first round of PCR (MJ Research PTC-100 DNA Engine; MJ Research Inc., Quebec, Canada) in a total volume of 20 µl (0.3 ml of LB1, 0.3 ml of APR1, 2 U of Takara r-Taq, 100 mM dNTPs, and 2 µl of 10× PCR buffer). The program was as follows: 3 min at 94°C followed by 35 cycles of 94°C for 30 s, 67°C for 45 s, and 72°C for 150 s, with a final elongation step at 72°C for 5 min. Next, 2 µl of a 1:50 dilution of the product was used for nested PCR in a total volume of 20 µl (0.5 µl of LB2, 0.5 µl of APR2, 0.8 U of r-Taq, 200 mM dNTPs, and 2 µl of 10× PCR buffer) under the same conditions. All of the products from the second round of PCR were loaded onto a 1.5% agarose gel for electrophoresis. Distinct DNA bands were recovered for direct sequencing using an ABI 3730xl DNA analyzer (Applied Biosystems, Life Technologies, Carlsbad, CA, USA), which has a sequencing capability of>1 kb. LB2 was used as the primer for sequencing according to the protocol included with the BigDye Terminator v3.1 Cycle Sequencing Kit (Life Technologies).

Co-segregating Genotype Analysis of *oscdm1* by PCR

An analysis of the *oscdm1* plants was performed by PCR using two gene-specific primers flanking the insertion site (P1 and P2) and another T-DNA LB-specific primer (P3). PCR was carried out in a total volume of 20 µl containing 20 ng of plant DNA, 10× PCR buffer, 0.2 mM dNTPs, 0.5 U of rTaq polymerase, and 1 mM each primer. The DNA was denatured at 95°C for 4 min, followed by 35 cycles of 94°C for 1 min, 58°C for 1 min, and 72°C for 2 min. The products were loaded onto a 1.5% agarose gel for electrophoresis.

OsSHMT1 cDNA Amplification, Complementation Analysis, and Overexpression Line Construction

Total RNA was extracted using TRIzol reagent (Invitrogen, Carlsbad, CA, USA) from wild-type plants. The RNA was quantified using a UV spectrophotometer (Beckman DU 800 Series UV/VIS Spectrophotometer; Beckman Coulter Inc., Brea, CA, USA). First-strand cDNA was synthesized by reverse transcription using a cDNA synthesis kit (Takara) in a total volume of 20 µl containing 1 µg of total RNA, 10 ng of oligo$(dT)_{14}$ primer, 2.5 mM dNTPs, 1 µl of AMV, and 0.5 µl of RNAsin. Amplification of the full-length cDNA sequence of *OsSHMT1* (1542 bp) was performed in a total of 20 µl containing a 1/20 aliquot of the cDNA reaction, 0.5 µM gene-specific primers (OP1, which contained a *Sal*I digestion site, and OP2, which contained a *Sma*I digestion site), 10 mM dNTPs, 1 U of rTaq DNA polymerase, and 2 µl of 10× reaction buffer. The reaction protocol was as follows: denaturation at 94°C for 3 min followed by 25 cycles of 94°C for 30 s, 60°C for 45 s, and 72°C for 1 min, with a final step at 72°C for 10 min. An aliquot (1 ţl) of the reaction was loaded onto a 1.5% agarose gel (regular; Biowest, Barcelona, Spain) and analyzed by electrophoresis. The PCR product was cloned into the pEASY-Blunt simple cloning vector (Beijing TransGen Biotech Co., Ltd., Beijing, China) and sequenced. For the complementation construct, full-length *OsSHMT1* cDNA was cloned into pCambia23A carrying the actin promoter to generate Actin::OsSHMT1. The constructs were introduced into *Agrobacterium tumefaciens* (strain EHA105) by electroporation and *Agrobacterium*-mediated transformation was performed using vigorously growing *oscdm1*calli derived from a segregating population. OsSHMT1-overexpressing (OV) lines were constructed by transforming Actin::OsSHMT1 into Nipponbare calli.

Generation of pUbi-RNAi309-SHMT1 Transgenic Rice

DNA corresponding to 307 bp of the 3′ end of *OsSHMT1* was amplified using the primers P6 and P7. The product was cloned into the pEASY-Blunt simple cloning vector (Beijing TransGen Biotech Co., Ltd.) and sequenced. The vector was inserted into pTCK309 using the *Sac*I, *Spe*I, *Kpn*I, and*Bam*HI sites in inverted orientations. The pUbi-RNAi309-SHMT1 construct was transferred to*Agrobacterium* strain AGL1 by electroporation and then introduced into wild-type rice calli through *A. tumefaciens*-mediated transformation. The regenerated T_0 plants were grown in a paddy field at the Chinese Academy of Agricultural Sciences in Beijing, China. Leaves of the pUbi-RNAi309-SHMT1 transgenic plants showing the *chlorina* phenotype were selected and analyzed.

Reverse Transcription (RT)-PCR and Quantitative Real-time RT-PCR (qRT-PCR)

To study the *OsSHMT1* expression level, total RNA was extracted using TRIzol reagent (Invitrogen) from the leaves of wild-type and *oscdm1* plants. Total RNA (2 μg) from each sample was reverse-transcribed with oligo(dT) primer and PrimeScript RT Enzyme (Takara) according to the manufacturer›s instructions. For RT-PCR, the primers used to amplify *OsSHMT1* were Q1f and Q1R, while the primers used to amplify actin were Actinf and Actinr. The PCR conditions were as follows: preincubation at 94°C for 2.5 min, then 30 cycles of 94°C for 30 s, 52°C for 30 s, and 72°C for 1 min. qRT-PCR was performed using an iQTM5 Muticolor Real-Time PCR Detection System (Bio-Rad, Hercules, CA, USA) with SYBR® Green Real-Time PCR Master Mix (Life Technologies). The primers used to amplify *OsSHMT1* were qF and qR.

Transmission Electron Microscopy (TEM)

Leaves were harvested from wild-type and *oscdm1* plants at the third leaf stage that had been grown in a greenhouse under a medium light intensity (~150 μmol of photons/m^{2}s). Leaf sections were fixed in 2% glutaraldehyde and further fixed in 1% OsO_4. The tissues were then stained with uranyl acetate, dehydrated in ethanol, and embedded in Spurr's medium prior to thin sectioning. The samples were then stained again and examined with a JEOL 100 CX electron microscope (JEOL Ltd., Tokyo, Japan). The total chlorophyll content was determined spectrophotometrically according to the method of Arnon and Whatley (1949). Leaves (300 mg) were ground to a powder in liquid nitrogen and then transferred to a 15-ml Falcon tube. Next, 5 ml of 80% acetone were added to the tube and the mixture was thoroughly combined then left in the dark overnight. Centrifugation was performed at 4°C for 15 min (3000 rpm). The supernatant was transferred to a new centrifuge tube and the chlorophyll absorbance (hereafter termed A) was measured by spectrophotometry. The chlorophyll concentrations were calculated as follows (using 80% acetone as a blank control): C_{a+b} (mg/g) = $[8.026A_{663} + 20.206A_{645}] \times V/1000 \times W$, where V is the volume of the extract (ml) and W is the fresh leaf weight (g).

Chlorophyll Fluorescence Measurement

The middle portion of the first leaves from wild-type and *oscdm1* plants at the third leaf stage were collected and used to measure the leaf photosynthetic efficiency. Chlorophyll fluorescence measurements were performed with an IMAGING-PAM 2000 chlorophyll fluorometer (Walz, Effeltrich, Germany). A small piece of the mutant leaf blade was removed and placed in the well of a PCR plate filled with water; after 15 min of dark preadaptation, the measurements were performed. Changes in the chlorophyll fluorescence of the chlorophyll-deficient mutants were documented based on F_t, the steady level of chlorophyll fluorescence; $F_v/F_m = (F_m-F_0)/F_m$, the maximum quantum yield of photosystem (PS)II photochemistry; $\Phi_{PSII} = (F'_m-F)/F'_m$, the effective quantum yield of PSII; $qP = (F'_m-F)/(F'_m-F'_0)$, photochemical quenching; and $qN = 1-([F'_m - F'_0]/[F_m - F_0])$, non-photochemical quenching. The measurements were recorded after 5 min under actinic light. F'_0 was calculated as $F'_0 = F_0/(F_v/F_m+F_0/F_m)$ because of the limitations of the fluorometer.

Blue Native (BN) Test

BN polyacrylamide gel electrophoresis for the isolation of membrane protein complexes in an enzymatically active form was performed mostly as described (Schagger et al., 1994; Yan et al., 2007). Thylakoids were resuspended in medium A (25 mM Bis-Tris HCl, pH 7.0, 20% [w/v] glycerol, and 0.25 mg/ml Pefabloc) to a final concentration of 0.5 mg/ml chlorophyll, and an equal volume of 2% (w/v) dodecyl β-D-maltoside (Sigma, St. Louis, MO, USA), freshly prepared in medium A, was added. The thylakoids were then solubilized on ice for 1 min and centrifuged at 18,000 × g at 4°C for 12 min. The supernatant was supplemented with a 1/10 volume of buffer (100 mM Bis-Tris HCl, pH 7.0, 0.5 M ε-amino-n-caproic acid, 30% [w/v] sucrose, and 50 mg/ml Serva Blue G) and loaded on a gel containing a 5–12% gradient of acrylamide in the separation gel. Electrophoresis was performed [using the Hoefer Mighty Small system from Amersham Biosciences (Little Chalfont, UK)] at 0°C for 3.5 h by gradually increasing the voltage from 75 to 200 V. The thylakoids photosynthetic complexes were separated.

Measurement of the H_2O_2 Levels

The H_2O_2 levels in the plants were measured according to the method of Dagmar et al. (2001). Briefly, 500 mg of leaves were homogenized in a cold mortar with a pestle and 0.2 g of silicon dioxide in pre-cooled acetone (5 ml). The homogenate was centrifuged at 12,000 × g for 5 min, after which 1 ml of the supernatant was mixed with 0.1 ml of 5% $Ti(SO_4)_2$ and 0.2 ml of 19% ammonia. After the formation of a precipitate, the reaction mixture was centrifuged at 12,000 × g for 5 min. The resulting pellet was dissolved in 3 ml of 2 M H_2SO_4 and the absorbance of the solution was recorded at 415 nm. The H_2O_2 content was calculated according to a standard curve of H_2O_2 ranging from 0 to 10 mM.

Photosynthetic Parameter Measurements

The photosynthetic rate, transpiration rate, and stomatal conductance in wild-type and OsSHMT1- OV plants were measured at the full heading stage using a portable photosynthetic LCPRO+instrument (ADC BioScientific, Hoddesdon, UK) with the following settings: 500 mmol/s flow velocity, 30°C leaf chamber, and 1800 umol/s light quantum flux density.

Phylogenetic Analysis

The BLAST search program (http://blast.ncbi.nlm.nih.gov/) was used to identify homologs of OsSHMT1 using its amino acid sequence as the query. The resulting sequences were aligned using ClustalX1.83 in multiple alignment modes, and a neighbor-joining phylogenetic tree was generated using MEGA 4.0 (Tamura et al., 2007). The bootstrap values for nodes in the phylogenetic tree were from 1000 replications. The handling gap option was pairwise deletion; the numbers at the branch points indicate the bootstrap values. The peptide identities among the different proteins were calculated using GeneDoc software. The OsSHMT1 nucleotide/amino acid sequence has been deposited in the NCBI database with accession number: Os03g0738400. The OsSHMT1 full cDNA clone AK063056 has been deposited in KOME.

Subcellular Localization

Full-length *OsSHMT1* cDNA was amplified using the primers Fp1 and Rp1, after which the product was cut with *Sma*I and *Xba*I and then fused with PAN580. The 35S::OsSHMT1-GFP expression construct was transfected into rice protoplasts as described previously (Zhang et al., 2011). Briefly, for each sample, 10 μg of plasmid DNA were mixed with 200 μl of protoplasts (~1 × 10^6 cells). Freshly prepared polyethylene glycol solution [220 μl; 40% (w/v) polyethylene glycol 4000, 0.2M mannitol, and 0.1 M $CaCl_2$] was added, and the mixture was incubated at room temperature for 20 min in the dark. After incubation, 1 ml of W5 solution was added slowly to the samples. The resulting solution was mixed well by gently inverting the tube, and the protoplasts were pelleted by centrifugation at 1500 rpm for 3min. The protoplasts were resuspended gently in 1 ml of W5 solution. Finally, the protoplasts were transferred to multi-well plates and cultured under light or in the dark at room temperature for 16 h. MitoTracker® Red CMXRos was purchased from Invitrogen. The samples were observed with a confocal laser scanning microscope (Leica TCS SP5; Leica Microsystems, Wetzlar, Germany).

Statistical Analysis

All the experiments were carried out at least in triplicate. The values shown in the figures are mean values ± SD. For multiple comparisons, means were compared by one-way analysis of variance and Duncan›s multiple range test with a 5% level of significance.

Primer Sequence

All the primers for gene cloning, complementation analysis, over-expression construction, sub-cellular localization, reverse transcription (RT)-PCR and quantitative real-time (qRT-PCR) and etc in our paper are listed.

RESULTS

Isolation of Chlorophyll-deficient Mutants from the T-DNA Insertion Population

To identify new genes modulating the photosynthetic mechanism in rice, we conducted a genetic screen of an enhancer trapping population containing more than 10,000 individual lines. In total, 420 lines with leaf blade color defects were identified, including 251 albino, 132 pale green, 35 chlorotic (yellow), 25 stripe, and 13 zebra. The mutant lines varied in phenotype from albino to pale green according to the degree of chlorophyll deficiency (Figure 1A). Of the 420 leaf color mutants tested, 34 lines were positive for GUS staining. However, only in nine lines were all of the mutants positive for GUS staining. These lines were selected for further study. To test for the co-segregation of GUS staining and the hygromycin resistance gene *HptII* in the T_1 generation, we amplified DNA from each mutant line using *HptII*-specific primers. The results were consistent, as expected, for all mutants of the nine lines. The GUS expression patterns in the mutant leaves are shown in Figure1B.

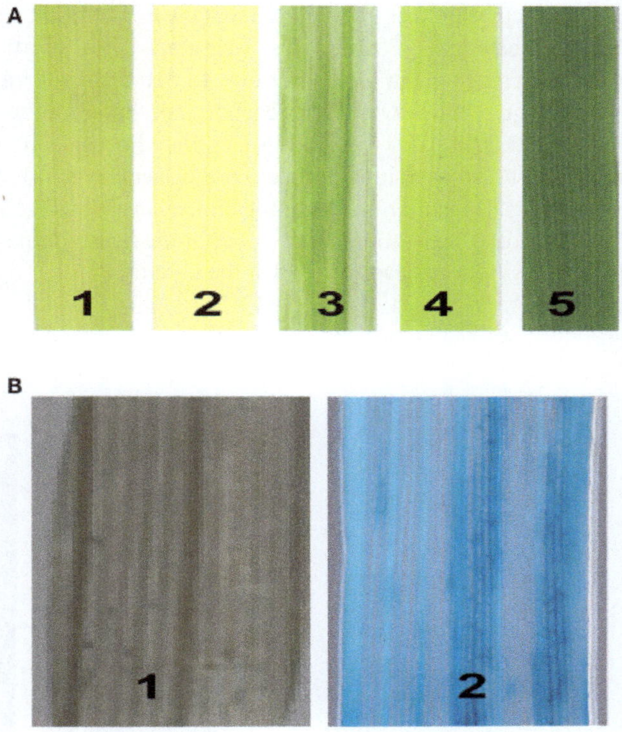

Figure 1: Isolation of chlorophyll-deficient mutants from the T-DNA-mutagenized population. (A) The phenotypes of the chlorophyll-deficient mutants: 1, yellow; 2, albino; 3, stripe; 4, pale green; and 5, wild type. (B) GUS expression patterns in the leaves: 1, negative; 2, vascular tissue-specific.

Analysis of T-DNA Insertion Sites in the Rice Genome and Prediction of Tagged Genes

PCR-based genome walking has been shown to be a highly efficient way of isolating T-DNA flanking sequences (Peng et al., 2005). Using this technique, 10 bands were amplified from the nine mutant lines. Of all of the bands sequenced, only six contained rice genomic sequences. The other four contained vector sequences, indicating border read-through or the termination of sequencing. Of the 10 flanking sequences rescued, six sequences originating from five mutants showed a unique hit to rice artificial chromosome (BAC/PAC) clones based on homology searches against sequences obtained from the NCBI and KOME databases. Of the six sequences, four flanking sequences were identified in genic regions and the other two in intergenic regions, based on a public annotation database for the rice genome. The locations of the T-DNA insertions are shown in Table 1.

Table 1: Locations of the T-DNA insertion and tagged putative genes.

Line	GenBank accession number	Insertion site	BLASTn *E*-value	Chr.	Insertion region (putative genic or intergenic)	Localization of putative protein
A33	AL731638	10450	2E-33	4	Intergenic region	
A167	AP002837	46557	9E-07	6	ORF of a putative Fe-SOD gene	Chloroplast
P239	AC117988	110492	9E-07	3	ORF of a putative glycine hydroxymethyltransferase gene	Mitochondrion
A304	AP004809	45787	3E-07	6	Intergenic region	
A401-1	AC125782	158205	3E-07	11	ORF of a putative GTP-binding gene	Chloroplast
A401-2	AP004182	53463	1E-17	7	ORF of a putative TPR-like domain-containing gene	No distinct localization

ORF, open reading frame; SOD, superoxide dismutase; TPR, tetratricopeptide repeat.

Identification of a Mutation in the SHMT Promoter

Sequence analyses using the NCBI database identified one line, P239, also named *oscdm1*, as having a T-DNA insertion upstream of a gene with high homology to SHMT genes from several plant species. *OsSHMT1* encodes the largest subunit of SHMT. *OsSHMT1* is 3962 bp in length and consists of 16 introns and 17 exons, with a complete coding sequence of 1542 bp encoding 513 amino acids and a molecular mass of 56.3 kDa; it encodes a putative serine methyltransferase domain, which spans from residues 51–449 at the C-terminus

Characterization of the *oscdm1* Mutant

There was no significant difference between 10-day-old (first leaf stage) *oscdm1* and wild-type plants (Figure 2A). However, the *oscdm1* plants appeared chlorotic at the third leaf stage (Figure2B), and they gradually died off from the fourth leaf stage onward. We measured the chlorophyll contents of both the *oscdm1* plants and wild-type plants at the third leaf stage. The ratio of chlorophyll *a* to chlorophyll *b* in the wild-type plants was about 3.5. The chlorophyll *a*, chlorophyll*b*, and total chlorophyll contents in the *oscdm1* plants at the third leaf stage were drastically lower than in the controls (Figure 2C). These results suggest that the chlorotic phenotype of the *oscdm1*seedlings was probably caused by a reduction in the total chlorophyll content, rather than a reduction in a particular pigment.

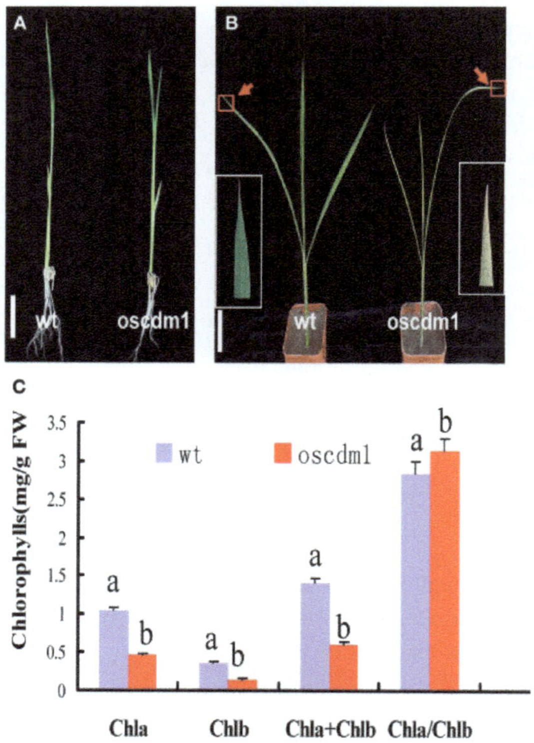

Figure 2: Phenotype comparison of *oscdm1* and wild-type plants. (A) Phenotypes of 10-day-old wild-type and mutant seedlings (bar = 1 cm). (B) Phenotypes of 1-month-old wild-type and mutant seedlings (bar = 1 cm). Red boxes denote the tips

of the mutant and wild-type leaves. The insets show magnified views of the areas in the red boxes. (C)Measurement of the chlorophyll content in the leaves of 1-month-old wild-type and *oscdm1*mutant plants. The data are the means ± standard deviation (SD). Bars represent the SDs (n= 5). Data were compared by one-way analysis of variance and Duncan's multiple range test. Different letters (a – b) indicate significant differences (P, 0.05) between lines. Chla, chlorophyll *a*; Chlb, chlorophyll *b*.

Light microscopic observations of cross-sections of a *chlorina* mutant leaf blade did not show any significant change in the size or number of mesophyll cells (data not shown). To examine whether the lack of chlorophyll in the *oscdm1* plants was accompanied by ultrastructural changes in the chloroplasts, we compared the ultrastructure of chloroplasts in the *oscdm1* and control plants at the third leaf stage using TEM. The analysis revealed no great change in chloroplast structure at the first leaf stage between the wild-type and *oscdm1* plants; however, at the third leaf stage, substantial changes in the number of chloroplasts, lamellar structures, and organization of the grana in the *oscdm1* chloroplasts (Figures 3B,D) were noted compared with wild type (Figures3A,C). In addition, the mutant chloroplasts exhibited vacuolation (Figure 3E). The BN test was used to assess thylakoid membrane integrity. A BN gel analysis of the thylakoid membrane showed some differences in five bands between the mutant and wild-type plants; in particular, bands III and IV were slightly more diffuse in the mutant plants. We concluded that the light harvesting complexes (LHCs) in the thylakoid membrane were altered and that chlorophyll synthesis was affected in the mutants beginning at the third leaf stage

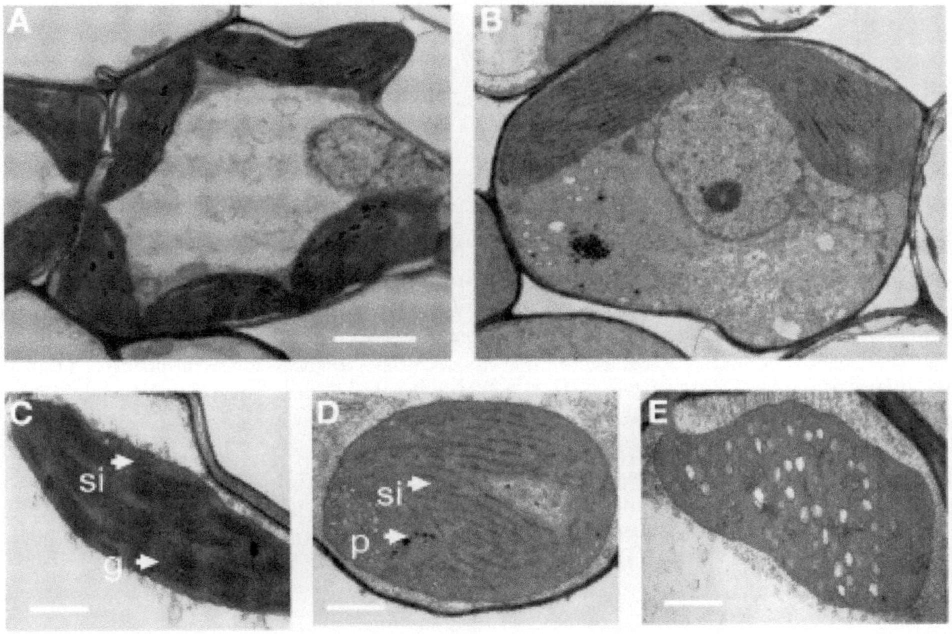

Figure 3: Transmission electron microscopic images of chloroplasts from 1-month-old wild-type and *oscdm1* plants. (A) Wild-type mesophyll cells. (B) The mesophyll cells of *oscdm1* plants. (C) Wild-type chloroplasts. (D,E) The chloroplasts of*oscdm1* plants. The chloroplasts of the wild-type plants had abundant, well-ordered membrane stacks, whereas the chloroplasts of the *oscdm1* mutant had almost no normal stacked membranous structures or grana (D). The mutant chloroplasts also exhibited vacuolation (E), p, plastoglobule; g, grana stack; sl, stroma lamellae; v, vacuolation. Bar = 1 μm in (A,B), and 100 nm in (C–E).

Co-segregation of the T-DNA and GUS Activity with the *Chlorina*Phenotype

The *oscdm1* plants appeared chlorotic at the third leaf stage. These plants eventually died both under greenhouse conditions and in a paddy field because of insufficient photosynthesis. In the T_2generation, 80 plants were wild type and 30 were *oscdm1*. In the T_3 generation, a heterozygous T_2plant generated 50 wild-type plants and 16 mutants, in accordance with our expectations for a single recessive locus. All of the chlorotic plants were GUS-positive, whereas only some of the wild-type plants were.

T_2 and T_3 seedlings were genotyped to test whether the chlorotic phenotype co-segregated with the T-DNA insertion (Figures 4A,B). Two randomly selected lines, 4 and 18, exhibited a wild-type phenotype, and T_3 seedlings of those lines were wild type in appearance. However, the progeny of plants 3 and 19 exhibited wild-type and mutant phenotypes at a ratio of 3:1. In contrast, plants 4 and 7 displayed the *chlorina* phenotype. These results demonstrate that the *chlorina* phenotype co-segregated with the T-DNA insertion. Furthermore, RT-PCR using RNA from the homozygous*oscdm1* plants did not detect full-length transcripts for *OsSHMT1*. Thus, the *oscdm1* plants were deemed to be loss-of-function mutants (Figure 4C).

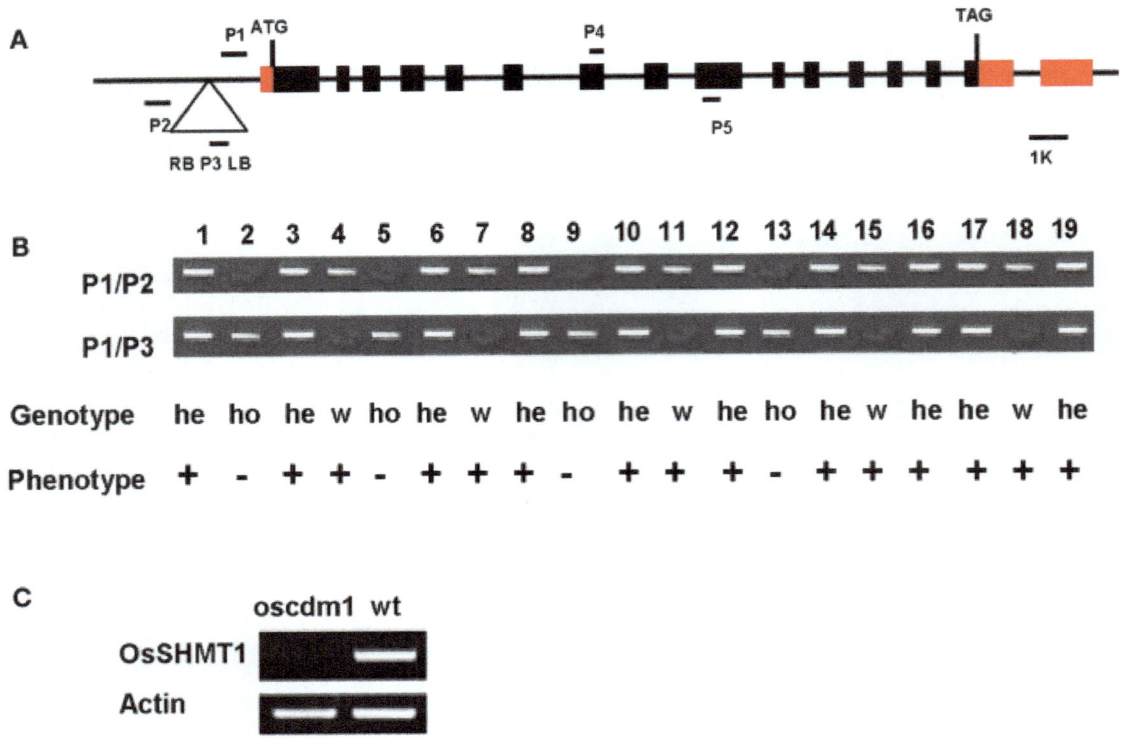

Figure 4: Co-segregation of the T-DNA with the *chlorina* phenotype. (A)Schematic representation of the T-DNA insertion upstream of *Os-SHMT1* (exons = black/red boxes; introns = black lines) and the domain organization of OsSHMT1 (http://www.uniprot.org/uniprot/Q6EPF0). ATG, start codon; TAG, termination codon; RB, right border; LB, left border. P1, P2, and P3 are the primers used for genotyping. (B) Co-segregation analysis of the genotype and phenotype in a segregating population. All plants homozygous for the T-DNA insertion showed a chlorotic phenotype, indicating that the recessive mutation was caused by T-DNA insertion. Numbers represent the different plants tested. P1/P2, PCR using primers P1 and P2; P1/P3, PCR using primers P1 and P3; He, hemizygous; Ho, homozygous; W, wild type. "+" represents the wild-type phenotype; "–" represents the mutant phenotype. Plants 2, 5, 9, and 13 were homozygous because only the 0.9-kb band was amplified using primers P1 and P2. Plants 4, 7, 11, 15, and 18 were designated as wild type because only the 0.7-kb band was amplified using primers P1 and P3. In contrast, both the 0.7- and 0.9-kb bands were amplified from plants 1, 3, 6, 8, 10, 12, 14, 16, 17, and 19, indicating that they were heterozygotic. (C) Analyses of OsSHMT1 expression in wild-type and *oscdm1* mutant plants using actin as a control. The RT-PCR primers P4 and P5 for *OsSHMT1* span introns 7 and 8. RT-PCR using RNA from homozygous *oscdm1* plants did not detect full-length *OsSHMT1* transcripts.

Functional Complementation of *osshmt1*

To further demonstrate that the *oscdm1* phenotype was due to insertion of the T-DNA into the candidate gene, functional complementation of the *oscdm1* mutant with wild-type *OsSHM*T1 was performed. Given the lethality of the homozygous mutation, homozygous mutant individuals were identified by genotyping the progeny of a heterozygous plant that was allowed to self-fertilize. These homozygous mutant individuals were used to initiate a callus in tissue culture, which was used for Agrobacterium-mediated transformation with the complementation construct. The sequence of *OsSHMT1* was used to search for the full-length cDNA sequence in the KOME database (http://cdna01.dna.affrc.go.jp/cDNA/). Fortunately, the full-length cDNA sequence of *OsSHMT1*was available. We cloned the full-length cDNA sequence by RT-PCR. The full-length cDNA sequence of *OsSHMT1*, which is 1542 bp in length, was found to be identical to that of the annotated gene Os03g0738400 in the TIGR database. This cDNA was cloned into a binary overexpression vector containing the actin promoter and subsequently transformed into *oscdm1*calli via *A. tumefaciens*-mediated transformation with G418 selection resistance (Figure 5A).

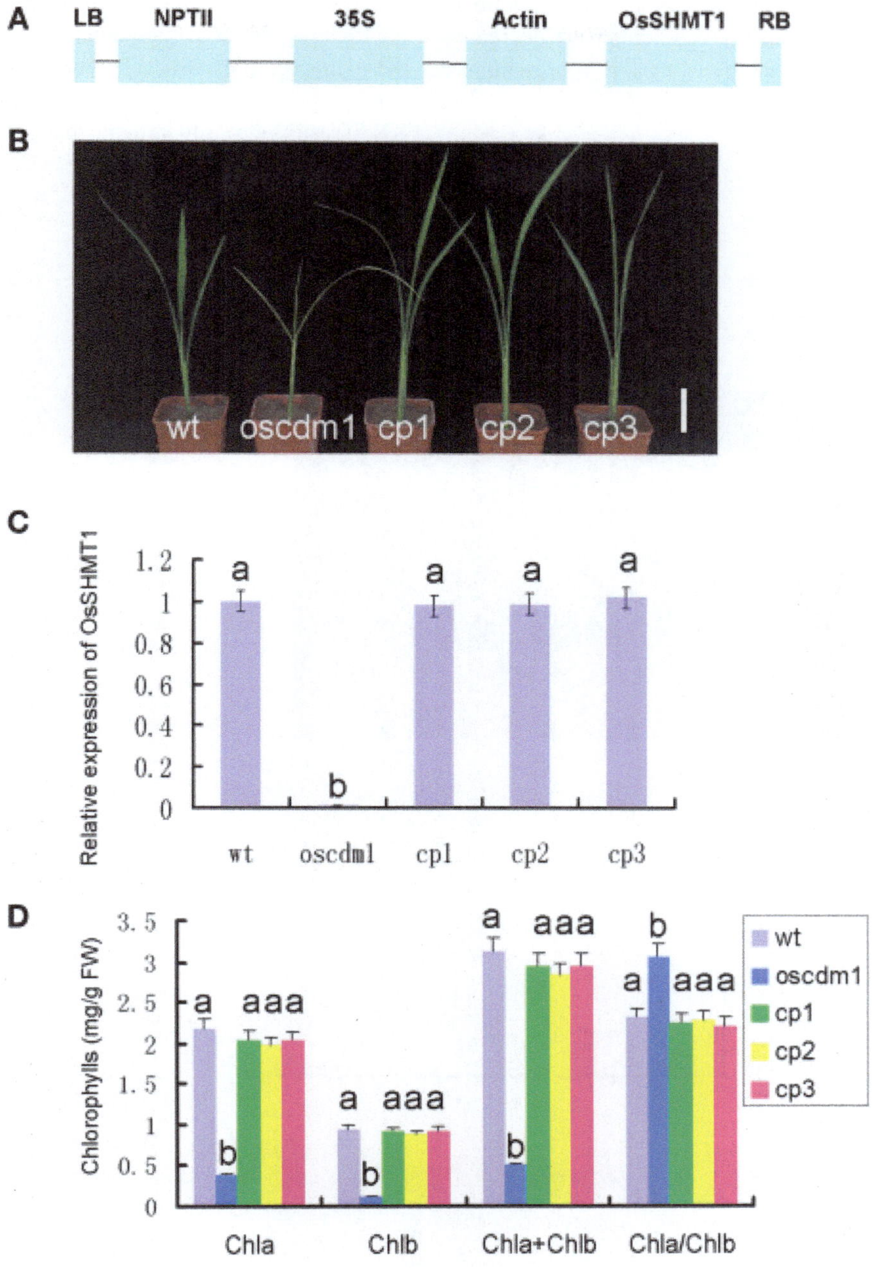

Figure 5: Complemented expression of *OsSHMT1* rescued the mutant phenotypes of *oscdm1*. (A) Schematic structure of the actin-OsSHMT1 complementation vector. NPTII was used as a marker in the transformation test. (B) Complementation of the*oscdm1* mutant by the wild-type gene. *Japonica* rice *oscdm1* callus was used for transformation (see the "Materials and Methods"). Images of the wild-type plants, *oscdm-1*mutant plants, and independent complemented plants (cp)1, cp2, and cp3 were taken after 1 month. Bar = 2 cm. (C) qRT-PCR analyses of *Os-SHMT1* expression in wild type, *oscdm1*, and in cp1, cp2, and cp3 showed that the expression level in cp1, cp2, and cp3 was near the wild-type level. The experiments are biological replicates with the SD. (D) The leaf pigment contents in wild type, *oscdm1*, and in cp1, cp2, and cp3 were measured. The pigment contents of the complemented lines were rescued to the wild-type level. Data were compared by one-way analysis of variance and Duncan's multiple range test. Different letters (a–b) indicate significant differences (P, 0.05) between lines.

The resulting G418-resistant lines were confirmed using specific PCR primers (p1, p2, and p3) and cDNA-amplifying primers (OP1 and OP2). In addition, quantitative PCR revealed that the exogenous *OsSHMT1* gene was overexpressed in 50 transgenic lines (CP1, CP2, and CP3) (Figure5C). Interestingly, these 50 lines displayed a wild-type phenotype even though their background was homozygous *oscdm1* (Figure 5B). The chlorophyll content in these lines was rescued to the wild-type level (Figure 5D).

RNAi Against *OsSHMT1* Reproduced the *oscdm1* Mutant Phenotype

To further confirm that *OsSHMT1* was the gene associated with the observed mutant phenotype, an*OsSHMT1* RNAi vector was constructed and transformed into *japonica* variety Nipponbare (Figure6A). About 400 independent transgenic lines were obtained and confirmed by PCR, and more than 100 of them displayed visibly chlorotic leaves during the early growth period (Figure 6B); however, after further growth and development the plants eventually died. Real-time PCR revealed that*OsSHMT1* transcription was decreased in the

RNAi transgenic plants compared to that in the Nipponbare controls (Figure 6D). The chlorophyll contents were decreased to 11 and 37% of that in Nipponbare using five representative T_0 lines (RNAi-1–5, respectively) (Figure 6E). Those RNAi plants showing a severe chlorotic phenotype eventually died in the field (Figure 6C). Conversely, other transgenic lines lacking the pUbi-RNAi309-SHMT1 vector (data not shown) failed to exhibit the chlorotic phenotype. These complementation and RNAi results strongly indicate that Os03g0738400 corresponds to *OsSHMT1*.

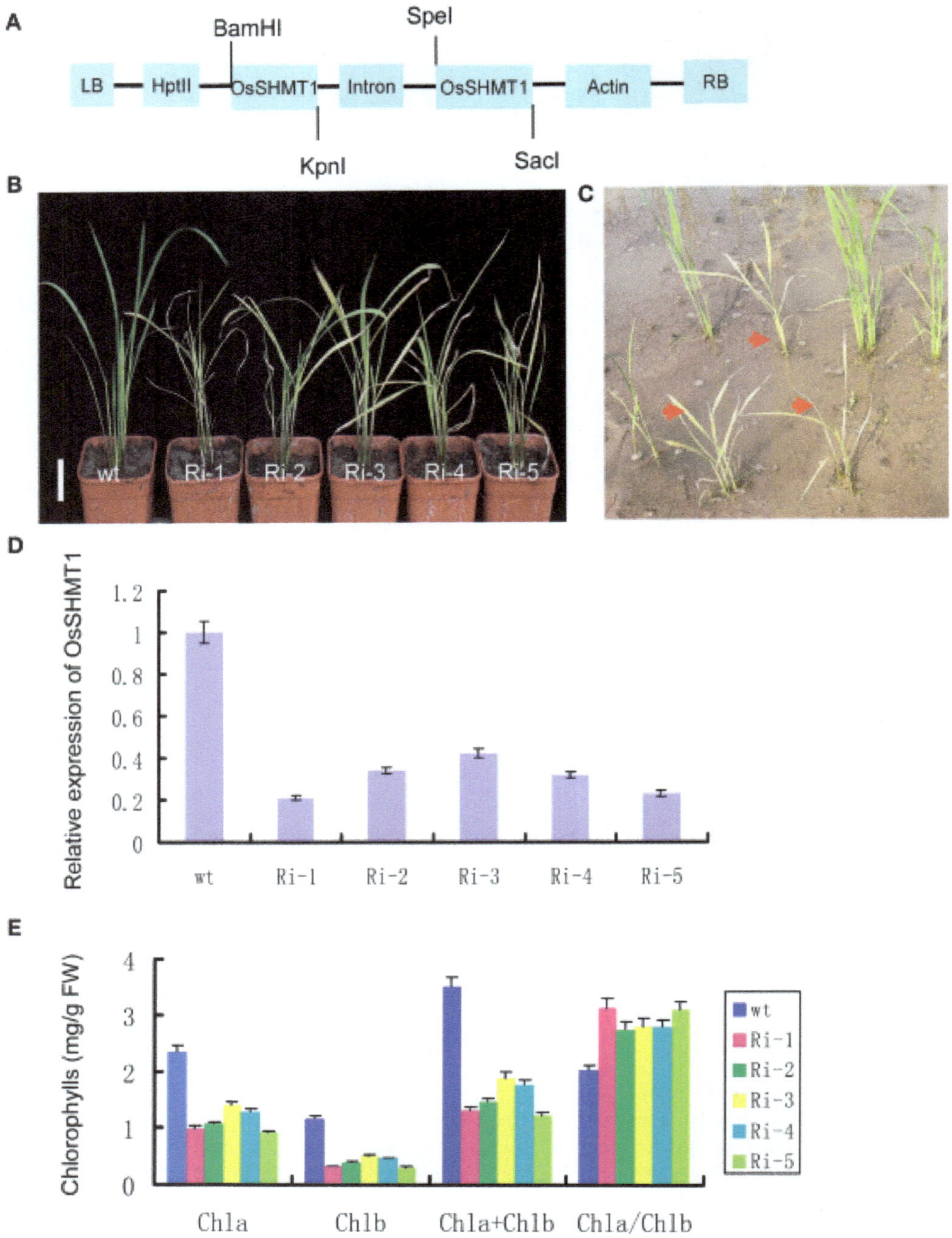

Figure 6: RNAi against *OsSHMT1* in wild type mimicked the mutant phenotypes of *oscdm1*. (A) Schematic structure of the pUbi-RNAi309-SHMT1 vector, including the inverted repeat at the 3′ end of the SHMT DNA sequence. (B) Phenotypes of 1-month-old wild-type plants and five independent RNAi lines (Ri-1–5). Bar = 2 cm. (C)Photographs of independent RNAi lines in the paddy field after 40 days. (D) qRT-PCR analyses of OsSHMT1 expression in wild type (WT-1 and -2) and five independent RNAi lines (Ri-1–5). The results indicate that the expression of OsSHMT1 in the five independent RNAi lines had decreased to some extent. The experiments are biological replicates with the SD.(E) The pigment contents in the leaves of wild-type plants and independent complemented lines (Ri-1–5) were measured.

Sequence Analysis Of *OsSHMT1*

BLAST searches of all available genomic sequences revealed that the rice genome contains five genes encoding an SHMT protein with strong homology to *OsSHMT1*. Close homologs of *OsSHMT1*were also identified in soybean and *Arabidopsis*. However, the functions of these genes are largely unknown. Notably, *OsSHMT1* and the *Arabidopsis* protein AtSHM1 (encoded by At4g37930) were found to possess a similar domain structure and to share the highest observed sequence homology across their entire lengths (87% amino acid identity). A neighbor-joining phylogenetic tree analysis based on the full-length protein sequences of SHMT family members from rice, soybean, and *Arabidopsis* showed that all of the SHMT protein sequences had a conserved pyridoxal phosphate binding site

OsSHMT1 contains an apparent mitochondrial targeting sequence of 65 amino acid residues at its N-terminus. To test whether OsSHMT1 is localized to mitochondria, the complete *OsSHMT1* coding region was fused to the gene encoding GFP under the control of the CaMV 35S promoter. Vectors to express the 35S::OsSHMT1-GFP fusion protein and 35S::GFP (as a control) were introduced into rice protoplasts. MitoTracker® Red was used as a positive control for mitochondria. The 35S::OsSHMT1-GFP fusion protein co-localized with the MitoTracker® Red signal (see the merged image), suggesting that OsSHMT1 is a mitochondrial protein (Figures 7E–H).

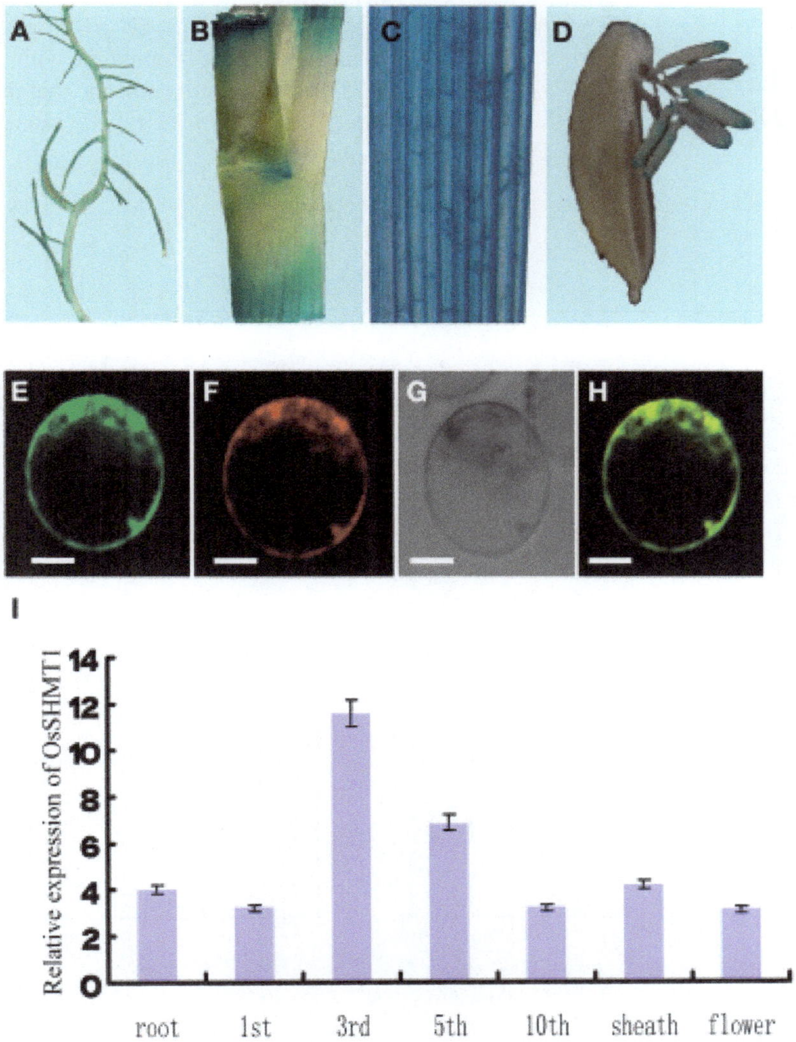

Figure 7: The expression pattern and subcellular localization of OsSHMT1. Different tissues were stained for GUS activity: (A) root; (B) internode; (C) blade; and (D)flower. The subcellular localization of OsSHMT1 was revealed in rice protoplasts as follows.(E) Subcellular localization of the OsSHMT1-GFP fusion protein in rice protoplasts. (F)MitoTracker® Red. (G) Bright field image of OsSHMT1-GFP. (H) Merged image of (E,F). Bars = 5 μm. (I) Quantitative PCR was performed to analyze the expression of OsSHMT1 in root, first leaf, third leaf, fifth leaf, tenth leaf, sheath (mature leaf), and flower.

OsSHMT1 Expression Pattern

An analysis by qRT-PCR revealed the expression of *OsSHMT1* mainly in young leaves, and not in root, leaf sheath, or flower, implying that *OsSHMT1* functions in seedlings or young leaves. To investigate the expression pattern of the gene in greater detail, total RNA was extracted from leaves from the first leaf stage to the mature stage. A qRT-PCR analysis of these RNA samples showed low-level expression of *OsSHMT1* at the first leaf stage; however, the expression level increased with leaf development and reached its highest level at the third leaf stage. Expression of the gene remained at a relatively high and stable level until the fifth leaf stage, suggesting that *OsSHMT1* is expressed at an early stage of growth and may play an important role in leaf development (Figure7I).

Because the GUS gene in the enhancer trapping vector and the *OsSHMT1* gene shared the same orientation, we expected that the expression pattern in *oscdm1* would be consistent with the *OsSHMT1* expression pattern. Thus, we analyzed heterozygous plant tissues from an *oscdm1* segregating population for GUS staining. In vegetative shoots, strong GUS activity was noted in young leaf and leaf ligule, whereas relatively weak GUS activity was detected in root and flower. This result indicates that *OsSHMT1* is expressed mainly in young leaves, consistent with our qRT-PCR results (Figures 7A–D,I).

Comparison of Chlorophyll Fluorescence Parameters between Wild-type and *oscdm1* Plants

Chloroplasts are the organelles that perform photosynthesis in plants; thus, they play an essential role in plant growth. To test whether the photosynthetic apparatus in the *oscdm1* plants was defective, we compared some key parameters of PSI and PSII between *oscdm1* and wild-type plants using an IMAGING-PAM fluorometer. The results of the comparison are shown in Table 2. Leaves from wild-type and *oscdm1* plants were dark-adapted for 10 min prior to fluorescence recording, and then a saturating light pulse was applied. F_v/F_m is widely considered to be a sensitive indicator of the maximum quantum efficiency of PSII. The maximum quantum yield (F_v/F_m) of PSII in the*oscdm1* plants was ~0.6, which is considerably lower than the value of 0.83 found in wild-type plants under a low light intensity, and which is the normal value for the mesophyll cells of most plant species. During illumination with actinic light, the maximal fluorescence yields (F'_m) were assessed in a time-dependent manner. The results indicate that the PSII quantum yield ratio [$\Delta F/F'_m = (F'_m - F)/F'_m$] was remarkably low in the *oscdm1* plants as compared with wild-type plants (Figures 8A,B). These results indicate that the *oscdm1* plants exhibited low photosynthetic activity because of damage to the reaction center of PSII.

Table 2: Comparison of fluorescence parameters between wild-type (WT) and*oscdm1* rice leaves ($n = 15$), the asterisks shows significantly different between the oscdm1 mutant and wild type at $P \leq 0.05$.

Parameter	57 μmol/m²s		103 μmol/m²s	
	WT	*oscdm1*	WT	*oscdm1*
F_v/F_m	0.83 ± 0.06	0.60 ± 0.05*	0.82 ± 0.08	0.53 ± 0.04*
Electron transport rate	6.21 ± 0.31	0.52 ± 0.04*	6.15 ± 0.25	0.25 ± 0.05*
Yield	0.73 ± 0.11	0.35 ± 0.06*	0.68 ± 0.14	0.15 ± 0.03*

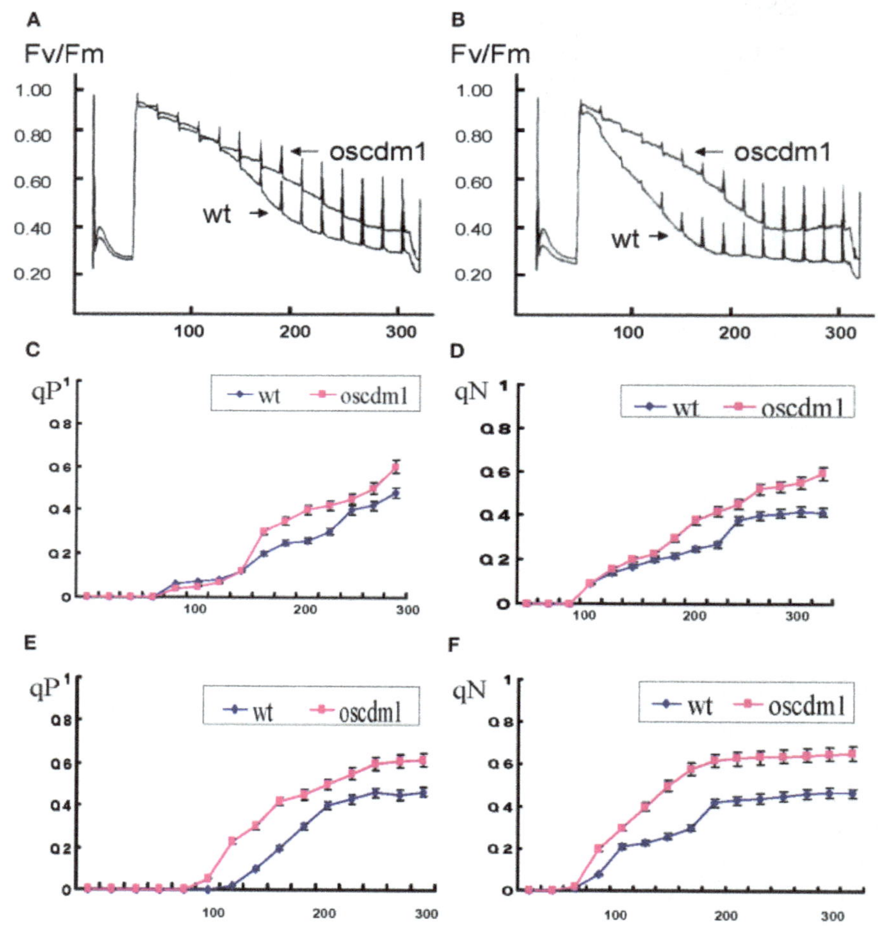

Figure 8: Chlorophyll fluorescence analysis in *oscdml* and wild-type plants. (A)Chlorophyll fluorescence induction curve was produced using third leaf stage plants under actinic light (A), low light intensity [57 μmol/m²s]; (B), high light intensity [103 μmol/m²s]. The induction curve did not differ between wild type and the mutant up to 120 s of treatment under a low light intensity, but after 120 s the wild-type plants exhibited a greater fluorescence quenching capacity than the *oscdm1* plants. Moreover, under a high light intensity this change occurred at an earlier time. (C–F) Changes in qP and qN in wild-type and *oscdm1* plants. (C,D), qP and qN changed under a low actinic light intensity (57 μmol/m²s). (E,F), qP and qN changed under a high actinic light intensity (103 μmol/m²s). The data show that qP and qN were not overly different

between the wild-type and *oscdm1*mutant plants up to 120 s of treatment at a low light intensity. However, under a high light intensity the change occurred earlier. This result suggests that the photochemical and non-photochemical quenching capacities of the mutant were affected and that the reaction speed became slower, especially under the higher light intensity. The data are the means ± SD ($n = 5$). These experiments were repeated more than three times with similar results.

The *oscdm1* Mutant was More Sensitive at a High Light Intensity

To detect the influence of different light intensities on the photosynthetic efficiency of the plants,*oscdm1* and wild-type plants were grown at a low (57 μmol/m^{2}s) or high light intensity (103 μmol/m^{2}s) and a chlorophyll fluorescence response curve was produced. Our results demonstrate no significant change in qP and qN between the wild-type and mutant plants up to 120 s of treatment at a low light intensity. However, at a high light intensity the change occurred earlier. This result suggests that the photochemical and non-photochemical quenching capacities of the mutant were affected, and that the reaction speed became slower, especially at a high light intensity. Similarly, a large change in qN was noted in the mutant plants at a high light intensity. The above results further demonstrate that the photosynthetic capacity of the mutant was abolished at a high light intensity (Figures 8C–F).

Oscdm1 is a Photorespiratory Mutant with a Chlorotic Phenotype Caused by Excessive H_2O_2

The *Arabidopsis* mutant *shm1-1* was one of the first photorespiratory mutants to be identified. The affected gene in *shm1-1* encodes an SHMT isozyme. The *shm1-1* mutant displays a lethal photorespiratory phenotype when grown at ambient CO_2 levels, but it is virtually unaffected by elevated CO_2 levels (Somerville and Ogren, 1981; Moreno et al., 2005; Jamai et al., 2009). Given the strong homology between *OsSHMT1* and *AtSHM1*, we examined whether the *oscdm1* mutant could survive exposure to a high (1%) CO_2 concentration. Thus, *oscdm1* mutant (leftmost plant in Figure 9A) plants were grown under ambient CO_2 levels while wild-type and *oscdm1*-hu (two rightmost plants in Figure 9A) plants were grown under elevated CO_2 levels in an artificial climate chamber. After 1 month at 1% CO_2, the *oscdm1* mutant plants were able to grow as well as wild type under ambient conditions (Figure 9A). Since the plants were kept in an artificial climate chamber for an extended period, some senescing leaves were observed on the wild-type and *oscdm1* plants. The pigment content of the *oscdm1*-hu plants was restored to the wild-type level (Figure 9B). This result is similar to that obtained for *Arabidopsis shm1-1*.

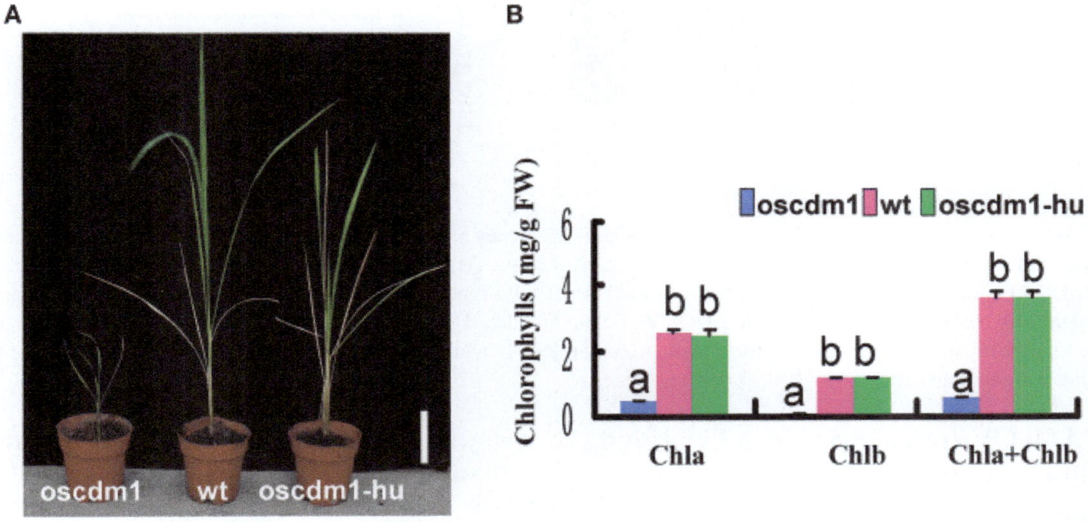

Figure 9: The *oscdm1* mutant exhibited a wild-type phenotype under a high CO_2 concentration. (A) The *oscdm1* mutant (leftmost plant in A) was grown at an ambient CO_2 level. The wild-type and *oscdm1*-hu plants (two rightmost plants in A) were grown under elevated CO_2 levels in an artificial climate chamber. The photographs were taken after about 2 months. Bar = 5 cm. (B) The pigment contents of leaves from wild-type,*oscdm1*, and *oscdm1*-hu plants were measured. The pigment contents of the *oscdm1*-hu plants were restored to the wild-type level. The data represent the means ± SD. Bars represent the SDs ($n = 10$). Chla, chlorophyll *a*; Chlb, chlorophyll *b*. Data were compared by one-way analysis of variance and Duncan's multiple range test. Different letters (a–b) indicate significant differences (P, 0.05) between lines.

Given that the *Arabidopsis* mutant *atshmt1-1* accumulates reactive oxygen species (ROS) under natural conditions (Moreno et al., 2005), we investigated the H_2O_2 contents among wild-type,*oscdm1* mutant, and OsSHMT1-OV plants. As shown in Figure 10A, for all of the lines the H_2O_2level in the *oscdm1* mutant plants was significantly higher than that in wild type at the fourth leaf stage. By contrast, the level of accumulation of H_2O_2 in the OsSHMT1-OV transgenic lines was slightly lower than that in wild type (Figure 10A). The above analyses indicate that *OsSHMT1*encodes an SHMT that is required for photorespiration and which plays a role in attenuating ROS production in chloroplasts to mitigate photoinhibition.

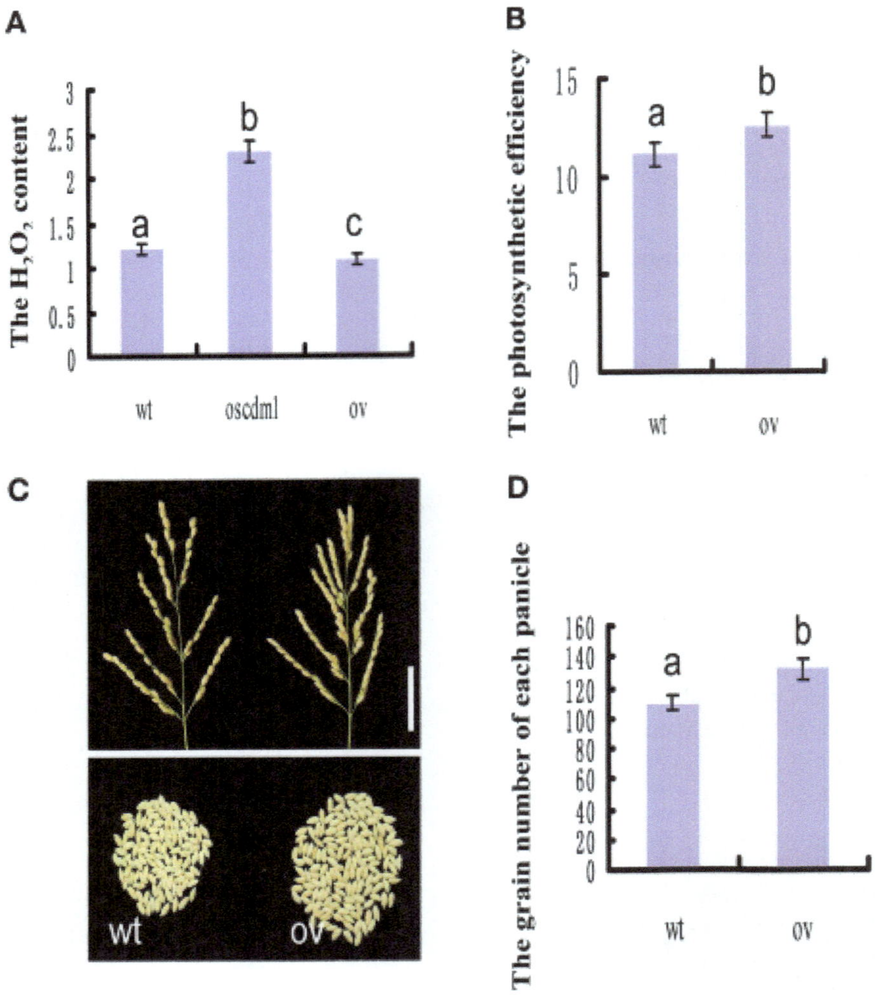

Figure 10: OsSHMT1-OV plants exhibited improved photosynthetic efficiency and productivity. (A) The H_2O_2 contents were measured among wild-type, *oscdm1*mutant, and OsSHMT1-OV plants at the fourth leaf stage. (B) The photosynthetic efficiency was measured between wild-type and OsSHMT1-OV plants at the mature stage. (C)Comparison of wild-type and OsSHMT1-OV panicles. (D) Comparison of the grain number per panicle between wild-type and OsSHMT1-OV panicles. The data are the means ± SD ($n = 5$). These experiments were repeated more than three times with similar results. Data were compared by one-way analysis of variance and Duncan's multiple range test. Different letters (a–c) indicate significant differences (P, 0.05) between lines.

The OsSHMT1-OV Transgenic Lines Exhibited Increased Photosynthetic Efficiency and Improved Productivity

To evaluate whether *OsSHMT1* could improve photosynthetic efficiency, several photosynthetic parameters were measured using a portable LCPRO+ photosynthesis measurement instrument (ADC BioScientific). The OsSHMT1-OV transgenic lines had a higher photosynthetic rate, stomatal conductance, intercellular CO_2 concentration, and transpiration rate than wild type at the heading stage (Figure 10B and Table 3). Interestingly, the productivity (grain number per plant) of the OsSHMT1-OV lines was 5% greater than that of wild type (Figures 10C,D). Thus, OsSHMT1 overexpression can improve the photosynthetic efficiency of plants and could be applied to breeding programs in the future.

Table 3: Comparison of photosynthesis parameters between WT and OsSHMT1-OV rice leaves ($n = 15$), the asterisks shows significantly different between the OsSHMT1-OV and wild type at $P \leq 0.05$.

Parameter	WT	OsSHMT1-OV lines
Stomatal conductance (mol/m^{2}s)	0.159 ± 0.03	0.194 ± 0.05*
Intercellular CO_2 concentration (μmol/mol)	237.1 ± 5.2	333.6 ± 7.3*
Transpiration rate (mmol/m^{2}s)	1.77 ± 0.21	2.62 ± 0.25*

DISCUSSION

Enhancer Trapping is a Valuable Tool for Gene Identification

GAL4/VP16-UAS, the improved binary T-DNA vector used to construct our study population (Wu et al., 2003; Peng et al., 2005), contains a promoter-less GUS reporter gene next to the RB. This enhancer trapping vector is designed to detect fusion between the GUS gene and an endogenous gene tagged by the T-DNA. Insertion of the construct not only destroys normal gene function, it also activates expression of the reporter gene. In rice, at least 40% of the insertion events led to GUS activation in various tissues (e.g., roots, leaves, flowers, and seeds) (Peng et al., 2005). Since T-DNA insertion lines do not all carry a single copy of the insert, positive GUS staining results do not fully confirm co-segregation between the mutant phenotype and gene. Still, this method has a number of advantages. First, it can be used to identify a candidate gene from several mutant lines in a short amount of time. Jung et al. (2003) used gene trapping technology to clone *OsCHLH*; since then, a number of genes, including *Wda1* (Jung et al., 2006), have been successfully cloned using this method. Second, enhancer trap vectors can be used to determine candidate gene expression patterns. GUS expression data can provide helpful information for functional gene analyses, including spatial-temporal expression specificity and clues as to a gene's function (Johnson et al., 2005; Peng et al., 2005; Liang et al., 2006). The cell division-related gene *CYCD3;2*was successfully cloned by reporter gene trapping. The GUS staining result in that case was similar but slightly different from the temporal patterns in T-DNA lines owing to the opposing orientation of the T-DNA insertion and *CYCD3;2*. This result may have been influenced by neighboring genes (Swaminathan et al., 2000). In the present study, the GUS gene was in the same orientation as*OsSHMT1*. Thus, the technique provided accurate *OsSHMT1* organ expression patterns and yielded helpful data for our bioinformatic analysis and prediction of gene function.

The *OsSHMT1* Gene has a Conserved Regulatory Function in Monocots and Dicots

In this report, a leaf color mutant with a *chlorina* phenotype was identified from a rice T-DNA-tagged pool and the gene was trapped by GUS staining. The mutated gene was identified as*OsSHMT1*, which encodes the largest subunit of SHMT. Rice OsSHMT1 is highly homologous to SHMT from *Arabidopsis*. The T-DNA was integrated into the *OsSHMT1* promoter, 99 bp away from the start codon (ATG). This generated a fusion transcript between the *OsSHMT1* promoter and GUS. Moreover, because of the insertion, the wild-type *OsSHMT1* transcript was not detectable in plants homozygous for the T-DNA. T-DNA insertion led to a loss-of-function mutation in*OsSHMT1*. The rice genome includes five genes that are homologous to *OsSHMT1*. RNAi against four other SHMT family genes caused no change in photorespiration compared with wild type (data not shown). This suggests that *OsSHMT1* is only one of multiple photorespiration-related genes in the rice SHMT family. SHMT in *Arabidopsis* is encoded by seven *SHM* genes, two of which encode mitochondrial isoforms (Jamai et al., 2009). However, only *SHM1* is necessary and sufficient to specify photorespiratory SHMT activity (Voll et al., 2006). The *oscdm1* plants produced chlorotic leaves, accumulated excessive H_2O_2, and subsequently died, similar to the mutant *Atshm1*. Moreover, the *oscdm1* plants were able to survive at a high concentration of CO_2 (1% CO_2). Therefore, SHMT family genes appear to have a conserved regulatory function in photorespiration in both monocots and dicots.

The *oscdm1* Mutant Exhibits Damage Caused by Photooxidation

Leaf color mutants may result from a variety of problems, including defects in chlorophyll synthesis, abnormal degradation pathways, and light damage. The *oscdm1* mutant is different from other reported leaf color mutants of rice. Almost all leaf color mutants show the mutant phenotype at the start of the second to the third leaf stage; at that stage, the *oscdm1* plants began to show slight chlorosis, which increased gradually. Structurally, the chloroplasts in the *oscdm1* plants developed well initially, but as time progressed the chloroplasts were destroyed and the plants eventually died. This chain of events is similar to that observed for photorespiration mutants of*Arabidopsis*. Photorespiration dissipates excess light energy as a protective process in plants. Based on our results, the phenotype of the *oscdm1* plants was due to photooxidation and changes in chloroplast structure. When the leaves of the mutant were subjected to stress, the chloroplasts exhibited ultrastructural changes, including irregularly stacked grana, and symptoms of vacuolization. The ultrastructural changes observed in the chloroplasts of the *oscdm1* plants are similar to the ultrastructural changes observed in the leaves under conditions of stress. When the plants were under stress, the level of ROS increased and caused cellular injury due to membrane oxidation. A number of physiological processes were affected, including photosynthesis, programmed cell death, hormone action, growth and development, and mitochondrial membrane structure. In higher plants, multiple SHMT genes predicted to localize to different compartments have been described (Zhang et al., 2010). Apart from its photorespiratory role in mitochondria, the physiological roles of SHMT are not well characterized. Given the finding that the OsSHMT1-OV lines had a lower H_2O_2 concentration compared with wild type, *OsSHMT1* may participate in biotic and abiotic stress responses. Genetic engineering using *OsSHMT1* represents a potential way to improve the tolerance of crop plants to stress.

Differences between the Current Study and a Recently Published Report

As we were submitting the present study for review, a report entitled "Characterization and molecular cloning of a serine hydroxymethyltransferase 1 (OsSHM1) in rice" in the *Journal of Integrative Plant Biology* (Wang et al., 2015). Both studies indicate that OsSHM1 is involved in photorespiration in rice, and that it has a conserved function in dicots and monocots. However, although the two mutants are allelic, there are many differences between the two studies. First, the*OsSHMT1* gene was cloned by two different methods: T-DNA tagging in our study and map-based cloning in the study of Wang et al. The corresponding mutant exhibited a loss-of-function point mutation (a single nucleotide change [T1300C] in LOC_ Os03g52840); however, for further analysis of the function of *OsSHMT1*, the *oscdm1* mutant obtained in the present study offers several advantages over the mutant described by Wang et al. The mutant oscdm1 not only provided a good material for photorespiration, but also *OsSHMT1* specical enhancer and promoter was further studied by oscdm1 mutant. Second, several additional color mutants and genes were isolated and presented in our study, including SOD (A167) and GTP (A401-1). Thus, the present report provides significant information and materials for further analysis of SOD (A167) and GTP (A401-1) function. Third, a greater number of photosynthetic parameters were measured in this study vs. that of Wang et al. In particular, the use of different light intensities and BN gel analysis in this study suggested that thylakoid membrane proteins (e.g., LHCs) were disrupted, extending the finding that photosynthesis was impaired. Fourth, and perhaps most interesting and/or valuable, our OsSHMT1-OV transgenic lines exhibited increased photosynthetic efficiency and improved productivity. These lines will be of great value for breeding high-efficiency cultivars of rice and other species.

The OsSHMT1-OV Lines Exhibited Increased Photosynthetic Efficiency and Improved Productivity

A particularly intriguing finding of our study is that *OsSHMT1* transcript accumulation increased the photosynthetic efficiency and productivity of the plants. This phenomenon can be explained as follows. The negative impact of photorespiration on plant growth and yield has been demonstrated by doubling the CO_2 concentration in a glasshouse environment; it can dramatically increase the performance of several crops. However, such a strategy cannot be implemented in the field, where photorespiratory losses are exacerbated by high temperatures and suboptimal water supplies. Under these conditions, the stomata close and the intercellular O_2 concentration increases through the release of O_2 from the light reactions of photosynthesis. Despite these disadvantages, photorespiration is important because it recovers 75% of the carbon from phosphoglycolate and efficiently removes potent inhibitors of photosynthesis. Moreover, photorespiration dissipates excess photochemical energy under high light intensities, thus protecting chloroplasts from overreduction. Kebeish et al. (2007) introduced the *Escherichia coli* glycolate catabolic pathway into *A. thaliana* chloroplasts to reduce the loss of fixed carbon and nitrogen that occurs in C3 plants when phosphoglycolate, an inevitable by-product of photosynthesis, is recycled by photorespiration. In our study, the transgenic OsSHMT-OV lines grew faster, produced more shoots and a greater panicle biomass, and contained more soluble sugars, reflecting reduced photorespiration and enhanced photosynthesis that correlated with an increased chloroplastic CO_2concentration in the vicinity of ribulose-1,5-bisphosphate carboxylase/oxygenase. The greater panicle biomass of the plants indicates that the overexpression of *OsSHMT1* enabled the plants to overcome at least some of the photorespiration-induced limitations on growth. This result indicates that the increase in biomass production of the transgenic plants coincided with improved photosynthesis.

In conclusion, we showed that the oscdm1 chloroplast thylakoid membrane damage may be from the excessive reactive oxygen (H_2O_2). The reactive oxygen (H_2O_2) was mainly produced in mitochondria. Mitochondria and chloroplasts are thought to have originated from prokaryotes during endosymbiotic evolution in eukaryotic cells. Intimate communication among organelles is necessary to coordinate their activities during growth, development, and other physiological processes. It is reasonable to assume that an intracellular mobile signal is transported from mitochondria to chloroplasts to trigger chloroplast change. This signal should be clarified in future investigations by taking advantage of *oscdm1* plants.

CONFLICT OF INTEREST STATEMENT

The authors declare that the research was conducted in the absence of any commercial or financial relationships that could be construed as a potential conflict of interest.

ACKNOWLEDGMENTS

This work was supported in part by National High-Tech Research and Development Project 863 "China Rice Functional Genomics" (2012AA10A304), Project 973 (2015CB150103) and Innovation project of Chinese Academy of Agricultural Science (to TL).

REFERENCES

1. Arnon, D. I., and Whatley, F. R. (1949). Is chloride a coenzyme of photosynthesis? *Science* 110, 554–556. doi: 10.1126/science.110.2865.554
2. Baker, R. F., and Braun, D. M. (2007). tie-dyed1 Functions non-cell autonomously to control carbohydrate accumulation in maize leaves. *Plant Physiol.* 144, 867–878. doi: 10.1104/pp.107.098814
3. Baker, R. F., and Braun, D. M. (2008). Tie-dyed2 functions with tie-dyed1 to promote carbohydrate export from maize leaves.*Plant Physiol.* 146, 1085–1097. doi: 10.1104/pp.107.111476
4. Baker, R. F., Slewinski, T. L., and Braun, D. M. (2013). The tie-dyed pathway promotes symplastic trafficking in the phloem.*Plant Signal. Behav.* 8:e24540. doi: 10.4161/psb.24540
5. Bashir, K., Ishimaru, Y., Shimo, H., Nagasaka, S., Fujimoto, M., Takanashi, H., et al. (2011). The rice mitochondrial iron transporter is essential for plant growth. *Nat. Commun.* 2, 322. doi: 10.1038/ncomms1326
6. Braun, D. M., Ma, Y., Inada, N., Muszynski, M. G., and Baker, R. F. (2006). tie-dyed1 Regulates carbohydrate accumulation in maize leaves. *Plant Physiol.* 142, 1511–1522. doi: 10.1104/pp.106.090381
7. Byrne, M., and Taylor, W. C. (1996). Analysis of Mutator-induced mutations in the iojap gene of maize. *Mol. Gen. Genet.* 252, 216–220. doi: 10.1007/s004389670025
8. Dagmar, P., Sairam, R. K., Srivastava, G. C., and Singh, D. V. (2001). Oxidative stress and antioxidant activity as the basis of senescence in maize leaves. *Plant Sci.* 161, 765–771. doi: 10.1016/S0168-9452(01)00462-9
9. Eisenhut, M., Hocken, N., and Weber, A. P. (2014). Plastidial metabolite transporters integrate photorespiration with carbon, nitrogen, and sulfur metabolism. *Cell Calcium.* 58, 98–104. doi: 10.1016/j.ceca.2014.10.007

10. Goh, C. H., Jung, K. H., Roberts, S. K., McAinsh, M. R., Hetherington, A. M., Park, Y. I., et al. (2004). Mitochondria provide the main source of cytosolic ATP for activation of outward-rectifying K+ channels in mesophyll protoplast of chlorophyll-deficient mutant rice (OsCHLH) seedlings. *J. Biol. Chem.* 279, 6874–6882. doi: 10.1074/jbc.M309071200

11. Gothandam, K. M., Kim, E. S., Cho, H., and Chung, Y. Y. (2005). OsPPR1, a pentatricopeptide repeat protein of rice is essential for the chloroplast biogenesis. *Plant Mol. Biol.* 58, 421–433. doi: 10.1007/s11103-005-5702-5

12. Govaerts, Y. M., Jacquemoud, S., Verstraete, M. M., and Ustin, S. L. (1996). Three-dimensional radiation transfer modeling in a dicotyledon leaf. *Appl. Opt.* 35, 6585–6598. doi: 10.1364/AO.35.006585

13. Gray, J., Janick-Buckner, D., Buckner, B., Close, P. S., and Johal, G. S. (2002). Light-dependent death of maize lls1 cells is mediated by mature chloroplasts. *Plant Physiol.* 130, 1894–1907. doi: 10.1104/pp.008441

14. Han, C. D., Coe, E. J., and Martienssen, R. A. (1992). Molecular cloning and characterization of iojap (ij), a pattern striping gene of maize. *EMBO J.* 11, 4037–4046.

15. Han, M. J., Jung, K. H., Yi, G., Lee, D. Y., and An, G. (2006). Rice Immature Pollen 1 (RIP1) is a regulator of late pollen development. *Plant Cell Physiol.* 47, 1457–1472. doi: 10.1093/pcp/pcl013

16. Huang, M., Slewinski, T. L., Baker, R. F., Janick-Buckner, D., Buckner, B., Johal, G. S., et al. (2009). Camouflage patterning in maize leaves results from a defect in porphobilinogen deaminase. *Mol. Plant* 2, 773–789. doi: 10.1093/mp/ssp029

17. Jamai, A., Salome, P. A., Schilling, S. H., Weber, A. P., and McClung, C. R. (2009). Arabidopsis photorespiratory serine hydroxymethyltransferase activity requires the mitochondrial accumulation of ferredoxin-dependent glutamate synthase. *Plant Cell* 21, 595–606. doi: 10.1105/tpc.108.063289

18. Jefferson, R. A., Kavanagh, T. A., and Bevan, M. W. (1987). GUS fusions: beta-glucuronidase as a sensitive and versatile gene fusion marker in higher plants. *EMBO J.* 6, 3901–3907.

19. Johnson, A. A., Hibberd, J. M., Gay, C., Essah, P. A., Haseloff, J., Tester, M., et al. (2005). Spatial control of transgene expression in rice (*Oryza sativa* L.) using the GAL4 enhancer trapping system. *Plant J.* 41, 779–789. doi: 10.1111/j.1365-313X.2005.02339.x

20. Jung, K. H., Han, M. J., Lee, D. Y., Lee, Y. S., Schreiber, L., Franke, R., et al. (2006). Wax-deficient anther1 is involved in cuticle and wax production in rice anther walls and is required for pollen development. *Plant Cell* 18, 3015–3032. doi: 10.1105/tpc.106.042044

21. Jung, K. H., Han, M. J., Lee, Y. S., Kim, Y. W., Hwang, I., Kim, M. J., et al. (2005). Rice Undeveloped Tapetum1 is a major regulator of early tapetum development. *Plant Cell* 17, 2705–2722. doi: 10.1105/tpc.105.034090

22. Jung, K. H., Hur, J., Ryu, C. H., Choi, Y., Chung, Y. Y., Miyao, A., et al. (2003). Characterization of a rice chlorophyll-deficient mutant using the T-DNA gene-trap system. *Plant Cell Physiol.* 44, 463–472. doi: 10.1093/pcp/pcg064

23. Kebeish, R., Niessen, M., Thiruveedhi, K., Bari, R., Hirsch, H. J., Rosenkranz, R., et al. (2007). Chloroplastic photorespiratory bypass increases photosynthesis and biomass production in *Arabidopsis thaliana*. *Nat. Biotechnol.* 25, 593–599. doi: 10.1038/nbt1299

24. Khandal, D., Samol, I., Buhr, F., Pollmann, S., Schmidt, H., Clemens, S., et al. (2009). Singlet oxygen-dependent translational control in the tigrina-d.12 mutant of barley. *Proc. Natl. Acad. Sci. U.S.A.* 106, 13112–13117. doi: 10.1073/pnas.0903522106

25. Knoetzel, J., Bossmann, B., and Grimme, L. H. (1998). Chlorina and viridis mutants of barley (*Hordeum vulgare* L.) allow assignment of long-wavelength chlorophyll forms to individual Lhca proteins of photosystem I *in vivo*. *FEBS Lett.* 436, 339–342. doi: 10.1016/S0014-5793(98)01158-2

26. Krol, M., Ivanov, A. G., Booij-James, I., Mattoo, A. K., Sane, P. V., and Huner, N. P. (2009). Absence of the major light-harvesting antenna proteins alters the redox properties of photosystem II reaction centres in the chlorina F2 mutant of barley.*Biochem. Cell Biol.* 87, 557–566. doi: 10.1139/O09-013

27. Lakshmanan, M., Mohanty, B., and Lee, D. Y. (2013). Identifying essential genes/reactions of the rice photorespiration by *in silico* model-based analysis. *Rice (N.Y.)* 6:20. doi: 10.1186/1939-8433-6-20

28. Lee, K. P., Kim, C., Lee, D. W., and Apel, K. (2003). TIGRINA d, required for regulating the biosynthesis of tetrapyrroles in barley, is an ortholog of the FLU gene of *Arabidopsis thaliana*. *FEBS Lett.* 553, 119–124. doi: 10.1016/S0014-5793(03)00983-9

29. Liang, D., Wu, C., Li, C., Xu, C., Zhang, J., Kilian, A., et al. (2006). Establishment of a patterned GAL4-VP16 transactivation system for discovering gene function in rice. *Plant J.* 46, 1059–1072. doi: 10.1111/j.1365-313X.2006.02747.x

30. Liu, S., Kandoth, P. K., Warren, S. D., Yeckel, G., Heinz, R., Alden, J., et al. (2012). A soybean cyst nematode resistance gene points to a new mechanism of plant resistance to pathogens. *Nature* 492, 256–260. doi: 10.1038/nature11651

31. Ma, Y., Baker, R. F., Magallanes-Lundback, M., DellaPenna, D., and Braun, D. M. (2008). Tie-dyed1 and sucrose export defective1 act independently to promote carbohydrate export from maize leaves. *Planta* 227, 527–538. doi: 10.1007/s00425-007-0636-6

32. Miyoshi, K., Ito, Y., Serizawa, A., and Kurata, N. (2003). OsHAP3 genes regulate chloroplast biogenesis in rice. *Plant J.* 36, 532–540. doi: 10.1046/j.1365-313X.2003.01897.x

33. Moreno, J. I., Martin, R., and Castresana, C. (2005). Arabidopsis SHMT1, a serine hydroxymethyltransferase that functions in the photorespiratory pathway influences resistance to biotic and abiotic stress. *Plant J.* 41, 451–463. doi: 10.1111/j.1365-313X.2004.02311.x

34. Mouillon, J. M., Aubert, S., Bourguignon, J., Gout, E., Douce, R., and Rebeille, F. (1999). Glycine and serine catabolism in non-photosynthetic higher plant cells:

their role in C1 metabolism. *Plant J.* 20, 197–205. doi: 10.1046/j.1365-313x.1999.00591.x

35. Murray, M. G., and Thompson, W. F. (1980). Rapid isolation of high molecular weight plant DNA. *Nucleic Acids Res.* 8, 4321–4325. doi: 10.1093/nar/8.19.4321
36. Peng, H., Huang, H., Yang, Y., Zhai, Y., Wu, J., Huang, D., et al. (2005). Functional analysis of GUS expression patterns and T-DNA integration characteristics in rice enhancer trap lines. *Plant Sci.* 168, 1571–1579. doi: 10.1016/j.plantsci.2005.02.011
37. Sallaud, C., Meynard, D., van Boxtel, J., Gay, C., Bes, M., Brizard, J. P., et al. (2003). Highly efficient production and characterization of T-DNA plants for rice (*Oryza sativa* L.) functional genomics. *Theor. Appl. Genet.* 106, 1396–1408. doi: 10.1007/s00122-002-1184-x
38. Sawers, R. J., Viney, J., Farmer, P. R., Bussey, R. R., Olsefski, G., Anufrikova, K., et al. (2006). The maize Oil yellow1 (Oy1) gene encodes the I subunit of magnesium chelatase. *Plant Mol. Biol.* 60, 95–106. doi: 10.1007/s11103-005-2880-0
39. Schagger, H., Cramer, W. A., and von Jagow, G. (1994). Analysis of molecular masses and oligomeric states of protein complexes by blue native electrophoresis and isolation of membrane protein complexes by two-dimensional native electrophoresis. *Anal. Biochem.* 217, 220–230. doi: 10.1006/abio.1994.1112
40. Siebert, P. D., Chenchik, A., Kellogg, D. E., Lukyanov, K. A., and Lukyanov, S. A. (1995). An improved PCR method for walking in uncloned genomic DNA. *Nucleic Acids Res.* 23, 1087–1088. doi: 10.1093/nar/23.6.1087
41. Silhavy, D., and Maliga, P. (1998). Mapping of promoters for the nucleus-encoded plastid RNA polymerase (NEP) in the iojap maize mutant. *Curr. Genet.* 33, 340–344. doi: 10.1007/s002940050345
42. Slewinski, T. L., Baker, R. F., Stubert, A., and Braun, D. M. (2012). Tie-dyed2 encodes a callose synthase that functions in vein development and affects symplastic trafficking within the phloem of maize leaves. *Plant Physiol.* 160, 1540–1550. doi: 10.1104/pp.112.202473
43. Slewinski, T. L., and Braun, D. M. (2010). The psychedelic genes of maize redundantly promote carbohydrate export from leaves. *Genetics* 185, 221–232. doi: 10.1534/genetics.109.113357
44. Somerville, C. R., and Ogren, W. L. (1981). Photorespiration-deficient mutants of *Arabidopsis thaliana* lacking mitochondrial serine transhydroxymethylase activity. *Plant Physiol.* 67, 666–671. doi: 10.1104/pp.67.4.666
45. Swaminathan, K., Yang, Y., Grotz, N., Campisi, L., and Jack, T. (2000). An enhancer trap line associated with a D-class cyclin gene in Arabidopsis. *Plant Physiol.* 124, 1658–1667. doi: 10.1104/pp.124.4.1658
46. Tamura, K., Dudley, J., Nei, M., and Kumar, S. (2007). MEGA4: Molecular Evolutionary Genetics Analysis (MEGA) software version 4.0. *Mol. Biol. Evol.* 24, 1596–1599. doi: 10.1093/molbev/msm092
47. Voll, L. M., Jamai, A., Renne, P., Voll, H., McClung, C. R., and Weber, A. P. (2006). The photorespiratory Arabidopsis shm1 mutant is deficient in SHM1. *Plant Physiol.* 140, 59–66. doi: 10.1104/pp.105.071399
48. Wan, S., Wu, J., Zhang, Z., Sun, X., Lv, Y., Gao, C., et al. (2009). Activation tagging, an efficient tool for functional analysis of the rice genome. *Plant Mol. Biol.* 69, 69–80. doi: 10.1007/s11103-008-9406-5
49. Wang, D., Liu, H., Li, S., Zhai, G., Shao, J., and Tao, Y. (2015). Characterization and molecular cloning of a *serine hydroxymethyltransferase 1 (OsSHM1)* in rice. *J. Integr. Plant Biol.* doi: 10.1111/jipb.12336. [Epub ahead of print].
50. Wu, C., Li, X., Yuan, W., Chen, G., Kilian, A., Li, J., et al. (2003). Development of enhancer trap lines for functional analysis of the rice genome. *Plant J.* 35, 418–427. doi: 10.1046/j.1365-313X.2003.01808.x
51. Wu, Z., Zhang, X., He, B., Diao, L., Sheng, S., Wang, J., et al. (2007). A chlorophyll-deficient rice mutant with impaired chlorophyllide esterification in chlorophyll biosynthesis. *Plant Physiol.* 145, 29–40. doi: 10.1104/pp.107.100321
52. Yan, L. J., Yang, S. H., Shu, H., Prokai, L., and Forster, M. J. (2007). Histochemical staining and quantification of dihydrolipoamide dehydrogenase diaphorase activity using blue native PAGE. *Electrophoresis* 28, 1036–1045. doi: 10.1002/elps.200600574
53. Yang, M., Wardzala, E., Johal, G. S., and Gray, J. (2004a). The wound-inducible Lls1 gene from maize is an orthologue of the Arabidopsis Acd1 gene, and the LLS1 protein is present in non-photosynthetic tissues. *Plant Mol. Biol.* 54, 175–191. doi: 10.1023/B:PLAN.0000028789.51807.6a
54. Yang, Y., Peng, H., Huang, H., Wu, J., Jia, S., Huang, D., et al. (2004b). Large-scale production of enhancer trapping lines for rice functional genomics. *Plant Sci.* 167, 281–288. doi: 10.1016/j.plantsci.2004.03.026
55. Yoshida, S., Forno, D. A., Cock, J. H., and Gomez, K. A. (1976). *Laboratory Manual for Physiological Studies of Rice, 3rd Edn.* Philippines: The International Rice Research Institute.
56. Zhang, G. H., Xu, Q., Zhu, X. D., Qian, Q., and Xue, H. W. (2009). SHALLOT-LIKE1 is a KANADI transcription factor that modulates rice leaf rolling by regulating leaf abaxial cell development. *Plant Cell* 21, 719–735. doi: 10.1105/tpc.108.061457
57. Zhang, H., Li, J., Yoo, J. H., Yoo, S. C., Cho, S. H., Koh, H. J., et al. (2006). Rice Chlorina-1 and Chlorina-9 encode ChlD and ChlI subunits of Mg-chelatase, a key enzyme for chlorophyll synthesis and chloroplast development. *Plant Mol. Biol.* 62, 325–337. doi: 10.1007/s11103-006-9024-z
58. Zhang, Y., Su, J., Duan, S., Ao, Y., Dai, J., Liu, J., et al. (2011). A highly efficient rice green tissue protoplast system for transient gene expression and studying light/chloroplast-related processes. *Plant Methods* 7:30. doi: 10.1186/1746-4811-7-30
59. Zhang, Y., Sun, K., Sandoval, F. J., Santiago, K., and Roje, S. (2010). One-carbon metabolism in plants: characterization of a plastid serine hydroxymethyltransferase. *Biochem. J.* 430, 97–105. doi: 10.1042/BJ20100566

CHAPTER 9 TILLING BY SEQUENCING TO IDENTIFY INDUCED MUTATIONS IN STRESS RESISTANCE GENES OF PEANUT (ARACHIS HYPOGAEA)

Yufang Guo, Brian Abernathy, Yajuan Zeng and Peggy Ozias-Akins

BACKGROUND

Peanut (*Arachis hypogaea* L.) is one of the world's most important oil seed crops. However, more than 70% of the peanut growing area is in arid and semi-arid regions. Abiotic stress caused by drought, as well as the co-occurring biotic stress due to *Aspergillus flavus* invasion of pods reduces plant productivity and the quality of seeds used for human consumption.

Using traditional breeding methods, several cultivars were developed for abiotic and biotic stress resistances, such as for nematode resistance [1] and tomato spotted wilt virus (TSWV) resistance [2-5]. However, due to the scarcity of drought and salt tolerance alleles in the peanut gene pool, the quantitative nature of abiotic stress tolerance, and the difficulty of selection, peanut stress resistance breeding can be a time consuming, labor-intensive, and difficult process [6].

Targeting Induced Local Lesions in Genomes (TILLING) is a powerful reverse genetics approach for functional genomics studies. It has been widely used to study gene function in various organisms including*Brassica napus* [7,8], *Brassica rapa* [9], *Lotus japonicus* [10], *Zea mays* [11], *Oryza sativa* [12], and*Drosophila* [13]. Traditional TILLING involves PCR amplification followed by digestion with CEL1 nuclease, a mismatch-specific nuclease that can recognize and cut heteroduplex DNA to identify single nucleotide substitutions or small insertions/deletions [14]. Alternatively, direct sequencing, high-performance liquid chromatography (HPLC), regular electrophoresis, high-resolution melting (HRM), and MALDI-TOF can also be used for TILLING applications [15]. Next-generation sequencing (NGS) has been reported as the latest mutation detection method to screen TILLING populations [16,17]. Previous reports using Illumina sequencing for high throughput TILLING not only confirmed known mutations, but also identified several mutants missed from previous heteroduplex detection assays [17]. In this study, we detected mutations in genes related to abiotic and biotic stress resistances in *Arachis hypogaea* using a high-throughput sequencing approach combined with two-dimensional pooling, and updated the TILLING bioinformatics pipeline by implementing either read percentage with non-reference nucleotide (single copy genes) or minimum variance multiplier to set mutation prediction parameters (multi-copy genes).

Lipoxygenases (LOX; EC1.13.11.12) comprise a family of structurally related non-heme iron-containing dioxygenases widely distributed among plants, animals, fungi and bacteria; they catalyse the addition of molecular oxygen to polyunsaturated fatty acids with a cis, cis-1,4-pentadiene structural unit. According to the dioxygenation position of the substrates, plant LOX enzymes can be classified into groups oxygenating polyunsaturated fatty acids at either C-13 (13-LOX) or C-9 (9-LOX) [18]. As a multi-gene family, LOX isoforms have an overall sequence identity of 25-40%. LOX proteins have conserved domains for catalytic iron binding. The catalytic iron is ligated in an octahedral arrangement by three conserved histidines, one*His*/*Asn*/*Ser*, and the C-terminal isoleucine. The LOX family has diversified functions in plants during vegetative growth and development, and contributes to formation of flavor and aroma compounds. In seeds, LOXs can function as storage proteins, while some can also play a role in plant-fungus interactions. In peanut, five lipoxygenase genes have been reported. The gene coding for *AhLOX1* (PnLOX1)

Citation: Yufang Guo, Brian Abernathy, Yajuan Zeng and Peggy Ozias-Akins, TILLING by sequencing to identify induced mutations in stress resistance genes of peanut (Arachis hypogaea), DOI: 10.1186/s12864-015-1348-0.

[19] is specifically expressed in immature cotyledons, and can be highly induced in mature seeds infected with*Aspergillus spp*., or by methyl jasmonate and wounding. Two other genes in peanut, *AhLOX2-3* [20], are also seed specifically expressed, with the highest expression level in mature embryo and immature cotyledons. Although the expression of *AhLOX2-3* was increased by wounding, in contrast to *AhLOX1*, both*AhLOX2* and *AhLOX3* were repressed upon *Aspergillus* infection of mature seed. Recently, *LOX4* and *LOX5*genes were described and suggested to also be involved in the response to *A. parasiticus* infection in peanut [21].

Phospholipase D (PLD, EC 3.1.4.4.) is a widely distributed ubiquitous eukaryotic enzyme participating in various cellular processes [22]. It hydrolyses membrane phospholipids to PA (phosphatidic acid) and a free head group such as choline in plants. PA is thought to be an effector in plant physiological processes such as secretion [23], DNA synthesis, and can be converted to second messengers such as diacyglycerol [24]. The active site of PLD consists of four conservative amino acid sequences. Motifs II and IV contain HxKx4Dx6G (G/S) (HDK) and are conserved in all organisms. A minority of PLD family members contain a single HKD motif, while most enzymes of this class contain two HDK motifs for bacterial, plant, yeast, and mammalian sources. Plant PLD normally contains a Ca^{2+} - dependent phospholipid - binding C2 domain and requires Ca^{2+} for activity (C2-PLD) [25], whereas the others may resemble mammalian PLD1 and PLD2, containing adjacent PX (phox homolog) and PH (pleckstrin homology) domains in the N-terminal region of the protein (PX/PH-PLD).

Recent studies have shown that PLD plays an important role in drought and stress tolerance [26-32]. It also plays a role in membrane degradation, seed germination, and acts in signal transduction cascades [25]. In peanut, two types of full-length PLD cDNA, *Ahpld1* and *Ahpld2* have been identified [33,34]. Up-regulation was observed in drought sensitive lines compared with drought tolerant lines of peanut. Southern blot analysis also indicated that PLD is a multigene family in cultivated peanut [33].

In this research, we targeted the above listed peanut stress related genes for cloning or amplification, set up an updated TILLING by sequencing pipeline, and used this pipeline to screen for mutations in an EMS induced population.

RESULTS AND DISCUSSION

Cloning of *Ah LOX* genes expressed in peanut roots

After screening a total of 1.75×10^5 colonies from a root cDNA library, 162 hybridization signals were detected, including 15 high-, 39 medium-, and 108 low- strength. A total of 80 signals including all high- and medium-, and 26 low-strength hybridization signals were selected for secondary screening. Colonies under a single signal were pooled and re-grown on a separate plate. Overall, 80 plates were grown for secondary screening and 61 were identified with hybridization signals, with 14 plates from the "high signal" group, 29 from the "medium signal" group, and 18 from the "low signal" group. A single colony was then picked under each hybridization signal from the secondary screening. Altogether, a total of 135 colonies were picked. Of those, 68 colonies contained recombinant plasmids with insert sizes > 2 kb; 32 colonies showed insert sizes of 1–2 kb. Overall, 77 clones (all 68 with insert sizes >2 kb, and the remaining with inserts between 1 kb and 2 kb) were picked for Sanger sequencing. Ten clones, ranging in size from 1,811 to 2,431 bp, showed similarity with LOX genes from *Arachis* or other species in plant EST databases (http://www.ncbi.nlm.nih.gov/nucest/). Based on these sequence similarities, the sequences were assigned to three genes.

The sequences from clones C05 (GenBank accession number KP710973) and E10 (KP710974) (hereafter named *AhLOX6),* had a complete open reading frame (ORF) of 2,604 bp, a 68 bp 5'-UTR (untranslated region) and 166 bp of 3'-UTR. The gene encodes a predicted 867 amino acids with the estimated molecular weight of 96.785 kDa. Clones C05 and E10 had 97.1% nucleotide sequence identity overall and their deduced amino acid sequences were 100% identical. A GenBank search found this gene was most similar to *Glycine max LOX9* (gb|ABS32275.1, gene ID: 100127399) and *Phaseolus vulgaris* (common bean) lipoxygenase (gb|AAB18970.2, gene ID: 18618345) (Additional file 1). *Glycine max LOX9* was most highly expressed in mature nodules and in roots, where it specifically expressed in the developing phloem [35]. In nodules, the expression of *LOX9* was correlated with the development of cells in the vasculature and lenticels. Histochemical analyses suggested that *Glycine max* LOX9 is involved in the growth and development of specific cells within these tissues. The *Phaseolus vulgaris* (common bean) lipoxygenase was also highly expressed in young, developing nodules [36]. Gene structure prediction by aligning the sequence to *Glycine max LOX3* (emb|X06928.1, gene ID: 547869) showed nine exons and eight introns for*AhLOX6*.

For the second gene, *AhLOX7* (GenBank accession number KP710975), two identical sequences (G02, H02) were most similar (82% identity) to *Glycine max* probable *linoleate 9S-lipoxygenase 5-like* mRNA (Gene ID: 100802887), *Glycine max LOX3* mRNA (U50081.1), which is highly expressed in seeds [35,37], and *Medicago truncatula* lipoxygenase mRNA (XM_003591072.1|) (Additional file 1). *AhLOX7* showed lower nucleotide similarity with *AhLOX2* and *AhLOX3*, 60% nucleotide sequence identity with *AhLOX4* (gb|EZ722311.1), and 97% with *AhLOX5* (gb|JR564445.1) with only three nucleotide differences in the ORF. The *AhLOX7* sequence has a complete open reading frame (ORF) of 2,583 bp, with a 90 bp 5'-UTR and 249 bp of 3'-UTR. The gene encodes 860 predicted amino acids with an estimated molecular weight of 96.961 kDa. The probable linoleate 9S-lipoxygenase 5 (UniGene Gma.24859, gene ID: 100802887) for *Glycine max* was predicted to be highly expressed in seed coat based on the EST profile from NCBI. The translated amino acid sequence showed 99.65% identity (Additional file 2) between *AhLOX7* and *AhLOX5*. Since peanut *LOX5* fell into the type I 9-LOX cluster, *AhLOX7* probably also belongs to the type I 9-LOXs [21]. Due to high predicted amino acid sequence identity, *AhLOX7* and *AhLOX5* may be different forms of the same gene, but given that*AhLOX2* and *AhLOX3* also had high identity between deduced amino acid sequences (99.42%) [20] we chose to assign a different name. This observation along with sequence variation among amplicons targeting *AhLOX7* described below (Additional file 3; Additional file 4), indicates that a more systematic study of peanut lipoxygenase genes in the peanut genome must be done to thoroughly characterize this multi-gene family. By aligning *AhLOX7* sequences with the soybean *LOX3* gene (emb|X06928.1 Gene ID: 547869), nine exons and eight introns were predicted. PCR spanning intron 1 indicated that the size of the first intron is greater than 3 kb, which is much larger than that in soybean *LOX3* (emb|X06928.1 Gene ID: 547869) (449 bp). The linoleate 9S-lipoxygenase 5-like sequence from soybean (gene ID: 100802887) has an intron 1 size of 1,956 bp, which is comparable with that of *AhLOX7*. Based on soybean lipoxygenase gene information [38,39], the corresponding locations of *AhLOX7* conserved histidine residues and fatty acid, iron, and oxygen binding sites were located in exons 7 to 9 (Table 1).

Table 1: Corresponding amino acid residue positions of *Glycine max LOX3* and *Arachis hypogaea AhLOX7* for conserved amino acids

Amino acid conserved region for *Glycine max*LOX3 [1]	**Fatty acid binding site**	**Iron binding site**		**Oxygen binding site**				
Amino acid abbreviation	**(H)**	**(H)**	**(H)**	**(H)**	**(H)**	**(H)**	**(N)**	**(I)**
Glycine max seed LOX 3 (emb|X06928.1|)	513	518	523	541	550	709	713	857
Glycine max exon	E7	E7	E8	E8	E8	E9	E9	E9
Arachis hypogaea AhLOX7	516	521	526	544	553	713	717	860
Arachis hypogaea exon	E7	E7	E8	E8	E8	E9	E9	E9

[1]The conserved histidine residues, the fatty acid-, iron-, and oxygen binding- residue in *Glycine max* are according to Reinprecht et al. [38].

For the third gene, *AhLOX8,* four sequences (D03, E03, H07, and A09) were most similar to *Glycine max*lipoxygenase-10 (*LOX10*) mRNA (Additional file 1). Cloned insert sizes ranged from 1,821 to 2,431 bp. The longest sequence (H07) (GenBank accession number KP710976) had an open reading frame (ORF) of 2,214 bp. The *Glycine max LOX10* was reported to have a lower expression level than *LOX9* in leaf and nodules but a similar expression level in roots [37]. Gene structure prediction by aligning these sequences to soybean *LOX3* (emb|X06928.1 gene ID: 547869) showed nine exons and eight introns.

Our RT-PCR results (Additional file 5) indicated that *AhLOX6* was expressed in roots and leaves, *AhLOX7*was expressed in seeds, roots and leaves, while *AhLOX8* also was expressed in seeds, roots, and leaves, but probably at lower levels than *AhLOX7*. Since *AhLOX7* was expressed at high levels in roots and seeds, and putative soybean ortholog (UniGene Gma.24859) was predicted to be highly expressed in the seed coat, seed tissues being the sites of *Aspergillus* invasion, *AhLOX7* was targeted for mutation discovery. Furthermore, differential expression of a highly similar lipoxygenase (*AhLOX5*) in genotypes resistant or susceptible to *Aspergillus* infection recently was documented [21]. Additional file 6 shows the phylogenetic relationships among all described peanut lipoxygenase genes.

Reference sequence amplification and characterization

For mutation detection, amplicon sequences from the genotype 'Tifrunner' (the wild type of the TILLING population) were used as references for alignment of short reads. Previously published peanut allergen genes *Ara h 1.01*, *Ara h 1.02*, *Ara h 2.01*, *Ara h 2.02*, and genes controlling oleic:linoleic acid ratio (O/L) -*ahFAD2A*, and *ahFAD2B* were also used in this study [40,41]. The "01" and "02" or "A" and "B" indicates that they are homeologous, with one copy in each subgenome.

For *AhLOX7* and *AhPLD*, the unique sequences obtained from amplicon sequencing were extracted as references. Because there was sequence variation for each gene from amplicon sequencing, more than one sequence per gene was used for reference (Additional file 3, 4, 7 and 8).

Two primer sets were designed to amplify the 5' and 3' ends of *AhLOX7*, aimed at identifying mutations with early transcription truncation or at functionally critical domains (Tables 1 and 2). The 5' amplicons (Figure 1a) spanned exons 2 (partial), 3, and 4 (partial) while the 3' amplicon (Figure 1b) covered exons 7 (partial), 8, 9 (partial), and the corresponding introns. Sequencing of the 5' amplicon revealed at least four copies with only three nucleotide positions differing overall (Additional file 3). Six sequence groups were obtained from the 3' amplicon with lengths of 1,532 bp, 1,544 bp, 1,698 bp, and 1,710 bp. The sequences can be divided into five categories based on nucleotide similarities. Totally there were 32 variant sites within the 3' end amplified region (Additional file 4). Based on the high sequence similarity between *AhLOX5* and*AhLOX7* (Additional file 2), the amplicon used for TILLING might be a mixture of both genes.

Table 2: Summary of amplicon sizes, mutation frequency, and primers used for the amplification

Gene	Amplicon (bp)	F primer no.	Sequence (5'-3')	R primer no.	Sequence (5'-3')
Ara h 1.01	1,865	1306	GAGCAATGAGAGGGAGGGTT	2079	TCTTCGTCTTC-GTCCTCCTCTTCTT
Ara h 1.02	1,666	1306	GAGCAATGAGAGGGAGGGTT	1309	CCTCCTCTTCTTCCCACTCTT G
Ara h 2.01	1,300	815	CGATTTACTCATGTACAATTAACAATAGAT	817	TCAAGATGGTTACAACTCTTG-CAGCAACA
Ara h 2.02	1,247	816	ATCACCTTAAATTTATACATATTTTCGG	817	TCAAGATGGTTACAACTCTTG-CAGCAACA
Ah FAD2A	1,241	1048	CTCTGACTATGCATCAG	1055	GATTACTGATTATTGACTT
Ah FAD2B	1,234	1048	CTCTGACTATGCATCAG	1101	CAGAACCATTAGCTTTG
AhLOX7_5'	1,713 1,714 1,715 1,716	2199	GCAGGAGAAGCAGCATTCACAGTTA	2184	CTCAAGAGGAACATTATCCC
AhLOX7_3'	1,532 1,544 1,698 1,710	2186	AAAGTCTACGGTGATCAAACCAGC	2188	AGCAGACACACCCATTGAAA
AhPLD1	1,271 1,272	2120	GACTTACGAACCTCAAAGATGCTGG	2121	TACTCTCCGTCCTTCTTCGCTT
AhPLD2	1,500	1991	AAGAACTGGGCACGTGGTGTTAGGAGT	1992	TCGACGGTTCTCCTGGGCTTT-TATGTA

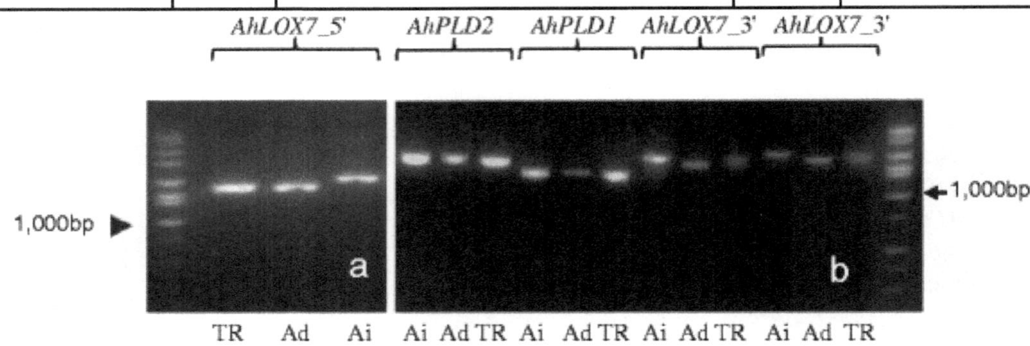

Figure 1: PCR amplicons of *AhLOX7 _5'*, *AhLOX7_3 '*, *AhPLD1* , and *AhPLD2*. PCR amplification of *AhLOX7_5'*, *AhLOX7_3'*,*AhPLD1*, and *AhPLD2*. **a**. Primers 2199/2184 amplify *AhLOX7_5'* on both progenitor genomes and the tetraploid. Amplicons cover partial E2, partial E4, entire E3, I2, and I3. **b**. Primers 1991/1992 amplified *AhPLD2* in tetraploid and diploid progenitor species; primers 2120/2121 amplified *AhPLD1* in both sub-genomes; primers 2187/2188 amplified *AhLOX7_3'* in both sub-genomes, amplicons contain partial E7, E9, entire E8, I7, and I8; primers 2186/2188 amplified *AhLOX7_3'* in both sub-genomes at a slightly different amplification start position resulted in similar amplification. DNA size standard: All-purpose Hi-Lo DNA marker, Bionexus, catalogue no. BN2050. TR=*A. hypogaea* cv. Tifrunner, Ad-*A. duranensis* (A genome), Ai=*A. ipaensis* (B genome).

The amplification of *AhPLD1* at exon 2 indicated at least four sequence groups. An overall sixteen single nucleotide differences were detected. Groups 1, 2, and 4 had only one or two nucleotide differences between each other and group 3 was most different from other three (Additional file 7). Amplification of diploid progenitor species *A. duranensis* (A genome) and *A. ipaensis* (B genome) with the primer set designed for *AhPLD1* did not show genome specific amplification at these regions (Figure 1b).

The amplicon sequences of *AhPLD2* indicated at least five copies for *AhPLD2* exon 3. There were five nucleotide differences in 1,500 bp. Sub-genome specific amplification was not detected from diploid progenitor species using the same set of primers (Figure 1b; Additional file 8).

Mutation detection, and validation for single copy genes

We obtained 367.8 million reads from one lane of an Illumina HiSeq run. After pre-processing, a total of 240 million reads (65%) remained. For the twenty libraries (12 from each column and 8 from each row), each library contained 3% to 7% of the total reads; the average read length was 91.44 bp. Over 85% of reads had the full read length of 94 bp.

The single copy genes included in this study were *Ara h 1.01*, *Ara h 1.02*, *Ara h 2.01*, *Ara h 2.02*, *AhFAD2A*, and *AhFAD2B*. The average depth of coverage for each sample for each gene per library at quality score cut-off from 10 to 25 was 76.01 (average depth of coverage for each sample for each gene across libraries = the total depth of coverage/(the number of genes × number of libraries × number of samples)). The average depth of coverage per sample for each gene per library was 76.5 at sequence quality score 10 and 75.2 at quality score 25. The average depth of coverage for each sample varied for different genes and libraries. The non-reference base percentage (percentage of reads containing a SNP compared with the reference sequence) was calculated for all six known mutants with sequence quality scores ranging from 10 to 21 to set up mutation detection parameters (Table 3; Additional file 9). Under these parameters (sequence quality scores from 10 to 21, and minimum read percentage with non-reference nucleotide from 0.058% to 0.37%), a set of unique mutants was detected (Table 4). We found that when the quality score was increased, the minimum read percentage with non-reference nucleotide had to be decreased to detect all six know mutations. However, with the combination of higher sequence quality score and lower minimum non-reference percentage, the number of predicted mutations increased. So, changing the minimum read percentage with non-reference nucleotide could be critical to control false positive predictions. When the sequence quality score increased to 22 and above, no known mutants could be detected, possibly due to a less than average number of mutations in the libraries caused by either uneven pooling and/or amplification efficiency. Because our purpose was to search for unknown single nucleotide mutations over a set of known sequences (amplicons) within a certain frequency range, the mutations that could be consistently detected across all sequence qualities (even at lower sequence quality cut-offs) was less likely due to sequencing errors.

Table 3: Mutation prediction parameters in various ranges for single copy genes

Min quality	Minimum read percentage with non-reference nucleotide [1] (%)	Background read percentage with non-reference nucleotide [1] (%)	Maximum read percentage with non-reference nucleotide (%)	Row non-reference % multiplier	All mutants (uniquely found [2])	Known mutants (uniquely found [2])	No. of new mutations validated
10	0.370	0.033	5.000	0.67	33	6	9
12	0.370	0.030	5.000	0.67	34	6	8
14	0.370	0.030	5.000	0.67	34	6	8
15	0.370	0.029	5.000	0.67	32	6	8
16	0.360	0.029	5.000	0.67	31	6	9
17	0.190	0.026	5.000	0.67	55	6	12
18	0.170	0.023	5.000	0.67	54	6	13
19	0.170	0.020	5.000	0.67	52	6	13
20	0.135	0.018	5.000	0.67	71	6	12
21	0.058	0.017	5.000	0.67	355	3	7

[1]The average non-reference nucleotide percentage at each quality cut-off.

[2]The mutations were only found once in each row and column.

Table 4: Summary of mutations identified in this study

Gene	Nucleotide change	Predicted AA change	Population	Plant ID	Amplicon length	SIFT score
Arahl.01	C321→T	Silent	08 F	213_1	1,865	
Arahl.01	C1524→T	T377→I	07JKEMS1	67	1,865	0.11
Arahl.01	C 1678→T	Silent	07JKEMS1	48	1,865	
Arahl.02	A 1258→G	silent	07JKEMS1	125	1,666	
Arahl.02	G72→T	Q 24→H	07JKEMS1	125	1,666	0.12
Arahl.02	C428→T	P143→L	08 F	216_1	1,666	0.16
Arahl.02	C 644→A	P215→H	08 F	221_5	1,666	0.00
Arahl.02	C765→T	Silent	07JKEMS1	125	1,666	
Arahl.02	G 891→A	Silent	07JKEMS1	2	1,666	
Arahl.02	A694→G	I232→V	07JKEMS1	125	1,666	1.00
Arahl.02	A 742→C	K248→Q	07JKEMS1	125	1,666	0.47
AhFAD2B	C 632→T	P 211→L	08 F	222_3	1,234	0.00
Arah2.02	C→T (upstream)	Silent	08 F	231_4	1,247	
AhLOX7_3'	T 1508→C	L503→P	07JKEMS1	69	1,532	0.00
AhLOX7_5'	C512→G	A171→G	07JKEMS1	125	1,714	0.09
AhLOX7_5'	A525→G	I 175→M	07JKEMS1	125	1,714	0.07
AhLOX7_5'	C532→G	L178→V	07JKEMS1	125	1,714	1.00
AhPLD1	C1328→T	S 443→F	08 F	201_4	1,272	0.00
AhPLD1	G1632→A	M 544→I	08 F	211_5	1,272	0.01
AhPLD2	C1727→T	P 576→L	07JKEMS	67	1,500	0.00

In theory, if the amplicons were mixed at equal quantity, in the row pool, the expected read percentage with non-reference nucleotide should be 1.04% (1/96 × 100%) for homozygous mutants and 0.52% (1/192 × 100%) for heterozygous mutants, while in the column pool, the non-reference reads should account for 1.56% (1/64 × 100%) and 0.78% (1/128 × 100%) for the homozygous and heterozygous mutants, respectively. The read percentage with non-reference nucleotide can be lower than the theoretical value because of sequencing errors, uneven amplicon pooling, etc. Decreasing the quality score can result in false-positive mutants due to sequencing errors (type I error), while increasing the quality score will decrease the mutation detection sensitivity (type II error). Thus the known mutants became very important for selecting the appropriate parameters that were sensitive enough to detect non-reference nucleotides and reliable enough to distinguish them from sequencing errors. When unique mutants were detected, all other nucleotides under the hypothetical mutation positions were also calculated; the hypothetical mutations should have significantly greater numbers of mutation-responsible nucleotides than other non-reference nucleotides. We found that the minimum read percentage with non-reference nucleotide needed to predict a mutation was 3–12 times the average non-reference nucleotide frequency.

With the above parameters, 14 mutants were found at all quality scores ("common group"), 34 detected only at sequence quality scores 17 to 20 ("high group"), and 15 detected at sequence quality scores 10 to 16 ("low group"), respectively. Subsequent mutation validation was done by several methods. Initially, CAPS (Cleaved Amplified Polymorphic Sequences) assay was carried out for all predicted mutations that were amenable. From the "common" mutation group, six were suitable for CAPS assay (three *Ara h 1.01* and three *Ara h 1.02*), and five were validated (two *Ara h 1.01* and three *Ara h 1.02*). From the "high" group, 13 were appropriate (two *AhFAD2B*, one *Ara h 1.01*, five *Ara h 1.02*, three *Ara h 2.01* and two *Ara h 2.02*), although only two *Ara h 1.02* mutations were confirmed. From the "low" group, seven putative mutants were screened (one *AhFAD2A*, one *AhFAD2B*, two *Ara h 1.02*, two *Ara h 2.02*, and one *Ara h 2.01*), but none of them was validated. Since the "common" group had the highest validation rate, SSCP (Single Strand Conformational Polymorphism) was then carried out for the remaining nine samples in this group; only one was validated. Because SSCP might have a higher false-negative rate, these nine samples were then tested by amplicon sequencing and four mutations were detected, including the previous one identified by SSCP. This might indicate that SSCP was not sufficiently sensitive for mutation detection in our study. Altogether, a total of 11 predicted mutations were confirmed by the combined CAPS/SSCP/amplicon sequencing approach.

For the remaining twenty samples in the "high" group and seven samples in the "low" group, five of them were predicted to have missense mutations (three from the "high" group and two from the "low" group) by Geneious software (http://www.geneious.com/) and thus were selected for amplicon sequencing. Two from the "high" group were validated. Altogether, 13 mutations were identified in the single copy genes. Comparing the validation results with the mutation detection parameters, optimum parameters of phred score = 18 or 19, with the minimum read percentage with non-reference nucleotide of 0.17% and maximum read percentage with non-reference nucleotide of 5% could predict the largest number of "true" mutations (Table 3).

In this experiment, we included 384 individuals from our previous TILLING study. In addition to the six previously identified mutants, eight more mutants were recognized. Five additional mutants were identified from the four new TILLING plates (384 individuals). This indicated that TILLING by sequencing is a more sensitive method for mutation detection than the CEL I/LI-COR heteroduplex detection method. One of the possible reasons is that for LI-COR gels, 200 bp are excluded from the amplicon to adjust for the 100 bp regions at the top and bottom of TILLING gel images that are difficult to analyse, which will shorten the actual TILLING target region. Another reason could be that CEL I preferentially recognizes certain mismatches, C/C ≥ C/A ~ C/T ≥ G/G, over others (A/C ~ A/A ~ T/C > T/G ~ G/T ~ G/A ~ A/G) which might decrease the sensitivity for detecting these mutations [42,43].

Three mutations were identified in *Ara h 1.01*. The C to T transition at bp positions 321 and 1687 are silent (Table 4). The third predicts a T 377 I change. The substitution was predicted to be tolerated according to SIFT analysis (http://sift.jcvi.org/) [44] (Table 4). Eight mutations were identified in *Ara h 1.02*. Three of the mutations, with a C to T transition at bp position 765, an A to G transition at bp position 1,258, and a G to A transition at bp position 891, are silent. The other five are predicted to induce amino acid changes: Q 24 H, P 143 I, P 215 H, I 232 V, and K 248 Q. The proline to isoleucine change at position 143 lies within epitopes 8 and 9 [45]. The proline to histidine change at position 215 was predicted to affect protein function by SIFT analysis ($p < 0.05$). Line 125 had five nucleotide changes in *Ara h 1.02*, two of which were silent.

Only known mutations were re-detected in *Ara h 2.01* and *AhFAD2A*. A new mutation was identified in *Ara h 2.02*. The C to T mutation was found upstream of the start codon, but doesn't appear to be located within any important promoter regions. One predicted amino acid change (P 211 L) that might affect protein function according to SIFT analysis was found in *AhFAD2B* (Table 4).

There are two possible reasons the predicted mutations were not validated. Firstly, because homeologous genes were used as TILLING references, in theory, the mutations predicted at identical regions for homeologous genes should be reported for both genes, while most of the time mutations from only one gene were reported due to the default BWA settings, which caused bias in the subsequent validation. Secondly, the false positives could arise from sequencing errors introduced by the combination of lower quality reads (especially for the mutations identified from the "low group") and the ambiguous mutation detection threshold. The minimum read percentage with non-reference nucleotide is set universally for each gene and each library once the quality cut-off is set, while the sequence depth for each gene and library may vary so that for a certain genes or libraries the true mutation calling may be mixed with noise.

Mutation prediction and validation in multi-copy genes

Seventeen validated mutations (eleven of those newly discovered from above and six known) were used as internal positive controls to detect mutations for multi-copy genes, with two major improvements. Firstly, the true mutation percentage and false mutation percentage for the reference genes at all quality scores were calculated and were used as reference to set the mutation detection parameter (Table 5). The improvement was made because when analyzing the first six known mutants, several mutants were relatively easier to detect, whereas one or two others were more difficult to detect. The more cryptic mutants required more relaxed parameters. These relaxed parameters introduced more noise into the results, and likely increased false positives. When using a larger set of reference mutations, and with multiple highly similar sequences, the problem of finding a relaxed value for read percentage with non-reference nucleotide that allowed detection of the entire mutation set was amplified. We thus updated our mutation detection strategy by maximizing efficiency (fewer false and unknown mutants) at the cost of decreasing the total number of true mutants. Secondly, a true/false mutant factor was added for mutation identification. While looking through the results for true mutants missed from the single copy gene prediction, we found that most missed mutants were typically counted ~10-20 times more in 1 row/column, when compared to variant counts in other rows and columns; however, the variant percentage was below the specified minimum non-reference percentage. Thus using minimum variance multiplier will be a more sensitive method for mutation prediction with multi-copy sequences. In addition, in Bowtie 2 alignments, options were set to report all alignments instead of in the previous experiment when the BWA default parameters were used (report first three alignments). This approach was taken because mutations detected in a conserved region for multi-copy genes could not be assigned to a specific copy without further validation (Table 5).

Table 5: The top ten mutant prediction parameters that produced the highest percent of true mutant and false mutant percentage difference using the known mutants as control

Minimum quality score	Minimum variance multiplier	Minimum variance percentage	Number of true mutants[1]	True %	Number of false mutants	False %	Total mutants predicted	Difference
19	12	0.05	11	68.8	1	6.2	16	62.6
19	14	0.05	10	66.7	1	6.7	15	60.0
20	12	0.05	12	63.2	1	5.3	19	57.9
21	12	0.05	12	63.2	1	5.3	19	57.9
16	10	0.05	11	68.8	2	12.5	16	56.3
22	10	0.05	11	61.1	1	5.6	18	55.5
23	8	0.05	11	61.1	1	5.6	18	55.5
21	10	0.05	12	60	1	5	20	55.0
15	10	0.05	10	66.7	2	13.3	15	53.4
16	12	0.05	10	66.7	2	13.3	15	53.4

[1]Based on the validated mutants and the known mutants detected from previous study [40].

For *AhLOX7*, seven out of the top ten predicted mutations had minimum quality scores above 19, and the minimum variance multipliers around 12. We selected six sets (minimum quality 19, 20, 21, with minimum variance multipliers 12 and 14, and the minimum read percentage with non-reference nucleotide of 0.05%) as candidate mutants. Seven of the hypothetical mutants were predicted to be missense and were chosen for validation, including one *AhLOX7_3'* and six *AhLOX7_5'*. The *AhLOX7_3'* and three *AhLOX7_5'* were validated by CAPS assay with expected digestion patterns. The *AhLOX7_3'* accounts for a leucine to proline change at bp position 1508. This mutation is located in exon 7, but not at the fatty acid binding site. The SIFT (Sorting Intolerant from Tolerant) prediction [44] indicated this amino acid change will affect protein function ($p<0.05$). All three mutations at the 5' end of *AhLOX7*, A 17 G, I 175 M, and L178 V, were found in exon 2 and in line 125 (Table 4). These amino acid changes were less likely to affect protein function according to SIFT ($p<0.05$).

For *AhPLD1* and *AhPLD2*, mutations were calculated with minimum quality scores from 10 to 25, and minimum variance multipliers from 2 to 20. Results showed the increase of minimum variance multiplier decreased the number of predicted mutations, as expected, while the increase of minimum quality had little effect on the predicted mutation numbers, which ranged from 4 to 40 when the parameters were changed. We chose twelve sets of parameters to keep the number of predicted mutations within a reasonable range, namely a minimum variance multiplier of 12 and 14 at quality 19, and minimum variance multipliers of 6, 8, 10, 12, and 14 at sequence quality of 20 and 21. All three mutations (all were missense) that were common to all parameter sets were selected for validation. Both sequencing and CAPS assay confirmed that all three predicted mutants were real. Predicted amino acid changes are serine to phenylalanine at position 433, methionine to isoleucine at position 544 for *AhPLD1* and proline to leucine at amino acid 576 for *AhPLD2*. The substitution at position 443 from S to F and the substitution at position 544 from M to I for *AhPLD1* and the substitution at position 576 for *AhPLD2* were predicted to affect protein function according to SIFT. None of them were located in the PLD active site HxKxxxxD (HKD) motif [46], or at the Ca^{2+}-dependent phospholipid binding C2 domain [46-48].

In total, 8.5 kb of the peanut genome was screened for mutations in single copy genes and 27.6 kb was screened for multi-copy genes. Twenty induced point mutations were identified. Single nucleotide substitution was identified in both coding and non-coding regions. Among the identified mutations, six were silent and fourteen were missense mutations; no nonsense mutations were found. The silent/missense mutation ratio is biased because we focused on validating non-silent mutations after the initial CAPS assay for single copy genes. The mutation frequency was calculated as follows: the size in base pairs of the region screened was multiplied by the total number of lines screened per total number of identified mutations. Thus, the overall mutation frequency was 1 SNP/1,066 kb ([(8.5 + 27.6)/ (20 + 6)] × 96 × 8). A total of thirteen mutations were newly identified for single copy genes in this study. The SNP frequency for single copy genes was 1 SNP/344 kb ([8.5/19] × 96 × 8) and the SNP frequency for multi-copy genes was 1 SNP/3,028 kb ([27.6/7] × 96 × 8). Our TILLING population was treated with 0.4% EMS for 12 hours, and compared with the previously reported mutation frequency of 1 SNP/1,067 kb for single copy genes for the same population, the current TILLING by sequencing method had a higher chance for mutation detection [40]. Eleven out of twenty nucleotide changes were G to A or C to T transitions, as expected for EMS-induced mutations. The other nine were unusual mutations. When looking closely at the unusual mutation types, eight out of nine were from the 07JKEMS1 population and out of these seven were from line 125. This line also had more than one base change detected in amplicons, for multiple genes. Moreover, some of the mutations were typical EMS-induced G/C to A/T transitions and others were not. All mutations in line 125 were heterozygous. Therefore line 125 might be interpreted as a suspicious contaminant (except for the high heterozygosity rate). If this sample is excluded, the mutation frequency would be 1 SNP/ 1,459 kb ([(8.5 + 27.6)/(26–7)] × 96 × 8) overall, and 1 SNP/466 kb ([8.5/(19–5)] × 96 × 8) for single copy genes, still higher than the previous report, and 1 SNP/5,299 kb ((27.6/4) × 96 × 8) for multi-copy genes. The mutation frequency decreased from multi-copy gene screening because multi-copy genes are highly similar in sequence and the short reads mapped couldn't be separated to a specific copy. Thus the multi-copy genes have a lower read percentage with non-reference nucleotide than single copy genes. For mutations in the conserved region to be effectively detected, perhaps a higher sequencing depth/minimum read percentage with non-reference nucleotide coverage needs to be achieved. Besides, thresholds need to be set to distinguish gene specific nucleotide differences from real mutations. By comparing with the previous TILLING by sequencing study [17], our mutation detection method has several differences. Firstly, we used a set of known mutations to adjust the parameters for mutation detection. When all known mutations can be detected, we set the threshold for minimum and maximum read percentage with non-reference nucleotide to detect new mutations for single copy genes. Secondly, we developed a method for detecting mutations in multi-copy, highly similar genes. After validating the single copy genes, our mutation detection pipeline was improved by setting the mutations at the ratio of non-reference nucleotide number counts of rows to columns at each position (variant multipliers), which ranged from 8 to 14 at the minimum non-reference nucleotide percentage of 0.05% and Phred quality scores ranging from 16 to 23. Although this method won't allow us to detect the exact gene in which the mutation resides for the highly conserved regions, the subsequent mutation validation showed it was effective in discerning real mutations from background in spite of gene copy number. Thirdly, due to the fact that mutation validation is time consuming, especially for multi-copy genes, and that the uneven pooling and PCR efficiency will all cause some mutations to be easier to detect than others, our mutation detection strategy focused on efficiency instead of detecting all possible mutations. After validating all the mutations, a few of the mutations were found to have a higher minimum non-reference coverage or a lower minimum non-reference coverage than the majority of the other known mutations. Those mutations will obscure the threshold and allow more false positive mutations to be predicted, thus will complicate the subsequent validation process. The updated pipeline set a threshold where the majority of mutations could be detected, which improved the efficiency.

CONCLUSIONS

By combining sequence quality score with minimum read percentage with non-reference nucleotide or non-reference variance multiplier, with the control being previously validated mutations, we refined our mutation discovery pipeline to identify mutations in both single copy and multiple copy genes. TILLING by sequencing identified previously reported genes as well as new mutations from the same population. We also provide evidence that the *AhLOX7* and *AhPLD* genes have multiple copies in the peanut genome. The mutations identified can be used to further study gene function.

METHODS

cDNA library screening and probe design

A cDNA library was constructed from developing roots of peanut (*Arachis hypogaea* L.) genotype C724-19-25 [49] (the isogenic line of TifGuard) with the In-Fusion®SMARTer® directional cDNA library construction kit (Clontech, Mountain View, CA) following the user's manual. Probes were designed from a conserved region identified by aligning EST sequences retrieved from GenBank by similarity to *AhLOX1-3* using the Align X feature with Vector NTI®Suite V6.0 (InforMax, Bethesda, MD). The probes were amplified by PCR using TifGuard as the template, with the amplicon sizes ranging from 150 bp to 300 bp. PCR products were purified with QIAquick Gel extraction kit (Qiagen Inc. Valencia, CA, USA), then were mixed in equal amounts for library screening. The cDNA library was plated on ten 150 × 15 mm Petri dishes (Fisher Scientific, Pittsburgh, PA), with around 2×10^4 colonies per plate. The colonies were transferred to Genescreen Plus nylon membrane (Perkin-Elmer, Boston, MA) following Sambrook and Russell [50]. The probes were labelled with α^{32}P-dCTP using the Random Primed DNA labelling kit (Roche, Indianapolis, IN). Unincorporated label was removed using Sephadex G-50 (Sigma, St. Louis, MO). Hybridization and washing followed Sambrook and Russell [50]. Hybridization signals were detected by exposing filters to film (X-U, Fuji Film, Tokyo). The corresponding clones were picked and plated for a secondary screening following the same protocol. Plasmid DNAs from colonies were cut with *Sac*I and *Xba*I (NEB catalogue No. R0156S, and R0145S) or *Eco*RI + *Bam*HI (NEB catalogue No. R010s and R0136s) to investigate the insertion sizes. Single clones with an insertion size > 2 kb were fully sequenced by primer walking at the Georgia Genomics Facility at the University of Georgia. RT-PCR was carried out by Superscript III first strand synthesis system for RT-PCR (Invitrogen, Carlsbad, CA).

The identified candidate LOX sequence that showed high expression in seeds was used for TILLING. Two amplicons of LOX were used for TILLING, one at the 5' end, starting at the second exon and ending at the fourth exon, and a second at the 3' end starting at exon 7 and ending at exon 9.

Amplification of phospholipase D genes

In peanut, two types of full-length cDNAs were reported, each encoding distinct PLD molecules with 794 (*AhPLD1*) and 807 (*AhPLD2*) amino acid residues, respectively [33,34,51]. Peanut PLD sequences from the above published sources were collected and sorted according to their sequence similarities. Intron-exon prediction was carried out by Splign (http://www.ncbi.nlm.nih.gov/sutils/splign/splign.cgi?textpage=online&level=form). *AhPLD1* has three exons and two introns, while *AhPLD2* has four exons and three introns. Domain prediction by InterProScan indicated three domains for both PLD1 and PLD2. The TILLING primers for *AhPLD1* and *AhPLD2* were designed by CODDLE (http://www.proweb.org/coddle/; The link is no longer available after Dec 31, 2013, suggest primer design with the following linkhttp://tilling.ucdavis.edu/index.php/Primer_Design_and_Testing_Guide). The amplicons were on the second exon for PLD1 and third exon for PLD2, respectively.

Amplicons from *AhPLD1*, *AhPLD2*, and *AhLOX7* were cloned with Zero-Blunt PCR cloning kit (Life Technologies, catalog no. K 2700–20) and colonies were sent to the Georgia Genomics Facility for Sanger sequencing. Sequences were analysed by Geneious software (http://www.geneious.com/).

Illumina library preparation and multiplexing

The mutagenized peanut population development and genomic DNA extraction from individual M_2 plants were previously described [40]. Equal amounts of DNAs were pooled following the bidimensional pooling strategy from Tsai et al. [17]. Briefly, eight 96-well plates were pooled into one 96-well plate according to their corresponding positions (micropools). Subsequently, 20 superpools were generated by combining all the wells in a row (a superpool containing 96 individuals) or in a column (a superpool containing 64 individuals) from the micropool. Primers for amplification of *Ara h 1*, *Ara h 2*, *AhFAD2A* and *AhFAD2B*genes were as described by Knoll et al. [40]. The PCR was carried out in a 20 μl final volume containing 20–30 ng gDNA, 0.4U iProof high fidelity DNA polymerase in 1 × PCR buffer (Bio-Rad, Hercules, CA), 0.2 μM each of dNTP, and 0.5 μM each forward and reverse primers. Reactions were conducted using a Gene Amp 9700 (Applied Biosystems, Carlsbad, CA) thermal cycler. The PCR conditions were as follows: denaturation at 98°C for 30 sec, then 30 to 35 cycles of 98°C for 10 sec, 61 to 65°C for 20 sec, and 72°C for 45 sec to 1 min.

All ten TILLING targets (Table 2) were amplified from the 20 superpools. PCR products were purified with QIAquick PCR purification kit (Qiagen, Valencia, CA) or Agencourt AMPure XP (Beckman Coulter, A63880), and then quantified using Quant-It PicoGreen dsDNA kit (Invitrogen, Carlsbad, CA). Amplicons from the same superpool were mixed for Illumina library preparation. Altogether, twenty five-base barcoded adapters were synthesized for Illumina sequencing [17]. The pooled amplicons were fragmented with NEBNext dsDNA fragmentase (NEB, catalog no. M0348S). Ends were repaired with NEB End Repair Module (catalog no. E6050S). The A bases were added to the 3›-end by NEB Klenow Fragment (NEB, catalog no. M0212s). Ligation was performed by NEB Quick Ligation kit (catalog no. M2200s). The size selection was done by QIAquick gel extraction kit (QIAGEN, catalog no. 287060). Products were enriched by PCR with Phusion High-Fidelity DNA polymerase (NEB, catalogue No. M0530s).

Mutation detection bioinformatics pipeline

The raw sequence was de-multiplexed and quality filtered by the TILLING pipeline from the Comai lab (http://comailab.genomecenter.ucdavis.edu/index.php/TILLING_by_Sequencing) with modifications. A custom Python script was written to replace the Ns in barcodes by checking their mate pair's barcode.

The analysis of single copy genes (*Ara h 1.02*, *Ara h 1.02*, *Ara h 2.01*, *Ara h 2.02*, *AhFAD2A*, *AhFAD2B*) was done by aligning the de-multiplexed sequences to reference sequences with BWA (Burrows-Wheeler Aligner) [52] using parameters –q 20 –k 1, which follows the parameter from the TILLING pipeline from the Comai lab. Samtools was used for creating Mpileup files for each library [53]. A custom Python script was written to pre-process the Mpileup files to create filtered pileups at the minimum Phred quality scores of 10, 12, 14, 15, 16, 17, 18, 19, and 20 for each alignment. The number of unique mutants was counted by a custom Python script when the maximum read percentage with non-reference nucleotide was 5%, and minimum read percentage with non-reference nucleotide ranged from 0.135% to 0.37%. Since the column super-pool had 2/3 the number of samples of the row super-pool, a dilution factor of 0.67 was used on non-reference percentages for column libraries when comparing with row libraries. Six known mutations from a previous study were used as internal controls to ensure that all six mutants were detected at all parameter sets selected and to determine the most efficient mutant calling parameters [40]. Unique mutants were filtered out from the mutant list for further validation. Mutants retrieved using the above parameters were compared and grouped into three categories according to the Phred quality score at which they were identified. Mutants found at all quality scores were placed in a "common" group, mutants found only at Phred scores 10 to 16 were in the "low" group, and mutants found only at Phred scores 17 to 20 were in the "high" group.

After validating the single copy genes, all validated mutants along with the six previously detected mutations were again used as internal controls for setting the parameters to detect multi-copy genes. The updated multi-copy gene mutation detection pipeline aligned all Illumina reads to reference sequences with Bowtie2, which can perform local alignments and exclude ends with low quality or similarity [54], with the parameters set as: perform local alignments allowing 1 seed alignment mismatch and report all alignments. Then the minimum variant percentage values were used to determine the cut-off value of a real mutant. The updated pipeline filtered mutants based on relative variance at the ratio of alternative nucleotide number counts of rows to columns at each position (variant multipliers), which ranged from 8 to 14 at the minimum non-reference count percentage of 0.05% and Phred quality scores ranging from 16 to 23 with a step of 1. With the updated known mutants as references, the numbers of known mutants detected and undetected were reported, respectively. By calculating the percentage of known true and false mutants, parameters resulting in the highest true mutation detection percentage and the lowest false mutation detection percentage were given priority; candidates detected from the above mentioned parameters were used for further validation.

Mutant Validation

All the candidate mutants were analysed by Geneious for potential amino acid changes and CAPS designer (http://solgenomics.net/tools/caps_designer/caps_input.pl) for their capability of CAPS assay. For multi-copy genes, only restriction enzymes with digestion patterns unique to the mutant gene were suitable for CAPS assay. The amplicons were purified by QIAquick PCR purification kit (Qiagen Inc. Valencia, CA, USA) or Agencourt AMPure XP (Beckman Coulter Inc A 63880), and digested with corresponding restriction enzymes. The digested fragments were separated by Ultra-Pure Agarose (Invitrogen) or NuSieve GTG agarose (Lonza Rockland, Atlanta, GA).

For samples not suitable for CAPS assay, SSCP was carried out for validation following the protocol of Zeng et al. [40]. Mutants identified from CAPS assay and SSCP were further confirmed by Sanger sequencing of amplicons (single copy genes) or performing PCR cloning with Zero-Blunt PCR cloning kit (Life Technologies, catalogue no. K 2700–20) prior to sequencing. The mutation effect was analysed by SIFT (Sorting Intolerant from Tolerant, http://sift.jcvi.org/) with default parameters [44]. Amino acids with probabilities <0.05 are predicted to affect protein function.

DECLARATIONS

Acknowledgements

This work was supported by the Georgia Peanut Commission, the National Peanut Board, and the Peanut Foundation. We thank Dr. Luca Comai and Helen Tsai from UC Davis for helpful discussions.

Competing interests

The authors declare that they have no competing interests.

REFERENCES

1. Holbrook C, Timper P, Culbreath AK, Kvien CK. Registration of ‘Tifguard’ Peanut. J Plant Reg. 2008;2:92–4.
2. Branch W. Registration of ‘Georgia Green’ peanut. Crop Sci. 1996;36:806.
3. Culbreath A, Todd J, Branch W, Brown S, Demski J, Beasley J. Effect of new peanut cultivar Georgia Browne on epidemics of spotted wilt. Plant Dis. 1994;78(12):1185–9.
4. Culbreath A, Todd J, Gorbet D, Branch W, Sprenkel R, Shokes F, et al. Disease progress of tomato spotted wilt virus in selected peanut cultivars and advanced breeding lines. Plant Dis. 1996;80(1):70–3.
5. Culbreath A, Tubbs R, Tillman B, Beasley J, Branch W, Holbrook C, et al. Effects of seeding rate and cultivar on tomato spotted wilt of peanut. Crop Prot. 2013;53:118–24.
6. Holbrook C, Stalker H. Peanut breeding and genetic resources. Plant Breed Rev. 2003;22:297–356.Google Scholar
7. Harloff H, Lemcke S, Mittasch J, Frolov A, Wu J, Dreyer F, et al. A mutation screeing platform for rapeseed (Brassica napus L.) and the detection of sinapine biosynthesis mutants. Theor Appl Genet. 2012;124:957–69.
8. Wang N, Wang Y, Tian F, King G, Zhang C, Long Y, et al. A functional genomics resource for *Brassica napus*: development of an EMS mutagenized population and discovery of *FAE1* point mutations by TILLING. New Phytol. 2008;180(4):751–65.
9. Stephenson P, Baker D, Girin T, Perez A, Amoah S, King J, et al. A rich TILLING resource for studying gene function in *Brassica rapa*. BMC Plant Biol. 2010;10:62.
10. Perry J, Brachmann A, Welham T, Binder A, Charpentier M, Groth M, et al. TILLING in Lotus japonicus identified large allelic series for symbiosis genes and revealed a bias in functionally defective ethyl methanesulfonate alleles toward glycine replacements. Plant Physiol. 2009;151(3):1281–91.
11. Till B, Reynolds S, Weil C, Springer N, Burtner C, Young K, et al. Discovery of induced point mutations in maize genes by TILLING. BMC Plant Biol. 2004;4:12.
12. Till B, Cooper J, Tai T, Colowit P, Greene E, Henikoff S, et al. Discovery of chemically induced mutations in rice by TILLING. BMC Plant Biol. 2007, 7(19).
13. Winkler S, Schwabedissen A, Backasch D, Bökel C, Seidel C, Bönisch S, et al. Target-selected mutant screen by TILLING in *Drosophila*. Genome Res. 2005;15:718–23.
14. McCallum C, Comai L, Greene E, Henikoff S. Targeted screening for induced mutations. Nat Biotechnol. 2000;18(4):455–7.
15. Sikora P, Chawade A, Larsson M, Olsson J, Olsson O. Mutagenesis as a tool in plant genetics, functional genomics, and breeding. Int J Plant Genomics. 2011;2011:314829.
16. Rigola D, van Oeveren J, Janssen A, Bonne A, Schneiders H, van der Poel H, et al. High-throughput detection of induced mutations and natural variation using KeyPoint TM technology. PloS One. 2009;4:e4761.
17. Tsai H, Howell T, Nitcher R, Missirian V, Watson B, Ngo K, et al. Discovery of rare mutations in populations: TILLING by sequencing. Plant Physiol. 2011;156(3):1257–68.
18. Siedow J. Plant lipoxygenase:structure and function. Annu Rev Plant Phys. 1991;42:145–88.
19. Burow G, Gardner H, Keller N. A peanut seed lipoxygenase responsive to *Aspergillus* colonization. Plant Mol Biol. 2000;42(5):689–701.
20. Tsitsigiannis D, Kunze S, Willis D, Feussner I, Keller N.

Aspergillus infection inhibits the expression of peanut 13S-HPODE-forming seed lipoxygenases. Mol Plant Microbe In. 2005;18(10):1081–9.

21. Müller V, Amé MV, Carrari F, Gieco JO, Asis R. Lipoxygenase activation in peanut seed cultivars resistant and susceptible to *Aspergillus parasiticus* colonization. Phytopathology. 2014;104(12):1340–8.
22. Exton J. Phospholipase D: Enzymology, mechanisms of regulation, and function. Physiol Rev. 1997;77(2):303–20.
23. Simmonds A, Halsey M. General and local-anesthetics perturb the fusion of phospholipid vesicles. Biochim Biophys Acta. 1985;813(2):331–7.
24. Nishizuka Y. Protein kinase C and lipid signaling for sustained cellular responses. FASEB J. 1995;9(7):484–96.
25. Wang X. Multiple forms of phospholipase D in plants: the gene family, catalytic and regulatory properties, and cellular functions. Prog Lipid Res. 2000;39(2):109–49.
26. Bargmann B, Munnik T. The role of *phospholipase D* in plant stress responses. Curr Opin Plant Biol. 2006;9(5):515–22.
27. Bargmann B, Laxalt A, ter Riet B, van Schooten B, Merquiol E, Testerink C, et al. Multiple *PLDs* required for high salinity and water deficit tolerance in plants. Plant Cell Physiol. 2009;50(1):78–89.
28. Hong Y, Zheng S, Wang X. Dual functions of phospholipase Dα1 in plant response to drought. Mol Plant. 2008;1(2):262–9.
29. Hong Y, Devaiah S, Bahn S, Thamasandra B, Li M, Welti R, et al. Phospholipase Dε and phosphatidic acid enhance Arabidopsis ni*trogen sign*aling and growth. Plant J. 2009;58(3):376–87.
30. Wang X. Phospholipase D in hormonal and stress signaling. Curr Opin Plant Biol. 2002;5(5):408–14.
31. Wang X. Lipid signaling. Curr Opin Plant Biol. 2004;7(3):329–36.
32. Wang X. Regulatory functions of phospholipase D and phosphatidic acid in plant growth, development, and stress responses. Plant Physiol. 2005;139(2):566–73.
33. Guo B, Xu G, Cao Y, Holbrook C, Lynch R. Identification and characterization of phospholipase D and its association with drought susceptibilities in peanut (*Arachis hypogaea*). Planta. 2006;223(3):512–20.
34. Nakazawa Y, Sato H, Uchino M, Takano K. Purification, characterization and cloning of phospholipase D from peanut seeds. Protein J. 2006;25(3):212–23.
35. Hayashi S, Gresshoff P, Kinkema M. Molecular analysis of lipoxygenases associated with nodule development in soybean. Mol Plant Microbe In. 2008;21(6):843–53.
36. Porta H, Rueda-Benítez P, Campos F, Colmenero-Flores J, Colorado J, Carmona M, et al. Analysis of lipoxygenase mRNA accumulation in the common bean (*Phaseolus vulgari*s L.) during development and under stress conditions. Plant Cell Physiol. 1999;40(8):850–8.
37. Yenofsky R, Fine M, Liu C. Isolation and characterization of a soybean (*Glycine max*) lipoxygenase-3 gene. Mol Gen Genet. 1998;211:215–22.
38. Reinprecht Y, Luk-Labey S, Yu K, Poysa V, Rajcan I, Ablett G, et al. Molecular basis of seed lipoxygenase null traits in soybean line OX948. Theor Appl Genet. 2011;122(7):1247–64.
39. Steczko J, Axelrod B. Identification of the iron-binding histidine residues in soybean lipoxygenase L-1. Biochem Biophys Res Commun. 1992;186(2):686–9.
40. Knoll J, Ramos M, Zeng Y, Holbrook C, Chow M, Chen S, et al. TILLING for allergen reduction and improvement of quality traits in peanut (Arachis hypogaea L.). BMC Plant Biol. 2011;11:81.
41. Ramos M, Huntley J, Maleki S, Ozias-Akins P. Identification and characterization of a hypoallergenic ortholog of Ara h 2.01. Plant Mol Biol. 2009;69(3):325–35.
42. Oleykowski CA, Bronson Mullins CR, Godwin AK, Yeung AT. Mutation detection using a novel plant endonuclease. Nucleic Acids Res. 1998;26(20):4597–602.
43. Yang B, Wen X, Kodali NS, Oleykowski CA, Miller CG, Kulinski J, et al. Purification, cloning, and characterization of the CEL I nuclease. Biochemistry. 2000;39(13):3533–41.
44. Ng P, Henikoff S. SIFT: Predicting amino acid changes that affect protein function. Nucleic Acids Res. 2003;31(13):3812–4.
45. Shin D, Compadre C, Maleki S, Kopper R, Sampson H, Huang S, et al. Biochemical and structural analysis of the IgE binding sites on Ara h1, an abundant and highly allergenic peanut protein. J Biol Chem. 1998;273(22):13753–9.
46. Ponting C, Kerr I. A novel family of phospholipase D homologues that includes phospholipid synthases and putative endonucleases: Identification of duplicated repeats and potential active site residues. Protein Sci. 1996;5(5):914–22.
47. Nalefski E, Falke J. The C2 domain calcium-binding motif: Structural and functional diversity. Protein Sci. 1996;5(12):2375–90.
48. Zheng L, Krishnamoorthi R, Zolkiewski M, Wang X. Distinct Ca^{2+} binding properties of novel C2 domains of plant phospholipase Dα and β. J Biol Chem. 2000;275(26):19700–6.
49. Holbrook C, Timper P, Dong W, Kvien C, Culbreath A. Development of near-isogenic peanut lines with and without resistance to the peanut root-knot nematode. Crop Sci. 2008;48(1):194–8.
50. Sambrook J, Russell D. Molecular cloning: A laboratory manual. Cold Spring Harbor, NY: Cold Spring Harbor Laboratory Press; 2001.
51. Dramé K, Clavel D, Repellin A, Passaquet C, Zuily-Fodil Y. Water deficit induces variation in expression of stress-responsive genes in two peanut (*Arachis hypogaea* L.) cultivars with different tolerance to drought. Plant Physiol Biochem. 2007;45(3–4):236–43.

52. Li H, Durbin R. Fast and accurate short read alignment with Burrows-Wheeler transform. Bioinformatics. 2009;25(14):1754–60.

53. Li H, Handsaker B, Wysoker A, Fennell T, Ruan J, Homer N, et al. The sequence alignment/map format and SAMtools. Bioinformatics. 2009;25(16):2078–9.

54. Langmead B, Trapnell C, Pop M, Salzberg S. Ultrafast and memory-efficient alignment of short DNA sequences to the human genome. Genome Biol. 2009;10(3):R25.

CHAPTER 10

INTERSPECIFIC AND INTERGENERIC HYBRIDIZATION AND CHROMOSOMAL ENGINEERING OF BRASSICACEAE CROPS

Yukio Kaneko and Sang Woo Bang

Keywords: Brassicaceae, alien gene introgression, alloplasmic, amphidiploid, chromosomal engineering, interspecific and intergeneric hybridization, monosomic alien chromosome addition

INTRODUCTION

The Brassicaceae comprise about 330 genera and 3700 species, of which two genera (*Brassica* and *Raphanus*) are widely grown for edible oils, vegetables, spices, ornamental flowers, and forage crops around the world. The genus *Brassica* contains three monogenomic diploid species, namely *B. rapa* L. (2n = 20, AA genome), *B. nigra* (L.) Koch (2n = 16, BB), and *B. oleracea* L. (2n = 18, CC), and their naturally produced allotetraploid species, *B. napus* L. (2n = 38, AACC), *B. juncea* (L.) Czern. & Coss. (2n = 36, AABB), and *B. carinata* A. Braun (2n = 34, BBCC). In contrast, the genus *Raphanus* contains only one species of agricultural importance, *R. sativus* (2n = 18, RR) (Mizushima 1950). Other crop species in this family include *Eruca sativa* L. (rocket salad), *Nasturtium officinale* R. Br. (water-cress), *Wasabia japonica* Matsumura (wasabi), *Matthiola incana* (L.) R. Br. (stock), and *Erysimum cheiri* (L.) Crantz (English wallflower). A number of wild relatives, on the other hand, have been evaluated as genetic resources in the development of new cultivars with biotic and abiotic stress resistance for use in agriculture in response to the diversification of ecotypes of diseases and pests, changing food preferences, adoption of advanced production technology, the use of new approaches such as *in vitro* breeding programs, and the need for economical production of F_1 seed (Warwick 1993).

Since *Raphanobrassica*, an intergeneric hybrid between *R. sativus* and *B. oleracea*, was reported by Karpechenko (1928), a number of intergeneric and interspecific hybridizations between *Brassica* species of the U triangle (U 1935) and *R. sativus* have been produced; the U triangle refers to the ancestral A, B, and C genomes that have given rise to modern *Brassica* crop species. For nearly a century, cytogenetic research based on their hybridization has been conducted, resulting in the discovery of excellent evidence for the cytogenetics and speciation of these species. Recently, interspecific and intergeneric hybridizations have been extensively performed between the cultivated species and between cultivated species and wild relatives to develop more potentially useful cultivars with improved biotic and abiotic stress tolerance (Harberd and McArthur 1980, Prakash *et al.* 1999).

However, cross-incompatibility in the interspecific and intergeneric hybridizations has hampered the development of these hybrids. Using knowledge gained from anatomical studies of the growth of hybrid embryos, Nishi *et al.* (1970)and Inomata (1977) produced interspecific and intergeneric hybrids through embryo and ovary culture techniques, respectively.

Matsuzawa *et al.* (1996) suggested a system for the use of interspecific and intergeneric hybridizations to develop five types of hybrid lines: synthetic amphidiploid lines, alien gene introgression lines, alloplasmic lines, monosomic alien chromosome addition lines, and monosomic alien chromosome substitution lines (Fig. 1). These hybrid lines would be valuable genetic resources both for breeding more productive cultivars with novel agronomic traits and for research to better understand each chromosome and gene in these hybrids. In each instance, it is first necessary to develop true F_1 hybrids and as many of their progeny as possible.

Citation: Yukio Kaneko and Sang Woo Bang, Interspecific and intergeneric hybridization and chromosomal engineering of Brassicaceae crops, doi:10.1270/jsbbs.64.14.

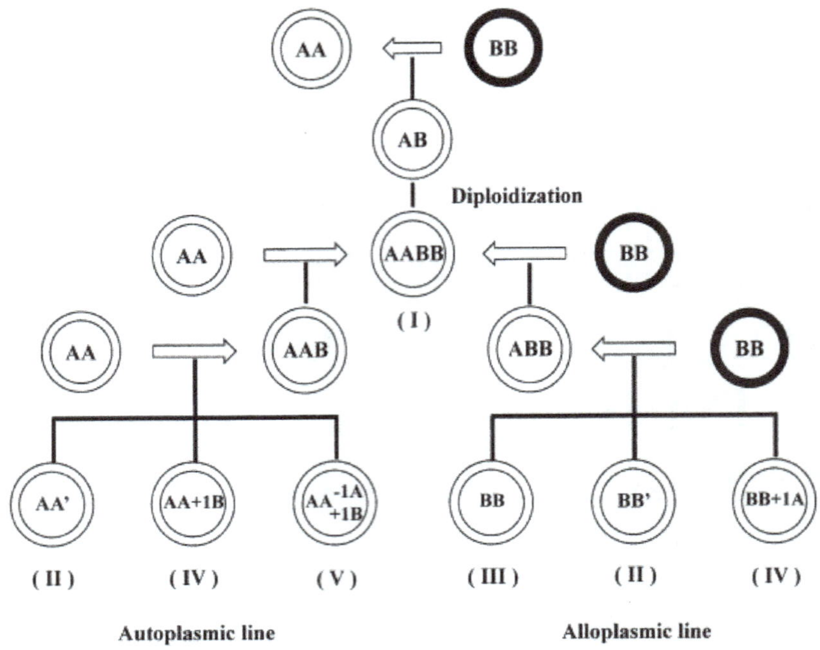

Figure 1: Schematic diagram of the distant hybridization breeding system between AA and BB genome species concerned (quoted fromMatsuzawa *et al.* 1996, revised). (I): Synthetic amphidiploid line (SADL), (II): Alien gene(s) introgression line (AGIL), (III): Alloplasmic line (ALPL), (IV): Monosomic alien chromosome addition line (MAAL), (V): Monosomic alien chromosome substitution line (MASL) 1) AB, AABB and AAB (ABB) show genomes for amphihaploid, amphidiploid and sesquidiploid, respectively. 2) A′ (B′) means some genetic modification via recombination between each complement of A and B genome. 3) AA + 1B(BB + 1A) means A (B) genome species added single chromosome of B (A) genome species. 4) ○ and ● show the difference in cytoplasmic background for A and B genome species, respectively. 5) A←B means the hybridization in which A and B genome species are pistillate and pollen parents, respectively. 6) In (V), a chromosome of A-genome is substituted by one of B-genome.

In this communication, we review current knowledge of the interspecific and intergeneric cross-incompatibility between *Brassica* crops and their wild relatives, and the production, characterization, and improvement of the five types of hybrid lines.

HYBRIDIZATION BARRIERS IN INTERSPECIFIC AND INTERGENERIC HYBRIDIZATIONS

The barriers to interspecific and intergeneric hybridization during sexual reproduction can be divided between those that operate before and after fertilization. Stebbins (1958) suggested that the pre-fertilization barriers might be due to failure of pollen germination, pollen tube growth, or pollen tube penetration of the ovule, whereas post-fertilization barriers would arise from degeneration of the hybrid embryo, male and female sterility in the hybrid plants, and lethality in the hybrid progeny. Khush and Brar (1992) surveyed the hybridization barriers in distant hybridization, and offered effective techniques to overcome both pre- and post-fertilization barriers.

Several researchers have investigated the nature of the pre-fertilization barriers for interspecific and intergeneric hybridizations in the Brassicaceae. For example, Matsuzawa (1983) evaluated the magnitude of the pre-fertilization barriers using a pollen germination index:

$$\text{P.G.I.} = (1b + 2c + 3d + 4e)/(a + b + c + d + e)$$

where $0 \leq \text{P.G.I.} \leq 4$, and *a*, *b*, *c*, *d*, and *e* represent the numbers of pistils in which no pollen grain is recognized on the stigma (with a score of 0), pollen grains do not germinate on the stigma (1), pollen grains germinate on the stigma but do not enter (2), pollen tubes reach the style tissue (3), and pollen tubes penetrate the style tissue to reach near or to the ovule (4), respectively. Kerlan *et al.* (1992) proposed a different equation for the index of pollination compatibility (*I*):

$$I = x + 2y + 3z$$

where *x*, *y*, and *z* are the corresponding scores for the numbers of pollen grains that germinate divided by the number that arrived on the stigma (with a score of 0, 1, 2 or 3 assigned to the ratio values of 0, 0–0.5, 0.5–0.7 and 0.7–1, respectively), of pollen tubes that grew into the style tissue (with a score of 0, 1 and 2 corresponding to 0, 1 to 5 and more than 6 pollen tubes, respectively), and of pollen tubes that penetrated the ovule (with a score of 0, 1 and 2 assigned to the same scores used for the y parameter), respectively.

In general, the pre-fertilization barriers between cultivated species and their wild relatives in interspecific and intergeneric hybridizations can be overcome by means of bud pollination and using wild species as the pistillate parent, although the proportion of successful outcomes may depend on the direction and combination in each crossing. In particular, when a self-compatible line was used as the pistillate parent, the pollen tubes of cultivated species grew well and penetrated into the ovule. Therefore, the prefertilization barriers in interspecific and intergeneric hybridizations seem to be similar to the self-incompatibility observed in Brassicaceae species, although genetic analysis using the F_2 population between an interspecific-incompatible line and a self-compatible cultivar could not confirm the responsibility of the line and the cultivar in the differences of the interspecific incompatibility phenotypes (Udagawa *et al.* 2010).

In contrast to pre-fertilization barriers, Wilmar and Hellendoorn (1968) evaluated post-fertilization barriers by means of an anatomical survey of the development and growth of embryos of Brussels sprouts (*B. oleracea*). The abortion of hybrid embryos in interspecific and intergeneric hybridizations resulted from abnormal development of the endosperm, and the underlying mechanisms have been explained by various hypotheses: the endosperm balance number (Johnston *et al.* 1980), activation of the polar nuclei (Nishiyama and Yabuno 1978), and genomic imprinting in the endosperm (Kinoshita 2007). Recently, Tonosaki *et al.* (2013) performed quantitative trait locus analysis during the formation of hybrid seed in intergeneric hybridization between *B. rapa* and *R. sativus*. Researchers have developed effective *in vitro* culture techniques using pollinated flowers, ovaries, ovules, and embryos to rescue hybrid embryos that would otherwise degenerate during the early stages of their development in interspecific and intergeneric hybridizations. Bang *et al.* (1996 to 2009) and Jeong *et al.* (2003 to 2009) inspected the post-fertilization barriers between cultivated species and wild relatives in intergeneric hybridizations, and produced many potential hybrid progeny through *in vitro* procedures for each of the five breeding strategies shown in Fig. 1.

PRODUCTION OF NOVEL F_1 HYBRID PLANTS BY MEANS OF EMBRYO RESCUE

Until the 1980s, F_1 hybrid plants had been produced from interspecific and intergeneric hybridizations mainly among *B. rapa*, *B. nigra*, *B. oleracea*, *B. napus*, *B. juncea*, *B. carinata* and *R. sativus*. Subsequently, a number of F_1hybrid plants were produced from interspecific and intergeneric hybridizations between *Brassica* crops and wild relatives in the genera *Brassica*, *Sinapis*, *Diplotaxis*, *Moricandia*, *Eruca*, and *Orychophragmus*. At least 45 species of wild relatives were reported in reviews of the literature on the production of F_1 hybrids (Harberd and McArthur 1980,Kaneko *et al.* 2009, Prakash *et al.* 1999). Numerous novel F_1 hybrids have been produced through the development of various embryo rescue techniques: embryo culture (Nishi *et al.* 1962, 1970), ovary culture (Inomata 1977, 1978a,1978b, Matsuzawa 1983), ovule culture (Momotaz *et al.* 1998) and placenta culture (Bang *et al.* 2007), as well as through combinations of these techniques, namely ovary culture followed by embryo or ovule culture, placenta culture followed by embryo culture, and successive ovary, ovule, and embryo culture. Ovary culture has been widely performed to overcome abortion of hybrid embryos during early stages of development, but this technique failed when a species such as *B. oleracea* with high regeneration ability was used as the pistillate parent. When *R. sativus*and *B. oleracea* are used as the pistillate parent, embryo culture is effective, although the hybrid embryo must develop to at least the heart-shaped stage before culture can begin. Placenta culture, in which placenta-attached ovules are removed and cultured starting about 15 days after pollination, may be more effective than ovule and embryo culture when *B. oleracea* is used as the pistillate parent.

SYNTHETIC AMPHIPLOID LINES

Synthetic amphidiploid lines have been developed by doubling the chromosome number of F_1 hybrid plants as well as by means of somatic cell fusion. Researchers have sporadically obtained amphidiploid plants in the F_2generation of amphihaploid F_1 hybrids through fusion of the unreduced F_1 female and male gametes. These amphidiploid lines show low pollen and seed fertility due to unequal chromosome segregation in pollen mother cells in initial generations. With advancing generations, however, the hybrids generally develop high pollen and seed fertility as a result of increasing meiotic stabilization (Howard 1938, Kato and Tokumasu 1976, Sarashima 1973). Six artificial amphidiploid *Brassica* lines, four *Raphanobrassica* lines and three *Brassicoraphanus* lines have been developed and maintained for use as breeding materials (Table 1 and Fig. 2).

Table 1: Self-pollinated seed fertility in artificial amphidipoids from reciprocal crosses among monogenomic *Brassica*crops as well as *Raphanus sativus*

Cross combination	Genome	Generations	No. of individuals observed	Seed fertility
B. rapa × *B. oleracea*	AACC	F_4	6	0.72 ± 0.63
B. oleracea × *B. rapa*	CCAA	F_4	10	1.19 ± 2.11
B. rapa × *B. nigra*	AABB	F_7	1	7.51
B. nigra × *B. rapa*	BBAA	F_6	3	2.21 ± 1.19
B. nigra × *B. oleracea*	BBCC	F_6	4	9.16 ± 2.48
B. oleracea × *B. nigra*	CCBB	F_7	5	1.55 ± 0.99
R. sativus × *B. oleracea*	RRCC	F_{10}	3	2.39 ± 0.88
B. oleracea × *R. sativus*	CCRR	F_4	5	0.76 ± 0.46
R. sativus × *B. rapa*	RRAA	F_7	–	1.3[a]
B. rapa × *R. sativus*	AARR	F_7	–	0.01[b]

B. napus 'N350' is 21.15 seeds, *B. juncea* 'Aso-Takana' 11.54, and *B. carinata*.

Ca115' 8.72, respectively.

[a] Data from Akaba *et al.* 2009b.

[b] Data from Matsuzawa *et al.* 2000.

Figure 2: Artificial synthetic amphidiploid plants. (A) Raphanobrassica(2n = 36, RRCC) derived from hybrid between R. sativus (2n = 18, RR) and B. oleracea (2n = 18, CC), (B) Brassicoraphanus(2n = 34, FFRR) derived from hybrid between B. maurorum (2n = 16, FF) and R. sativus (2n = 18, RR), (C) Brassicoraphanus(2n = 50, FFFFRR) derived from hybrid between B. fruticulosa(2n = 32, FFFF) and R. sativus (2n = 18, RR), (D)Brassicoraphanus (2n = 36, OORR) derived from hybrid between B. oxyrrhina (2n = 18, OO) and R. sativus (2n = 18, RR).

Synthetic fertile amphidiploid lines derived from hybrids between cultivated crops and their wild relatives have been used as genetic stocks in breeding programs, and as bridging materials for the transfer of desirable traits from wild species into cultivated ones. The fertile amphidiploid lines obtained through doubling of the chromosome number of sterile hybrids or through somatic hybridization between a cultivated crop and wild relatives have led to novel crops, including 'Hakuran' (*Brassica campestris* × *B. oleracea*), 'Radicole' (*Raphanus sativus* × *B. oleracea*), and 'Raparadish' (*B. campestris* × *R. sativus*) (Namai 1987). As a result, several novel cultivars, 'Gifu-Green' (Takada 1985), 'Senpousai No. 1 and No. 2' (Nagano 1988), 'Hanakkori' (Matsumoto *et al.* 1997), and 'F_1-Hosoda-Wase' (Takada 2006), were developed from amphidiploids between *B. rapa* and *B. oleracea*, and have been released by Japanese seed companies. Recently, we produced a number of amphidiploid lines between crop cultivars and wild relatives: *Brassica maurorum* × *R. sativus* (Bang *et al.* 1997, 1998, Fig. 2B), *Brassica fruticulosa* × *R. sativus* (Bang *et al.* 1997, 2000, Fig. 2C), *Brassica oxyrrhina* × *R. sativus* (Bang *et al.* 1997, Matsuzawa *et al.* 1997, Fig. 2D), *B. oxyrrhina* × *B. rapa* (Bang, unpublished data), *B. oxyrrhina* × *B. oleracea* (Bang, unpublished), *B. rapa* × *Diplotaxis tenuifolia* (Jeong *et al.* 2009), and *B. oleracea* × *D. tenuifolia* (Bang, unpublished). *Diplotaxis tenuifolia* has considerable potential as a healthy leafy salad vegetable because of its content of bioactive photochemicals, such as those that are involved in intermediate C_3–C_4 photosynthetic activity, and is commonly eaten as "rocket", a name it shares with some *Eruca* species (Martínez-Sánchez *et al.* 2007). The amphidiploid line of *B. rapa* × *D. tenuifolia* has desirable morphological characteristics, vigorous growth, and the attractive fragrance of *D. tenuifolia*, but its seed fertility was insufficient for commercial production as a crop. However, the line produced abundant seeds that grew into sesquidiploid plants, which inherited desirable traits when they were backcrossed with commercial cultivars of *B. rapa* (Jeong *et al.* 2009). Therefore, the amphidiploid lines of *B. rapa* × *D. tenuifolia* and *B. oleracea* × *D. tenuifolia*could both be useful genetic resources for the breeding of new leafy salad vegetables.

MONOSOMIC ALIEN CHROMOSOME ADDITION LINES

Chromosome addition lines, such as monosomic alien chromosome addition lines (MAALs), have been examined for genetic analysis in breeding programs for agronomic traits, and in studies of genes that are assumed to be located on the added chromosome. The advantages of using MAALs involve the possibility of assigning species-specific genes or characteristics to particular chromosomes, and the potential to transfer desirable agronomic traits between species (Matsuzawa *et al.* 1996, McGrath and Quiros 1990, Namai 1987, Prakash and Chopra 1990). Since Kaneko*et al.* (1987) produced MAALs that combined *R. sativus* with *B. oleracea* and Quiros *et al.* (1987) produced MAALs that combined *B. rapa* with *B. oleracea*, several MAALs have been bred through interspecific and intergeneric hybridizations between crop species and wild relatives. These include combinations of *B. campestris* with *Brassica alboglabra* (Chen *et al.* 1992), *R. sativus* with *E. sativa* (Bang 1996), *R. sativus* with *Sinapis arvensis* (Bang 1996), *B. campestris* with *B. oxyrrhina* (Srinivasan *et al.* 1998), *R. sativus* with *Moricandia arvensis* (Bang *et al.* 2002), *B. napus*with *Sinapis alba* (Wang *et al.* 2005), *B. napus* with *R. sativus* (Akaba *et al.* 2009b), and *B. napus* with *B. juncea*(Takashima *et al.* 2012).

To produce and classify MAALs, the cross-incompatibility that exists in successive backcrosses with the background species (as the recurrent parent) must be overcome, and it is essential to use specific markers for the chromosome from the donor species to assist in selection. The cross-incompatibility that often appears in successive backcrosses of the amphidiploid (F_1) progeny with the background species seems to depend upon the genetic affinity between the background species and the chromosome donor. These barriers in the first backcross may be reflected in male and female sterility of the amphidiploid as well as in pre- and post-fertilization barriers (Table 2). In the intergeneric hybridization between *M. arvensis* and *R. sativus* (Bang *et al.* 1996a), the amphidiploid plants showed more stable chromosome association at metaphase I in pollen mother cells and higher pollen fertility than the amphihaploid plants. In the backcross of the amphidiploid with the parent *R. sativus*, there was no pre-fertilization barrier, but a post-fertilization barrier appeared in the form of abortion of the hybrid embryos, although this was overcome using an embryo rescue technique. Occasionally, male and female sterility of the BC_1 hybrid plants (sesquidiploids) and subsequent pre- and post-fertilization barriers also appear in successive back-crossing (usually by BC_2). On the other hand, these barriers in the first and second backcrosses were not observed in other cross-combinations to *R. sativus* × *S. arvensis* (Bang 1996), *R. sativus* × *E. sativa* (Bang 1996), and *B. campestris* × *B. oxyrrhina* (Srinivasan *et al.* 1998). However, whole MAALs have not been developed from any cross-combinations in the Brassicaceae thus far. The genetic system that controls hybridization barriers in interspecific and intergeneric hybridizations seems to function between the nuclear and cytoplasmic genomes in the recurrent parent's genetic background and a few chromosomes from the donor genome.

Table 2: Seed fertility by backcrossing between amphidiploid plants and their parents

Pistillate parent	Genome	Male parent	Genome	Seed fertility
Artificial *B. napus*	AACC	*B. rapa*	AA	2.50 ± 1.55
Natural *B. napus*	AACC	ditto		9.47 ± 3.42
Artificial *B. napus*	CCAA	ditto		3.06 ± 3.13
Artificial *B. juncea*	AABB	*B. rapa*	AA	0.26
Natural *B. juncea*	AABB	ditto		9.01 ± 3.40
Artificial *B. juncea*	BBAA	ditto		2.25 ± 0.62
Artificial *B. carinata*	BBCC	*B. oleracea*	CC	1.72 ± 0.58
Natural *B. carinata*	BBCC	ditto		0.37 ± 0.04
Artificial *B. carinata*	CCBB	ditto		0.01 ± 0.01
Raphanobrassica	RRCC	*R. sativus*	RR	0.12 ± 0.06[a]
Raphanobrassica	RRAA	*R. sativus*	RR	0.03 ± 0.03
Brassicoraphanus	AARR	*B. rapa*	AA	0.95 ± 1.59

[a] Data from Kaneko *et al.* (1987).

These MAALs have been effectively exploited to analyze various agronomic and genetic traits to homoeologous relationships in the genus *Brassica* (Kaneko *et al.* 2002); disease resistance to blackleg (*Leptosphaeria maculans*;Chevre *et al.* 1996), turnip mosaic virus (TuMV; Kaneko *et al.* 1996), beet cyst nematode (*Heterodera schachtii*;Peterka *et al.* 2004), and clubroot (*Plasmodiophora brassicae*; Akaba *et al.* 2009a); and photorespiratory characteristics (Bang *et al.* 2009); and in restoring pollen fertility (Akaba *et al.* 2009b, Budahn *et al.* 2008) and avoiding lethality in an alloplasmic line (Tsutsui *et al.* 2011).

ALLOPLASMIC LINES

The production of F_1 hybrid seed of *Brassica* crops is currently based on either self-incompatibility or cytoplasmic male sterility (CMS). The self-incompatibility controlled by *S* alleles is not sufficiently stable because it is affected by various environmental factors (Horisaki and Niikura 2004). CMS is a maternally inherited trait, and has been ascertained to be more successful because of the stable expression of pollen sterility without any obvious changes in vegetative growth or female fertility. Alloplasmic lines have been developed by exchanging the cytoplasm through inter-specific and intergeneric hybridizations followed by successive backcrossing to the nuclear genome's donor species. These hybrids acquire the advantages of CMS and useful agronomic traits, such as herbicide resistance, modification of flavor and nutrient content, and morphogenetic potential (Namai 1987).

Sarashima *et al.* (1990a, 1990b, 1990c) produced many kinds of alloplasmic lines of monogenomic Brassicaceae crops (*B. rapa*, *B. oleracea*, and *R. sativus*), but most of them did not exhibit any agronomically useful traits, except for CMS of *B. oleracea*, which carried *R. sativus* cytoplasm but no *R. sativus* nuclear genetic material. The CMS in the latter alloplasmic line may have been induced by a lack of genetic affinity between the mitochondrial genome and the nuclear genome. The cytoplasm from several wild relatives has been introduced as a source of CMS into several*Brassica* crops (*B. rapa* [*B. campestris*], *B. juncea*, and *B. napus*) through interspecific and intergeneric hybridizations followed by successive backcrossings. These wild relatives include *Arabidopsis thaliana*, *Diplotaxis berthautii*,*Diplotaxis muralis*, *Diplotaxis catholica*, *Diplotaxis siifolia*, *Diplotaxis erucoides*, *B. oxyrrhina*, *Brassica tournefortii*,*Enarthrocarpus lyratus*, *E. sativa*, *Eruca vesicaria*, *M. arvensis*, and *Trachystoma ballii* (reviewed by Kaneko *et al.*2009, Prakash *et al.* 1999).

Recently, some new CMS sources have been introduced into several *Brassica* crops in addition to *R. sativus*, including *B. maurorum* (Bang *et al.* 2011), *B. oxyrrhina* (Shim *et al.* 2010, 2011a, 2012), *D. tenuifolia* (Shim *et al.* 2011b), and *B. fruticulosa* (Tsutsui 2013).

When alloplasmic lines are produced through interspecific and intergeneric hybridizations followed by successive backcrossings, there are various affinity natures that can be observed in cross-combinations and in different generations. For example, two alloplasmic *B. juncea* lines were produced using amphidiploid lines of *D. erucoides* ×*B. campestris* and *D. berthautii* × *B. campestris* as cytoplasmic donor plants (Malik *et al.* 1999). Alloplasmic *B. napus*carrying *A. thaliana* cytoplasm was obtained from six backcrosses of the somatic hybrid to *B. napus* (Leino *et al.*2003). Alloplasmic *B. campestris* with *E. sativa* cytoplasm was produced via a sesquidiploid plant obtained from an amphihaploid F_1 hybrid by open pollination (Matsuzawa *et al.* 1996), and alloplasmic *R. sativus* with *E. vesicaria*cytoplasm was produced via the same method (Bang 1996). Alloplasmic *R. sativus* with *M. arvensis* cytoplasm was also produced via a sesquidiploid plant generated from backcrossing of an amphidiploid F_1 hybrid to *R. sativus*, followed by embryo rescue (Bang *et al.* 1996a, 2002). As Namai (1976) indicated, sesquidiploid plants can be useful as a bridge plant to generate an alloplasmic line with no fertilization barriers in various interspecific and intergeneric hybridizations.

Occasionally, male or female sterility of sesquidiploid plants and subsequent pre- and post-fertilization barriers also appear in successive backcrossings. When this happens, an alloplasmic line cannot be generated. The pre- and post-fertilization barriers can often be overcome by using bud-pollination and embryo rescue, respectively. The female sterility of sesquidiploid plants, which is a primary cause of hybrid breakdown, can also be overcome by using F_1 cultivars as the pollen parents and subsequently as the recurrent parent because male and female gametes with diverse genotypes can be confirmed in each generation. In sesquidiploid hybrids between *Brassica* crop species and their wild relatives, Tsutsui *et al.* (2011) found that many progeny involving alloplasmic lines could be generated through repeated pollination without embryo rescue, but that no progeny could be obtained by bud pollination and subsequent embryo rescue techniques owing to severe chlorosis. It is likely that the female sterility in the sesquidiploid plants was caused by delayed maturation of the egg cells. As mentioned above, using synthetic allopolyploids between different combinations as bridge plants to transfer cytoplasm from wild relative species and ascertaining the cause of sexual incompatibility barriers in the hybrid plants as accurately as possible may provide practical ways to improve the efficiency of producing alloplasmic lines.

ALIEN GENE INTROGRESSION LINES

Until the 1980s, alien gene introgression lines had been produced by hybridization mainly between the Brassicaceae crops, resulting in the development of novel cultivars with desirable agronomic traits. Examples include Chinese cabbage (*B. rapa*) cultivars resistant to bacterial soft rot (Shimizu *et al.* 1962), early-maturing cultivars of *B. napus* (Namai 1976, 1987), and virus-resistant cultivars of*B. oleracea* (Namai 1976). Warwick (1993) described wild genera of the tribe Brassiceae as the sources of various agronomically interesting traits, such as hairiness, resistance to pod shattering, photosynthesis, soil adaptation and disease resistance. A large collection of wild relative species of*Brassica* crops have been screened as sources of resistance to blackleg (also called stem canker), alternaria leaf spot (*Alternaria* spp.), and clubroot. Interspecific and intergeneric hybrids between wild relatives and *Brassica* crop species, and their progeny, would certainly provide valuable breeding materials for the introgression of agronomically useful traits. Siemens (2002) reviewed the possibility of using interspecific and intergeneric hybridizations between *B. napus* and some wild relatives in the tribe Brassiceae to introduce resistance to important fungal pathogens. AsNamai (1987) indicated, the introgression of desirable genes from one species to another in the Brassiceae has been accomplished through meiotic allosyndesis in *Brassica* crops. Bang (1996) described the homoeologous pairing between an *R. sativus* chromosome and an *S. arvensis* chromosome in an *S. arvensis* monosomic addition line of *R. sativus* (MAAL-d type). A pollen fertility restoration gene from the cytoplasmic donor species has been transferred to alloplasmic *Brassica* oilseed crops through homoeologous pairing between the AB and M chromosomes in an *M. arvensis* MAAL of *B. juncea* (Prakash *et al.* 1999), and between the A and E chromosomes in an *E. lyratus* MAAL of *B. rapa* (Banga *et al.* 2003). Recently, resistance to the beet cyst nematode and clubroot has been transferred from *R. sativus* to *B. napus* using *R. sativus* MAALs of *B. napus* (Akaba *et al.* 2009a, Budahn *et al.* 2008). Analysis of genes and genomes and the efficient induction and identification of recombination for introgression might be prerequisites for the development of an alien gene introgression line. The genetic linkage maps that have been recently constructed for some *Brassica* crops will offer useful tools to support genetic introgression.

MONOSOMIC ALIEN CHROMOSOME SUBSTITUTION LINES

Many disomic alien chromosome substitution lines (DASLs) have been bred in wheat cultivars. In these lines, one member of a chromosome pair is replaced by a homoeologous chromosome from *Secale*, *Hordeum*, and *Aegilops*(Khush 1973). Such DASLs can be used to study the genetic effects of individual chromosomes, and to estimate the number of genes that control a given trait and their linkage relationships. These lines can be generally developed in the progeny after self-fertilization of a monosomic alien chromosome substitution line (MASL). In this breeding system, the MASLs can be generated only when the two corresponding chromosomes are homoeologous and complement each other, especially within diploid species. In the Brassicaceae, DASLs and MASLs generated using the technique described above have not been found, as far as we know. DASLs are known to occur spontaneously during successive backcrossings after interspecific hybridization. By analyzing the meiotic configuration, Banga (1988) reported the development of C genome chromosome substitution lines in *B. juncea* that arose spontaneously in an interspecific hybridization between *B. juncea* and *B. napus*. Although phylogenetic relationships have been suggested in *Brassica* (Armstrong and Keller 1981, 1982, Röbbelen 1960) as well as between *Brassica* and allied genera (Quiros *et al.* 1988), we cannot currently confirm the generation of MASLs among segregated amphiploid lines, as techniques are not currently available for the critical analysis and identification of each chromosome. Chen*et al.* (1997) produced *B. alboglabra* chromosome substitution *B. rapa* lines. In this context, *in situ* hybridization procedures, the production of somatic chromosome maps by various imaging methods, and the development of species-specific markers would strongly support the development and analysis of MASLs.

CONCLUSIONS

Five categories of hybrid progeny can be produced through interspecific and intergeneric hybridization: synthetic amphidiploid lines, alien gene introgression lines, alloplasmic lines, monosomic alien chromosome addition lines, and monosomic alien chromosome substitution lines. These approaches can provide materials that are useful both for developing practical breeding strategies and for studying the genetic effects of individual chromosomes on plant traits. This includes estimating the number of genes that control a trait, describing their linkage relationships, and using this knowledge for genetic improvement of Brassicaceae crops. For these research programs to be successful, it will be necessary to produce and maintain as many true lines as possible of these five hybrid types.

REFERENCES

1. Akaba, M., Y. Kaneko, K. Hatakeyama, M. Ishida, S.W. Bang and Y. Matsuzawa (2009a) Identification and evaluation of clubroot resistance of radish chromosome using a *Brassica napus-Raphanus sativus* monosomic addition line. Breed. Sci. 59: 203–206.
2. Akaba, M., Y. Kaneko, Y. Ito, Y. Nakata, S.W. Bang and Y. Matsuzawa (2009b) Production and characterization of*Brassica napus-Raphanus sativus* monosomic addition lines mediated by the synthetic amphidiploids "*Raphanobrassica*". Breed. Sci. 59: 109–118.
3. Armstrong, K.C. and W.A. Keller (1981) Chromosome pairing in haploids of *Brassica campestris*. Theor. Appl. Genet.59: 49–52.
4. Armstrong, K.C. and W.A. Keller (1982) Chromosome pairing in haploids of *Brassica oleracea*. Can. J. Genet. Cytol.24: 735–739.
5. Bang, S.W. (1996) Studies on the intergeneric crossability and production of hybrid progenies between radish (*Raphanus sativus* L.) and allied genera. PhD Thesis. Univ. of Tokyo Agric. and Techni., Tokyo, pp. 1–276.
6. Bang, S.W., Y. Kaneko and Y. Matsuzawa (1996a) Production of intergeneric hybrids between *Raphanus* and*Moricandia*. Plant Breed. 115: 385–390.
7. Bang, S.W., Y. Kaneko and Y. Matsuzawa (1996b) Production of intergeneric hybrids between *Raphanus* and *Sinapis*and the cytogenetics of their progenies. Breed. Sci. 46: 45–51.
8. Bang, S.W., D. Iida, Y. Kaneko and Y. Matsuzawa (1997) Production of new intergeneric hybrids between *Raphanus sativus* and *Brassica* wild species. Breed. Sci. 47: 223–228.
9. Bang, S.W., Y. Kaneko and Y. Matsuzawa (1998) Cytogenetical stability and fertility of an intergeneric amphidiploid line synthesized from *Brassica maurorum* Durieu and *Raphanus sativus* L. Bull. Coll. Agric., Utsunomiya Univ.17: 23–29.
10. Bang, S.W., Y. Kaneko and Y. Matsuzawa (2000) Cytogenetical stability and fertility in an intergeneric amphidiploid line synthesized from *Brassica fruticulosa* Cyr. spp. *mauritanica* (Coss.) Maire and *Raphanus sativus* L. Bull. Coll. Agric., Utsunomiya Univ. 17: 67–73.
11. Bang, S.W., Y. Kaneko, Y. Matsuzawa and K.S. Bang (2002) Breeding of *Moricandia arvensis* monosomic chromosome addition lines (2n = 19) of alloplasmic (*M. arvensis*) *Raphanus sativus*. Breed. Sci. 52: 193–199.
12. Bang, S.W., Y. Mizuno, Y. Kaneko, Y. Matsuzawa and K.S. Bang (2003) Production of intergeneric hybrids between the C_3-C_4 intermediate species *Diplotaxis tenuifolia* (L.) DC. and *Raphanus sativus* L. Breed. Sci. 53: 231–236.
13. Bang, S.W., K. Sugihara, B.H. Jeong, R. Kaneko, E. Satake, Y. Kaneko and Y. Matsuzawa (2007) Production and characterization of intergeneric hybrids between *Brassica oleracea* and a wild relative *Moricandia arvensis*. Plant Breed. 126: 101–103.
14. Bang, S.W., O. Ueno, Y. Wada, S.K. Hong, Y. Kaneko and Y. Matsuzawa (2009) Production and photorespiratory characteristics of *Raphanus sativus* (C_3)-*Moricandia arvensis* (C_3-C_4 intermediate) monosomic and disomic addition lines in each parental cytoplasmic background and their photorespiratory characteristics. Plant Prod. Sci.12: 70–79.
15. Bang, S.W., K. Tsutsui, S.H. Shim and Y. Kaneko (2011) Production and characterization of the novel CMS line of radish (*Raphanus sativus*) carrying *Brassica maurorum* cytoplasm. Plant Breed. 130: 410–412.
16. Banga, S.S. (1988) C-genome chromosome substitution lines in *Brassica juncea* (L.) Coss. Genetica 77: 81–84.
17. Banga, S.S., P.B. Bhaskar and I. Ahuja (2003) Synthesis of intergeneric hybrids and establishment of genomic affinity between *Diplotaxis catholica* and crop *Brassica* species. Theor. Appl. Genet. 106: 1244–1247.
18. Budahn, H., O. Schrader and H. Peterka (2008) Development of a complete set of disomic rape-radish chromosome-addition lines. Euphytica 162: 117–128.
19. Chen, B.Y., V. Simonsen, C. Lanner-Herrer and W.K. Heneen (1992) A *Brassica campestris-alboglabra* addition line and its use for gene mapping, intergenomic gene transfer and generation of trisomics. Theor. Appl. Genet. 84: 592–599.
20. Chen, B.Y., R.B. Jorgensen, B.F. Cheng and W.K. Heneen (1997) Identification and chromosomal assignment of RAPD markers linked with a gene for seed color in a *Brassica campestris-alboglabra* addition line. Hereditas 126: 133–138.
21. Chevre, A.M., F. Ever, P. This, P. Barret, X. Tanguy, H. Brun, M. Delseny and M. Renard (1996) Characterization of*Brassica nigra* chromosomes and of blackleg resistance in *B. napus-B. nigra* addition lines. Plant Breed. 115: 113–118.

22. Harberd, D.J. and E.D. McArthur (1980) Meiotic analysis of some species and genus hybrids in the Brassiceae. *In*: Tsunoda, S., K. Hinata and C. Gomez-Campo (eds.) *Brassica* Crops and Wild Allies, Biology and Breeding, Jpn. Sci. Soc. Press, Tokyo, pp. 65–87.
23. Horisaki, A. and S. Niikura (2004) Effectiveness of insect-pollination to evaluate the level of self-incompatibility and genetic variation in *Brassica rapa* L. Breed. Sci. 54: 291–295.
24. Howard, H.W. (1938) The fertility of amphidiploids from the cross *Raphanus sativus* × *Brassica oleracea*. J. Genet. 36: 239–273.
25. Inomata, N. (1977) Production of interspecific hybrids between *Brassica campestris* and *Brassica oleracea* by culture in vitro of excised ovaries. I. Effects of yeast extract and casein hydrolysate on the development of excised ovaries. Jpn. J. Breed. 27: 295–304.
26. Inomata, N. (1978a) Production of interspecific hybrids between *Brassica campestris* and *B. oleracea* by culture in vitro of excised ovaries. II. Development of excised ovaries in various culture media with coconut milk and casein hydrolysate. Jpn. J. Genet. 53: 1–11.
27. Inomata, N. (1978b) Production of interspecific hybrids between *Brassica campestris* and *B. oleracea* by culture in vitro of excised ovaries. III. Development of excised ovaries in the crosses of various cultivars. Jpn. J. Genet. 53: 161–173.
28. Jeong, B.H., S.W. Bang, S. Niikura, Y. Kaneko and Y. Matsuzawa (2003) Wide cross-compatibility and production of hybrid progenies between *Brassica* crops and wild relatives. Breed. Res. 5 (Suppl 2): 282.
29. Jeong, B.H., S.W. Bang, S. Niikura, Y. Kaneko and Y. Matsuzawa (2004) Intergeneric cross-compatibility and production of hybrid progenies between *Moricandia arvensis* and the inbred lines of *Brassica* crops. Breed. Res. 6(Suppl 2): 194.
30. Jeong, B.H., S.W. Bang, S. Niikura, Y. Kaneko and Y. Matsuzawa (2006a) Progenies of intergeneric hybrid between*Moricandia arvensis* and *Brassica* crops. Breed. Res. 8 (Suppl 1): 208.
31. Jeong, B.H., S.W. Bang, S. Niikura, Y. Kaneko and Y. Matsuzawa (2006b) Intergeneric cross-compatibility and production of hybrid progenies between *Diplotaxis* spp. and the inbred lines of *Brassica* crops. Breed. Res. 8(Suppl 2): 136.
32. Jeong, B.H., T. Saga, K. Okayasu, G. Hattori, Y. Kaneko and S.W. Bang (2009) Production and characterization of an amphidiploid line between *Brassica rapa* and a wild relative *Diplotaxis tenuifolia*. Plant Breed. 128: 536–537.
33. Johnston, S.A., T.P.M. Nijs, S.J. Peloquin and R.E. Hanneman (1980) The significance of genic balance to endosperm development in interspecific crosses. Theor. Appl. Genet. 57: 5–9.
34. Kaneko, Y., Y. Matsuzawa and M. Sarashima (1987) Breeding of the chromosome addition lines of radish with single kale chromosome. Jpn. J. Breed. 37: 438–452.
35. Kaneko, Y., T. Natsuaki, S.W. Bang and Y. Matsuzawa (1996) Identification and evaluation of turnip mosaic virus (TuMV) resistance gene in kale monosomic addition lines of radish. Breed. Sci. 46: 117–124.
36. Kaneko, Y., N. Nagasawa, S.W. Bang and Y. Matsuzawa (2002) Homoeological relationships between the f chromosome of *Brassica rapa* and the e chromosome of *Brassica oleracea*. Plant Breed. 121: 171–173.
37. Kaneko, Y., S.W. Bang and Y. Matsuzawa (2009) 11 Distant Hybridization. *In*: Gupta, S.K. (ed.) Biology and Breeding of Crucifers, Taylor & Francis Group, New York, pp. 207–247.
38. Karpechenko, G.D. (1928) Polyploid hybrids of *Raphanus sativus* L. × *Brassica oleracea* L. Z. Ind. Abst. Vererbgsl. 48: 1–85.
39. Kato, M. and S. Tokumasu (1976) The mechanism of increased seed fertility accompanied with the chance of flower color in *Brassicoraphanus*. Euphytica 25: 761–767.
40. Kerlan, M.C., A.M. Chevre, F. Eber, A. Baranger and M. Renard (1992) Risk assessment of outcrossing of transgenic rapeseed to related species: I. Interspecific hybrid production under optimal conditions with emphasis on pollination and fertilization. Euphytica 62: 145–153.
41. Khush, S.G. (1973) Cytogenetics of alien addition and substitution. *In*: Cytogenetics of Aneuploids, Academic Press, New York, pp. 238–248.
42. Khush, S.G. and D.S. Brar (1992) Overcoming the barriers in hybridization. *In*: Kalloo, G. and J.B. Chowdhury (eds.) Distant hybridization of crop plants, Springer-Verlag, Berlin, pp. 47–61.
43. Kinoshita, T. (2007) Reproductive barrier and genomic imprinting in the endosperm of flowering plants. Genes Genet. Syst. 82: 177–186.
44. Leino, M., R. Teixeira, M. Landgren and K. Glimelius (2003) *Brassica napus* lines with rearranged *Arabidopsis*mitochondria display CMS and a range of developmental aberrations. Theor. Appl. Genet. 106: 1156–1163.
45. Malik, M., P. Vyas, N.S. Rangaswamy and K.R. Shivanna (1999) Development of two new cytoplasmic male-sterile lines in *Brassica juncea* through wide hybridization. Plant Breed. 118: 75–78.
46. Martinez-Sánchez, A., L. Rafael, I.G. Maria and F. Federico (2007) Identification of new flavonoid glycosides and flavonoid profiles to characterize rocket leafy salads (*Eruca vesicaria* and *Diplotaxis tenuifolia*). J. Agric. Food Chem. 55: 1356–1363.
47. Matsumoto, O., Y. Okafuzi, K. Kaneko and S. Katakawa (1997) Breeding of new vegetable 'Hanakkori' by ovule culture. Bull. Yamaguchi Agric. Expt. Stn. 48: 21–24.
48. Matsuzawa, Y. (1983) Studies on the interspecific and intergeneric crossability in *Brassica* and *Raphanus*. Special Bull. Coll. Agric., Utsunomiya Univ. 39: 1–86.
49. Matsuzawa, Y., Y. Kaneko and S.W. Bang (1996) Prospects of the wide cross for genetics and plant

breeding in*Brassicaceae*. Bull. Coll. Agric., Utsunomiya Univ. 16: 5–10.

50. Matsuzawa, Y., T. Minami, S.W. Bang and Y. Kaneko (1997) A new *Brassicoraphanus* (2n = 36); the true-breeding amphidiploid line of *Brassica oxyrrhina* Coss. (2n = 18) × *Raphanus sativus* L. (2n = 18). Bull. Coll. Agric., Utsunomiya Univ. 16: 1–7.

51. Matsuzawa, Y., T. Funayama, M. Kamibayashi, M. Konnai, S.W. Bang and Y. Kaneko (2000) Synthetic *Brassica rapa-Raphanus sativus* amphidiploids lines developed by reciprocal hybridization. Plant Breed. 119: 357–359.

52. McGrath, J.M. and C.F. Quiros (1990) Generation of alien chromosome addition lines from synthesis *Brassica napus*: morphology, cytology, fertility, and chromosome transmission. Genome 33: 374–383.

53. Mizushima, U. (1950) Karyogenetic studies of species and genus hybrids in the tribe *Brassiceae* of *Cruciferae*. Tohoku J. Agri. Res. 1: 1–14.

54. Momotaz, A., M. Kato and F. Kakihara (1998) Production of intergeneric hybrids between *Brassica* and *Sinapis*species by means of embryo rescue techniques. Euphytica 103: 123–130.

55. Nagano, K. (1988) Senpousai No. 1 and No. 2. *In*: Hujii, K. (ed.) The monograph 'New varieties of vegetable crops': 10. Seibundo-Shinkosha, Tokyo, p. 129.

56. Namai, H. (1976) Cytogenetic and breeding studies on transfer of economic characters by means of interspecific and intergeneric crossing in the tribe *Brassiceae* of *Cruciferae*. Memoirs Fac. Agric., Tokyo Univ. Education 22: 101–171.

57. Namai, H. (1987) Inducing cytogenetical alterations by means of interspecific and intergeneric hybridization in brassica crops. Gamma Field Symp. 26: 41–89.

58. Nishi, S., J. Kawada and M. Toda (1962) Studies on the embryo culture in vegetable crops. Part 2. Breeding of interspecific hybrids between cabbage varieties and Chinese cabbage varieties through the application of embryo culture techniques. Bull. Hort. Res. Sta., Japan, Ser. A1: 111–156.

59. Nishi, S., M. Toda and T. Toyoda (1970) Studies on the embryos culture in vegetable crops. III. On the conditions affecting to embryo culture of interspecific hybrids between cabbage and Chinese cabbage. Bull. Hort. Res. Sta., Japan, Ser. A9: 73–88.

60. Nishiyama, I. and T. Yabuno (1978) Causal relationships between the polar nuclei in double fertilization and interspecific cross-incompatibility in *Avena*. Cytologia 43: 453–466.

61. Peterka, H., H. Budahn, O. Schrader, R. Ahne and W. Schutze (2004) Transfer of resistance against the beet cyst nematode from radish (*Raphanus sativus*) to rape (*Brassica napus*) by monosomic chromosome addition. Theor. Appl. Genet. 109: 30–41.

62. Prakash, S. and V.L. Chopra (1990) Male sterility caused by cytoplasm of *Brassica oxyrrhina* in *B. campestris* and *B. juncea*. Theor. Appl. Genet. 79: 285–287.

63. Prakash, S., Y. Takahata, P.B. Kirti and V.L. Chopra (1999) Cytogenetics. *In*: Gomez-Campo, C. (ed.) Biology of*Brassica* coenospecies. Elsevier, Amsterdam, pp. 59–106.

64. Quiros, C.F., O. Ochoa, S.F. Kianian and D. Douches (1987) Analysis of the *Brassica oleracea* genome by the generation of *B. campestris-oleracea* chromosome addition lines: characterization by isozymes and rDNA genes. Theor. Appl. Genet. 74: 758–766.

65. Quiros, C.F., O. Ochoa and D.S. Douches (1988) Exploring the role of x = 7 species in *Brassica* evolution: hybridization with *B. nigra* and *B. oleracea*. J. Hered. 79: 351–358.

66. Röbbelen, G. (1960) Beiträge zur analyse des *Brassica*-genoms. Chromosoma (Berl.) 11: 205–228.

67. Sarashima, M. (1973) Studies on the breeding of artificially synthesized forage rape (*Brassica napus* ssp. *oleifera*) by means of interspecific crosses between *B. campestris* and *B. oleracea*. Special Bull. Coll. Agric., Utsunomiya Univ.29: 1–117.

68. Sarashima, M., M. Matsuzawa and S. Suto (1990a) Synthesis of alloplasmic *Brassica campestris* L. Bull. Coll. Agric., Utsunomiya Univ. 14: 1–10.

69. Sarashima, M., M. Matsuzawa and S. Suto (1990b) Synthesis of alloplasmic *Brassica oleracea* L. Bull. Coll. Agric., Utsunomiya Univ. 14: 11–26.

70. Sarashima, M., M. Matsuzawa and S. Suto (1990c) Synthesis of alloplasmic *Raphanus sativus* L. Bull. Coll. Agric., Utsunomiya Univ. 14: 27–34.

71. Shim, S., W. Yamada, S.W. Bang and Y. Kaneko (2010) Production of interspecific hybrids between *Brassica* wild species and *B. oleracea*. Breed. Res. 12 (Suppl 2): 239.

72. Shim, S., S.W. Bang and Y. Kaneko (2011a) Production and characterizations of *Brassica oleracea* carrying *Diplotaxis*cytoplasm. Breed. Res. 13 (Suppl 1): 208.

73. Shim, S., S.W. Bang and Y. Kaneko (2011b) Production and versatility of the CMS *Brassica rapa* line carrying *B. oxyrrhina* cytoplasm. Breed. Res. 13 (Suppl 2): 186.

74. Shim, S., S.W. Bang and Y. Kaneko (2012) Production and characterizations of CMS *Brassica napus* carrying*Brassica oxyrrhina* cytoplasm. Breed. Res. 14 (Suppl 2): 202.

75. Shimizu, S., K. Kanazawa and T. Kobayashi (1962) Studies on the breeding of Chinese cabbage for resistance to soft rot. Part III. The breeding of the resistant variety "Hiratsuka No. 1" by interspecific crossing. Bull. Hort. Res. Sta., Japan, Ser. A1: 157–174.

76. Siemens, J. (2002) Interspecific hybridization between wild relatives and *Brassica napus* to introduce new resistance traits into the oilseed rape gene pool. Czech J. Genet. Plant Breed. 38: 155–157.

77. Srinivasan, K., V.G. Malathi, P.B. Kirti, S. Prakash and V.L. Chopra (1998) Generation and characterization of monosomic chromosome addition lines of *Brassica campestris-B. oxyrrhina*. Theor. Appl. Genet. 97: 976–981.

78. Stebbins, G.L. (1958) The inviability, weakness and sterility of interspecific hybrids. Adv. Genet. 9: 147–215.

79. Takada, M. (1985) Gihu-Gtreen. *In*: Hujii, K. (ed.) The monograph 'New varieties of vegetable crops': 9, Seibundo-Shinkosha, Tokyo, p. 144.

80. Takada, M. (2006) F_1-Hosoda Wase (Hakuran). *In*: Ito, T. (ed.) The monograph 'New varieties of vegetable crops': 16, Seibundo-Shinkosha, Tokyo, p. 88.

81. Takashima, M., S.W. Bang and Y. Kaneko (2012) Production and characterization of monosomic addition lines of autoplasmic and alloplasmic *Brassica napus* with each B-genome chromosome of *Brassica juncea*. Breed. Res.14: 95–105.

82. Tonosaki, K., K. Michiba, S.W. Bang, H. Kitashiba, Y. Kaneko and T. Nishio (2013) Genetic analysis of hybrid seed formation ability of *Brassica rapa* in intergeneric crossings with *Raphanus sativus*. Theor. Appl. Genet. 126: 837–846.

83. Tsutsui, K. (2013) Breeding utilization of cytoplasmic male sterile lines in cruciferae carrying cytoplasm of their related species. PhD Thesis. Univ. of Tokyo Agric. and Techni., Tokyo, pp. 1–205.

84. Tsutsui, K., B.H. Jeong, Y. Ito, S.W. Bang and Y. Kaneko (2011) Production and characterization of an alloplasmic and monosomic addition line of *Brassica rapa* carrying the cytoplasm and one chromosome of *Moricandia arvensis*. Breed. Sci. 61: 373–379.

85. U, N. (1935) Genome-analysis in *Brassica* with special reference to experimental formation of *B. napus* and peculiar mode of fertilization. Jpn. J. Bot. 7: 389–452.

86. Udagawa, H., Y. Ishimaru, F. Li, Y. Sato, H. Kitashiba and T. Nishio (2010) Genetic analysis of interspecific incompatibility in *Brassica rapa*. Theor. Appl. Genet. 121: 689–696.

87. Wang, Y.P., X.X. Zhao, K. Sonntag, P. Wehling and R.J. Snowdon (2005) Behaviour of *Sinapis alba* chromosome in a*Brassica napus* background revealed by genomic *in-situ* hybridization. Chromosome Res. 13: 819–826.

88. Warwick, S.I. (1993) Guide to the wild germplasm of *Brassica* and allied crops. Part IV. Wild species in the tribe*Brassiceae* (*Cruciferae*) as sources of agronomic traits. Centre for Land and Biological Resources Research, Research Branch, Agriculture Canada, Ottawa, Ont. Technical Bulletin 1993-17E. 1–19.

89. Wilmar, J.C. and M. Hellendoorn (1968) Embryo culture of Brussels sprouts for breeding. Euphytica 17: 28–37

CHAPTER 11

CENH3: AN EMERGING PLAYER IN HAPLOID INDUCTION TECHNOLOGY

Anne B. Britt and Sundaram Kuppu

Keywords: Uniparental inheritance, Centromere, plant breeding, CENH3, haploid induction

THE VARIANT HISTONE CENH3 IS REQUIRED FOR CENTROMERE LOCALIZATION

Eukaryotic DNA is wrapped around nucleosomes. This wrapping, and subsequent orders of supercoiling of the nucleosome/DNA complex, serves to compact the chromosome, allowing the rather long linear genome in flowering plants, between 61 Mb (*Genliseatuberosa*) (Fleischmann et al., 2014) and 160 Gb (*Paris japonica*) (Pellicer et al., 2010) to fit into a rather small compartment- the nucleus of the cell. Compaction of the entire genome or discrete sections within it varies with both the stage of the cell cycle and the function of the DNA. The degree of supercoiling of any particular region of the chromosome is determined by the presence or absence of particular histone variants that make up each nucleosome and by posttranslational modifications of these variants (Gibcus and Dekker, 2013; Cutter and Hayes, 2015). The pattern of histone deposition and modification is maintained after DNA replication, and in some cases across generations, through a variety of mechanisms (Mattiroli et al., 2015). This semi-stable modification of polymerase access through the modification of nucleosomal histone content, often heritable as a pseudoallele, has come to be referred to as "epigenetic" regulation of gene expression and heritable forms of this modification are termed epigenetic inheritance.

An exceptional example of epigenetic determination of chromatin function is the role of histone content in the localization of the centromere. While centromeric DNAs in eukaryotic species often share general features (long arrays of tandem repeats, often interspersed with transposable elements) recent findings have demonstrated that these features are not essential to centromere localization; some eukaryotes lack long tandem arrays entirely, some have different repeat arrays on different chromosomes, and neocentromeres can form at unique sequences (Marshall et al., 2008). Rather than being determined by DNA sequence the location of the centromere is thought to be determined by the presence of nucleosomes carrying the variant histone CENH3 (Allshire and Karpen, 2008; Stimpson and Sullivan, 2010; Henikoff and Smith, 2015; Westhorpe and Straight, 2015; also known as CID in *Drosophila*, CENP-A in mammals and Cse4 in yeast; Talbert et al., 2002; Sekulic and Black, 2012; Lermontova et al., 2014; Steiner and Henikoff, 2015), although in some exceptional cases (notably *Saccharomyces cerevisiae*) this histone variant does have sequence-specific DNA binding affinity (and so centromeres will form at specific DNA sequences). Thus in the majority of eukaryotes studied, centromere positioning appears to be an epigenetic, rather than sequence-based, phenomenon (Liu et al., 2015).

CENH3, like other histones, carries an N-terminal tail (which protrudes from the nucleosome and is a target for posttranslational modification) and C-terminal Histone Fold Domain (which interacts with DNA and other histones to form the nucleosome). Unlike conventional histones, CENH3 is rapidly evolving. The N-terminal tail of CENH3 barely alignable even among closely related species whereas the histone fold domain is relatively well conserved (Malik and Henikoff, 2003).

Because CENH3 determines the position of the centromere, the presence of CENH3 is essential for segregation of chromosomes. Alteration of CENH3 in several organisms has been shown to have adverse consequences leading to chromosome segregation errors and lethality.

Citation: Britt AB and Kuppu S (2016) Cenh3: An Emerging Player in Haploid Induction Technology. Front. Plant Sci. 7:357. doi: 10.3389/fpls.2016.00357.

In budding yeast mutations in *CENH3* cause chromosome non-disjunction and cell cycle arrest at mitosis (Stoler et al., 1995). In *Caenorhabditis elegans*, knockout of CENH3 induces missegregation and lethality (Buchwitz et al., 1999). In *cenpA* null mice, early disruption of centromeric chromatin organization has been observed (Howman et al., 2000). Reduction in the level of CENPA in human cell lines induces p53-dependent cellular senescence (Maehara et al., 2010), presumably to prevent cells from undergoing error-prone mitosis. Down-regulation of *CENH3* in *Arabidopsis* leads to reduced mitosis and increased meiotic segregation errors (Lermontova et al., 2011). Homozygous *cenh3*-null mutants of *Arabidopsis* exhibit early embryonic lethality (Ravi et al., 2010).

Given the importance of maintaining one- and only one- centromere per chromosome, the loading of CENH3-containing nucleosomes at a single locus per chromosome is a critical step in genome maintenance. As we will discuss below, haploid induction (HI) in CENH3-modified lines may result from errors in reloading of CENH3 at centromeres derived from CENH3-modified lines. Although the reloading process has been extensively investigated in animals and fungi (Müller and Almouzni, 2014), we are only beginning to learn about this process in plants (Lermontova et al., 2013; Liu et al., 2015).

THE IMPORTANCE OF HAPLOID INDUCTION

Doubled haploids- plants carrying a genome derived from a single (haploid) gamete- provide enormous timesaving in the production of true-breeding lines, literally reducing the time to acceptable levels of homozygosity by an order of magnitude. Haploid lines, and "haploid genetics" can accelerate a variety of genetic constructions and is useful in the research laboratory also (Ravi et al., 2014). The spontaneous occurrences of haploids have been reported in several species reviewed in (Dunwell, 2010; Dwivedi et al., 2015). In some species haploids have been generated through interspecies and intraspecies hybridization. (Guha and Maheshwari, 1964) showed that haploid plants could be generated from anther culture in some plant species. A variety of other protocols have been developed to generate haploids, reviewed in detail in Wędzony et al. (2009). Though all of these techniques have proved to be useful over the last 50 years, they are limited to few crop species and/or varieties. In this review we will highlight the recent advances in HI technology via modification of CENH3. Because CENH3 is present in all plants there is promise that this technology can be translated to the majority of crop species.

CENH3 MEDIATED HAPLOID INDUCTION

In a breakthrough discovery, Ravi and Chan (2010) studying the structure/function relationship of Histone 3 variants, found that a *cenh3–/– Arabidopsis* null mutant, when complemented by altered version of *CENH3*, could induce haploids. A remarkably effective haploid inducer, termed "green fluorescent protein *(GFP)-tailswap*," carried GFP fused to the N-terminal tail domain of an H3 variant (At1g13370), which then replaced the N-terminal tail of CENH3 (see Figure 1). They found, surprisingly, that the CENH3 mitotic functionality was maintained in spite of the "tailswap," though meiotic function was compromised- the plants were almost entirely male sterile, with some female sterility also. On self-pollination, *GFP-tailswap* produces normal diploid seeds (though again, with reduced fertility). However, when crossed as a female by plants carrying wild-type CENH3, *GFP-tailswap* produces a remarkable 25–45% paternal haploid progeny.

Transgenic variant	Complementation	Haploids	Aneuploids
At Tail | At Histone Fold Domain	Y	N	N
GFP | At Tail | At Histone Fold Domain	Y	Y	Y
At Tail | At Histone Fold Domain | GFP	N	-	-
At Histone Fold Domain	N	-	-
At H3.3Tail | At Histone Fold Domain	Y	Y	Y
GFP-tailswap: GFP | At H3.3Tail | At Histone Fold Domain	Y	Y	Y
Zm Tail | Zm Histone Fold Domain	Y	ND	ND
GFP | Zm Tail | Zm Histone Fold Domain	N	-	-
Lo Tail | At Histone Fold Domain	Y	Y	Y
At Tail | Lo Histone Fold Domain	Y	N	Y
Lo Tail | Lo Histone Fold Domain	Y	Y	Y
Br Tail | Br Histone Fold Domain	Y	Y	Y
Vv Tail | Vv Histone Fold Domain	Y	ND	ND

Figure 1: Effect of various modifications of transgenic *CENH3* **variants in** an*Arabidopsis thaliana cenh3* **null.** Table on the right summarizes complementation (of viability and fertility), haploid, and aneuploid induction by these lines. *GFP-tailswap* is the commonly used nomenclature in published literature for this construct.

The maternal (*GFP-tailswap*) genome is lost in these haploids (Ravi and Chan, 2010). The almost total male sterility observed in *GFP-tailswap* plants is ascribed to its meiosis-specific loading abnormalities (Ravi et al., 2011). The authors also observed some, though less frequent, HI in plants carrying a simple N-terminal addition of GFP to CENH3 (no "tailswap"). The three independently derived lines of *GFP-tailswap* analyzed did not display significant differences in their level of HI frequencies (Tan et al., 2015). This suggests that HI is dependent on the modification of CENH3 rather than on expression levels- although it must be noted that a certain minimum, and perhaps maximum (Marshall et al., 2008) level of expression is selected for in these transgenic experiments, which are performed in a *cenh3–/–* background.

The initial findings with *GFP-CENH3* and *GFP-tailswap* suggested that addition of a bulky tag to CENH3 might interfere with CENH3 recognition, and so produce a centromere that is less competitive for the reloading of CENH3 or subsequent components of the centromere when confronted with competing wild-type centromeres. Working from this hypothesis, we will refer to these modified CENH3's, and their resulting centromeres, as "weak" or "non-competitive" vs. wild-type CENH3. Addition of GFP on its own presumably results in a slightly defective protein, and modification of the N-terminal tail (tailswap) further degrades the strength of the resulting centromere.

The addition of adducts such as GFP, and changes in the sequence of CENH3, can act additively to affect protein function. For example, as described above, GFP-CENH3, in *Arabidopsis*, is a haploid inducer, but otherwise fully functional. The *Arabidopsis* null mutant can also be complemented (mitotically and meiotically) by CENH3 from a wide range of angiosperms (Maheshwari et al., 2015), including the monocot *Zea mays*. However, addition of GFP to the N-terminus of this distantly related protein compromised its function to the extent that while it could localize to centromeres, it could not complement the lethality of the *cenH3–/–* mutant (Ravi et al., 2010; Figure 1).

CENH3 FUNCTION IS CONSERVED ACROSS WIDE VARIETY OF SPECIES

In pioneering study with human cell lines, it was shown that yeast *CENH3* (*Cse4*) was able to complement a human defect in *CENH3* (*CENPA*) function (Wieland et al., 2004). Similarly, Moraes et al. (2011), found that EYPF-tagged versions of CENH3 from monocentric plant species like,*Arabidopsis lyrata, A. arenosa, Capsella bursa-pastoris*, and *Z. mays* were able to target the *A. thaliana* centromere. In contrast, an EYPF tagged version of *CENH3* from holocentric plant *Luzula nivea* failed to target *A. thaliana* centromeres. Ravi et al. (2010) found that GFP-tagged versions of CENH3 from *A. aeronosa, A. lyrata*, and *Z. mays* were able to target *A. thaliana* centromere. However, the authors also found that GFP-tagged versions of *CENH3* from other kingdoms (*S. cerevisiae, C. elegans* and *Homo sapiens*) did not target to *A. thaliana* centromeres. In the same study it was shown that GFP tagged versions of CENH3 from *Brassica rapa* and *Z. mays* were able to localize to *A. thaliana* centromere, but were unable to complement *A.thaliana cenh3* null function. This result was surprising given ability of the yeast protein to complement the human defect, which suggests that CENH3 function is highly conserved across kingdoms. This discrepancy was explained when Maheshwari et al. (2015), in a subsequent study employed untagged versions of CENH3 from *B. rapa, Lepidium oleracuem, Vitis Vinifera*, and *Z. mays* found that all of these untagged proteins were able to complement *A. thaliana cenh3* null function. In another intriguing note, when the GFP is attached C-terminally to the *AtCENH3* protein, it is able to localize to the centromere but unable to complement *cenh3* null function in *Arabidopsis* (Ravi et al., 2010), while the N-terminally tagged protein restores viability and full fertility to this mutant. Thus GFP-tagging of CENH3 often places restrictions on its functionality. From the above studies using heterologous, untagged CENH3, it's clear that the essential functions of *CENH3* are well conserved across a wide variety of species.

A ROLE FOR N-TERMINAL DOMAIN MUTATIONS IN HAPLOID INDUCTION

The Ravi and Chan (2010) study described above not only indicated that substantial alterations to CENH3 can result in HI, it also demonstrated, surprisingly, that the N-terminal domain of CENH3 is not required for its mitotic function. In a study to designed to determine whether naturally occurring divergence in CENH3 sequence can result in functional constraints, Maheshwari et al. (2015) found that CENH3 function is well conversed across species on a broad evolutionary scale. *Arabidopsis cenh3-1* mutants transformed with exotic *CENH3*s from *B. rapa, L. oleraceum, V. vinifera*, and *Z. mays* were viable and fertile. However, although these CENH3s are fully functional in terms of mitosis, meiosis, and in self-pollinating crosses, chromosomes derived from lines expressing the alien but closely related CENH3s from *B. rapa* and *L. oleraceum* proved non-competitive when crossed by wild-type plants, producing aneuploid and haploid offspring that retained the wild-type parent's genome. Furthermore, the authors demonstrated that replacement of the *Arabidopsis* N-terminal tail with the N-terminal tail from *L. oleraceum* produced plants that, while fully viable and fertile on self-pollination, exhibited extensive seed death, as well as induction of haploidy and aneuploidy, when crossed by wild-type lines. The reverse tailswap, carrying the At-NTD and the LoHFD, did not induce haploids at the scale tested (approximately 100 progeny) though some aneuploids were reported. Thus the sequence of the N-terminal tail affects the "strength" of CENH3 in a competitive cross. The difference in sequence between the N-terminal tails of *L. oleraceum* and that of *A. thaliana* is substantial- they are of similar lengths but differ at 29/82 positions (see alignment in Figure 2). Similarly, the N-terminal tail of histone H3.3 (see alignment in Figure 2) has very little similarity to that of CENH3, but it complements CENH3's mitotic and meiotic functions and is, in the absence of the GFP tag, a weak haploid inducer. Whether subtle differences in the N-terminal tail- for example, single amino acid substitutions- can possibly affect the "strength" of CenH3 remains to be determined.

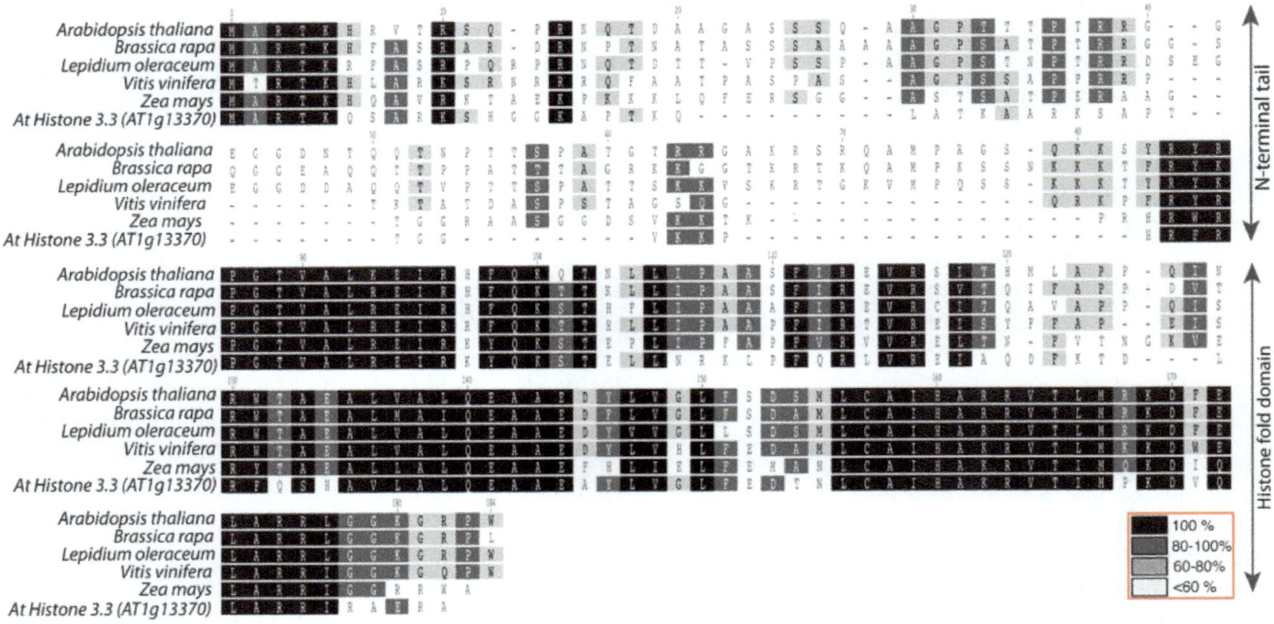

Figure 2: Alignment of *Arabidopsis thaliana* **histone3.3 vs. CENH3 sequences from** *A. thaliana, Brassica rapa, Lepidium oleraceum, Vitis vinifera*, and *Zea mays*. Top two rows represent N-terminal tail with less conservation and the bottom three columns represent relatively well-conserved histone fold domain.

Point Mutations in CENH3 Histone Fold Domain Induce Haploids

The tail region of CENH3 is rapidly evolving in both length and sequence whereas the histone fold domain is well conserved. This suggests that subtle changes in the HFD might have substantial effects on its activity. Recently, it has been shown that point mutations resulting in single amino acid substitutions induced uniparental genome elimination, resulting in haploidy or aneuploidy, with retention of the wild-type chromosomes (Karimi-Ashtiyani et al., 2015; Kuppu et al., 2015). Remarkably, changes in the majority of conserved residues tested resulted in plants, which were viable, fertile, and, to various degrees, haploid-inducing. None of the tested residues were essential to mitotic or meiotic function. The single amino acid substitutions P82S, G83E, A86V, A132T, L130F, and A136T (*Arabidopsis* CENH3) induced haploids (among 0.6–12 percent of progeny, depending on the mutation), while P102S and G173E appeared to be fully competitive with wild-type CENH3. Aneuploid progeny were also obtained from among the progenies of these plants. Because these mutations could be obtained in other crop plants by ethyl methane sulfonate (EMS) mediated mutagenesis (the sites tested were conserved among crops), this discovery suggests that non-transgenic haploid inducers could be induced- and may already exist, undetected- in EMS mutagenized crop species.

SHATTERED CHROMOSOMES AND MINICHROMOSOMES ARE PRODUCED DURING CENH3- MEDIATED GENOME ELIMINATION

Genome loss presumably results from the inability of one parent's chromosomes to efficiently attract reloading of CENH3 and/or additional constitutive centromeric components; either of these defects would result in a compromised ability to attract outer kinetochore components and a failure to attach to the spindle. The mixed haploid, diploid, and aneuploid progeny of CENH3-based HI crosses indicates that the individual HI-derived chromosomes have independent fates. In some cases, individual chromosomes derived from HI crosses appear to have undergone substantial rearrangements, duplications and/or truncations. These events have been particularly well characterized in progeny of *GFP-tailswap* crosses in *Arabidopsis* (Tan et al., 2015). The authors of this study employed low-pass whole-genome sequencing to bin and count genomic sequences from hundreds of progeny of an HI × wt cross. This method allows the researcher to detect imbalances in copy number (balanced translocations are not detectable via this procedure). The great majority of the plants characterized were chosen from among the 37% of all progeny that resembled previously characterized (Steinitz-Sears, 1963) "numerical" aneuploids- plants carrying a diploid genome plus one or more entire additional chromosome.

Analysis of these putative aneuploids revealed that plants carrying $2n + 1$ (or more) chromosomes were indeed the most commonly encountered class of aneuploids. Because the analysis was sequence-based, and the crosses were between two different ecotypes, the parental origin of each chromosome could also be scored. In all cases the supernumerary chromosomes were derived from the *GFP-tailswap* (here also female) parent. The second most commonly encountered class of aneuploids were diploids carrying an additional truncated chromosome (again *GFP-tailswap*-derived). In a very small fraction of aneuploids, during the process of genome elimination, a chromosome from the *GFP-tailswap* parent undergoes "shattering," revealed by copy number discrepancies within a single supernumerary chromosome. The investigators observed the encapsulation of chromatin into micronuclei, and hypothesize that one or more laggard *GFP-tailswap* chromosomes left behind during segregation can be compartmentalized into micronuclei, which occasionally rejoin the nucleus proper. At some point during this excursion, the chromosomes undergo endonucleolytic degradation (referred to as "shattering") and subsequent rejoining, resulting in variety of intra- or interchromosomal rearrangements. Retention of– or fusion to- a centromeric region would result in heritability. Meiotically heritable mini- and interchromosomal rearrangements were observed and confirmed cytologically. The resulting mini-chromosomes are potentially of value for plant breeding, if they can be designed to carrying a one or more useful traits in a single, simple package that can be easily transferred between varieties.

RECOMBINATION OF SHATTERED CHROMOSOMES OCCURS VIA NON-HOMOLOGOUS END JOINING (NHEJ)

The source of the double strand breaks that result in shattered chromosomes is unclear. Breaks might be induced by prolonged exposure to the cytoplasmic environment, or through the deregulation of replication vs. repair in micronuclei. Rearrangements could theoretically be generated either via non-homologous end joining or through ectopic homologous recombination (HR) of repetitive DNAs. Tan et al. (2015) cloned and sequenced 12 randomly selected junctions derived from a single shattered chromosome, and found that all displayed junction characteristic of NHEJ events, including the presence of microhomologies at the joint and the insertion of sequences of unknown origin. The authors then went on to investigate the effects of deficiencies in DNA ligase IV, an enzyme routinely involved in NHEJ. Crossing *GFP-tailswap* by plants homozygous for a null mutation in *Lig4* resulted in an increased frequency of haploids (doubling from 40% of progeny to 80%) at the expense of both diploid and aneuploid progenies. This suggests that *both* aneuploids *and* diploids experience *LIG4*-dependent chromosome rescue. This in turn suggests that the *GFP-tailswap*-derived chromosomes in diploids may have undergone extensive rearrangement and/or point-like mutations that are undetectable via the phenotypic screen employed to collect aneuploids for sequence-based characterization.

Fascinatingly, the *LIG4* status of the *GFP-tailswap* (maternal) parent has no effect on HI frequency; the observed *LIG4*-dependent rescue of aneuploids and diploids depended entirely on the (wild-type *CENH3*) paternal genome. A potential explanation for this phenomenon is that chromosomes- even chromosomes that will eventually produce diploids- are frequently being rescued into what is essentially a haploid, paternal nucleus, and that the (perhaps extranuclear) *GFP-tailswap*-derived genome is not effectively expressed during this process. In mammalian systems, it has also been shown that unrepaired DNA damage increased the frequency of uniparental genome elimination (Wang et al., 2014). This finding clearly shows that DNA repair mutants could be used in CENH3-mediated genome elimination system to enhance the frequency of HI.

CENH3 MEDIATED GENOME ELIMINATION IS MOST EFFICIENT WHEN THE HI LINE IS THE FEMALE PARENT

Three different classes of modifications of CENH3 have been shown to produce haploids: Domain swapping with addition of fluorescent tag, complementing with CENH3 from different species, and point mutations in the highly conserved Histone Fold Domain. It is strikingly evident from all three cases that the HI works well when the CENH3 mutants are used as female parent, but not as a male. In case of *GFP-tailswap* the haploid inducing frequency is around 40 percent when it is used as a female parent, but drops down to 5% when used as a male (Ravi and Chan, 2010). Karimi-Ashtiyani et al. (2015) found that the point mutation L130F induced haploids only used as female parent (at the scale tested). This does not rule out the possibility of using CENH3 mediated HI system as a male, but it is clear that modification of CENH3 has a much stronger haploid-inducing effect when used as a female. The biological significance of this effect is unclear; possibly a higher level of wt CENH3 transcript in the female gametophyte helps to negate the effect of modified CENH3s on loading of additional centromeric factors. However, see below for a discussion of the phenomenology of CENH3 loading and active unloading in the gametophyte and zygote.

SEED ABORTION IS CORRELATED WITH HAPLOID INDUCTION IN CENH3 MEDIATED GENOME ELIMINATION

A common feature observed in all CENH3 mediated haploid inducing systems is seed abortion (the presence of small, wrinkled, dark, inviable seeds in the silique, amongst plump viable seeds) in the HI cross. All variants of CENH3 (*GFP-tailswap*, CENH3's from distant species and point mutants) that complement endogenous *cenh3-1* null do not display seed death on self-pollination. Though*GFP-tailswap* is nearly sterile, with poor seed set per silique, those seeds that are produced are plump and healthy. In contrast, when these lines are hybridized to wild-type plants they display substantial seed death, and the frequency of death is positively correlated with the frequency of HI.

In angiosperms, embryogenesis involves double fertilization and triple fusion where one sperm nuclei fuses with the egg to form the zygote and one sperm fuses with two already fused central cells to form the triploid endosperm (a tissue that supports embryonic and seed development). Euploidy of both the zygote and endosperm are important for seed development; defects in chromosomal balance in either tissue type could result in seed death. It makes sense that loss of the HI genome would occur in both the embryo and endosperm, but this is not necessarily true, as CENH3 loading and active unloading rates are developmentally specific (Ingouff et al., 2010). Thus the effects of CENH3 modification on the genome of the endosperm require additional investigation.

In an effort to facilitate screening of haploids at seed level, Ravi et al. (2014) expressed GPF under the control of seed storage protein 2S3 promoter (At2S3: GFP), and transformed this construct into*GFP-tailswap*. This marker is expressed in both embryo and endosperm, and although endosperm tissue is limited in the mature *Arabidopsis* seed, the marker can be visualized through the seed coat of mature, dried seeds in both the seedling (where it is highly expressed) and in the residual endosperm. Seeds derived from *GFP-tailswap* p23S:GFP HI crosses fell into two phenotypic classes: fully GFP+ (endosperm vs. embryonic expression was not distinguished in this class), and mottled GFP expression (in the endosperm, but not the embryo). Mottled seeds always gave rise to haploid (paternal, GFP-) seedlings, while fully green seeds produced diploids or aneuploids. Perhaps the most significant result, aside from the success of the marker for identification of haploid seeds, is that there were no GFP-free seeds, suggesting that fertilization events resulting in paternal "monoploid" endosperm are either very rarely produced (both copies of the HI-inducer's genome are rarely lost) or generally lead to seed abortion, or particularly lead to abortion of haploids.

OBSERVATION OF CENH3 DEPLETION IN THE ZYGOTE AND EMBRYO

In a classical barley HI system involving crosses between *Hordeum vulgare* × *Hordeum bulbosum*, the *H. bulbosum* genome is lost during embryogenesis (Kasha and Kao, 1970). Sanei et al. (2011)were able to cytologically distinguish between the *vulgare* and *bulbosum* genomes via FISH, while simultaneously following deposition of CENH3 via immunolocalization (employing an antibody which does not distinguish between the CENH3s encoded by each species). In their visualization of anaphase mitoses in early embryos, the authors found that lagging chromosomes were, unsurprisingly, *bulbosum*-derived. More notably, these laggards lacked detectable CENH3, which is easily observed via immunolocalization in normal barley chromosomes. This indicates that there is-minimally- a failure of CENH3 to reload onto the *bulbosum* chromosomes, rather than (or perhaps in addition to), a failure of more distal centromeric components to recognize the *bulbosum*-derived centromeres.

It is not clear in the barley system described above whether CENH3 is lost from the *bulbosum*chromosomes due to dilution (with a failure to reload) after replication, or due to active unloading of CENH3 in the zygote. Interestingly, Ingouff et al. (2010) have shown- in *Arabidopsis*- that neither *CENH3* transcripts nor CENH3-GFP (tagged at the C terminus) were detectable in the*Arabidopsis* egg cell. Although the authors observed localization of CENH3-GFP to sperm centromeres, this mark was rapidly lost after fertilization of wild-type eggs, prior to further cell division, in what is apparently the active removal of this mark. Localization of CENH3-GFP was visible later in embryogenesis. This very provocative observation raises the possibility that CENH3 is not present in the zygote. If this were true for CENH3, as well as for CENH3-GFP, this observation would challenge the model that centromeres are determined simply by the presence or absence of CENH3. It remains to be determined, however, whether wild-type CENH3 is similarly actively depleted from the zygote. CENH3-GFP may behave differently than CENH3; it does not, for example, complement the *cenH3–/–* mutation (Ravi et al., 2010).

The observations with point mutants that the behavior of chromosomes in the zygote is determined by the amino acid sequence of the CENH3 protein present in the parental lines- is most easily explained by assuming that significant levels of parental CENH3 are retained in both the gametes and the zygote. However, it is also conceivable that CENH3 is truly absent, and that modification of CENH3 results in some sort of modified- and therefore "weak"- zygotic signal, or "footprint," at the centromeric locus. It is possible that CENH3 is not effectively reloaded onto these weak footprints.

REGULATORY ISSUES AND POSSIBLE ALTERNATIVES

In a *CENH3*-mediated HI system, all chromosomes in the haploid product are, from a molecular point of view, non-transgenic, regardless of whether the haploid-inducing parent carried a transgenically modified *CENH3* allele. From a regulatory perspective, however, it is interesting to note that HI through modification of *CENH3* could be considered GMO or non-GMO based on interpretation of the law in the geographical region for which it is being considered. In some parts of the world, the genetically modified organisms are labeled based on the product- the genotype of the plant itself- whereas in other parts GMOs may be identified based on the process employed to generate that plant (Hartung and Schiemann, 2014). *GFP-tailswap* based HI would be considered non-GMO if the product only is taken into account whereas it would be considered GMO if process of generation is an important criterion. Similarly, if CENH3 from other species were introduced through molecular tools, the resulting new varieties would be considered GMO if process of generation is considered but non-GMO if the product is considered instead. But if the same alien gene were introgressed by traditional crossing approach it would be considered non-GMO under both conditions. Defining a variety as "GMO" precipitates a costly and time-consuming approval process.

Single amino acid substitutions obtained through point mutations generated by EMS is currently accepted as non-GMO under both interpretations of the law. Therefore, under existing regulations, chemically-induced point mutations of *CENH3* offer a non-transgenic process for generation of haploids. Additionally, because alien *CENH3s* or *GFP-tailswap* are only functional in a *cenH3–/–*homozygote, identification of point-mutation-based haploid inducers may be faster than other two approaches if robust TILLING populations are already available (Figure 3).

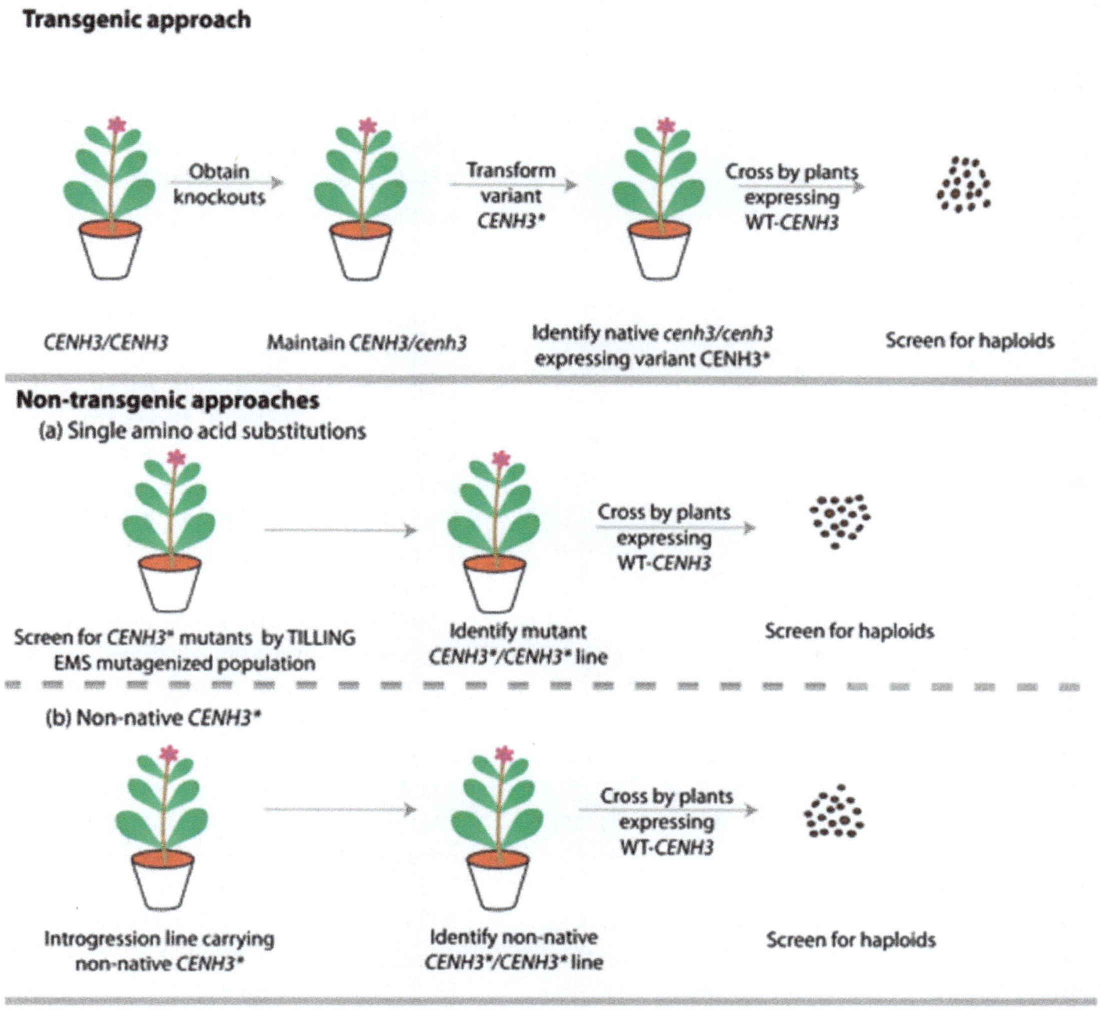

Figure 3. Schematic representation comparing the steps involved in generating haploids by transgenic vs. non-transgenic approaches. The transgenic approach involves generating *CENH3* knockouts by TILLING or genome editing, with transgenic addition of point mutant, non-native or *GFP-tailswap* alleles (*CENH3**). The non-transgenic approach involves novel alleles, created either by chemical mutagenesis or though introgression of non-native *CENH3*. This later non-native approach is hypothetical and is yet to be demonstrated in plants. *CENH3** represents variant *CENH3* (Point mutant*CENH3*/ Non-native *CENH3*/ *GFP-tailswapCENH3*).

CONCLUSION

Haploid induction technologies- and CENH3-based HI is only one among many- significantly reduces the time and labor required to derive new true-breeding varieties. Within the short period of its discovery, CENH3-mediated genome eliminations have also been shown to be useful in trait mapping (Seymour et al., 2012), reverse breeding (Wijnker et al., 2012), and a variety of other applications (Ravi et al., 2014), all of which can be applied to crop breeding. In spite of the great success with various applications in model plant *Arabidopsis*, CENH3-mediated genome elimination needs to be tested in other crop species– its application to maize, described in this volume (Kelliher et al., 2016) is very encouraging. The possible delay in implementing this technology (first published in 2010) may be due to the lack of *CENH3* knockouts in other species. The recent development of efficient CRISPR-based gene targeting conveniently addresses this issue. *GFP-tailswap* is a simple and straightforward approach for plants with well-annotated and sequenced genomes. The alternative approaches to CENH3-mediated HI may also be employed. Non-GMO point mutants can be obtained by EMS mutagenesis, and may already be present in existing populations. The recent findings of complementation and HI by CENH3 from related species (Maheshwari et al., 2015) offers an alternative non-GMO approach for species in which EMS mutagenesis is inefficient.

The mechanism, through which CENH3 modification affects competitive crosses, without affecting self-pollinations, mitosis, or meiosis, remains a mystery. The elucidation of this process will not only provide insight into CENH3 function and the mechanism of chromosome assortment, but may reveal additional targets that will, upon modification, provide the next generation of haploid inducers.

CONFLICT OF INTEREST STATEMENT

The authors declare that the research was conducted in the absence of any commercial or financial relationships that could be construed as a potential conflict of interest.

The reviewer JK and handling Editor declared their shared affiliation, and the handling Editor states that the process nevertheless met the standards of a fair and objective review.

ACKNOWLEDGMENTS

The authors gratefully acknowledge the Rijk Zwaan Zaadteelt en Zaadhandel B.V., research agreement 201016851, and the National Science Foundation (BREAD award #1109882) for their support of our research in this area.

REFERENCES

1. Allshire, R. C., and Karpen, G. H. (2008). Epigenetic regulation of centromeric chromatin: old dogs, new tricks? *Nat. Rev. Genet.* 9, 923–937. doi: 10.1038/nrg2466
2. Buchwitz, B. J., Ahmad, K., Moore, L. L., Roth, M. B., and Henikoff, S. (1999). Cell division: a histone-H3-like protein in *C. elegans*. *Nature* 401, 547–548. doi: 10.1038/44062
3. Cutter, A. R., and Hayes, J. J. (2015). A brief review of nucleosome structure. *FEBS Lett.* 589, 2914–2922. doi: 10.1016/j.febslet.2015.05.016
4. Dunwell, J. M. (2010). Haploids in flowering plants: origins and exploitation. *Plant Biotechnol. J.* 8, 377–424. doi: 10.1111/j.1467-7652.2009.00498.x
5. Dwivedi, S. L., Britt, A. B., Tripathi, L., Sharma, S., Upadhyaya, H. D., and Ortiz, R. (2015). Haploids: constraints and opportunities in plant breeding. *Biotechnol. Adv.* 33, 812–829. doi: 10.1016/j.biotechadv.2015.07.001
6. Fleischmann, A., Michael, T. P., Rivadavia, F., Sousa, A., Wang, W., Temsch, E. M., et al. (2014). Evolution of genome size and chromosome number in the carnivorous plant genus *Genlisea* (Lentibulariaceae), with a new estimate of the minimum genome size in angiosperms. *Ann. Bot.* 114, 1651–1663. doi: 10.1093/aob/mcu189
7. Gibcus, J. H., and Dekker, J. (2013). The hierarchy of the 3D genome. *Mol. Cell* 49, 773–782. doi: 10.1016/j.molcel.2013.02.011
8. Guha, S., and Maheshwari, S. C. (1964). In vitro production of embryos from anthers of datura. *Nature* 204, 497–497. doi: 10.1038/204497a0
9. Hartung, F., and Schiemann, J. (2014). Precise plant breeding using new genome editing techniques: opportunities, safety and regulation in the EU. *Plant J.* 78, 742–752. doi: 10.1111/tpj.12413
10. Henikoff, S., and Smith, M. M. (2015). Histone variants and epigenetics. *Cold Spring Harbor Perspect. Biol.* 7:a019364. doi: 10.1101/cshperspect.a019364
11. Howman, E. V., Fowler, K. J., Newson, A. J., Redward, S., MacDonald, A. C., Kalitsis, P., et al. (2000). Early disruption of centromeric chromatin organization in centromere protein A (Cenpa) null mice. *Proc. Natl. Acad. Sci. U.S.A.* 97, 1148–1153. doi: 10.1073/pnas.97.3.1148
12. Ingouff, M., Rademacher, S., Holec, S., Šoljić, L., Xin, N., Readshaw, A., et al. (2010). Zygotic resetting of the HISTONE 3 variant repertoire participates in epigenetic reprogramming in *Arabidopsis*. *Curr. Biol.* 20, 2137–2143. doi: 10.1016/j.cub.2010.11.012
13. Karimi-Ashtiyani, R., Ishii, T., Niessen, M., Stein, N., Heckmann, S., Gurushidze, M., et al. (2015). Point mutation impairs centromeric CENH3 loading and induces haploid plants. *Proc. Natl. Acad. Sci. U.S.A.* 112, 11211–11216. doi: 10.1073/pnas.1504333112
14. Kasha, K. J., and Kao, K. N. (1970). High frequency haploid production in barley (*Hordeum vulgare* L.). *Nature* 225, 874–876. doi: 10.1038/225874a0
15. Kelliher, T., Starr, D., Wang, W., McCuiston, J., Zhong, H., Nuccio, M. L., et al. (2016). Maternal haploids are preferentially induced by CENH3-tailswap transgenic complementation in maize. *Front. Plant Sci.* 7:414. doi: 10.3389/fpls.2016.00414
16. Kuppu, S., Tan, E. H., Nguyen, H., Rodgers, A., Comai, L., Chan, S. W. L., et al. (2015). Point mutations in centromeric histone induce post-zygotic incompatibility and uniparental inheritance. *PLoS Genet.* 11:e1005494. doi: 10.1371/journal.pgen.1005494
17. Lermontova, I., Koroleva, O., Rutten, T., Fuchs, J., Schubert, V., Moraes, I., et al. (2011). Knockdown of CENH3 in *Arabidopsis*reduces mitotic divisions and causes sterility by disturbed meiotic chromosome segregation. *Plant J.* 68, 40–50. doi: 10.1111/j.1365-313X.2011.04664.x
18. Lermontova, I., Kuhlmann, M., Friedel, S., Rutten, T., Heckmann, S., Sandmann, M., et al. (2013). *Arabidopsis* KINETOCHORE NULL2 is an upstream component for centromeric histone H3 variant cenH3 deposition at centromeres. *Plant Cell* 25, 3389–3404. doi: 10.1105/tpc.113.114736
19. Lermontova, I., Sandmann, M., and Demidov, D. (2014). Centromeres and kinetochores of Brassicaceae. *Chromosome Res.* 22, 135–152. doi: 10.1007/s10577-014-9422-z
20. Liu, Y., Su, H., Zhang, J., Liu, Y., Han, F., and Birchler, J. A. (2015). Dynamic epigenetic states of maize centromeres. *Front. Plant Sci.* 6:904. doi: 10.3389/fpls.2015.00904
21. Maehara, K., Takahashi, K., and Saitoh, S. (2010). CENP-A reduction induces a p53-dependent cellular senescence response to protect cells from executing defective mitoses. *Mol. Cell. Biol.* 30, 2090–2104. doi: 10.1128/MCB.01318-09
22. Maheshwari, S., Tan, E. H., West, A., Franklin, F. C. H., Comai, L., and Chan, S. W. L. (2015). Naturally occurring differences in CENH3 affect chromosome segregation in zygotic mitosis of hybrids. *PLoS Genet.* 11:e1004970. doi: 10.1371/journal.pgen.1004970
23. Malik, H. S., and Henikoff, S. (2003). Phylogenomics of the nucleosome. *Nat. Struct. Mol. Biol.* 10, 882–891. doi: 10.1038/nsb996
24. Marshall, O. J., Chueh, A. C., Wong, L. H., and Choo, K. A. (2008). Neocentromeres: new insights into centromere structure, disease development, and karyotype evolution. *Am. J. Hum. Genet.* 82, 261–282. doi: 10.1016/j.ajhg.2007.11.009
25. Mattiroli, F., D'Arcy, S., and Luger, K. (2015). The right place at the right time: chaperoning core histone variants. *EMBO Rep.*16:e201540840. doi: 10.15252/embr.201540840

26. Moraes, I. R., Lermontova, I., and Schubert, I. (2011). Recognition of *A. thaliana* centromeres by heterologous CENH3 requires high similarity to the endogenous protein. *Plant Mol. Biol.* 75, 253–261. doi: 10.1007/s11103-010-9723-3

27. Müller, S., and Almouzni, G. (2014). A network of players in H3 histone variant deposition and maintenance at centromeres.*Biochim. Biophys. Acta (BBA)-Gene Regul. Mech.* 1839, 241–250.

28. Pellicer, J., Fay, M. F., and Leitch, I. J. (2010). The largest eukaryotic genome of them all? *Bot. J. Linn. Soc.* 164, 10–15. doi: 10.1111/j.1095-8339.2010.01072.x

29. Ravi, M., and Chan, S. W. L. (2010). Haploid plants produced by centromere-mediated genome elimination. *Nature* 464, 615–618. doi: 10.1038/nature08842

30. Ravi, M., Kwong, P. N., Menorca, R. M. G., Valencia, J. T., Ramahi, J. S., Stewart, J. L., et al. (2010). The rapidly evolving centromere-specific histone has stringent functional requirements in *Arabidopsis thaliana*. *Genetics* 186, 461–471. doi: 10.1534/genetics.110.120337

31. Ravi, M., Marimuthu, M. P. A., Tan, E. H., Maheshwari, S., Henry, I. M., Marin-Rodriguez, B., et al. (2014). A haploid genetics toolbox for *Arabidopsis thaliana*. *Nat. Commun.* 5:5334. doi: 10.1038/ncomms6334

32. Ravi, M., Shibata, F., Ramahi, J. S., Nagaki, K., Chen, C., Murata, M., et al. (2011). Meiosis-specific loading of the centromere-specific histone CENH3 in *Arabidopsis thaliana*. *PLoS Genet.* 7:e1002121. doi: 10.1371/journal.pgen.1002121

33. Sanei, M., Pickering, R., Kumke, K., Nasuda, S., and Houben, A. (2011). Loss of centromeric histone H3 (CENH3) from centromeres precedes uniparental chromosome elimination in interspecific barley hybrids. *Proc. Natl. Acad. Sci. U.S.A.* 108, E498–E505. doi: 10.1073/pnas.1103190108

34. Sekulic, N., and Black, B. E. (2012). Molecular underpinnings of centromere identity and maintenance. *Trends Biochem. Sci.* 37, 220–229. doi: 10.1016/j.tibs.2012.01.003

35. Seymour, D. K., Filiault, D. L., Henry, I. M., Monson-Miller, J., Ravi, M., Pang, A., et al. (2012). Rapid creation of*Arabidopsis*doubled haploid lines for quantitative trait locus mapping. *Proc. Natl. Acad. Sci. U.S.A.* 4227–4232. doi: 10.1073/pnas.1117277109

36. Steiner, F. A., and Henikoff, S. (2015). Diversity in the organization of centromeric chromatin. *Curr. Opin. Genet. Dev.* 31, 28–35. doi: 10.1016/j.gde.2015.03.010

37. Steinitz-Sears, L. M. (1963). Chromosome studies in *Arabidopsis thaliana*. *Genetics* 48, 483–490.

38. Stimpson, K. M., and Sullivan, B. A. (2010). Epigenomics of centromere assembly and function. *Curr. Opin. Cell Biol.* 22, 772–780. doi: 10.1016/j.ceb.2010.07.002

39. Stoler, S., Keith, K. C., Curnick, K. E., and Fitzgerald-Hayes, M. (1995). A mutation in CSE4, an essential gene encoding a novel chromatin-associated protein in yeast, causes chromosome nondisjunction and cell cycle arrest at mitosis. *Genes Dev.* 9, 573–586. doi: 10.1101/gad.9.5.573

40. Talbert, P. B., Masuelli, R., Tyagi, A. P., Comai, L., and Henikoff, S. (2002). Centromeric localization and adaptive evolution of an *Arabidopsis* histone H3 variant. *Plant Cell Online* 14, 1053–1066. doi: 10.1105/tpc.010425

41. Tan, E. H., Henry, I. M., Ravi, M., Bradnam, K. R., Mandakova, T., Marimuthu, M. P., et al. (2015). Catastrophic chromosomal restructuring during genome elimination in plants. *eLife* 4:e06516. doi: 10.7554/eLife.06516

42. Wang, Z., Yin, H., Lv, L., Feng, Y., Chen, S., Liang, J., et al. (2014). Unrepaired DNA damage facilitates elimination of uniparental chromosomes in interspecific hybrid cells. *Cell Cycle* 13, 1345–1356. doi: 10.4161/cc.28296

43. Wędzony, M., Forster, B. P., Żur, I., Golemiec, E., Szechyńska-Hebda, M., Dubas, E., et al. (2009). "Progress in doubled haploid technology in higher plants," in *Advances in Haploid Production in Higher Plants*, eds A. Touraev, B. Forster, and S. M. Jain (Amsterdam: Springer), 1–33. doi: 10.1007/978-1-4020-8854-4_1

44. Westhorpe, F. G., and Straight, A. F. (2015). The centromere: epigenetic control of chromosome segregation during mitosis.*Cold Spring Harb. Perspect. Biol.* 7:a015818. doi: 10.1101/cshperspect.a015818

45. Wieland, G., Orthaus, S., Ohndorf, S., Diekmann, S., and Hemmerich, P. (2004). Functional complementation of human centromere protein A (CENP-A) by Cse4p from *Saccharomyces cerevisiae*. *Mol. Cell. Biol.* 24, 6620–6630.

46. Wijnker, E., van Dun, K., de Snoo, C. B., Lelivelt, C. L. C., Keurentjes, J. J. B., Naharudin, N. S., et al. (2012). Reverse breeding in *Arabidopsis thaliana* generates homozygous parental lines from a heterozygous plant. *Nat. Genet.* 44, 467–470. doi: 10.1128/MCB.24.15.6620-6630.2004

CHAPTER 12

IMPROVED EFFICIENCY OF DOUBLED HAPLOID GENERATION IN HEXAPLOID TRITICALE BY IN VITRO CHROMOSOME DOUBLING

Tobias Würschum, Matthew R Tucker, Jochen C Reif and Hans Peter Maurer

BACKGROUND

Doubled haploid (DH) technology is a valuable tool in modern breeding programs since it allows for the production of completely homozygous lines within a few months and dramatically reduces the time required to establish new cultivars [1]. In addition, many genomic approaches such as association or QTL mapping benefit greatly from the use of DH populations [2, 3, 4]. DH plants are routinely generated in crops such as barley [5]. However, due to high costs and low efficiency, DH production is yet to be established in applied breeding programs for other cereals such as triticale. Of the different methods available for DH production, microspore embryogenesis (androgenesis) shows the greatest potential due to the abundance of microspores per spike and consequently the higher frequency of DH output as compared to other approaches such as wide crosses. Factors limiting the application of microspore culture at a commercial level include the rate of embryogenesis and regeneration, the frequency of albinism among regenerants and the frequency of chromosome doubling required to obtain fertile DH plants [6]. In barley, the rate of spontaneous chromosome doubling is high and ranges from 70-90 % [6]. By contrast, the rate of spontaneous doubled haploids reported for triticale ranges from 6-38 % [7, 8, 9], which similar to most crops is too low and variable to omit a specific step for induced chromosome doubling. The most common doubling agent is colchicine which is traditionally applied *in vivo* [10]. This application of colchicine at the seedling stage, however, has several disadvantages: large amounts of colchicine solution are required which imparts high costs and increased exposure to this toxic substance, a high rate of mortality of seedlings owing to the treatment, and production of mixoploid or chimeric plants which results in low seed set and consequently the requirement of additional seed multiplication before field testing [11].

The application of colchicine during early stages of androgenesis could alleviate some of these problems [6] and has been applied to different cereals including wheat [11, 12], durum wheat [13] and rice [14]. In wheat, the incorporation of colchicine during the first hours of anther or microspore culture improved the doubling rate [11, 12]. The application of doubling agents, however, not only increases the percentage of chromosome doubling, but also affects the whole androgenetic process. The optimum concentration and time of application for chromosome doubling may have negative effects on embryogenesis and regeneration rate, as well as on the percentage of green plants [6]. Thus, *in vitro* chromosome doubling must be investigated and optimized in light of the entire androgenetic process.

In triticale, *in vitro* application of colchicine has been applied in anther culture approaches [7] but not for microspore culture which is emerging as the method of choice. The objective of this work was to investigate the suitability of *in vitro* chromosome doubling during microspore culture for efficient doubled haploid production in triticale. More specifically, the aim was to determine the optimum concentration and incubation period of colchicine for *in vitro* chromosome doubling while considering the androgenetic response as a whole.

Citation: Tobias Würschum, Matthew R Tucker, Jochen C Reif and Hans Peter Maurer, Improved efficiency of doubled haploid generation in hexaploid triticale by in vitro chromosome doubling, DOI: 10.1186/1471-2229-12-109.

RESULTS

Visual scoring of Petri dishes containing microspores in embryo induction revealed no obvious effect of the colchicine incubation on the number of regenerated embryos 4 weeks after microspore isolation (Figure 1). In addition, the embryos were of similar high quality and almost no calli were formed.

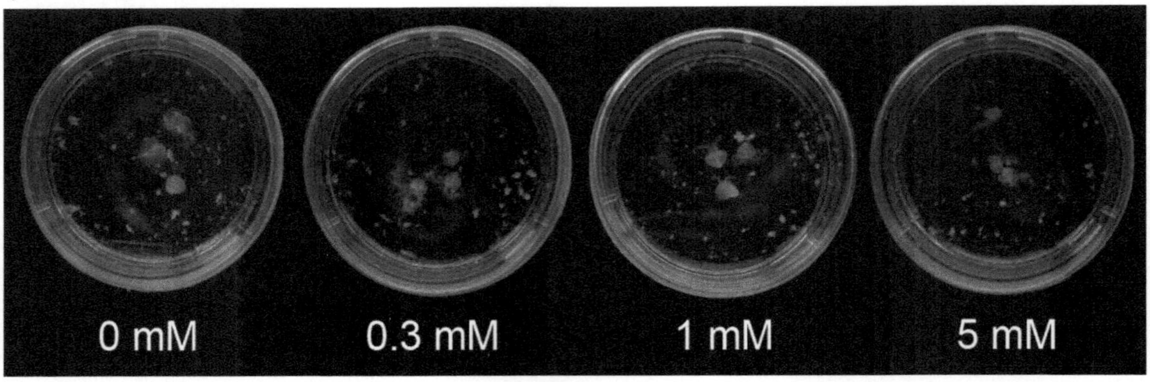

Figure 1: Effect of colchicine on embryo formation. Effect of different colchicine concentrations applied for 72 h on embryo formation 4 weeks after microspore isolation.

The regeneration of plants from the transferred embryos averaged 42.7% for the negative controls without colchicine (Table 1) and similar results were obtained for most combinations of colchicine concentration and incubation duration. Lower values were found for the 5 mM concentration at 48 h and 72 h and for 1 mM at 72 h. No clear trend could be observed relating to the percentage of green plants. Whereas colchicine increased the percentage of green plants for the 24 h and 72 h periods, it decreased green plant formation for the 48 h colchicine treatments. In comparison, the *in vivo* approach showed a higher regeneration rate (58.8%).

Table 1: Effect of *in vitro* versus *in vivo* colchicine treatment

Colchicine treat-ment	Transferred embryos	Regeneration rate	Green plants [%]	Survival rate [%]
24 h - 0 mM	330	42.4 a	54.3 a	85.5 a
24 h - 0.3 mM	240	39.2 a	61.7 ab	81.0 ab
24 h - 1 mM	240	41.7 a	72.0 ab	83.3 ab
24 h - 5 mM	180	41.7 a	65.3 ab	71.4 ab
48 h - 0 mM	210	43.3 a	75.8 b	62.3 bc
48 h - 0.3 mM	240	47.1 a	59.3 ab	82.1 ab
48 h - 1 mM	240	47.1 a	69.0 ab	43.6 cd
48 h - 5 mM	120	24.2 b	41.4 a	50.0 ab
72 h - 0 mM	210	42.4 a	48.3 a	60.5 b
72 h - 0.3 mM	240	43.8 a	69.5 ab	68.5 ab
72 h - 1 mM	240	33.3 a	62.5 ab	64.0 ab
72 h - 5 mM	270	26.3 b	71.8 ab	62.7 ab
in vivo	835	58.8 c	70.9 b	44.0 d

Survival rate refers to the rate of plants that survived the transfer from culture medium to soil and grew to maturity. Means of columns followed by different letters indicate significant differences at the $\alpha = 0.05$ level.

The survival rate for plants derived from colchicine-treated seedlings was 44% i.e. approximately half of the green plantlets that were transferred from culture medium could be grown to maturity (Table 1). The remainder died, predominantly within 4 weeks of the colchicine treatment and subsequent transfer to soil. Compared to this *in vivo* chromosome doubling treatment, the survival rate was generally higher for the *in vitro* colchicine approach with an average survival rate of 67.9%. No negative effect was observed for the different colchicine concentrations as compared to the respective controls (Table 1).

The rate of fertile doubled haploid plants from the colchicine treatment at the seedling stage was 52.3% (Table2), while the percentage of fertile plants in the negative controls, which is equivalent to the spontaneous doubling rate, averaged 32.1%. There was a clear effect of the *in vitro* colchicine treatment on the proportion of fertile plants (Table 2, Figure 2). The highest values, ranging from 56.2% to 66.7%, were obtained for 1 mM and 5 mM concentrations and incubation times of 48 h and 72 h. These values roughly doubled that of the spontaneous doubling rate and were even higher than that obtained by the *in vivo* procedure. Consistent with this, we observed significant variance components of colchicine concentration and incubation time for the proportion of fertile plants (Additional file 1: Table S1). The plants produced mainly fertile tillers as compared to the plants from the *in vivo* approach for which, on average, half of the tillers were sterile (Table 2). There appeared to be no effect on the number of seeds per fertile tiller.

Table 2: Effect of the chromosome doubling procedure on plant fertility

Colchicine treatment	Plants	Fertile plants [%]	Fertile plants per 100 embryos	Mean sterile tillers[1]	Mean fertile tillers[2]	Mean seeds per fertile tiller[3]
24 h - 0 mM	65	29.2 a	5.8 ab	0.0 a	5.6 cdef	29.1 bcde
24 h - 0.3 mM	47	27.7 a	5.4 ab	0.4 a	6.9 abcd	33.8 a
24 h - 1 mM	60	50.0 ab	12.5 b	0.5 a	4.5 fg	24.2 e
24 h - 5 mM	35	45.7 ab	8.9 ab	0.3 a	4.9 efg	24.1 cde
48 h - 0 mM	43	32.6 a	6.7 ab	0.7 a	7.7 ab	19.8 bcde
48 h - 0.3 mM	55	50.9 ab	11.7 b	0.9 a	7.0 abc	22.4 bcde
48 h - 1 mM	34	61.8 b	8.8 ab	0.5 a	5.7 defg	33.7 abcd
48 h - 5 mM	6	66.7 ab	3.3 a	0.3 a	7.0 a	26.6 ab
72 h - 0 mM	26	38.5 a	4.8 ab	0.9 a	6.3 abcde	32.0 ab
72 h - 0.3 mM	50	52.0 ab	10.8 ab	0.5 a	5.0 defg	31.3 bcde
72 h - 1 mM	32	59.4 ab	7.9 ab	0.9 b	6.5 abcde	27.4 abc
72 h - 5 mM	32	56.2 ab	6.7 ab	0.2 a	6.3 bcde	36.7 a
in vivo	153	52.3 ab	9.6 ab	2.9 b	3.7 g	31.0 de

[1] mean number of sterile and [2] fertile tillers per seed-setting plant, and [3] mean number of seeds per fertile or partially fertile tiller. Means of columns followed by different letters indicate significant differences at the $\alpha = 0.05$ level.

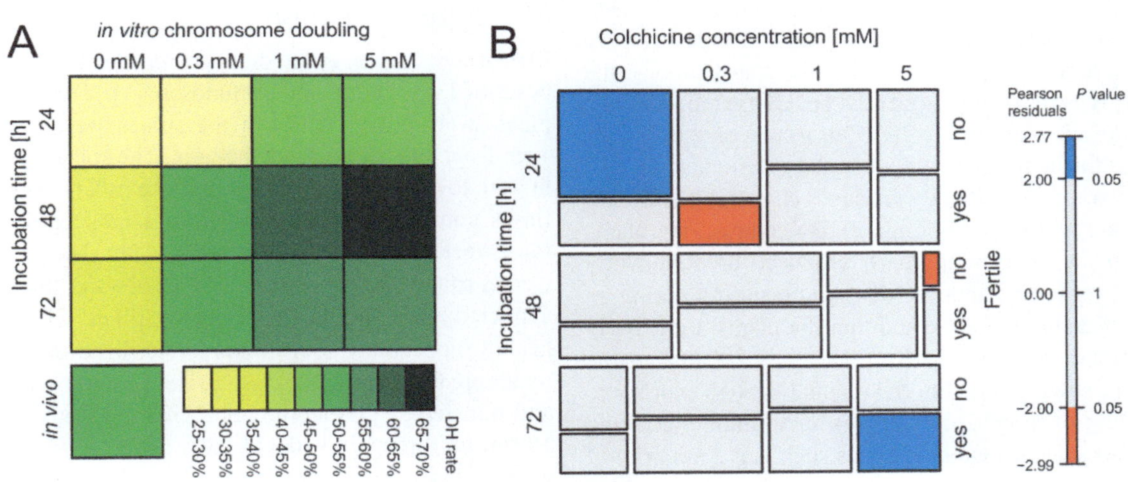

Figure 2: *In vitro* chromosome doubling. (A) Heatplot for the rate of fertile doubled haploid plants among regenerated green plants obtained by the tested *in vitro* methods and by the conventional *in vivo* approach. **(B)** Mosaic plot visualizing the contingency table. The area of each tile is proportional to the corresponding cell entry and the colors reflect the residuals (fit or lack of fit) of the loglinear model.

DISCUSSION

For scientists as well as for commercial breeders, the establishment of a routinely applicable doubled haploid (DH) protocol is of significant interest. For triticale, microspore culture has been shown to provide a high frequency of DH plants, and thus is becoming the method of choice for DH production [8, 9]. Key issues remain, however, that prevent the widespread application of this method. A major limitation relates to the low frequency of spontaneous DH plants produced in triticale, which necessitates chemically-induced doubling of the chromosomes to obtain fertile plants. In this study we compared the traditionally employed *in vivo*approach to those where colchicine is applied *in vitro* during the first stage of microspore culture.

The *in vitro* application of colchicine has been reported to induce positive as well as negative effects on the whole androgenetic pathway, influencing the events of embryogenesis, regeneration rate and the percentage of green plants [15]. In contrast to Zhou et al. [16], who found an increased frequency of embryogenesis in*Brassica* by the application of colchicine, we observed no effect of the *in vitro* colchicine treatment on embryo formation in the experiments described here (Figure 1) nor in our routine DH program (Additional file 1: Figure S1). We did, however, observe a reduced regeneration rate for the treatments with high colchicine concentrations and extended incubation time (Table 1). A similar trend was observed by Soriano et al. [11] for the wheat cultivar Chris, but not for two other cultivars, indicating that this process is likely genotype dependent.

A critical point for a DH procedure is the survival rate, which gives a measure of how many green plants regenerated on culture medium survive until maturity. We found that this survival rate was only 44% for the *in vivo* approach mainly due to a high mortality after the colchicine treatment of the seedlings. Colchicine is an anti-microtubule agent that not only doubles the chromosomes in the meristem but also has severe effects on the health status of the seedlings, possibly explaining the high mortality [6]. By contrast, the *in vitro* approach resulted in much higher survival rates (Table 1) as the seedlings are not subjected to the deleterious colchicine treatment. In this approach, the survival rate of seedlings appeared to be largely dependent upon the transfer from culture medium to soil.

We found that the rate of spontaneous doubled haploids was 32.1%, which is consistent with previous reports for triticale [8, 9], and confirms that an additional chemically-induced chromosome doubling step is required for triticale DH programs. The *in vitro* colchicine method tested here (Table 2) resulted in improved frequency, with the best treatments more than doubling the rate of fertile plants as compared to the spontaneous doubled haploid rate. More importantly, this method also outperformed the *in vivo* approach. The highest rates of fertile plants were observed for the 1 mM and 5 mM colchicine concentrations applied for 48 or 72 h (Figure 2A). The increasing proportion of fertile plants among the regenerated green plants produced after *in vitro* treatment can be seen in the mosaic plot (Figure 2B). Our results on the number of sterile and fertile tillers per regenerated green plant suggest that the *in vivo* approach mainly produces chimeric plants with sectors that are still haploid (Table 2). By contrast, the plants derived from the *in vitro* approach produced predominantly fertile tillers suggesting that the doubling occurred during an early stage of embryogenesis, and that the plants are entirely doubled haploid. The few sterile tillers may be attributable to inadvertent stress applied to the plants (through transfer to soil or bagging of tillers for example), or to doubling taking place at a later stage of embryogenesis such that a minority of the recovered plants are also mosaic for haploid and doubled haploid sectors.

Whereas the enhancement of the proportion of fertile plants is an important goal, the number of fertile plants derived from a defined number of transferred embryos is the ultimate measure of DH procedure success. In this study we found that this is maximized by three treatment combinations: 1 mM colchicine applied for 24 h, or the lower concentration of 0.3 mM colchicine for 48 or 72 h. The variation in optimal incubation time is advantageous because it gives the experimenter more flexibility, especially in applied high-throughput DH programs. With regard to the rate of fertile plants produced from a standard number of transferred embryos, the *in vivo* approach was only slightly less effective than the *in vitro* approach. This similar rate, however, is misleading since a fair comparison must consider the effort required to reach this value. Culturing of seedlings and fostering of treated plants to maturity, including the bagging required to prevent cross-pollination, are resource and time consuming steps. For the *in vivo* approach, many plants are maintained throughout the entire method but do not yield any DH seeds. By contrast, the higher rate of fertile plants and the higher survival rate observed for the *in vitro* approach more than compensate for the reduced regeneration rate as compared to the *in vivo* approach.

CONCLUSION

In conclusion, we have shown that an *in vitro* approach for chromosome doubling performs well for microspore culture of triticale. The higher rate of fertile plants produced per number of transferred embryos and the more economical employment of resources make the *in vitro* approach an attractive alternative to the traditionally used *in vivo* colchicine treatment of seedlings.

METHODS

Plant material

This study was based on winter triticale (*Triticosecale* Wittmack L.) F_1 plants derived from a cross between the two cultivars Witon and Corino. Seeds of donor plants were sown in pots in September and kept outside the greenhouse for vernalization until February. Plants were then transferred to the greenhouse with initial temperatures of 12°C for 3 days and subsequently grown at 24°C with a photoperiod length of 16 h.

Microspore isolation

Microspores were isolated following the general procedure described by Eudes and Amundsen [9]. Tillers from donor plants were harvested when the spikes were approximately 1 cm from emergence from the boot. The spikes were wrapped in foil to maintain humidity and the tillers were stored in flasks containing distilled water in a dark cold room (4°C) for 3 weeks as a cold pretreatment. After this time the tillers were surface sterilized with 70% EtOH and the spikes were extracted under sterile conditions. Spikes with a generally healthy appearance were surface sterilized again and, after removing the awns with scissors, were cut into 2 cm fragments and transferred to a sterile cold 250 ml Waring Blender cup. 100 ml cold mannitol (0.3 M) was added to the blender cup, which was sealed with sterile parafilm, and spikes were blended twice for 7 seconds at 18,000 rpm. The suspension was filtered through sterile mesh (100 μm) into a sterile glass beaker and then transferred to two 50 ml centrifuge tubes. The microspores were pelleted by centrifugation at 100 g for 5 minutes in a cooled centrifuge (4°C), resuspended in 45 ml 0.3 M mannitol and centrifuged again. The pellets were subsequently resuspended in 2 ml 0.3 M mannitol each. The microspore solution was then carefully layered on 5 ml of 21% maltose solution in a 15 ml tube and centrifuged at 100 g for 5 minutes at 4°C. The interphases from both tubes containing viable microspores at the appropriate developmental stage were removed by a Pasteur pipette and transferred to a fresh 15 ml centrifuge tube containing NPB-99 medium. Following a washing step (centrifugation at 100 g for 5 minutes at 4°C) the supernatant was removed and approximately 1 ml NPB-99 medium added to the tube. The microspore concentration was determined using a hemacytometer and the final concentration was set at 33,000 microspores per ml medium.

Embryo induction and plant regeneration

For embryo induction, microspores were cultured in 35 mm Petri dishes with ~100,000 microspores per Petri dish in 3 ml medium. The induction medium was NPB-99 supplemented with 10 mg/l arabinogalactan-proteins (gum arabic from acacia tree, Sigma-Aldrich G9752) and 10 mg/l arabinogalactan (AG) Larcoll ((+)-Arabinogalactan, Sigma-Aldrich 851361) [17], 200 μg/ml cefotaxime, and 120 g/l Ficoll-400. Three ovaries from sterilized spikes were added to each Petri dish. The dishes were sealed with Parafilm and incubated in the dark at 25°C for 4 weeks.

Regenerated embryos were removed from the Petri dishes and placed on regeneration medium (NPB-99 medium, supplemented with 3.5 g/l gelrite and 1 g/l charcoal) in sterile plastic boxes. Plantlets were regenerated at 20°C with a 16 h light period. Once the plants reached the 2–3 leaf stage they were transplanted to pots with soil and grown under the same conditions. Plants were grown to maturity, several tillers per plant were isolated with bags to prevent cross-pollination and seed set was determined after ripening. The plants with seed set were scored as fertile doubled haploid plants as these determine the success of the procedure. The sterile plants contain haploids, aneuploids, mixoploids, and sterile doubled haploids.

Colchicine treatment

For the *in vitro* colchicine treatment, isolated microspores (~33,000 per ml; i.e. same concentration as for the embryo induction, thus 3 ml for each embryogenesis Petri dish) were transferred to 15 ml tubes containing induction medium (minus Ficoll to allow for washing and centrifugation) and supplemented with either 0.3 mM, 1 mM, or 5 mM colchicine. The negative control was subjected to the same procedure without colchicine treatment. Three incubation periods were tested: 24 h, 48 h, and 72 h. Three replications were used for each combination of colchicine concentration and incubation time and the presented results are averages over all replicates. The tubes were kept in the dark at 25°C and placed on a shaker to keep the microspores in solution and prevent them from settling. After the incubation time the microspores were pelleted by centrifugation (100 g, 5 minutes), washed once with medium without colchicine, and then transferred to Petri dishes for embryo induction as described above.

As a control for the *in vitro* colchicine treatment, we included an *in vivo* colchicine treatment according to Arzani and Darvey [7]. This step was done with seedlings, which were approximately at the 3-4-leaf stage and which were derived from ten replications (after isolation, microspores were directly placed in the Petri dishes for embryogenesis). Seedlings were submerged with their roots and the meristem in a 0.06% colchicine solution supplemented with 1.5% DMSO and Tween 20 (pH 5.5) for 6 h at room temperature and light. Plants were subsequently rinsed in water for 2 h and then planted in soil. Plants were grown to maturity as described above.

Statistical analyses

Statistical analyses were conducted using the statistical software R [18]. Mosaic plots were drawn using the vcd package for R. Variance components were determined by the restricted maximum likelihood (REML) method. Significance for variance component estimates was tested by model comparison with likelihood ratio tests in a full random model where the halved P values were used as an approximation [19]. Calculations were performed using the software ASReml 2.0 [20]. Mean comparisons in Tables 1 and 2 were obtained for the binominal traits by calculating confidence intervals with the R package binom (Pearson-Klopper method) or by Fisher's least significant difference (LSD) tests following ANOVA for the traits with a normal distribution.

ACKNOWLEDGEMENTS

We greatly acknowledge the lab work of Alexandra Appel, Sylvia Kaiser, Barbara Renz, and Angela Harmsen. We greatly appreciate the helpful comments and suggestions of two anonymous reviewers.

REFERENCES

1. Wedzony M, Forster BP, Zur I, Golemice E, Szechynska-Hebda , Dubas E, Gotebiowska G: Progress in doubled haploid technology in higher plants.In Edited by: Touraev A, Forster BP, Mohan Jain S. Advances in Haploid Production in Higher Plants; 2009: 1-33.
2. Alheit KV, Reif JC, Maurer HP, Hahn V, Weissmann EA, Miedaner T, Würschum T: Detection of segregation distortion loci in triticale (x Triticosecale Wittmack) based on a high-density DArT marker consensus genetic linkage map. BMC Genomics. 2011, 12: 380-10.1186/1471-2164-12-380.
3. Reif JC, Maurer HP, Korzun V, Ebmeyer E, Miedaner T, Würschum T: Mapping QTLs with main and epistatic effects underlying grain yield and heading time in soft winter wheat. Theor Appl Genet. 2011, 123: 283-292. 10.1007/s00122-011-1583-y.
4. Würschum T: Mapping QTL for agronomic traits in breeding populations. Theor Appl Genet. 2012, DOI 10.1007/s00122-012-1887-6.
5. Maluszynski M, Kasha KJ, Szarejko I: Published doubled haploid protocols in plant species. Doubled haploid production in crop plants, a manual. Edited by: Maluszynski M, Kasha KJ, Forster BP, Szarejko I.Dordrecht, The Netherlands, Kluwer Academic Publishers;2003: 309-335.
6. Castillo AM, Cistué L, Vallés MP, Soriano M: Chromosome doubling in monocots. Forster BP. Edited by: Touraev A. Mohan Jain S (ed) Advances in Haploid Production in Higher Plants:Springer;2009: 329-338.
7. Arzani A, Darvey NL: The effect of colchicine on triticale anther-derived plants: Microspore pre-treatment and haploid-plant treatment using a hydroponic recovery system. Euphytica. 2001, 122: 235-241. 10.1023/A:1012966506030.
8. Oleszczuk S, Sowa S, Zimny J: Direct embryogenesis and green plant regeneration from isolated microspores

of hexaploid triticale (x Triticosecale Wittmack) cv. Bogo. Plant Cell Rep. 2004, 22: 885-893.

9. Eudes F, Amundsen E: Isolated microspore culture of Canadian 6x triticale cultivars. Plant Cell Tiss Org Cult. 2005, 82: 233-241. 10.1007/s11240-005-0867-9.
10. Kasha KJ: Chromosome doubling and recovery of doubled haploid plants. Biotechnology in Agriculture and Forestry 56. Edited by: Palmer CE, Keller WA, Kasha KJ. Berlin/Heidelberg: Haploids in Crop Improvement II. Springer;2005; 123-152.
11. Soriano M, Cistué L, Vallés MP, Castillo AM: Effects of colchicine on anther and microspore culture of bread wheat. Plant Cell Tiss Org Cult. 2007, 91: 225-234. 10.1007/s11240-007-9288-2.
12. Hansen NJP, Andersen SB: In vitro chromosome doubling with colchicine during microspore culture in wheat (Triticum aestivum L.). Euphytica. 1998, 102: 101-108. 10.1023/A:1018348816205.
13. Tersi M, Xynias IN, Gouli-Vavdinoudi E, Roupakias DG: Anther culture response of F-1 durum X bread wheat hybrids after colchicine application. Plant Breed. 2006, 125: 457-460. 10.1111/j.1439-0523.2006.01285.x.
14. Alemanno L, Guiderdoni E: Increased doubled haploid plant regeneration from rice (Oryza sativa L.) anthers cultured on colchicine supplemented media. Plant Cell Rep. 1994, 13: 432-436.
15. Shariatpanahi ME, Bal U, Heberle-Bors E, Touraev A: Stresses applied for the re-programming of plant microspores towards in vitro embryogenesis. Physiol Plant. 2006, 127: 519-534. 10.1111/j.1399-3054.2006.00675.x.
16. Zhou WJ, Hagberg P, Tang GX: Increasing embryogenesis and doubling efficiency by immediate colchicine treatment of isolated microspores in spring Brassica napus. Euphytica. 2002, 128: 27-34. 10.1023/A:1020687517888.
17. Letarte J, Simion E, Miner M, Kasha KJ: Arabinogalactans and arabinogalactan-proteins induce embryogenesis in wheat (Triticum aestivum L.) microspore culture. Plant Cell Rep. 2006, 24: 691-698. 10.1007/s00299-005-0013-5.
18. R Development Core Team: R: a language and environment for statistical computing, R foundation for statistical computing. 2010, http://www.R-project.org,
19. Stam DO, Lee JW: Variance components testing in longitudinal mixed effects models. Biometrics. 1994, 50: 1171-1177. 10.2307/2533455.
20. Gilmour AR, Gogel BJ, Cullis BR, Thompson R: ASReml User Guide Release 2.0.UK: VSN International Ltd., Hermel Hempstead; 2006:

CHAPTER 13

COMPARATIVE MAPPING IN THE POACEAE FAMILY REVEALS TRANSLOCATIONS IN THE COMPLEX POLYPLOID GENOME OF SUGARCANE

Karen S Aitken, Meredith D McNeil, Paul J Berkman, Scott Hermann, Andrzej Kilian, Peter C Bundock and Jingchuan Li

Keywords: Saccharum, Poaceae, Synteny, Orthology, Translocations, SNP markers

BACKGROUND

Sugarcane is a member of the Poaceae family that contributes 75% of the worlds sugar supply and is being heralded as an important source as an alternative biofuel [[1]]. However, knowledge of the genetic architecture of this crop currently lags behind that of other crops of the Poaceae family such as wheat, barley and rice. It is important to uncover the genome structure of a particular crop to guide informed decisions in the identification of important genes/traits that can be utilized during breeding and germplasm introgression.

Modern sugarcane cultivars (2n = 100-120) have complex polyploid, aneuploid genomes that make classical genetic, molecular genetic, and breeding studies difficult to interpret, as information on the structure and organisation of the genome is incomplete [[2]]. They are essentially hybrids between the polyploid domesticated species, Saccharum officinarum L (2n = 80) and the polyploid wild species, *S. spontaneum* L (2n = 40-128). The cultivated hybrids have undergone a series of backcrosses to *S. officinarum* as the female parent to recover agronomically adapted genotypes with a high sugar yield. Using in situ hybridisation, [[3]] determined that the basic chromosome number of *S. officinarum* is x = 10 and for *S. spontaneum* is x = 8, with 10-15% of the genome of modern sugarcane cultivars contributed by *S. spontaneum*, 80% *S. officinarum* and 11-15% recombinant chromosomes between the two species [[4]]. Studies have shown that chromosome pairing is mostly as bivalents, although an assortment of meiotic irregularities can occur and contribute to the formation of aneuploid gametes [[5],[6]]. Sugarcane clones are propagated vegetatively and are stable over many years some cultivars have been grown commercially for over 20 years [[7]].

Genetic studies of sugarcane have been difficult due to the polyploid and highly heterozygous nature of the genome. For example, molecular markers such as restriction fragment length polymorphism (RFLPs) and simple sequence repeats (SSRs) can reveal several markers or alleles at one locus. They can be single dose (ie single copy at a particular locus), double dose (two copies) or multi dose where it is often not possible to know how many copies are at a particular locus. To date, a number of genetic maps have been constructed using single dose (SD) markers. Single dose markers are used in coupling to construct linkage groups, and multi allelic markers and markers in repulsion are used to sort these linkage groups into homology groups. The genetic maps generated to date have utilized the following markers; random amplified polymorphic DNA (RAPDs) [[8],[9]], RFLPs [[10]-[13]] and amplified fragment length polymorphism (AFLPs) [[14]-[16]] and simple sequence repeats (SSRs) [[17],[18]]. Even the most extensive of these genetic maps still only covers approximately half of the sugarcane genome [[15]]. Recently, the development of a high-throughput marker system, diversity arrays technology (DArTs), has enabled the creation of medium-density genetic maps for plants with complex genomes and limited sequence information available [[19]]. DArT arrays generate whole genome fingerprints by scoring the presence versus absence of DNA fragments in genomic representations generated from

Citation: Karen S Aitken, Meredith D McNeil, Paul J Berkman, Scott Hermann, Andrzej Kilian, Peter C Bundock and Jingchuan Li, Comparative mapping in the Poaceae family reveals translocations in the complex polyploid genome of sugarcane, DOI: 10.1186/s12870-014-0190-x.

genomic DNA samples through the process of complexity reduction [[20]]. DArT maps have now been generated for many plants in the grass family, including rice [[20]], barley [[19]], wheat [[21]] and sorghum [[22]]. In an effort to increase the marker density for sugarcane, a diversity array was developed using a combination of *S. officinarum* and *S. spontaneum* accessions and sugarcane cultivars [[23]]. This has resulted in a medium density genetic map for sugarcane with enough sequence information to allow comparative mapping [[24]].

Comparative genomics is a powerful tool that can provide a starting point for accelerating progress in revealing the genomic structure of crops that are lacking in the necessary genomic tools [[25]]. Aligning markers or partial sequences to fully sequenced genomes enables researchers to deduce information about gene content and order of a particular crop of interest. Sugarcane is a member of the Andropogoneae tribe, within the grass family Poaceae, along with other tropical grasses such as sorghum (*Sorghum bicolor* L. Moench), maize (*Zea mays*) and rice (*Oryza sativa*). Evolutionary studies have speculated that sorghum and maize may have diverged from a common ancestor about 29 million years ago [[26]] after which maize underwent polyploidation and subsequent reorganisation of the genome. In contrast, rice and the maize/sorghum lineage may have diverged from a common ancestor around 46 million years ago and show much greater levels of chromosome structural rearrangement [[26]]. Whilst, sugarcane and sorghum may have shared a common ancestor less than 10 million years ago and retain very similar gene order [[27],[28]]. This close evolutionary relationship has led to the use of sorghum as a model crop for comparative genetic studies to gain a greater understanding of the genome of sugarcane [[25]]. The availability of the complete genome sequence of rice, maize and more recently sorghum, and the close evolutionary history of these species, has allowed some powerful comparative genomic studies [[25],[26],[29]] to be undertaken to gain an understanding of the genetic architecture of other crops where genomic sequence is lacking.

In this current study we have taken advantage of the development of a new, high-throughput marker system (DArTs) to increase the marker density combined with the addition of SNP and EST-SSR markers. This has led to an increase in genome coverage which will aid in the identification of QTL regions containing agronomically important genes of interest for sugarcane breeders. Furthermore, the availability of the genome sequence of maize, rice and more specifically sorghum has allowed a detailed study of the syntenic relationships between these crops. The generation of this denser genetic map will ultimately lead to a greater understanding of the genetic architecture of the sugarcane genome which had previously been lacking. This understanding will now allow the interrogation of important regions and identification of target genes for breeding.

RESULTS

Assembly of Saccharum linkage groups into homo(eo)logous groups (HGs)

The genetic linkage map of Q165 presented in Additional file 1 was arranged into eight HGs which represents the lowest *Saccahrum* basic chromosome number ($x=8$). Initially this was carried out using the multiple allelic loci detected by single SSR, RFLP or SNP markers. The location of the LGs was then verified using sequence information from all the 677 markers with a known sequence that had a primary correspondence to the sorghum genome at a significance level of $P<e^{-20}$. This allowed the 160 LGs to be condensed into the eight HGs (Table 1). Cultivar Q165 has a total of 110 chromosomes of which 15–17 are inherited from *S. spontaneum*, 82–83 are inherited from *S. officianrum* and 11–12 are recombinant between the two species [[30]].

Table 1: Number of single dose DArT markers mapped in each HG and length of HG in the Q165 linkage map and aligned to the sorghum genome (see Additional file 1 for details)

HG	Sorghum chromosome	No. LG	No. DART	Total number of markers mapped	Length of HG in cM
4	Sb1	19	125	308	1217.2
8	Sb8 Sb2	24	134	297	1319.2
3	Sb3	20	75	218	1243.6
1	Sb4	19	125	282	1253.4
2	Sb5 Sb6	23	198	475	1818.0
5	Sb7	13	94	222	846.6
6	Sb9	18	111	232	678.3
7	Sb10	17	90	204	1020.5
unassigned		18	21	45	395.9
total		168	976	2283	9792.7

Alignment to sorghum

The sequence information from the DArT, SNP, RFLP and the EST derived SSR markers were used to align the sugarcane LGs to the sorghum genome sequence (Additional file 1, Table 2). This was achieved by comparing the DNA sequence of the 959 markers with a known sequence to the genome sequence of sorghum using BLASTN homology searching. At the significance threshold of $P<1\times10^{-20}$ there were 677 primary correspondences and a further 175 secondary and tertiary correspondences where the sequence aligned to a further one or two sites at $P < 1 \times 10^{-20}$ but a lower significance level than the primary correspondence (Table 2). In combination with the sequence information from 677 markers, 282 alleles from SSR, SNP and RFLP loci gave a total of 959 links between the sugarcane genetic map and the sorghum genome. Using these 959 informative markers the majority of the sugarcane linkage groups could be assigned to a single sorghum chromosome with high confidence ($P < 1 \times 10^{-20}$) at the macrosyntenic level. As *S. spontaneum* has a basic chromosome number of x = 8, the 8 HGs of *Saccharum* are aligned to the ten chromosome sequences of sorghum. A composite map was produced for each HG to increase the number of markers that could be used for alignment between the *Saccharum* HGs and the sorghum genome. The sugarcane LGs within a HG where possible were condensed into single consensus LGs. This was only possible where the LGs shared two or more markers in common. The consensus LG for each HG was compared with the sorghum genome (Figure 1). In some cases the LGs could not be included as there were not enough markers in common between the LGs within the HG. For example HG6 aligned to Sb9 and 3 using the consensus sequence but when all the LGs are considered parts of LGs also aligned to Sb8 and Sb5 (Table 3). In total over 92% of the sorghum genome was covered by these composite LGs.

Table 2: Number of sugarcane loci with BLASTN ($P < 1 \times 10^{-20}$) correspondences in ten sorghum chromosomes (Sb1-10)

***Sorghum bicolor*-chromosomes**	**Primary correspondences**	**All correspondences**	**No. of incongruous loci**	**No. with no hit to sorghum**
Sb 1	109	150	16	15
Sb 2	69	80	16	4
Sb 3	79	101	13	2
Sb 4	98	138	32	10
Sb 5	78	96	6	13
Sb 6	68	85	8	12
Sb 7	44	53	41	13
Sb 8	25	29	5	2
Sb 9	55	62	37	10
Sb 10	52	58	21	6
Total	677	852	195	87

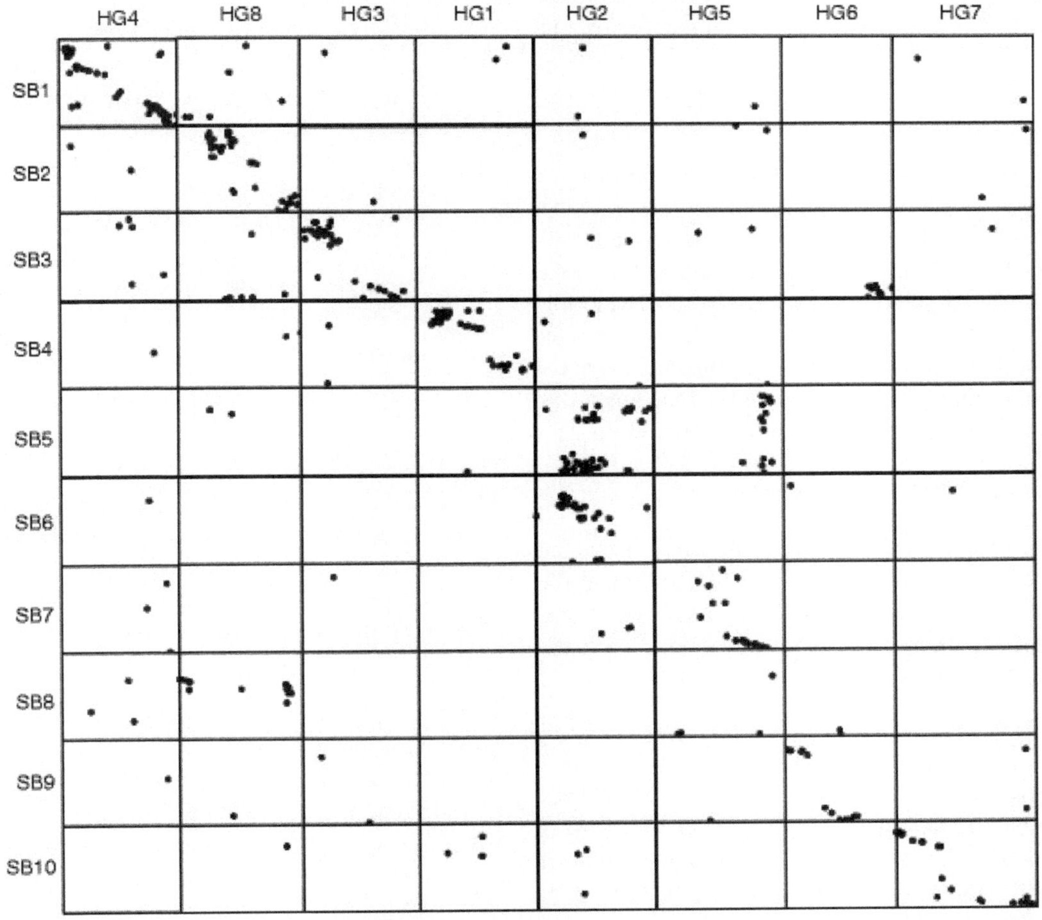

Figure 1: Global distribution of synteny between the 10 sorghum chromosome sequence (SB1-10) and the 8 composite linkage groups from *Saccharum* cultivar Q165. Loci showing homology between the two genomes at $P < 1e^{-20}$significance threshold are indicated by dots.

Table 3: Distance covered in base pairs of the sorghum genome sequence by the mapped markers

Sugarcane HG	Sorghum chromosome	Distance covered (bp)	Percent of sorghum chromosome covered	Translocation chromosome(s)	Distance covered (bp)	Percent of chromosome covered by markers
HG1	**4**	61181-68024049	99.0%	10	49989490-68023941	29.6%
HG2	**5**	1921299-62103178	96.5%	**6**	1820265-61619875	96.1%
HG3	**3**	42138-74366564	99.8%			
HG4	**1**	175578-73559859	99.4%			
HG5	**7**	724155-63641050	97.8%	5	380277-61719143	95.3%
HG6	**9**	407317-58839141	98.0%	3	463958-38224709	50.1%
				8	633124-7753209	12.8%
				5	31121104-62153593	49.8%
HG7	**10**	1611962-60601826	96.7%			
HG8	**2**	652374-77373466	98.4%	**8**	33432881-55441494	40%

Interchromosomal rearrangements between Saccharum and sorghum chromosomes

HG1

When all the LGs within this HG are combined 99% of Sb4 is covered by this HG (Table 3). This HG is highly syntenic to Sb4 with 80% of the markers with sequence hits of e-values $\leq^{-20}$ aligning to Sb4. Only on LG8 did there appear to be an interchromosomal rearrangement with 11 markers that had homology to Sb10 inserted at the end of the LG (Additional file 1; Table 3), this same LG has an intrachromosomal rearrangement.

HG2

HG2 aligned to Sb5 and Sb6 with 8 LGs containing markers aligned to both chromosomes. Of all the markers with a known sequence, over 90% hit these chromosomes indicating a high level of synteny. When a composite LG was created from the LGs over 96% of Sb5 and Sb6 were covered by this HG (Table 3). The LGs that demonstrated synteny with both chromosomes could be explained in the most part by one translocation event although LG1a exhibited a more complex rearrangement that was supported by gene sequence markers (Additional file 1).

HG3

HG3 was highly syntenic to Sb3 and the consensus LG covered 99.8% of this sorghum chromosome (Table 3, Additional file 1). Of the markers with known sequence only LG 10 contained markers that had significant homology to another sorghum chromosome (Sb4).

HG4

HG4 was syntenic to Sb1 and the consensus LG covered 99.4% of this chromosome (Table 3, Additional file 1). Only LG1b contained any significant markers that had homology to other chromosomes and as they were at the end of the LG could be due to spurious linkage.

HG5

HG5 aligned to two sorghum chromosomes Sb7 and Sb5. The coverage was over 90% for both chromosomes. Five of the 13 LGs contained significant homology to both chromosomes and LGs 46, 47, 52 and 84 all represented a single translocation event at a similar position in the LG (Additional file 1, Table 3). LG52 also contain the lower part of Sb8 but more markers are needed to verify this.

HG6

HG6 aligned to Sb9 but contained a large number of markers that aligned to different chromosomes. Of the markers that had known sequence 67% mapped to other chromosomes than Sb9, including Sb3, Sb8 and Sb5. When the composite LG was considered, 98% of Sb9 was covered, 50% of Sb3 and Sb5 and 12% of Sb8 (lower part) (Table 3, Additional file 1). Although the majority of the LGs seemed to contain a backbone of Sb9, LG 20, 17 and 1 all contained a region that had homology to Sb8. In addition LG20 contained a group of markers with homology to Sb3 which appeared from a single translocation event at the end of Sb9.

HG7

This HG aligned to Sb10 with the composite LG covering 97% of the chromosome (Table 3, Additional file 1). Although a number of the markers did align to different chromosomes (38%) there was no real pattern observed. This could be due to low marker coverage as this HG contained the smallest number of markers of any of the HGs (Additional file 1).

HG8

This HG aligned to Sb8 and Sb2 and the composite LGs covered approximately 98% of Sb2 and 40% of Sb8 (Table 3, Additional file 1). Only LG3, 99 and 26 had markers with homology to both chromosomes. The majority of the LG (15/24) aligned only to Sb2 which is the largest sorghum chromosome. Sb8 is the smallest sorghum chromosome and only 6 LGs aligned to this chromosome. LG 99 and 26 both had translocations at the end of Sb2 but LG3 contained at least two rearrangements. Less than half of Sb8 was present in this HG, which could be due to low polymorphism levels of this chromosome or it is possible that the rest of the chromosome is present in other HGs.

Rearrangements between Saccharum and Sorghum chromosomes

Four major rearrangements were identified from syntenic analysis between the sugarcane genetic map and the sorghum genome (Figure 2A-D). Two of these account for the reduction in basic chromosome number from x = 10 in sorghum and *S. officinarum* to x = 8 in *S. spontaneum*. The complicated nature of the hybrid sugarcane genome means that only a few of the LGs within a HG have the translocation, for example in HG2, 7 of the LGs display the translocation, the other LGs align only to one of the two chromosomes (Additional file1). This recombination results in the two sorghum chromosomes combining into one sugarcane HG. This translocation event reduces the basic chromosome number of sugarcane (Figure 2A). The other translocation event is in HG8 which aligns to Sb8 and Sb2 and again reduces the chromosome number to its final basic chromosome number of x = 8 (Figure 2B). The other major rearrangements had no effect on basic chromosome number but involved rearrangements between chromosomes. HG6 aligns in the main to Sb9 but one LG contains a significant translocation from Sb3 covering approximately 50% of Sb3 (Table 3, Figure 2C) The last rearrangement is in HG5 which aligns to Sb7 and Sb5 (Figure 2D); this appears to be a simple translocation between Sb7 and half of Sb5 and is present in 5 LGs within this HG (Additional file 1).

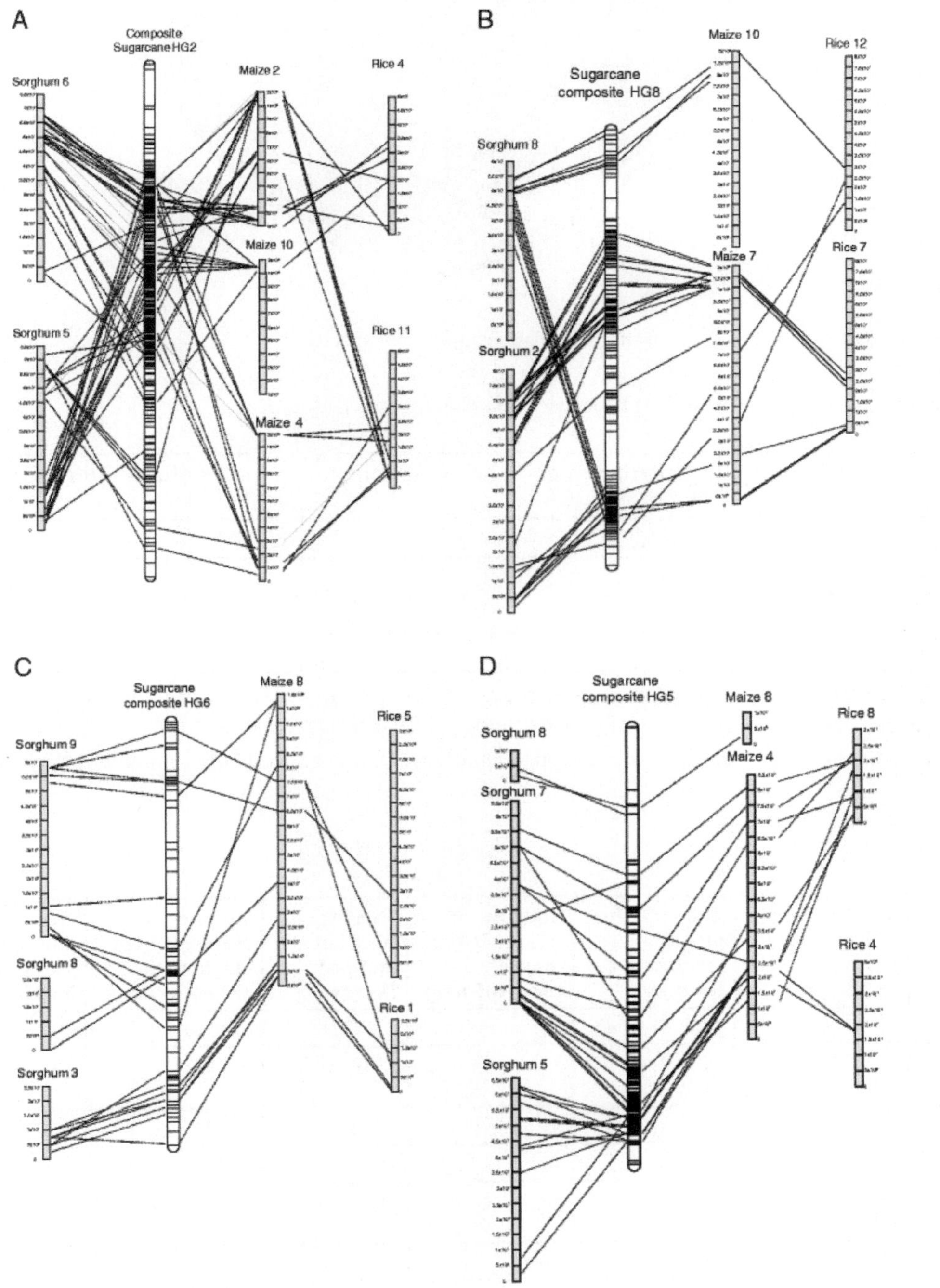

Figure 2: Alignment of the composite sugarcane LGs to sorghum, maize and rice genomic sequences (A-D). The bars on the sugarcane LG represent markers from Additional file 1. These are aligned using the BLASTN algorithm ($P < e^{-20}$) and the position indicated by lines to the other chromosomes. The scale of the chromosomes is in bp.

Analysis of the DArT sequences

In total, 7846 DArT clones were sequenced and a proportion of these were also used to generate the genetic map. These sequences produced a total sequence length of 3,602,170 bp with an average length of 465 bp. These DArT sequences were assembled into a total of 5278 sequences, comprised of 1501 contigs and 3777 singletons. Of these 5278 unique sequences, 1750 hit against the *Saccharum* EST dataset, 978 to the *S. bicolor*EST dataset, 1226 to the *Z. mays* EST dataset and 825 to the *O. sativa* EST dataset. A Venn diagram representing the assembled DArT sequences that aligned to each of the relevant datasets is displayed in Figure 3. Table 4 contains the number of ESTs in each dataset from the different genomes and the total number of assembled DArT sequences that aligned to each dataset.

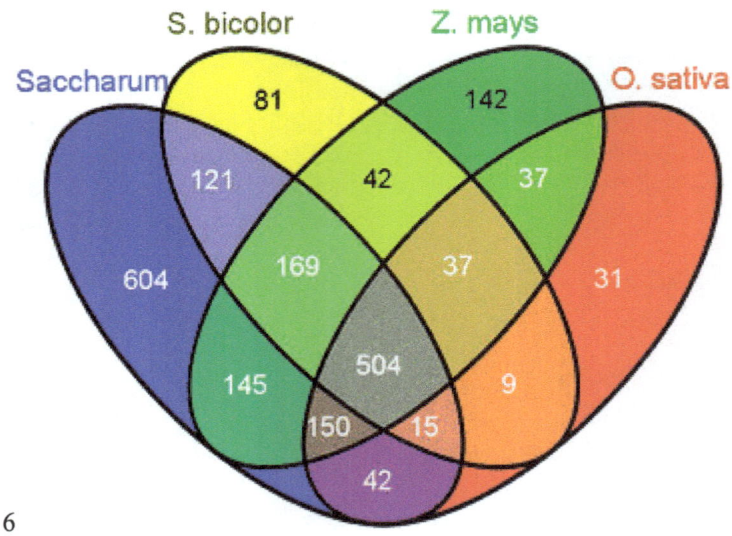

6

Figure 3: Venn diagram of DArTs with hits against the EST datasets from four species.

Table 4: Number of ESTs and total number of DArT sequences aligning for each EST dataset

EST dataset species of origin	Number of ESTs in NCBI dataset	Number of assembled DArT sequences aligned
Saccharum	283,254	1,750
Sorghum bicolor	209,828	978
Zea mays	2,019,114	1,226
Oryza sativa	1,252,989	825

The DArT markers aligned to all ten sorghum chromosomes with the largest number having homology to Sb1 (Table 5). Of the 7846 DArT markers 976 were mapped into the Q165 linkage map. All HGs contained DArT markers, 772 were from unique sequences which resulted in 18% redundancy of the mapped markers. Of the mapped DArT markers 34% had hits to genes (Table 5).

Table 5: Total number of DArT markers mapped in sugarcane and aligned to each sorghum chromosome using the BLASTN algorithm

Sorghum bicolour chromosomes	Total number of DArT markers	Total number of DArT markers on the map	Total no. of unique DArT markers on the Q165 map	Hits to NCBI of mapped DArTs	Number of mapped DArT markers that are not genes	Number with no hit to NCBI	Percent of DArT markers mapped that are genes
Sb1	553	128	105	23[b] (32)[a]	62 (76)	20	27
Sb2 and Sb8	426 (Sb2) 250 (Sb8)	134	113	33 (45)	56 (65)	24	37
Sb3	449	75	61	20 (28)	38 (44)	3	34
Sb4	395	126	99	38 (54)	52 (63)	9	42
Sb5 and Sb6	247 (Sb5) 307 (Sb6)	198	154	28 (30)	100 (142)	26	22
Sb7	252	94	82	24 (27)	47 (56)	11	34
Sb9	252	107	86	27 (40)	51 (59)	8	35
Sb10	320	89	72	23 (37)	35 (38)	14	40
No hits	1783						

Super contigs	44						
Total	5278	951	772	216	441	115	34

[a]total number of individual DArT markers.

[b]total number of unique markers.

Genome sequence comparison

A results table of the DArT sequence comparison against the genomes of *S. bicolor*, *Z. mays*, and *O. sativa* is provided in the Additional file 2. A total of 1755 assembled DArT sequences aligned to an annotated gene in the *S. bicolor* genome, while a similar number (1783) did not align to the genome at all. A small proportion of the DArT sequences aligned to repeats (Figure 4). The categories of repeats against which the DArT sequences aligned on the *S. bicolor* genome are displayed in Figure 5.

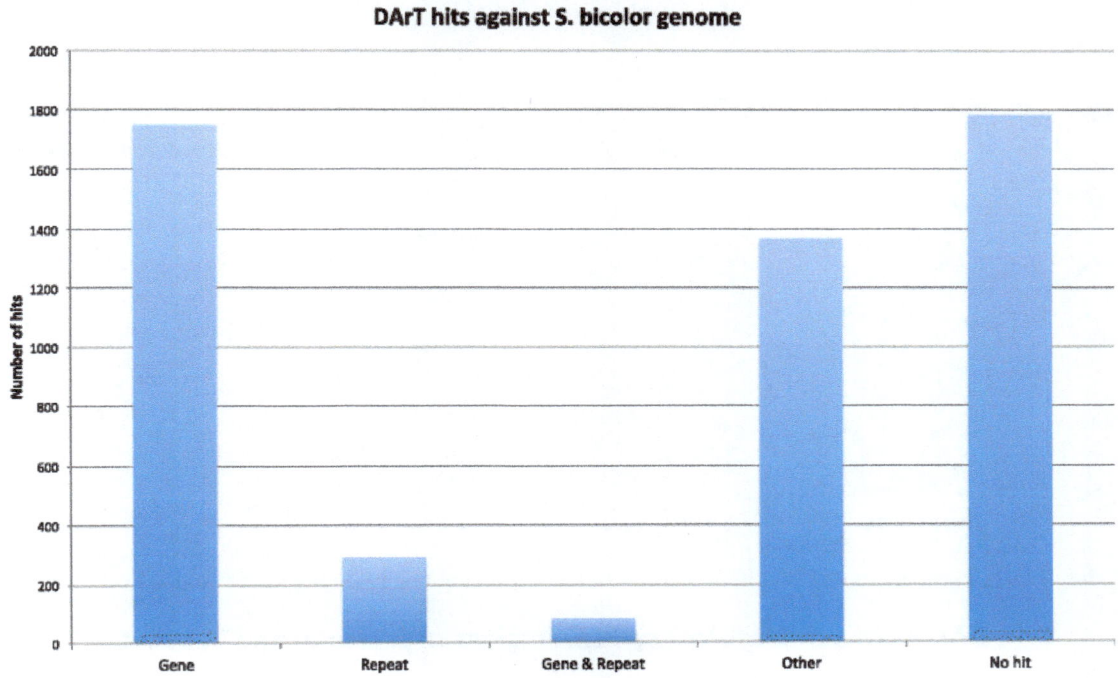

Figure 4: Annotations of assembled DArT sequence alignments to different parts of the annotated S . bicolor genome including genes, repeats, the boundary between a repeat and genes and other which are regions that are not annotated.

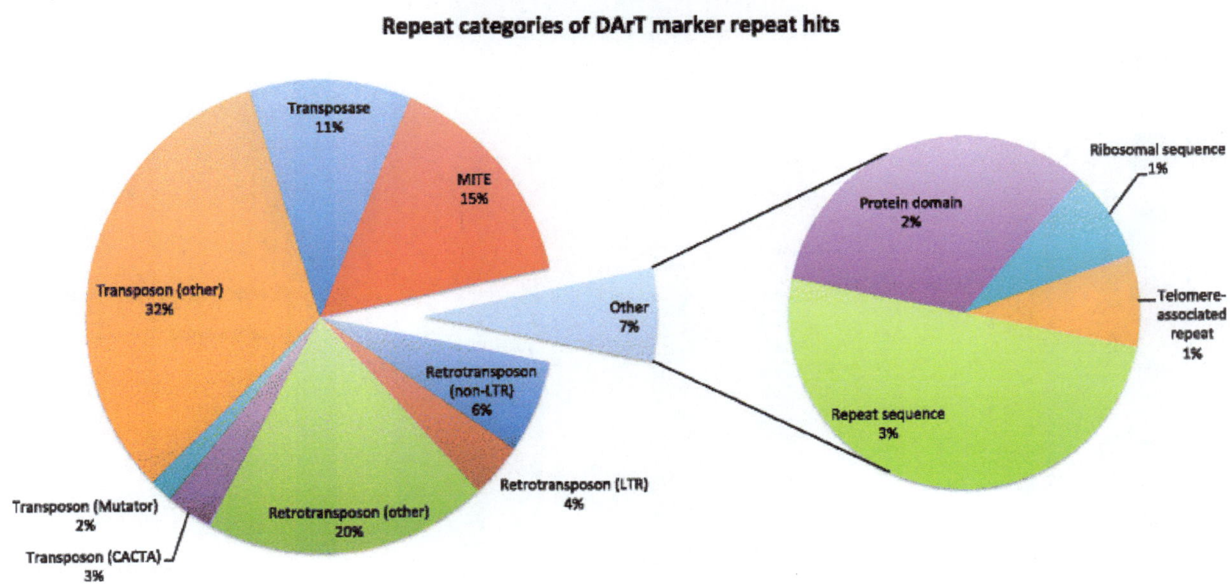

Figure 5: Categories of repeats in the *S. bicolor* genome against which assembled DArT sequences aligned.

In total, 2124 assembled DArT sequences which hit against the *S. bicolor* genome sequence also hit against the*Z. mays* genome sequence, of which 1888 (88.9%) aligned to regions syntenic between these two species [[31]]. In contrast, only 846 assembled DArT sequences hitting against the *S. bicolor* genome also hit against the *O. sativa* genome sequence, of which only 671 (79.3%) aligned in regions syntenic between these two species. The ratios of syntenic to non-syntenic region alignments for each category of annotation are displayed in

Figures 6 and 7. In the case of both *Z. mays* and *O. sativa*, DArT sequences aligning to annotated genes most frequently aligned to syntenic regions (92.2% and 83.0% respectively).

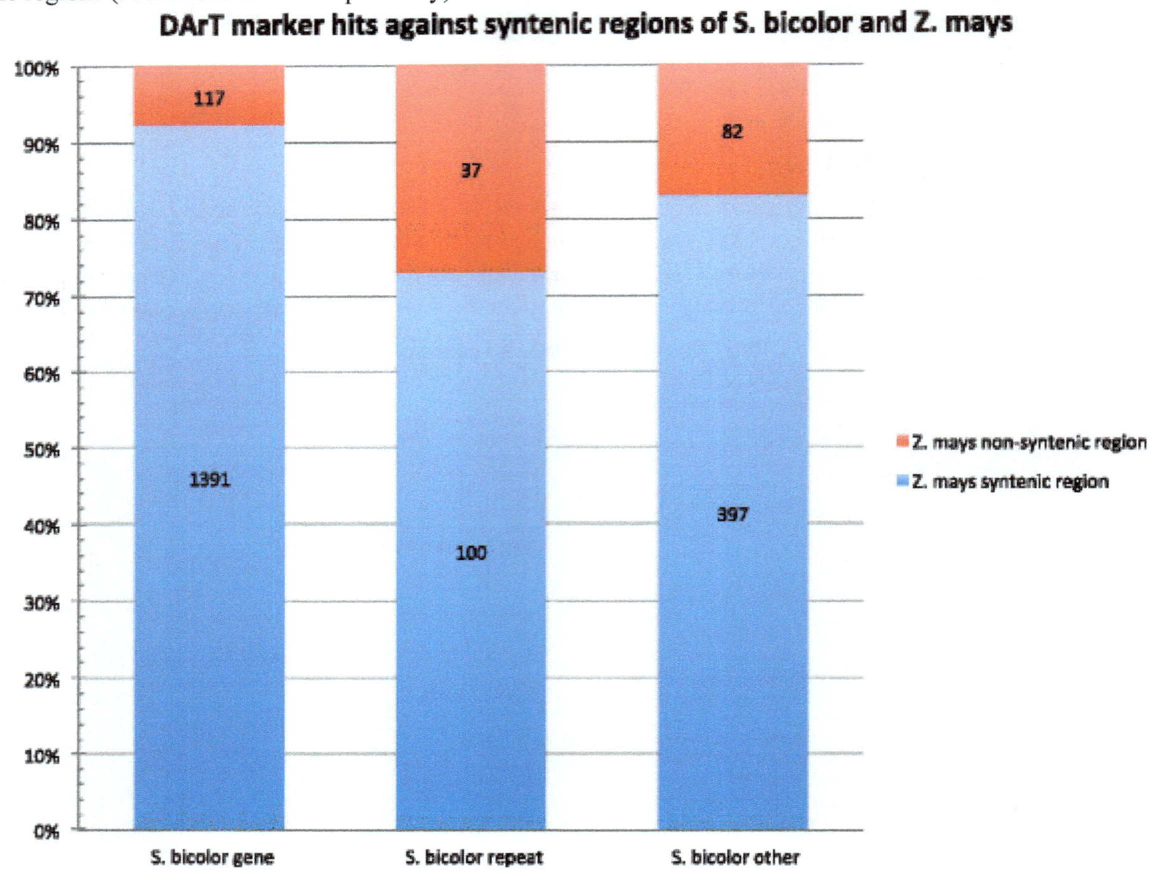

Figure 6: Ratio of DArT sequences aligning in *S. bicolor* and *Z. mays* syntenic to non-syntenic regions.

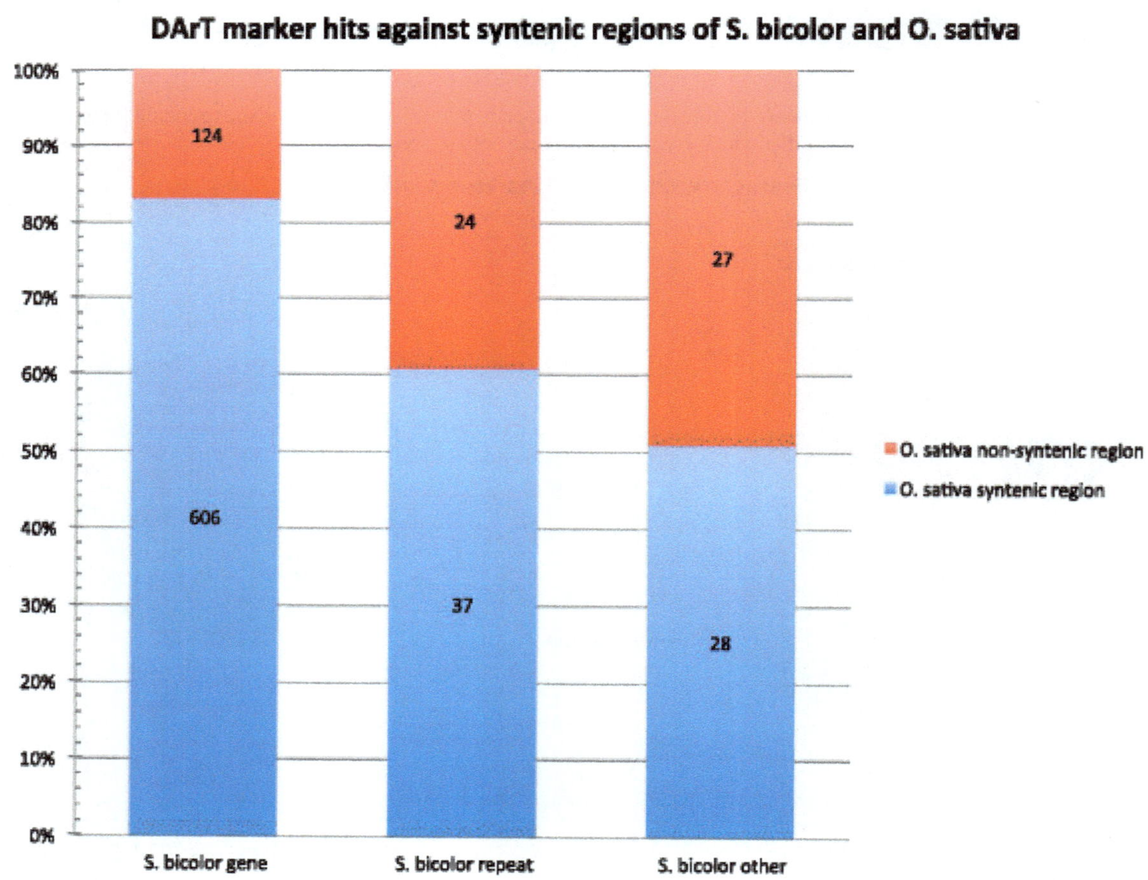

Figure 7: Ratio of DArT sequences aligning in *S. bicolor* and *O. sativa* syntenic to non-syntenic regions.

DISCUSSION

Comparison of the sugarcane genetic map to the sorghum genome

To date only low resolution maps have been used to compare the sugarcane genetic map to the sorghum genome [[12],[32],[33]]. These maps were constructed with low-copy RFLP markers that were selected for their ability to provide a signal in cross-hybridisation thereby limiting the ability to assess orthologous and paralogous relationships in gene families, since comparative mapping by RFLP often identifies paralogous rather than orthologous sequences [[29]]. The large number of markers with known sequence in this study made it possible for the first time to align a significant part of the sugarcane genetic map to the sorghum genome sequence (Table 3, Additional file 1). Although with 954 sequences this analysis is at the macro level and only the major structural changes are identified, at the micro level there could be many more rearrangements but more markers and sequence information is needed to verify these. To use as many markers as possible in the comparison to sorghum a composite linkage map was generated where possible for each HG. This was then aligned to the sorghum genome sequence (Figure 1). Six HGs aligned to one sorghum chromosome and four HGs aligned in the main to two sorghum chromosomes. For all chromosomes there were regions that appeared to contain markers from other chromosomes (Additional file 1, Table 2) but to be considered for further analysis a number of linked markers from the same chromosomes showing homology to gene sequences had to be identified. This left four regions that showed strong likelihood of being due to rearrangement. A major translocation was identified in HG8, where LG3 aligned to Sb8 and Sb2. This result agrees with previous attempts to align sugarcane to sorghum [[32],[33]] although the greater number of markers in this map means a higher level of resolution of the alignment. Two sugarcane populations were studied by [[12]] and they determined that no more than four or five major interchromosomal translocations had occurred since the *Saccharum*-Sorghum divergence. They also identified a translocation between Sb2 and Sb8 in both *S. spontaneum* parents and the hybrid parents of their populations. They identified additional possible translocations between sorghum chromosomes 8 and 7 in the hybrid parent and 6 and 7 and 2 and 4 in the *S. spontaneum* parent, these were not verified in this study. They did not identify the Sb5/Sb6 translocation that was identified in this analysis and by [[32],[33]]. These two major translocations resulted in both cases in the condensation of two chromosomes into one. This process is known in many lineages and is demonstrated within the grass family where all grasses have evolved from an ancestral genome of 12 chromosomes [[29],[34]]. A further 2 HGs contained translocated chromosomes although these did not result in reduced chromosome number. HG5 aligned in the main to Sb7 but 5 of the 13 LGs also aligned to Sb5 indicating a translocation between Sb7 and Sb5. An interchromosomal rearrangement between LGs that aligned to Sb7 and Sb5 was also identified by [[12]] in the hybrid parent of the cross studied. The additional interchromosomal rearrangement in HG6 LG20 which aligns to Sb9 and Sb3 has not previously been identified (Table 3, Additional file 1).

Translocations

A more detailed comparison of the four major interchromosomal rearrangements (Figure 2A-D) revealed that in some cases the rearrangement of the grass ancestral genome that has occurred in maize is the same as in sugarcane [[29]]. The chromosomal rearrangement seen in HG6 on LG20, which aligns to Sb9 and part of Sb3, corresponds to maize chromosome 8 (Figure 2C). Maize diverged from sorghum approximately 12 million years ago (mya) followed by allopolyploidisation approximately 5 mya [[29]]. The present maize chromosomes are a result of numerous rearrangements which reduced the chromosome number from 20 to 10 [[29]]. Recent work by [[35]] has shown that the whole genome duplication (WGD) which occurred in the ancestor of both maize and sorghum 29 mya has resulted in large duplicated blocks in their genomes. Sb9 of sorghum shares a duplicated block with Sb3 these two chromosomes have recombined to form maize m8. The same rearrangement has occurred in sugarcane as seen in LG20 in HG6 (Figure 2C). The translocation identified in HG2 which aligns to Sb5 and Sb6 is also seen in the top of maize m2 (Figure 2A). Parts of HG2 also align to m10 and m4 which are the duplicated ancestral regions [[26]]. The rearrangement in HG5 which aligns to Sb7 and Sb5 again is present in maize and aligns to m4 (Figure 2D). HG8 aligns to sorghum Sb8 and Sb2 which also aligns to maize m1 and m7. Interestingly the ancestral fusion event in Sb2 is retained in both maize and sugarcane although a further rearrangement of this chromosome has occurred uniquely in sugarcane (Figure 2C).

Saccharum spontaneum

Sugarcane being a polyploid, is more tolerant of gene deletions and rearrangements than a diploid. It would appear that the rearrangements in HG2 and HG8 are major chromosome rearrangements to account for the smaller number of basic chromosomes in *S. spontaneum*. A sparse RFLP map was generated by [[10]] of a *S. spontaneum* clone SES208 using 234 single dose RFLP markers. From comparative mapping of the RFLP clones to sorghum and maize, HG4 of this map contains clones UMC130 and UMC102 which align to Sb8 and SG298 and UMC139 which align to Sb2 and are on the same LG. This is comparable to the Q165 HG8 which aligns to Sb8 and Sb2. Similarly RFLP markers in HG2 of SES208 align to Sb5 and Sb6 as does the Q165 HG2. The translocations in HG2 and HG8 appear to be inherited from the *S. spontaneum* part of the genome of Q165 but with the current data it is not possible to distinguish *S. spontaneum* chromosome from *S. officinarum*chromosomes. However, as *S. officinarum* is x = 10, the assumption is that at least the two rearrangements in HG2 and HG8 are inherited from the *S. spontaneum* part of the genome. The other rearrangements in HG5 and HG6 could be inherited from either or both species.

Comparison within the Poaceae

Recently a detailed genetic map was constructed for another member of the Andropogoneae: Poaceae,*Miscanthus sinensis*. This species has a diploid inheritance patterns (2 n = 38) and when aligned to sorghum contains two sub-genomes, each syntenic to the entire sorghum genome with one major structural rearrangement which resulted in the condensation of two chromosomes into one and a reduction in chromosome number to x = 19 [[36]]. In this case the translocated LG aligned to Sb7 and Sb4 a translocation not seen in *Saccharum* but again seen in maize on m4. Interestingly [[36]] found that the *M. sinensis* LGs that were syntenic to Sb5 had the lowest percentage of markers that mapped to sorghum, a similar finding to switchgrass. The data suggests that linkages syntenic to Sb5 have encountered more evolutionary changes than other chromosomes after the species diverged from each other. Similarly HG2 in *Sacharrum* which is syntenic to Sb5 has undergone more rearrangement than other HGs.

Recent work by [[37]] comparing sequenced genomes of grasses has demonstrated that the chromosome number variation/reduction from the n = 12 common paleo-ancestor to the n = 10 of sorghum could be due to the insertion of a chromosome in the centromeric region of another chromosome although these reductions could also come about through pericentric inversions followed by reciprocal translocations [[38]]. The comparison of the Q165 *Saccharum* genetic map to the sorghum and maize genomes demonstrates*Saccharum* contains the same two ancestral fusion events. Moreover, out of the 4 major translocations identified, 3 of them have also occurred in maize (Figures 2 and 8), one of which resulted in a reduction in chromosome number. *S. officinarum* is an octoploid and *S. spontaneum* ranges from 2n = 5x = 40 to 16x = 128 [[39],[40]] in chromosome number. It would seem probable that after the whole genome duplication that also occurred in maize the *Saccharum* ancestor underwent some of the same rearrangements as maize then diverged and under went further WGD events. More sequence analysis is needed to determine how these events occurred but in evolutionary terms they would have been recent (less than 5 mya).

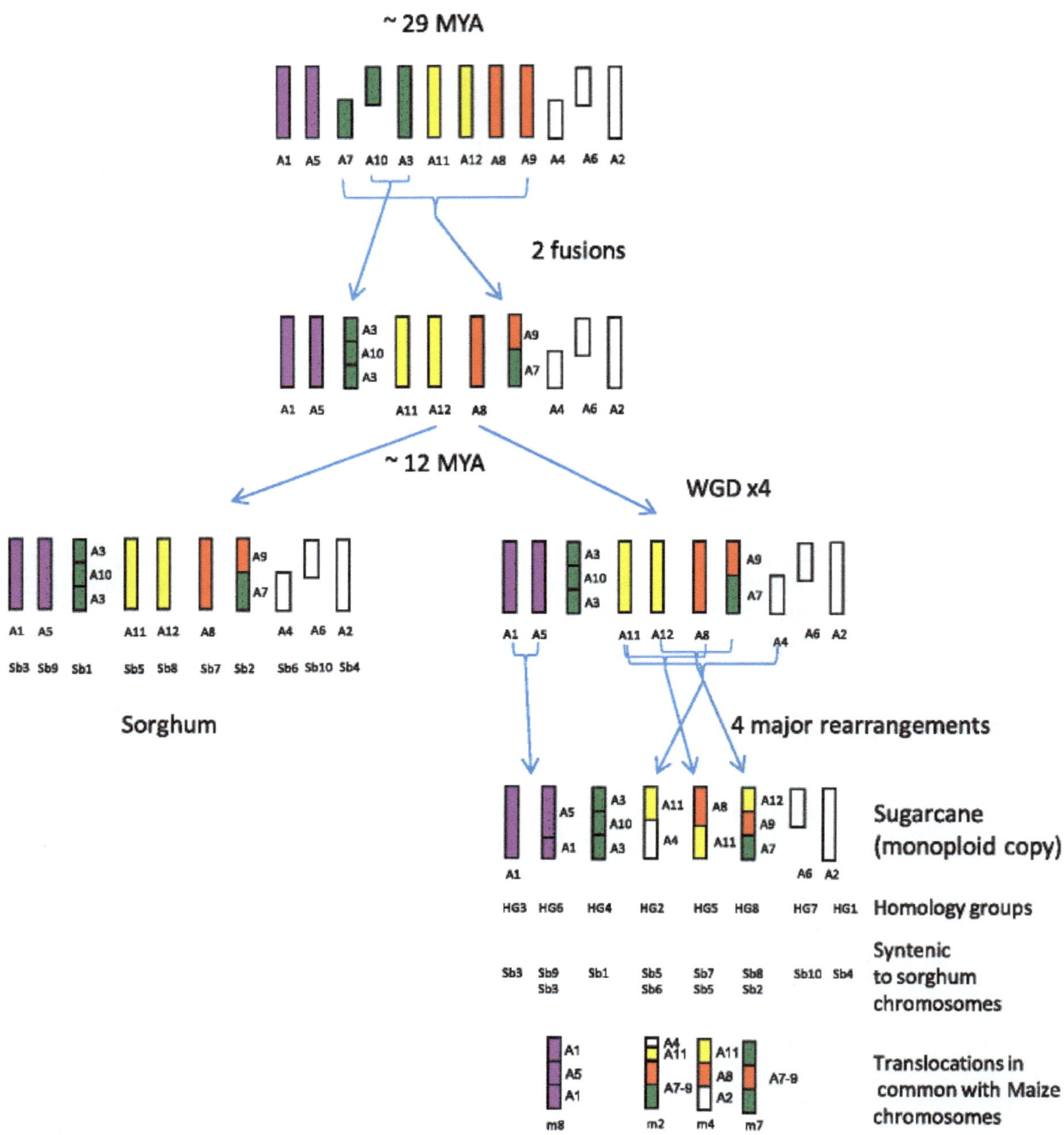

Figure 8: Model for the structural evolution of the sugarcane monoploid genome (modified from [[26]]). As in [[26]] the ancestral chromosomes are colour coded to represent the original 5 chromosomes (and labelled A1-12) which after a whole genome duplication (~90 MYA) and breakage/fusion formed the 12 ancestral chromosomes of sorghum.

Whole genome duplication and the rapid and massive structural changes which have occurred in the *Poaceae*and to a greater extent in the *Saccharum* might explain the ability of these polyploids to adapt and survive environmental conditions that are not tolerated by their diploid ancestors. Within the *Saccharum* there are no known diploids as all species are polyploid. *S. spontaneum* has the widest distribution from Japan and Indonesia through the Indian subcontinent to the Mediterranean and Africa [[39]] and the largest range of ploidy levels (2n=40-128). It has been reported that polyploidy provides increased vigour [[41]] and provided an advantage for colonizing a wider range of environments [[42]].

Dart sequence comparison to EST and genome sequences

It is to be expected that the highest number of DArTs would align to ESTs from *Saccharum* as the genome of origin. Interestingly, a higher number of DArTs align to the maize EST dataset (1226) than to the sorghum (978) EST dataset, even though sorghum is generally thought to be a much closer relative to sugarcane [[25]] (Figure 3). This may be attributable to the higher volume of repetitive DNA in the maize genome [[25],[43]], some of which may be expressed and represented in the EST database but missing from the sorghum EST database which is much smaller than the maize database. As would be expected, the least number of DArT sequences aligned to the rice EST dataset, which is the most distant relative of the species surveyed.

A proportionally low number of DArT sequences aligned to annotated repeats (7.7%) in the sorghum genome compared to the number of DArTs aligned to annotated genes (35.1%), given that 55% of the sorghum genome is known to be repetitive and the repetitive content of sugarcane has been demonstrated to be similar to that of sorghum [[25],[44]]. This suggests that the enrichment for unique regions applied in the production of the DArT markers has largely succeeded. Interestingly, 1783 (33.8%) of the DArT marker sequences can be observed in Figure 2 to display no significant alignment to the sorghum genome, suggesting that these originate from divergent regions of sugarcane.

CONCLUSION

Both the analysis of the sequence data and the comparative mapping has indicated that sugarcane is closely related to both maize and sorghum. Three out of the four translocations identified from the sugarcane genetic map are also present in the maize genome (Figure 8). These events occurred in maize after a WGD which also occurred in the ancestors of sugarcane indicating its possible sugarcane diverged after the initial rearrangements occurred. This study has highlighted the need for generation of a denser genetic map and ultimately the sugarcane genome sequence to allow the determination of the evolution of these three closely related species, sorghum, maize and sugarcane.

METHODS

Plant material

The mapping population used to generate the Q165 genetic map consisted of 227 progeny derived from a cross between a *S. officinarum* clone IJ76-514 (2n=80) as the female parent and Q165 (2n=115) an Australian sugarcane cultivar and elite parent as the male parent.

Sequencing of the DArT clones

A total of 7846 DArT clones, the majority of which were present on the DArT array used to genotype the mapping populations were isolated and DNA purified using QIAGEN plasmid purification kit. Each of these clones were sequenced using the ABI PRISM BigDye Terminator Cycle Sequencing Ready reaction kit (Perkin-Elmer, Wellesley, MA) and analyzed on an ABI 3700 sequencer. Sequencing was done with both M13 Forward and Reverse primers. The sequences were analyzed using the Lasergene software program Seqman (DNAstar Inc., Madison, WI). The vector and adapter sequences were removed and the original restriction sites restored. The DArT sequences can be obtained on request from the corresponding author.

DArT sequence clustering and analysis

The 7846 raw DArT sequences were assembled using Sequencher with a minimum overlap length of 50 nucleotides. The EST sequences available in the NCBI EST database as at 27th October 2011 were downloaded to produce four species-specific EST datasets for *Saccharum*, *Sorghum bicolor*, *Zea mays*, and *Oryza sativa*. For each DArT contig, coding DNA sequences (CDS) were predicted by comparing the DArT sequence to these databases using TBLASTX [[45]] with an e-value cutoff of $\leq 10^{-20}$, and the best hit for each DArT sequence identified by e-value.

Comparison with public genome sequences

Genome sequences for *S. bicolor*, *Z. mays* and *O. sativa* as well as the gene and RepeatMasker annotations for*S. bicolor* were downloaded from Phytozome version 7.0 [[46]]. The 5278 assembled DArT sequences were compared to each genome sequence using BLASTN [[40]] with an e-value cutoff of $\leq 10^{-20}$, identifying the best hit for each DArT sequence along with the chromosomal location of these hits.

We compared the position of each top DArT marker hit on the *S. bicolor* genome sequence against the annotations for genes and repeats using a custom Python script to identify whether a DArT sequence could be 'annotated' as one of four categories; aligned to an annotated gene, aligned to an annotated repeat, aligned on the boundary between an annotated repeat and an annotated gene, or in a region without an annotation. Assembled DArT sequences without a BLAST hit against the *S. bicolor* genome were also recorded and DArT hits against repeats were categorised.

We then compared the position of hits against the *S. bicolor* genome with the position of hits against the genomes of *Z. mays*, and *O. sativa* to identify whether DArT sequences hit against syntenic regions in these genomes based on the syntenic regions identified by [[47]].

Marker analysis and genotyping

The DNA extraction and marker data were generated as in

[[15]]. This consisted of the original data from [[15]] to which were added further AFLP, SSR, RFLP, SNP and DArT markers. Generation of further SSR, and *Pst* I/*Mse*I AFLP marker data were as reported by [[15]]. The SNP data was generated as in [[48]] and RFLP data as in [[49]]. The DArT data was generated as in [[23]]. The genetic map was produced using 2283 simplex markers that were present as a single copy in Q165, the map is reported in detail in [[24]]. Composite LGs were generated using JoinMap V4 [[50]].

Comparison to the sorghum genome

Sequence information from the DArT, SNP and EST - SSR markers were used to align the Q165 linkage map to the genome sequence of *Sorghum bicolour*. This was carried out using BLASTN homology search; sequences had to show at least 80% nucleotide identity for at least 50% of the marker sequence length with an e-value cutoff of $\leq 10^{-20}$ to be used for comparative analysis. The results were visualised using Sigmaplot. This high stringency was used as it is difficult to infer orthologous (derived from a common ancestor by speciation) and paralogous (derived by duplication within one genome) relationships from sequence comparisons using a less stringent e-value.

ACKNOWLEDGEMENTS

We thank the Cooperative Research Centre for Sugar Industry Innovation through Biotechnology (CRC SIIB) for providing funding for this work. We thank Anne Rae and Graham Bonnett for helpful discussions and comments on the manuscript.

COMPETING INTERESTS

Andrzej Kilian is Director of Diversity Arrays Technology who provide DArT array commercial genotyping services for a range of crops.

REFERENCES

1. Carruthers SP: Solid biofuels: fuel crops. In *Crops for Industry and Energy 1994 CAS Report.* Edited by Carruthers SP, Miller PA, Vaughan CMA. Centre for Agricultural Strategy, University of Reading; 1994:168–180.
2. Grivet L, Arruda P: Sugarcane genomics: depicting the complex genome of an important tropical crop. Curr Opin Plant Biol. 2001, 5: 122-127. 10.1016/S1369-5266(02)00234-0.
3. D'Hont A, Ison D, Alix K, Roux C, Glaszmann JC: Determination of basic chromosome numbers in the genus *Saccharum*by physical mapping of ribosomal RNA genes. Genome. 1998, 41: 221-225. 10.1139/gen-41-2-221.
4. D'Hont A, Grivet L, Feldmann P, Rao PS, Berding N, Glauszmann JC: Characterisation of the double genome structure of modern sugarcane cultivars (Saccharum spp.) by molecular cytogenetics. Mol Gen Genet. 1996, 250: 404-413.
5. Sreenivasan TV, Ahloowalia BS, Heinz DJ: Cytogenetics. Sugarcane improvement through breeding. Edited by: Heinz DJ. 1987, Elsevier Press, New York, 211-253. 10.1016/B978-0-444-42769-4.50010-2.
6. Burner DM, Legendre BL: Chromosome transmission and meiotic stability of sugarcane (*Saccharum* spp.) hybrid derivatives. Crop Sci. 1993, 33: 600-606. 10.2135/cropsci1993.0011183X003300030036x.
7. Aitken KS, Li JC, Jackson P, Piperidis G, McIntyre CL: AFLP analysis of genetic diversity within *Saccharum officinarum* and comparison with sugarcane cultivars. Aust J Agric Res. 2006, 57: 1167-1184. 10.1071/AR05391.
8. Al-Janabi SM, Honeycutt RJ, McClelland M, Sobral BWS: A genetic linkage map of *Saccharum spontaneum* L. 'SES 208'. Genetics. 1993, 134: 1249-1260.
9. Mudge J, Andersen WR, Kehrer RL, Fairbanks DJ: A RAPD genetic map of *Saccharum officinarum*. Crop Sci. 1996, 36: 1362-1366. 10.2135/cropsci1996.0011183X003600050046x.
10. Da Silva JAG, Honeycutt RJ, Burnquist W, Al-Janabi SM, Sorrells ME, Tanksley SD, Sobral BWS: *Saccharum spontaneum* L. 'SES 208' genetic linkage map combining RFLP and PCR-based markers. Mol Breeding. 1995, 1: 165-179. 10.1007/BF01249701.
11. Grivet L, D'Hont A, Roques D, Feldmann P, Lanaud C, Glaszmann JC: RFLP mapping in cultivated sugarcane (*Saccharum*spp.): genome organisation in a highly polyploid and aneuploid interspecific hybrid. Genetics. 1996, 142: 987-1000.
12. Ming R, Liu S-C, Lin Y-R, da Silva J, Wilson W, Braga D, van Deynze A, Wenslaff TF, Wu KK, Moore PH, Burnquist W, Sorrells ME, Irvine JE, Paterson AH: Detailed alignment of saccharum and sorghum chromosomes: comparative organisation of closely related diploid and polyploidy genomes. Genetics. 1998, 150: 1663-1682.
13. Ming R, Del Monte TA, Hernandez E, Moore PH, Irvine JE, Paterson AH: Comparative analysis of QTLs affecting plant height and flowering among closely-related diploid and polyploidy genomes. Genome. 2002, 45: 794-803. 10.1139/g02-042.
14. Hoarau JY, Offmann B, D'Hont A, Risterucci AM, Roques D, Glaszmann JC, Grivet L: Genetic dissection of a modern sugarcane cultivar (*Saccharum* spp.). 1. Genome mapping with AFLP markers. Theor Appl Genet. 2001, 103: 84-97. 10.1007/s001220000390.
15. Aitken KS, Jackson PA, McIntyre CL: A combination of AFLP and SSR markers provides extensive map coverage and identification of homo(eo)logous linkage groups in a sugarcane cultivar. Theor Appl Genet. 2005, 110: 789-801. 10.1007/s00122-004-1813-7.
16. Andru S, Yoong-Bao P, Thongthawee S, Burner DM, Kimbeng CA: Genetic analysis of the sugarcane (Saccharum spp.) cultivar 'LCP 85-384'. I. Linkage mapping using AFLP, SSR and TRAP markers. Theor Appl Genet. 2011, 123: 77-93. 10.1007/s00122-011-1568-x.
17. Rossi M, Araujo PG, Paulet F, Garsmeur O, Dias VM, Chen H, Van Sluys MA, D'Hont A: Genomic distribution and characterisation of EST-derived resistance gene analogs (RGAs) in sugarcane. Mol Gen Genomics. 2003,

269: 406-419. 10.1007/s00438-003-0849-8.

18. Oliveira KM, Pinto LR, Marconi TG, Margarido GRA, Pastina MM, Teixeira LHM, Figueira AV, Ulian EC, Garcia AAF, Souza AP: Functional integrated genetic linkage map based on EST-markers for a sugarcane (Saccharum spp.) commercial cross. Mol Breed. 2007, 20: 189-208. 10.1007/s11032-007-9082-1.

19. Wenzl P, Carling J, Kudrna D, Jaccoud D, Huttner E, Kleinhofs A, Kilian A: Diversity arrays technology (DArT) for whole-genome profiling of barley. Proc Natl Acad Sci U S A. 2004, 101: 9915-9920. 10.1073/pnas.0401076101.

20. Jaccoud D, Peng K, Feinstein D, Kilian A: Diversity arrays: a solid state technology for sequence information independent genotyping. Nucleic Acids Res. 2001, 29: e25-10.1093/nar/29.4.e25.

21. Akbari M, Wenzl P, Caig V, Carling J, Xia L, Yang SY, Uszynski G, Mohler V, Lehmensiek A, Kuchel H, Hayden MJ, Howes N, Sharp P, Vaughan P, Rathmell B, Huttner E, Kilian A: Diversity arrays technology (DArT) for high-throughput profiling of the hexaploid wheat genome. Theor Appl Genet. 2006, 113: 1409-1420. 10.1007/s00122-006-0365-4.

22. Mace ES, Xia L, Jordan DR, Halloran K, Parh DK, Huttner E, Wenzl P, Kilian A: DArT markers: diversity analyses and mapping in Sorghum bicolor. BMC Genomics. 2008, 9: 26-37. 10.1186/1471-2164-9-26.

23. Heller-Uszynska K, Uszynski G, Huttner E, Evers M, Carlig J, Caig V, Aitken K, Jackson P, Piperidis G, Cox M, Gilmour R, D'Hont A, Butterfield M, Glaszmann JC, Kilian A: Diversity arrays technology effectively reveals DNA polymorphism in a large and complex genome of sugarcane. Mol Breed. 2010, 28: 37-55. 10.1007/s11032-010-9460-y.

24. Aitken KS, McNeil MD, Hermann S, Bundock PC, Kilian A, Heller-Usznska K, Henry RJ, Li J: A comprehensive genetic map of sugarcane that provides enhanced map coverage and integrates high throughput Diversity Array Technology (DArT) markers. BMC Genomics. 2014, 15: 152-10.1186/1471-2164-15-152.

25. Paterson AH, Bowers JE, Bruggmann R, Dubchak I, Grimwood J, Gundlach H, Haberer G, Hellsten U, Mitros T, Poilakov A, Schmutz J, Spannagl M, Tang HB, Wang XY, Wicker T, Bharti AK, Chapman J, Feltus FA, Gowik U, Grigoriev IV, Lyons E, Maher CA, Martis M, Narechania A, Otillar RP, Penning BW, Salamov AA, Wang Y, Zhang LF, Carpita NC, et al: The Sorghum bicolor genome and the diversification of grasses. Nature. 2009, 457: 551-556. 10.1038/nature07723.

26. Salse J, Bolot S, Throude M, Jouffe V, Piegu B, Masood Quraishi U, Calcagno T, Cooke R, Delseny M, Feuillet C: Identification and characterisiation of shared duplications between rice and wheat provide new insight into grass genome evolution. Plant Cell. 2008, 20: 11-24. 10.1105/tpc.107.056309.

27. Jannoo N, Grivet L, Chantret N, Garsmeur O, Glaszmann JC, Arruda P, D'Hont A: Orthologous comparison in a gene-rich region among grasses reveals stability in the sugarcane polyploidy genome. Plant J. 2007, 50: 574-585. 10.1111/j.1365-313X.2007.03082.x.

28. Garsmeur O, Charron C, Bocs S, Jouffe V, Samain S, Couloux A, Droc G, Zini C, Glaszmann JC, Van Sluys MA, D'Hont A: High homologous gene conservation despite extreme autoplyploid redundancy in sugarcane. New Phytol. 2010, 189: 629-642. 10.1111/j.1469-8137.2010.03497.x.

29. Salse J, Abrouk M, Murat F, Quraishi UM, Feuillet C: Improved criteria and comparative genomics tool provide new insights into grass paleogenomics. Brief Bioinform. 2009, 2009 (10): 619-630. 10.1093/bib/bbp037.

30. Piperidis G, Piperidis N, D'Hont A: Molecular cytoogenetic investigation of chromsome composition and transmission in sugarcane. Mol Gen Genomics. 2010, 284: 65-73. 10.1007/s00438-010-0546-3.

31. Swigonova Z, Lai JS, Ma JX, Ramakrishna W, Llaca V, Bennetzen JL, Messing J: Close split of sorghum and maize genome progenitors. Genome Res. 2004, 14: 1916-1923. 10.1101/gr.2332504.

32. Dufour P, Deu M, Grivet L, D'Hont A, Paulet F, Bouet A, Lanaud C, Glaszmann JC, Hamon P: Construction of a composite sorghum genome map and comparison with sugarcane, a related complex polyploidy. Theor Appl Genet. 1997, 94: 409-418. 10.1007/s001220050430.

33. Glaszmann JC, Dufour P, Grivet L, D'Hont A, Deu M, Paulet F, Hamon P: Comparative genome analysis between several tropical grasses. Euphytica. 1997, 96: 13-21. 10.1023/A:1002987620250.

34. Tang H, Bowers JE, Wang X, Ming R, Alam M, Paterson P: Synteny and collinearity in plant genomes. Science. 2008, 320: 486-488. 10.1126/science.1153917.

35. Wang X, Tang H, Bowers JE, Paterson AH: Comparative inference of illegitimate recombination between rice and sorghum duplicated genes produced by polyploidization. Genome Res. 2009, 19: 1026-1032. 10.1101/gr.087288.108.

36. Ma X-F, Jensen E, Alexandrov N, Troukhan M, Zhang LP, Thomas-Jones S, Farrar K, Clifton-Brown J, Donnison I, Swaller T, Flavell R: High resolution genetic mapping by genome sequencing reveals genome duplication and tetraploid genetic structure of the diploid *Miscanthus sinesis*. PLoS ONE. 2012, 7: e33821-10.1371/journal.pone.0033821.

37. Murat F, Xu J-H, Tannier E, Abrouk M, Guilhot N, Pont C, Messing J, Salse J: Ancestral grass karyotype reconstruction unravels new mechanisms of genome shuffling as a source of plant evolution. Genome Res. 2010, 2010 (20): 1545-1557. 10.1101/gr.109744.110.

38. Schubert I, Lysak MA: Interpretation of karyotype evolution should consider chromsome structural constraints. Trends Genet. 2011, 27: 207-216. 10.1016/j.tig.2011.03.004.

39. Panje R, Babu C: Studies in *Saccharum spontaneum*. Distribution and geographical association of chromosome numbers. Cytologia. 1960, 25: 152-172. 10.1508/cytologia.25.152.

40. Fan LN, Deng HH, Luo QW, He HY, Li Y, Wang QN, Huang ZX, Wu JT, Li SM, Qi YW: Genetic diversity of

Saccharum spontaneum from geographical regions of China assessed by simple sequence repeats. Genetics and Molecular Research. 2013, 12: 5916-5925. 10.4238/2013.November.26.1.

41. Hegarty M, Hiscock S: Polyploidy: doubling up for evolutionary success. Curr Biol. 2007, 2007 (17): 927-929. 10.1016/j.cub.2007.08.060.
42. Van de Peer Y, Maere S, Meyer A: The evolutionary significance of ancient genome duplications. Nat Rev Genet. 2009, 10: 725-732. 10.1038/nrg2600.
43. Schnable PS, Ware D, Fulton RS, Stein JC, Wei FS, Pasternak S, Liang CZ, Zhang JW, Fulton L, Graves TA, Minx P, Reily AD, Courtney L, Kruchowski SS, Tomlinson C, Strong C, Delehaunty K, Fronick C, Courtney B, Rock SM, Belter E, Du FY, Kim K, Abbott RM, Cotton M, Levy A, Marchetto P, Ochoa K, Jackson SM, Gillam B, et al: The B73 maize genome: complexity, diversity, and dynamics. Science. 2009, 326: 1112-1115. 10.1126/science.1178534.
44. Wang J, Roe B, Macmil S, Yu Q, Murray JE, Tang H, Chen C, Najar F, Wiley G, Bowers J, Van Sluys MA, Rokhsar DS, Hudson ME, Moose SP, Paterson AH, Ming R: Microcollinearity between autopolyploid sugarcane and diploid sorghum genomes. BMC Genomics. 2010, 11: 261-10.1186/1471-2164-11-261.
45. Altschul SF, Gish W, Miller W, Myers EW, Lipman DJ: Basic local alignment search tool. J Mol Biol. 1990, 1990 (215): 403-410. 10.1016/S0022-2836(05)80360-2.
46. Goodstein DM, Shu S, Howson R, Neupane R, Hayes RD, Fazo J, Mitros T, Dirks W, Hellsten U, Putnam N, Rokhsar DS: Phytozome: a comparative platform for green plant genomics. Nucleic Acids Res. 2012, 40: 1178-1186. 10.1093/nar/gkr944.
47. Bolot S, Abrouk M, Masood-Quraishi U, Stein N, Messing J, Feuillet C, Salse J: The 'inner circle' of the cereal genomes. Curr Opin Plant Biol. 2009, 12: 119-125. 10.1016/j.pbi.2008.10.011.
48. Bundock PC, Eilot FG, Ablett G, Benson AD, Casu RE, Aitken KS, Henry RJ: Targeted single nucleotide polymorphism (SNP) discovery in a highly polyploidy plant species using 454 sequencing. Plant Biotechnol J. 2009, 2009 (7): 347-354. 10.1111/j.1467-7652.2009.00401.x.
49. Chao S, Sharp PJ, Worland AJ, Warham EJ, Koebner RMD, Gale MD: RFLP-based genetic maps of wheat homologous group-7 chromosomes. Theor Appl Genet. 1989, 78: 495-504. 10.1007/BF00290833.
50. Van Ooijen JW: JoinMao 4, Software for the calculation of genetic linkage maps in experimental populations. Kyazma BV, Wageningen, Netherlands 2006.

CHAPTER 14 GENOME EVOLUTION IN ALPINE OAT-LIKE GRASSES THROUGH HOMOPLOID HYBRIDIZATION AND POLYPLOIDY

Grit Winterfeld, Alexandra Wolk and Martin Roser

Keywords: Alps, Helictotrichon, hybridization, karyotype, Poaceae, polyploidy, Topoisomerase VI

INTRODUCTION

Grasses, especially the cereals, have long been investigated as model organisms for studies in chromosome organisation, genome dynamics and molecular evolution. The grasses are the most important plant family on earth and represent the fifth largest family in number of species. The systematics and taxonomy of oat-like grasses (tribe Aveneae Dumort.), one of the major groups of grasses, has a long and partly complicated history. Since 1823, when Dumortier (1823) separated the annuals with large and hanging spikelets (*Avena* L. s.str.) from the perennials with more or less small and upright spikelets, many different characters have been used for classification and taxonomy. The instability of the taxonomic treatment of this complex is furthered by the occurrence of intermediate anatomical and morphological characters and combinations thereof, caused by extensive events of hybridization and polyploidization which can lead to newly homoploid or polyploid species (e.g. Matthews *et al.*2015).

One of the large hybrid-polyploid systems is formed by the perennial oat-like genus *Helictotrichon* Besser, recently newly circumscribed by Wölk and Röser (2013, 2014). In the current valid classification the genus*Helictotrichon* comprises four species complexes distributed in the Mediterranean and the Eurasian mountains, which may have evolutionarily originated in parallel (Röser 1996). One of them is the endemic alpine*Helictotrichon parlatorei* group, which forms a clearly defined monophyletic group within the genus (cf. Heuchert and Röser 2004; Röser 1996), sharing a common derived character of diaspores consisting not of single florets, but a compound structure formed by all florets of a spikelet. This is due to a spikelet axis that is not disarticulating at maturity, a character not found in the other taxa of this genus (Röser 1989, 1996). The *H. parlatorei* species complex comprises three species. These are *H. parlatorei* (Woods) Pilg., *H. sempervirens* (Vill.) Pilg. and *Helictotrichonsetaceum* (Vill.) Henrard with two subspecies *setaceum* and *petzense* (H. Melzer) Röser.

H. parlatorei is the most widespread taxon of the species group. Based on observations from some isolated populations in the French western Alps, the species seems to be especially abundant mainly along the calcareous mountains of the southern Alps and more scattered in the northern Alps to the southeastern Alps (Karavanke Range, Steiner Alps in Slovenia). However, the species is rare in the dominantly crystalline ranges of the Inner Alps (Aichinger 1933; Gervais 1973; Pignatti 1982 Vollmann 1914).*H. parlatorei* is diploid ($2n = 2x = 14$) across its distribution (Gervais 1966,1972; Polatschek 1966; Röser 1989, 1996; Sauer 1971; Winterfeld *et al.*2009a) except for three tetraploid populations ($2n = 4x = 28$) from the Piedmont in the valley of the Chisone River (Gervais 1973; Röser 1989,1996).

The two subspecies of *H. setaceum* (Vill.) Henrard are distributed in the Southern Alps. The southwesterly distributed subsp. *setaceum* occurs mainly in the French Alps and the Preálpes, but there are additional stands in the Alpes Maritimes. Subspecies *petzense* (H. Melzer) Röser, which

Citation: Grit Winterfeld, Alexandra Wölk and Martin Röser, "Genome evolution in alpine oat-like grasses through homoploid hybridization and polyploidy," doi: 10.1093/aobpla/plw039.

was originally described as a separate species, *H. petzense* H. Melzer (Melzer 1967), is distributed on the Petzen and on a few neighboring mountains in the Karavanke Range, the border between Carinthia and Slovenia in the southeastern Alps (Hartl *et al.* 1992; Melzer 1967). Thus, the partial ranges of both subspecies are separated by approximately 600 km. Chromosome numbers of subsp. *setaceum* (Gervais 1965, 1972, 1973;Litardière 1950; Röser 1989) and subsp. *petzense* (Sauer 1971) are diploid ($2n=2x=14$). The tetraploid number of $2n=28$ previously reported for a population of subsp. *setaceum* (Röser 1989, 1996) was due to a sample switching error.

On Petzen Mountain, where *H. setaceum* subsp. *petzense* and *H. parlatorei*grow together, individuals with exactly intermediate morphological characters between these two species occur. These plants have been recognized as a hybrid taxon with *H. setaceum* subsp. *petzense* and *H. parlatorei* as its parental species, named as *Helictotrichon ×krischae* byMelzer (1967). The hybrid status was partly confirmed with karyotaxonomical characters by Sauer (1971) and in micromorphological, anatomical and morphometric studies by Heuchert and Röser (2004). In addition to these strictly intermediate hybrids, intermediates between the (F1) hybrids and the parental taxa exist (named *H.* cf. *×krischae*), displaying a mosaic of morphological characters in the hybrid zone of the Petzen Mountain populations.

The only consistently polyploid taxon of the *H. parlatorei* group, *H. sempervirens* (Vill.) Pilg., is hexaploid with $2n=6x=42$ (Gervais 1966,1973; Litardière 1950; Röser 1989, 1996). It is endemic to the southwestern Alps, namely the French Limestone Alps and their Préalpes, but occurs in only a few stands in the main range of the Alps between France and Italy (Röser 1989, 1996). The distribution of *H. sempervirens*largely corresponds with that of *H. setaceum* subsp. *setaceum*. Both are comparatively rare and only locally frequent (Röser 1989).

In the present study an attempt was made to ascertain the parentage of the *H. parlatorei* species complex by a molecular and cytogenetic survey. Because the commonly used nuclear ribosomal (ITS) or plastid DNA marker sequences (e.g. *matK*) were invariant within this species group (unpublished data), Topoisomerase VI sequences were used for phylogenetic analyses. The single copy nuclear gene Topoisomerase VI shows a highly variable intron structure and seems to be less susceptible to intergenomic concerted evolution. Therefore, the Topoisomerase VI sequences can be used to analyze relationships amongst closely related taxa and offer a suitable resource for phylogenetic studies in polyploids (Brassac *et al.* 2012; Jakob and Blattner 2010; Wölk and Röser 2014).

Combined karyotype studies are the ideal basis for the identification of basic chromosome sets in parental taxa (see Winterfeld and Röser 2007a,b). The present study physically maps the 5S and 45S rDNA as well as two satellite DNAs (CON1, CON2) on chromosomes of the hybrid and proposed parental species using FISH, enabling the detection of hybridization events or chromosomal rearrangements (Abd El-Twab and Kondo 2007; Heslop-Harrison 2000; Jiang and Gill 2006; Kotseruba *et al.*2003; Mishima *et al.* 2002; Vaio *et al.* 2005). We then map the presence of base pair-specific heterochromatin by differential fluorochrome banding with chromomycin and 4′,6-diamidino-2-phenylindole (DAPI; cf.Schweizer 1976; Schweizer and Ambros 1994; Sumner 1990).

These chromosome mapping methods were used to (i) identify hybrid taxa contributions of the different parental genomes at the DNA sequence and chromosome level, in order to detect a potential continuous transition between *H. setaceum* subsp. *petzense* and *H. parlatorei* and to clarify the origin of polyploid *H. sempervirens*, (ii) obtain insights on chromosomal changes that occurred after hybridization and polyploidization and to assess the role and extent of 'concerted evolution', i.e. the homogenization of different DNA repeat types and (iii) reconstruct biogeographical and ecological properties of the different taxa in the hybrid zone of the Petzen and across the entire distribution area in the Alps.

METHODS

Plant material

Plants were collected in the field from natural populations, either as living plants or as seeds that were later grown in the greenhouse. One sample of each taxon was used for chromosome and molecular survey. Taxon samples and voucher information are shown in Table 1. Vouchers of the accessions studied are housed in the herbarium of the Martin Luther University Halle-Wittenberg (HAL).

Table 1: Voucher information of the*Helictotrichon* samples used for chromosome and molecular study, their ploidy level and GenBank accession numbers, collectors (M. Röser, B. Heuchert, G. Winterfeld) and collection number, location of voucher specimens

Taxon	Provenance and collector	Ploidy level	Topoiso-merase 6 clones	GenBank accession number
H. ×*krischae* H. Melzer (= *H. parlatorei* × *H. setaceum* subsp. *petzense*)	Austria, Karavanke Mountains; M. Röser 10648 (HAL)	2*x*	KRI$_3$	HG797232[b]
			KRI$_5$	HG797233[b]
			KRI$_7$	HG797234[b]
			KRI$_8$	HG797235[b]
H. cf. ×*krischae*	Austria, Karavanke Mountains; M. Röser 10649 (HAL)	2*x*		
H. parlatorei (Woods) Pilg.	Austria, Karavanke Mountains; M. Röser 10647 (HAL)	2*x*		HG797240[b2]
	Italy, Mt. Baldo; G. Winterfeld 11	2*x*		
	Austria, Karavanke Mountains; B. Heuchert 1108 (HAL)			LN884300
	Italy, Piedmont; M. Röser 2344 (HAL)	4*x*[a]		LN589918[c]
H. sempervirens (Vill.) Pilg.	Cultivated in the Botanical Garden of Leipzig	6*x*		
	France, Dépt. Drôme; M. Röser 2429 (HAL)		SEM$_1$	HG797243[b]
			SEM$_2$	HG797244[b]
			SEM$_3$	HG797245[b]
			SEM$_4$	HG797246[b]
			SEM$_5$	HG797247[b]
			SEM$_6$	HG797248[b]
			SEM$_7$	HG797249[b]
			SEM$_8$	HG797250[b]
			SEM$_9$	HG797251[b]
			SEM$_{10}$	HG797252[b]
H. setaceum (Vill.) Henrard subsp. *setaceum*	France, Mt. Ventoux; M. Röser 10631 (HAL)	2*x*		
	France, Montagne de Clairet; M. Röser 2420 (HAL)			HG797254[b]
H. setaceum subsp. *petzense* (H. Melzer) Röser	Austria, Karavanke Mountains; M. Röser 10646 (HAL)	2*x*		HG797253[b]

[a] Röser (1989).
[b] Wölk & Röser (2014).
[c] Wölk et al. (2015).

Molecular phylogenetic analyses

DNA extraction, amplification and sequencing

Total genomic DNA was isolated from 20 to 45 mg dried leaf material using the NucleoSpin Plant Kit (Macherey-Nagel, Düren, Germany), following the manufacturer's instructions. The amplified region spans exon 17 to exon 19 of Topoisomerase VI (Topo6), numbered according to the gene structure of the published *Oryza sativa* sequence in EBI (2011). Most Topo6 sequences were amplified using primers Top6-15F and Top6-17R (Jakob and Blattner 2010), binding in exon 17 and exon 19. In some cases, the additional primers TopoHe-18F and TopoHe-18R (Wölk and Röser, 2014), binding in exon 18, were used to amplify the Topo6 region in two fragments. Amplifications were carried out on an Eppendorf Mastercycler (Eppendorf, Hamburg, Germany) using a reaction mix including 2 µl of 10× PCR buffer, 1.9 mM $MgCl_2$, 0.8–1 U Taq DNA polymerase (all maximum parsimony (MP) Biomedicals, Heidelberg, Germany), 5% DMSO (AppliChem, Darmstadt, Germany), 0.5 µM of forward and reserve primer, 100 µM dNTPs (GeneCraft, Lüdinghausen, Germany), ~20 ng template DNA and an aliquot of purified water to obtain a final volume of 20 µl. The PCR program for the amplification using Top6-15F and Top6-17R primers was 3 min at 94 °C, followed by 35 cycles of 30 s at 94 °C, 1 min at 50 °C, 2 min at 72 °C and the final extension at 72 °C for 10 min. We reduced the annealing time of the PCR protocol to 1 min for amplifications using the primer combinations TopoHe-18F/Top6-17R and Top6-15F/TopoHe-18R. PCR products were column-purified with the NucleoSpin Extract II Kit (Macherey-Nagel) and were used for direct sequencing and cloning. The sequencing reaction was carried out with 75 ng DNA and 0.3 µM forward or reverse primer and an aliquot of purified water to obtain a final volume of 15 µl. The reaction mix was sequenced by Eurofins MWG Operon (Ebersberg, Germany).

Cloning of PCR products

Cloning of PCR products was carried out using the pGEM-T Easy Vector System of Promega (Mannheim, Germany). Ligation and transformation were performed according to the technical manual. We used the GeneJET Plasmid Miniprep Kit (Fermentas, St Leon-Rot, Germany) to isolate the plasmid, according to the manufacturer's protocol. Plasmid DNA was eluted in 40 μl nuclease-free water. Presence of the insert was checked by digestion with the restriction enzyme *Eco*RI according to the manufacturer's protocol (Promega) or by PCR. The restriction product was visualized by gel electrophoresis. The sequencing reaction was carried out by Eurofins.

Data and phylogenetic analyses

Sequence editing and the generation of a multiple sequence alignment were both done manually in Sequencher 4.6 (Gene Codes Corporation, Ann Arbor, USA) and ambiguous positions were coded according to the IUPAC Code. Potentially parsimony-informative indels were encoded as binary data and gaps were treated as missing data. We checked rare nucleotide substitutions against the chromatograms to exclude errors in the manual editing. Sequences from *Hordeum marinum* subsp. *gussoneanum* were used as the outgroup according to Wölk *et al.* (2015).

Parsimony analyses and bootstrap

A MP analysis was conducted using PAUP* version 4.0b10 (Swofford 2002) and the following options: heuristic search, characters and character-state changes equally weighted, TBR branch swapping, 10 000 maxtrees, RANDOM addition and 100 replications. The heuristic search resulted in several most parsimonious trees, from which a strict consensus was generated. To test the statistical support of the clades a bootstrap analysis with 500 replications, 1000 maxtrees, TBR branch swapping, and closest addition was performed.

Bayesian analyses

The best-fit model of nucleotide substitution for the sequence data was performed using MrModeltest 2.3 (Nylander 2004), which selected HKY + G (hLRT) and GTR + G (AIC) as the optimal models for the Topo 6 dataset. We used the more general GTR + G model for the Bayesian inference (BI) analysis, implemented in MrBayes 3.1.2 (Ronquist and Huelsenbeck 2003). The BI analysis was executed for 3 million generations, which were sufficient for an average standard deviation of split frequencies <0.01, with trees sampled every 1000 generations. The first 10% of trees were excluded as burn-in.

Molecular cytogenetic analyses

Chromosome preparation

Young growing root tips were treated in iced water for 24 h, fixed in ethanol:acetic acid (3:1) for 3 h and stored in absolute ethanol at −20 °C. Before preparation the root tips were softened in 1% cellulase (w/v) and 10% pectinase (v/v) in citric acid-sodium citrate buffer pH 4.8 at 37 °C (cf.Schwarzacher *et al.* 1980) and squashed in 45% propionic acid with 2% carmine. After removal of the cover slips by freezing the slides at −90 °C, immersing in absolute ethanol for 5 min and air drying, they were stored at −20 °C until *in situ* hybridization and fluorochrome banding.

DNA probes for in situ hybridization

DNA probes used for *in situ* hybridization were 5S rDNA (GenBankAJ390195; Röser *et al.*, 2001), parts of 45S rDNA (GenBank Z96856,Z96857; Grebenstein *et al.* 1998), satellite DNA CON1 (GenBank Z68761- 68765) and satellite DNA CON2 (GenBank Z68772-Z68775; Grebenstein *et al.* 1995, 1996). The probes were labeled with fluorescein-11-dUTP using the 'random primed labeling kit' of Roche Diagnostics, according to the manufacturer's instructions.

FISH, fluorochrome banding and signal detection

For detection of the ribosomal and satellite DNA probes, FISH was done successively according to the protocols of Leitch (1994) with modifications of the procedures described in detail in Winterfeld and Röser (2007a). Unhybridized DNA was counterstained with propidium iodide (1.5 mg ml^{-1}) dissolved in fluorescence antifade solution (Vector Laboratories). Metaphase chromosomes on the slides were screened with a Zeiss Axioskop 2 epifluorescence microscope equipped with filter block 09 for simultaneous detection of fluorescein and propidium iodide fluorescence. To reveal GC or AT-rich chromosome segments by fluorochome banding the slides were cleaned in distilled water, air-dried, and stained with a 0.03% solution of chromomycin A_3 (Sigma, St Louis, MO, USA) in McIlvaine's buffer pH 7.0 containing 5 mM $MgCl_2$ for 1.5 h before embedding in antifade solution (Vector Laboratories) containing 2 mg ml^{-1}DAPI. The slides were stored in the dark until examination by epifluorescence microscopy with the Zeiss filter blocks 05 for chromomycin and 02 for DAPI. Photographs were taken using a computer-assisted cooled CCD camera (Zeiss Axiocam HRC) employing Zeiss Axiovision software.

Chromosome analyses and graphical representation

The chromosome sizes, localization of primary constrictions (centromere), secondary constrictions (satellite region; SAT = localization of nucleolus organizer regions, NORs) and extension of chromosome bands (5S rDNA, 45S rDNA, chromomycin and DAPI bands) were measured digitally using Zeiss Axiovision and Adobe PhotoShop 6.0 software. Measurement data was input into Microsoft Excel for calculations. Chromosomes were numbered according to their individual length. Complete chromosome idiograms were constructed using CorelDraw software. In the idiograms the chromosomes were arranged into groups of presumable homologues or homoeologues (indicated by Roman numerals in lines I–VII in Fig. 1) according to FISH and fluorochrome banding signals. Satellite chromosomes representing a secondary constriction, 45S rDNA band or chromomycin positive and DAPI negative bands (cf. Winterfeld and Röser 2007a) were arranged in lines I and II (Fig. 1). The terminology for chromosome shape according to the centromere position (metacentric or submetacentric) followed Levan *et al.* (1964).

Figure 1: Idiograms of chromosome complements of diploid*Helictotrichon* species and hexaploid *H. sempervirens*on a strict consensus tree derived from Topoisomerase IV gene sequences. Bootstrap and Bayesian support are shown on the left and right side above the branches. Chromosomes are numbered below according to their length and arranged into groups of presumable homologues or homoeologues (lines I–VII) according to FISH with 45S and 5S rDNA probes and fluorochrome banding signals. Chromosome bands arranged above or below the chromosomes are subtelomeric, unless an intercalary position is indicated by a square bracket at the side of the respective chromosome. Chromosomes with a NOR bearing the 45 S rDNA bands or chromomycin-positive staining are indicated by a secondary constriction in the chromosome idiogram additional to their centromeric constriction (= satellite chromosomes; cf. Winterfeld and Röser 2007b). Genome-specific DNA and chromosome variants are shown in light blue for *H. parlatorei* and in pink for *H. setaceum*. Possible mechanisms of chromosome and genome evolution as well as geographical distribution of the species are listed below. B = B chromosomes.

RESULTS

Molecular phylogenetics

DNA-sequences

Due to nucleotide additivity in direct sequencing of Topo6 in 2*x H.*×*krischae* and 6*x H. sempervirens*, the PCR products of these taxa were cloned and up to ten individual clones were sequenced. The alignment of 20 sequences from direct sequencing (three accessions of *H. parlatorei*, two of *H. setaceum*) and from cloned fragments (*H.* ×*krischae*, *H. sempervirens*) comprised 922 nucleotide positions. 53 of 180 variable characters were parsimony-informative. The sequences obtained from this study are available in GenBank (see Table 1).

Partial alignment

In total, 48 of the parsimony-informative nucleotide sequence variants and their positions are shown in the partial alignment depicted in Table 2. Clone 8 of *H.* ×*krischae* and clones 1, 3, 6 and 7 of *H. sempervirens*conform as far as possible (85%) with the sequences of the two subspecies of *H. setaceum* (marked in pink in Table 2). However, five variable nucleotides, 193 (A/T polymorphism), 226 (T/A), 295 (A/G), 384 (T/G) and 650 (A/G), in clone 8 of *H.* ×*krischae* agree with the repeat types of the two subsp. of *H. setaceum* but differ from clones 1, 3, 6 and 7 of *H. sempervirens*. At positions 75 and 76 an additional polymorphism is found (G/A), in which clone 8 of *H.* ×*krischae* matches *H. setaceum* subsp.*petzense*. Clones 3, 5 and 7 of *H.* ×*krischae* and clones 2, 4, 5, 8, 9 and 10 of *H. sempervirens* closely match sequences of *H. parlatorei* (marked in light blue in Table 2). Despite 180 variable characters (point mutations) in the whole alignment (not shown) single base pair changes occur in the partial alignment of Table 2 at positions 15 (clone 3 of *H.* ×*krischae*: A/G polymorphism), 43 (clone 5 of *H.* ×*krischae*: T/- polymorphism), 242 (*H. parlatorei* accession H 1108: G/- polymorphism), 295 (clone 7 of *H. sempervirens*: A/G polymorphism), 429 (clones 4, 5 of *H. sempervirens*and *H. parlatorei* accession M. Röser 2344: T/C polymorphism), and 683 (*H. parlatorei* B. Heuchert 1108: C/- polymorphism).

Table 2: Variable nucleotide positions of Topoisomerase 6 in direct sequencing of PCR products in 2*x H. parlatorei* (blue shading), *H. setaceum* subsp. *setaceum*and subsp. *petzense* (red shading) and cloned sequences of 2*x H.* ×*krischae* and 6*x H. sempervirens.*

sequence data / alignment position	15	43	62	72	73	74	75	76	193	226	242	247	268	277	284	294	295	307	308	312	319	360	361	373	380	384	386	423	426	429	588	590	606	609	624	625	642	649	650	683	688	691	704	710	712	724	727	733
H. ×krischae clone 8	G		A	C	T	A	G	G	A	T	G	C	C	G	A	A	A			A	C	T	A	T	T	T	G	A	A	G	G	T	T	T	T	C	T	T	A	C	T	T	C	C	G	G	A	A
H. setaceum subsp. *petzense*	G		A	C	T	A	G	G	A	T	G	C	C	G	A	A	A			A	C	T	A	T	T	T	G	A	A	G	G	T	T	T	T	C	T	T	A	C	T	T	C	C	G	G	A	A
H. setaceum subsp. *setaceum*	G		A	C	T	A	A	A	A	T	G	C	C	G	A	A	A			A	C	T	A	T	T	T	G	A	A	G	G	T	T	T	T	C	T	T	A	C	T	T	C	C	G	G	A	A
H. sempervirens clone 1	G		A	C	T	A	A	A	T	A	G	C	C	G	A	A	G			A	C	T	A	T	T	G	G	A	A	G	G	T	T	T	T	C	T	T	G	C	T	T	C	C	G	G	A	A
H. sempervirens clone 3	G		A	C	T	A	A	A	T	A	G	C	C	G	A	A	G			A	C	T	A	T	T	G	G	A	A	G	G	T	T	T	T	C	T	T	G	C	T	T	C	C	G	G	A	A
H. sempervirens clone 6	G		A	C	T	A	A	A	T	A	G	C	C	G	A	A	G			A	C	T	A	T	T	G	G	A	A	G	G	T	T	T	T	C	T	T	G	C	T	T	C	C	G	G	A	A
H. sempervirens clone 7	G		A	C	T	A	A	A	T	A	G	C	C	G	A	A	A			A	C	T	A	T	T	G	G	A	A	G	G	T	T	T	T	C	T	T	G	C	T	T	C	C	G	G	A	A
H. ×krischae clone 5	A		G	T					A	T		T	T	T	C	T	A	T	T	C	G	A	T	C	A	T	C	G	T	C	C	C	G	G	C	G	C		A		C	C	G	T	A	A	G	C
H. ×krischae clone 3	G	T	G	T					A	T		T	T	T	C	T	A	T	T	C	G	A	T	C	A	T	C	G	T	C	C	C	G	G	C	G	C		A		C	C	G	T	A	A	G	C
H. ×krischae clone 7	A	T	G	T					A	T		T	T	T	C	T	A	T	T	C	G	A	T	C	A	T	C	G	T	C	C	C	G	G	C	G	C		A		C	C	G	T	A	A	G	C
H. parlatorei (Rö 10647)	A	T	G	T					A	T		T	T	T	C	T	A	T	T	C	G	A	T	C	A	T	C	G	T	C	C	C	G	G	C	G	C		A		C	C	G	T	A	A	G	C
H. parlatorei (Rö 2344)	A	T	G	T					A	T		T	T	T	C	T	A	T	T	C	G	A	T	C	A	T	C	G	T	T	C	C	G	G	C	G	C		A		C	C	G	T	A	A	G	C
H. parlatorei (H 1108)	A	T	G	T					A	T	G	T	T	T	C	T	A	T	T	C	G	A	T	C	A	T	C	G	T	C	C	C	G	G	C	G	C		A	C	C	C	G	T	A	A	G	C
H. sempervirens clone 2	A	T	G	T					A	T		T	T	T	C	T	A	T	T	C	G	A	T	C	A	T	C	G	T	C	C	C	G	G	C	G	C		A		C	C	G	T	A	A	G	C
H. sempervirens clone 4	A	T	G	T					A	T		T	T	T	C	T	A	T	T	C	G	A	T	C	A	T	C	G	T	T	C	C	G	G	C	G	C		A		C	C	G	T	A	A	G	C
H. sempervirens clone 5	A	T	G	T					A	T		T	T	T	C	T	A	T	T	C	G	A	T	C	A	T	C	G	T	T	C	C	G	G	C	G	C		A		C	C	G	T	A	A	G	C
H. sempervirens clone 8	A	T	G	T					A	T		T	T	T	C	T	A	T	T	C	G	A	T	C	A	T	C	G	T	C	C	C	G	G	C	G	C		A		C	C	G	T	A	A	G	C
H. sempervirens clone 9	A	T	G	T					A	T		T	T	T	C	T	A	T	T	C	G	A	T	C	A	T	C	G	T	C	C	C	G	G	C	G	C		A		C	C	G	T	A	A	G	C
H. sempervirens clone 10	A	T	G	T					A	T		T	T	T	C	T	A	T	T	C	G	A	T	C	A	T	C	G	T	C	C	C	G	G	C	G	C		A		C	C	G	T	A	A	G	C

Phylogenetic Analyses

A heuristic search found a single tree island consisting of 6956 equally parsimonious trees with a length of 201 steps. All trees in the island have a high consistency and retention index (CI = 0.965, RI = 0.9727) which indicates a low level of homoplasy in the data matrix. MP and Bayesian analysis resulted in similar tree topologies. A strict consensus of all most-parsimonious trees is shown in Figure 1, with posterior probability (PP, on the right side above the branches in Fig. 1 values of the Bayesian analysis (BS, on the left side above the branches in Fig. 1). Two clades were maximally supported. The first clade shows support for relatedness of both subspecies of *H. setaceum* and clone 8 of *H.* ×*krischae* (BS 82%, PP 1.00) that are closely related to a strongly supported clade comprising clones 1, 3, 6 and 7 of *H. sempervirens* (BS 100%, PP 1.00; copy type SET; cf. Wölk *et al.* 2015; marked in pink in Fig. 1). The second clade includes all populations of *H. parlatorei* and clones 2, 4, 5, 8, 9 and 10 of *H. sempervirens* (BS 100%, PP 1.00; copy type PAR; marked in light blue inFig. 1); for the most part relationships among these sequences are unresolved within this clade.

Molecular cytogenetic analyses

Karyotypes

Figure 2 shows mitotic metaphases of *H. setaceum*, *H. parlatorei*, *H.*×*krischae* and *H. sempervirens* after *in situ* hybridization and fluorochome staining. The respective idiograms of all taxa investigated are shown (Fig. 1). The karyotypes are constituted by seven pairs of chromosomes (Roman numerals; Fig. 1) in 2*x* *H. setaceum*, *H. parlatorei* and *H.* ×*krischae*, and three sets of seven chromosome pairs in 6*x* *H. sempervirens*. All chromosomes are metacentric with the exception of one pair in *H.* cf. ×*krischae* (chromosomes 3 and 4), which are submetacentric.

Satellite chromosomes

In all taxa the presence of four terminal SATs per diploid chromosome complement was observed at the distal end of the short arms identified by secondary constrictions, 45S rDNA bands and/or chromomycin-positive and DAPI-negative bands (Fig. 2). One exception was found in *H. setaceum* subsp. *setaceum*, which displayed only three SATs (obviously loss of one NOR). Additionally, subtelomeric minor 45S rDNA bands were found in *H. parlatorei* (Winterfeld 11; chromosomes 3, 5, 8, 9, 14), *H.*×*krischae* (chromosomes 2, 12) and *H. sempervirens* (chromosomes 1, 4, 14, 15, 30, 42).

5S rDNA-bearing chromosomes

Bands of 5S rDNA are intercalary in non-satellite chromosomes, either in the long or in the short arms, with two, three or four loci per diploid complement. Two loci occur in chromosomes 1 and 2 of *H. setaceum*subsp. *setaceum* and chromosomes 7, 8, 9, 12, 21 and 22 of *H. sempervirens*. Three loci occur in chromosomes 3, 4 and 5 of *H. setaceum*subsp. *petzense*, chromosomes 3, 4 and 5 of *H.* ×*krischae* and 6, 7 and 9 of *H.* cf. ×*krischae*. Four loci occur in chromosomes 1, 2, 4 and 7 of *H. parlatorei* (M. Röser 10647) and 1, 4, 10 and 11 of *H. parlatorei*(Winterfeld 11).

Bands of DAPI positive heterochromatin and satellite DNAs

DAPI-positive bands, if present, are subtelomeric either in both arms or only in one arm. They are often, but not always, correlated with the loci of the satellite DNAs CON1 and CON2.

In the chromosome complement of *H. setaceum* subsp. *setaceum* broad DAPI bands occur in the long arms of all SAT chromosomes (5, 12, 14), in both arms of the 5S rDNA-bearing chromosomes (1, 2), in both arms of two non rDNA-bearing chromosome pairs (chromosomes 6, 7, 9, 10) and in one arm of an additional pair (chromosomes 3, 4). They are correlated mostly (chromosomes 3, 4, 5, etc.) but not always (e.g. chromosomes 7, 10) with the bands of satellite DNAs CON1 and CON2 (Fig. 1). In three chromosomes (8, 11, and 13) no chromosome band was found.

In *H. setaceum* subsp. *petzense* only one of the four satellite chromosomes had a subtelomeric DAPI band (chromosome 13), two showed broad bands of satellite DNA CON2 (11, 14) and one had no DAPI or satellite DNA band. As seen in subsp. *setaceum* the 5S rDNA-bearing chromosomes 3 and 4 had subtelomeric DAPI bands in both arms, which are correlated with bands of the satellite DNAs CON1 and CON2. Additional DAPI bands were found in both arms of chromosomes 6 and 7 and in one arm of chromosomes 1, 2, 5, 8, 9 and 12, partly corresponding with the bands of satellite DNAs CON1 and/or CON2 (Fig. 1).

Karyotype structures and banding pattern in both putative hybrid accessions *H.* ×*krischae* (M. Röser 10548) and *H.* cf. ×*krischae* (M. Röser 10549) were very similar. Two satellite chromosomes showed subtelomeric DAPI bands in the non-satellite arms (chromosomes 10 and 11 in accession M. Röser 10548; 11 and 12 in M. Röser 10549) but the two other SAT chromosomes 8 and 9 (M. Röser 10548) and 13 and 14 (M. Röser 10549) were devoid of DAPI or satellite DNA bands. Two of the 5S rDNA-bearing chromosomes 4 and 5 (M. Röser 10548) had small DAPI bands in only one arm, the non 5S rDNA-bearing arm, while the presumable homoeologous chromosomes 6 and 7 of the accession M. Röser 10549 are devoid of DAPI bands. The third 5S rDNA-bearing chromosome of both accessions had, like both subspecies of *H. setaceum*, DAPI bands in both arms, which are correlated with bands of the satellite DNAs CON1 and/or CON2. In the remaining chromosomes, DAPI bands occurred in both arms (1, 2 in M. Röser 10548; 5, 8 in M. Röser 10549), in only one arm (12, 13 in M. Röser 10548; 1, 2 in M. Röser 10549) or were absent (chromosomes 6, 7, 14 in M. Röser 10548; 3, 4, 10 in M. Röser 10549), which is partly correlated with the bands of the satellite DNAs CON1 and CON2 (see Fig. 1).

In the hexaploid chromosome complement of *H. sempervirens* in four of twelve satellite chromosomes subtelomeric DAPI bands occur in the arms lacking satellites. All 5S rDNA-bearing chromosomes reveal DAPI bands in the arms without the 5S rDNA loci. Five chromosomes (5, 10, 16, 17, 38) had DAPI bands in both arms; two chromosomes were devoid of DAPI bands (14, 36). The remaining chromosomes showed DAPI bands in one arm, partly correlated with the satellite DNAs CON1 and CON2 (Fig. 1).

H. parlatorei M. Röser 10647 from Petzen Mountain is very similar to the population G. Winterfeld 11 from Monte

Baldo with respect to DAPI banding. In both populations, only one of the four satellite chromosomes had DAPI bands near the telomeres (chromosome 5 in M. Röser 10647, 8 in G. Winterfeld 11). Two of four 5S rDNA-bearing chromosomes show DAPI bands in one arm (1, 2 in M. Röser 10647; 4, 10 in G. Winterfeld 11). In other 5S rDNA-bearing chromosomes DAPI bands were absent (4, 7 in M. Röser 10647; 11 in G. Winterfeld 11), while chromosome 1 of G. Winterfeld 11 revealed DAPI bands in both arms. In the remaining chromosomes DAPI bands were localized either in one arm (11, 12, 13, 14 in M. Röser 10647; 5, 6, 9, 12 in G. Winterfeld 11), in both arms (13 in M. Röser 10647; 2, 7 in G. Winterfeld 11) or were absent (9, 10 in M. Röser 10647).

DISCUSSION

Structures of parental genomes at sequence and chromosomal level reveal hybrid origins: biogeography and evolution of the *H.Parlatorei* group

Hybrid evidence at sequence level

Sequence variation in the nuclear single copy gene Topo6 and resulting trees show the occurrence of two different copies, in taxa of the *H. parlatorei* group (SET and PAR, Fig. 1; Wölk *et al.* 2015). In the diploid taxa, SET type copies occur in both subspecies of *H. setaceum*, whereas PAR type copies were found the three populations of *H. parlatorei*investigated (see Table 2). Both the diploid hybrid populations of *H.×krischae* and *H.* cf. *×krischae* and hexaploid *H. sempervirens* have two distinct variants of Topo6 sequences, which correspond to SET and PAR, respectively. The resulting gene phylogeny (Fig. 1) corroborates the hybrid character of *H. ×krischae* and *H.* cf. *×krischae*, in accordance with the systematic treatment based on morphological and anatomical characters by Melzer (1967), the classical karyological investigation by Sauer (1971)and micromorphological, anatomical and morphometric studies byHeuchert and Röser (2004). Hexaploid *H. sempervirens* is shown to be an allopolyploid hybrid species between *H. setaceum* and *H. parlatorei* due to the occurrence of their two Topo6 gene copies.

Detection of marker chromosomes and reconstructing of hybrid origins

The presence of four satellite chromosomes per $2n=2x=14$ chromosomes is a consistent karyotype found in the *H. parlatorei* group diploids. The only exception is *H. setaceum* subsp. *setaceum* with only three satellite chromosomes, most likely due to the loss of one NOR in the specimen sampled. However, karyotypes of *H. parlatorei* and *H. setaceum*are distinguished by differences in the amount and distribution of heterochromatin indicated by DAPI bands and also by bands of the satellite DNA s CON1 and CON2 (cf. Winterfeld and Röser 2007b), especially in the 5S rDNA-bearing and the SAT chromosomes. The occurrence and localization of the heterochomatin bands in 5S rDNA-bearing chromosomes provide the identification of parental species marker chromosomes involved in the formation of hybrid taxa between *H. parlatorei* and *H. setaceum*.

The karyotypes of the two populations of *H. parlatorei* studied are very similar, characterized by the occurrence of two chromosome pairs with intercalary 5S rDNA bands. These pairs occur in two slightly different variants; with small subtelomeric bands of satellite DNAs and DAPI in the arm without the 5S rDNA, termed marker chromosome par I (light blue triangle in Fig. 1), or without additional bands, termed par II (light blue circle in Fig. 1). The two DAPI bands in both arms of chromosome 1 in line IV (Fig. 1) of the Monte Baldo population (Winterfeld 11) are the result of a translocation from the homologous chromosome 11, which is devoid of subtelomeric heterochromatin and considerably shorter. As a result, the chromosomes of this pair are par II. In both populations only one of the four SAT chromosomes of *H. parlatorei* has subtelomeric DAPI bands in the arm without the satellite. Modifications of both karyotypes were found in the different number of eight (Petzen population) and 13 (Monte Baldo population) DAPI heterochromatin bands which correlate partially with the satellite DNAs CON1 and CON2. Based on the hypothesis of heterochromatin accumulation during evolution (Guerra *et al.* 2000; Ikeda 1988; Morawetz and Samuel 1989; Röser 1994; Winterfeld and Röser 2007b), the population of *H. parlatorei* from Monte Baldo might be considered younger.

In contrast to *H. parlatorei*, both subspecies of *H. setaceum* were characterized by subtelomeric CON1 and CON2 satellite DNA and DAPI bands on both arms of the 5S rDNA-bearing chromosomes, termed 'marker chromosomes set' (pink square in Fig. 1), making these chromosomes distinguishable from those of *H. parlatorei*. However, in addition to the varying number of satellite chromosomes (three in subsp.*setaceum* and four in subsp. *petzense*), the subspecies differ in the number of 5S rDNA bands (two in subsp. *setaceum* and three in subsp. *petzense*) and the number of DAPI-positive heterochromatin bands and their localization (15 in subsp. *petzense* and 17 in subsp. *setaceum*). All in all, chromosomal characters of subsp. *setaceum* seem to be derived from subsp. *petzense* by loss of one NOR (cf. Winterfeld and Röser 2007a) and 5S rDNA sequences (cf. Wendel *et al.* 1995) or their inactivation (cf. Chen and Pikaard 1997; Pikaard, 1999) and by changes in the DAPI heterochromatin banding pattern (cf. Guerra *et al.* 2000; see Fig. 1).

The presumed hybrid plants of *H. ×krischae* and *H.* cf. *×krischae*investigated here were found to be diploid like their parents (cf. Sauer 1971). Their karyotypes contain marker chromosomes (set, par I, par II) as well as satellite chromosomes from both *H. parlatorei* and *H. setaceum*. Therefore, at the chromosome level, they are intermediate between their presumable parental taxa. Their common geographical distribution as well as co-possession of B chromosomes in *H. ×krischae* and *H. setaceum*subsp. *petzense* supports the latter as one parent of the hybrid *H.×krischae*. This is in agreement with intermediate morphological characters, especially noted in the laminas of *H. xkrischae* leaves (Heuchert and Röser 2004; Melzer 1967).

Interestingly, the localization of DAPI positive heterochromatin and satellite DNA CON1 and CON2 bands in the chromosomes of *H. ×krischae*and *H.* cf. *×krischae* are not fully intermediate between their parents. Obviously either translocations between single chromosomes of the parental taxa occurred, or these hybrid plants are not immediate F_1hybrids, indicating that other intermediates between *H. parlatorei* and *H. setaceum* subsp. *petzense* occur. It appears the examined plants belong to a complex hybrid swarm in which the parental plants are in a continuous process of crossing and backcrossing due to the lack of a barrier to gene flow. This result supports the morphometric findings (Heuchert and Röser 2004), in which hybrid plants display a mosaic of characters of their parental taxa characteristic of many other natural hybrid plants (Arnold 1997; Rieseberg and Ellstrand 1993).

The hexaploid cytotype of *H. sempervirens* shares the karyotype and the marker chromosomes par I with diploid *H. parlatorei*, especially of the population from Monte Baldo, but is missing the characteristic chromosome pattern of *H. setaceum*, which would be expected based on the Topo6 data. Which subspecies of *H. setaceum* is the second parent of*H. sempervirens* cannot be determined from the molecular and cytogenetic methods used, because the Topo6 sequences of subsp. *petzense* and subsp. *setaceum* are highly similar and their chromosomal differences, which are widely restricted to subtelomeric heterochromatin bands, are rather weak. The close relationship of *H. sempervirens* to the diploids *H. parlatorei* and *H. setaceum*, which was taxonomically implied (Röser 1989,1996), is corroborated by molecular data. Deviations in the karyotype of *H. sempervirens* which shows a reduction of 5S rDNA loci, because the number of 5S rDNA bands are not a multiple of their alleged diploid parental taxa, is presumably caused by unidirectional homogenization consistently at the cost of repeats from the *H. setaceum* genome. This result represents an example of a beginning diploidization in the course of polyploidization, found also in other plant groups (e.g. Ma and Gustafson 2005; Ozkan and Feldman 2009; Renny-Byfield *et al.* 2013).

Biogeography and evolution of the *H. parlatorei* group

The current occurrence of the alpine *H. parlatorei* group seems to be influenced by glaciations during the last ice age period: the distribution of widespread *H. parlatorei*, known mostly from the southern calcareous flank of the Alps (Pignatti 1982) and minor refuge areas in the Bavarian and Austrian Alps of the northern flank, lies in regions of the Alps that remained unglaciated during the last glacial period (cf. Merxmüller 1952;Pawlowski 1970; Schönswetter *et al.* 2005). The disjunct east-west geographic occurrence of the two subspecies of *H. setaceum* implies that their distribution was also shaped during glacial periods, with the southwestern and southeastern Alps serving as important refuge areas (cf.Merxmüller 1952).

Cytogenetic evidence, e.g. the presumably older karyotype of *H. setaceum*subsp. *petzense* (see above), points to a primary center of speciation in the eastern region of the Alps. Processes of hybridization with *H. parlatorei* likely took place when the parental taxa became sympatric. As a result, the homoploid hybrid *H.* ×*krischae* and backcross offspring with the parents (*H.* cf. ×*krischae*) have become established on the Petzen Mountain.

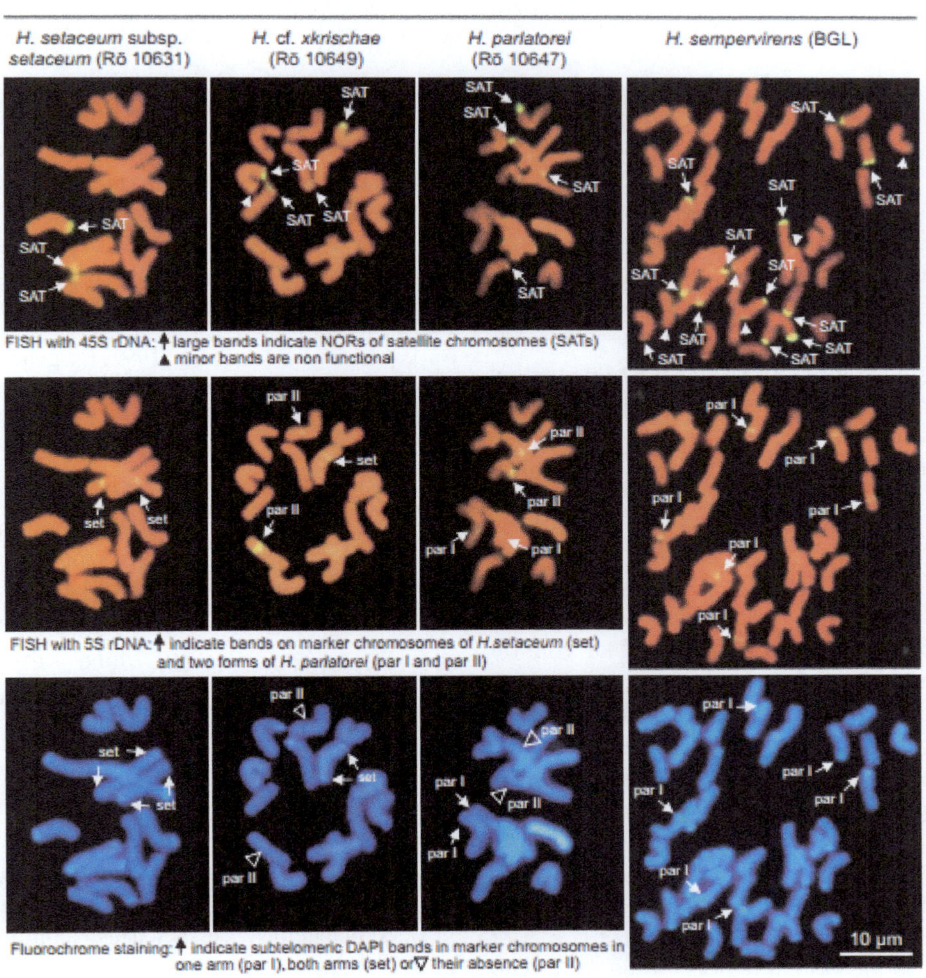

Figure 2: Somatic metaphase plates of*Helictotrichon* species after FISH and fluorochome staining.

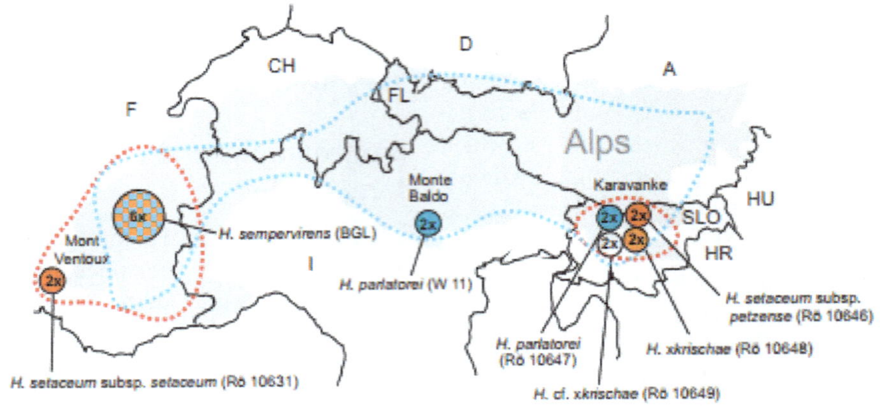

Figure 3: Collection sites of studied taxa of the *H. parlatorei*group in the Alps and simplified general geographical distribution of*H. parlatorei* (blue dotted line) and disjunct areas of *H. setaceum* subsp. *setaceum*in the West and subsp. *petzense* in the East (red dotted lines). Circles filled with checkerboard pattern indicate hybrid character in molecular and cytogenetic analyses.

Westward expansion of this diploid hybrid (or hybrid swarm) initially resulted in a large southern alpine distribution area. Most probably during the ice age glaciations this range became fragmented into two distinct areas in the eastern and the western Alps with a gap in distribution of ca. 600 km. It is possible that in the western area harboring currently *H. setaceum* subsp. *setaceum* and *H. sempervirens* (whose genome also indicates a hybrid origin), polyploidization took place in the latter species after isolation of the glacial refuges. Another hypothesis implies that hybridization took place at least two times in parallel in the western and the eastern Alps followed by polyploidization only in western *H. sempervirens* and genome diploidization (Fig. 3).

Recently, the hybrid plants *H.* ×*krischae* and *H.* cf. ×*krischae* but also the polyploid *H. sempervirens* have narrower overall geographical distributions than their putative diploid parents *H. parlatorei* and *H. setaceum* (see Fig. 3). This is a clear example where genetic introgression by homoploid hybridization on the one hand and maybe followed by polyploidization on the other hand of more widely distributed diploid species like *H. parlatorei* and/or *H. setaceum* resulted in taxa with smaller distribution areas. The same pattern has been reported in other polyploid series of other species, e.g. of the genus *Helictochloa* (= *Helictotrichon*; cf.Winterfeld *et al.* 2009b), which was discussed in relation to the allopolyploid origin of these taxa. This contrasts to the general opinion, that polyploid plants are more vital and efficient colonizers than their diploid ancestors.

CONCLUSIONS

The cytogenetic and molecular phylogenetic data confirm that the endemic Karavanke Range *H.* ×*krischae* and *H.* cf. ×*krischae* in the southeastern Alps are homoploid hybrids, resulting from a cross between the widespread alpine *H. parlatorei* and the local endemic *H. setaceum* subsp.*petzense*. Hexaploid *H. sempervirens* from the southwestern Alps, the French Alps and their Préalps is also suspected to be of hybrid origin involving the *H. parlatorei* and *H. setaceum* genomes. Cytogenetic evidence, e.g. the potentially older karyotype of *H. setaceum* subsp.*petzense*, might suggest a primary center of species establishment in the eastern regions of the Alps, with the westward expansion and subsequent ice age interruption resulting in a recently disjunct distribution between populations of the southwestern and southeastern Alps.

REFERENCES

1. Abd El-Twab MH, Kondo K. 2007. Isolation of chromosomes and mutation in the interspecific hybrid between Chrysanthemum boreale and C. vestitum using fluorescence in situ hybridization and genomic in situ hybridization. Chromosome Botany 2:19–24.
2. Aichinger E. 1933. Vegetationskunde der Karawanken. Jena: G. Fischer.
3. Arnold ML. 1997. Natural Hybridization and Evolution. New York: Oxford University Press.
4. Brassac J, Jakob SS, Blattner FR. 2012. Progenitor-derivative relationships of Hordeum polyploids (Poaceae, Triticeae) inferred from sequences of Topo6, a nuclear low copy gene region. Public Library of Science. One 7:e33808. doi: 10.1371/journal.pone.0033808.
5. Dumortier BCJ. 1823.Observations sur les Graminéesde la Flore belgique. Tournay.
6. Chen ZJ, Pikaard CS. 1997. Epigenetic silencing of RNA polymerase I transcription. A role for DNA methylation and histone modification in nucleolar dominance. Genes and Development 11:2124–2136.
7. European Bioinformatics Institute (EBI). [Ensemble plants release]. http://plants.ensembl.org/Oryza_sativa/Transcript/Summary?db=core;g=LO_Os010770;r=9:5873972-5882020;t=LOC_Os09g10770.1 (accessed 11 September 2011).
8. Heslop-Harrison JS. 2000. Comparative genome organization in plants: from sequence and markers to chromatin and chromosomes. Plant Cell 12:617–636.
9. Gervais C. 1965. Nombres chromosomiques chez quelques graminées des Alpes. Bulletin De La Société Neuchâteloise Des Sciences Naturelles 88:61–64.
10. Gervais C. 1966. Nombres chromosomiques chez quelques graminées alpines. Bulletin De La Société Neuchâteloise Des Sciences Naturelles 89:87–100.
11. Gervais C. 1972. Nouvelles déterminations de nombres chromosomiques chez les avoines vivaces. I. Bulletin

De La Société Neuchâteloise Des Sciences Naturelles 95:57–61.

12. Gervais C. 1973. Contribution à l›étude cytologique et taxonomique des avoines vivaces. Denkschrift Der Schweizerischen Naturforschenden Gesellschaft 88:1–166.

13. Grebenstein B, Röser M, Sauer W, Hemleben V. 1998. Molecular phylogenetic relationships in Aveneae (Poaceae) species and other grasses as inferred from ITS1 and ITS2 rDNA sequences. Plant Systematics and Evolution 213:133–150.

14. Grebenstein B, Grebenstein O, Sauer W, Hemleben V. 1995. Characterization of a highly repeated DNA component of perennial oats (Helictotrichon, Poaceae) with sequence similarity to a A-genome-specific satellite DNA of rice (Oryza). Theoretical and Applied Genetics 90:1101–1105.

15. Grebenstein B, Grebenstein O, Sauer W, Hemleben V. 1996. Distribution and complex organization of satellite DNA sequences in Aveneae species. Genome 39:1045–1050.

16. Guerra M, Galva K, Dos Santos B, Barros e Silva AE, Ehrendorfer F. 2000. Heterochromatin banding patterns in Rutaceae-Aurantioideae – a case of parallel chromosomal evolution. American Journal of Botany 87:735–747.

17. Hartl H, Kniely G, Leute GH, Niklfeld H, Perko M. 1992. Verbreitungsatlas der Farn- und Blütenpflanzen Kärntens. Klagenfurt: Naturwissenschaftlicher Verein.

18. Heuchert B, Röser M. 2004. The Carinthian grass hybrid Helictotrichon ×krischae (Poaceae tribe Avenaeae) – epitome of a hybrid zone. Schlechtendalia 12:69–83.

19. Ikeda H. 1988. Karyomorphological studies on the genus Crepis with special reference to C-banding pattern. Journal of Science, Hiroshima University, Series B, Division 2. 22: 65–117.

20. Jakob SS, Blattner FR. 2010. Two extinct diploid progenitors were involved in allopolyploid formation in the Hordeum murinum (Poaceae: Triticeae) taxon complex. Molecular Phylogenetics and Evolution 55:650–659.

21. Jiang J, Gill BS. 2006. Current status and the future of fluorescence in situ hybridization (FISH) in plant genome research. Genome 49:1057–1068.

22. Kotseruba V, Gernand D, Meister A, Houben A. 2003. Uniparental loss of ribosomal DNA in the allotetraploid grass Zingeria trichopoda (2n = 8). Genome 46:156–163.

23. Leitch AR, Schwarzacher T, Jackson D, Leitch IJ. 1994. In situ hybridization: a practical guide. Oxford: BIOS Scientific Publishers.

24. Levan A, Fredga K, Sandberg AA. 1964. Nomenclature for centromeric position on chromosomes. Hereditas 52:201–220.

25. Litardière R. 1950. Nombres chromosomiques de quelques graminées. Boletim Da Sociedade Broteriana 24: 79–87.

26. Ma XF, Gustafson JP. 2005. Genome evolution of allopolyploids: a process of cytological and genetic diploidization. Cytogenetic and Genome Research 109:236–249.

27. Matthews A, Emelianova K, Hatimy AA, Chester M, Pellicer J, Ahmad KS, Guignard MS, Rouhan G, Soltis DE, Soltis PS, Leitch IJ, Leitch AR, Mavrodiev EV, Buggs RJA. 2015. 250 years of hybridization between two biennial herb species without speciation. AoB PLANTS 7: plv081; doi:10.1093/aobpla/plv081.

28. Melzer H. 1967. Helictotrichon petzense, spec. nova – ein neu entdeckter Endemit der südöstlichen Kalkalpen. Österreichische Botanische Zeitschrift 114:309–319.

29. Merxmüller H. 1952. Untersuchungen zur Sippengliederung und Arealbildung in den Alpen. I. Jahrbuch Des Vereins Zum Schutze Der Alpenpflanzen und -Tiere. 1: 96–133.

30. Mishima M, Ohmido N, Fukui K, Yahara T. 2002. Trends in site-number change of rDNA loci during polyploid evolution in Sanguisorba (Rosaceae). Chromosoma 110:550–558.

31. Morawetz W, Samuel MRA. 1989. Karyological patterns in the Hamamelidae. In: Crane PR,, Blackmore S, eds. Evolution, systematics and fossil history of the Hamamelidae, vol. 1, Introduction and "Lower Hamamelidae," Oxford: Clarendon Press, 129–154.

32. Nylander JAA, Ronquist F, Huelsenbeck JP, Nieves Aldrey JL. 2004. Bayesian phylogenetic analysis of combined data. Systematic Biology 53:47–67.

33. Ozkan H, Feldman M. 2009. Rapid cytological diploidization in newly formed allopolyploids of the wheat (Aegilops-Triticum) group. Genome 52:926–934. doi: 10.1139/g09-067.

34. Pawlowski B. 1970. Remarques sur l'endémisme dans la flore des Alpes et des Carpates. Vegetatio 21:181–243.

35. Pignatti S. 1982. Flora d'Italia, Bd. 3. Bologna: Edagricole.

36. Pikaard CS. 1999. Nucleolar dominance and silencing of transcription. Trends in Plant Science 4:478–483.

37. Polatschek A. 1966. Cytotaxonomische Beiträge zur Flora der Ostalpenländer, II. Österreichische Botanische Zeitschrift 113:101–147.

38. Renny-Byfield S, Kovarik A, Kelly LJ, Macas J, Novak P, Chase MW, Nichols RA, Pancholi MR, Grandbastien M-A, Leitch AR. 2013. Diploidization and genome size change in allopolyploids is associated with differential dynamics of low- and high-copy sequences. The Plant Journal 74:829–839. doi:10.1111/tpj.12168.

39. Rieseberg LH, Ellstrand NC. 1993. What can molecular and morphological markers tell us about plant hybridization? CRC. Critical Reviews in Plant Sciences 12:213–241.

40. Ronquist F, Huelsenbeck JP. 2003. MrBayes 3: Bayesian phylogenetic inference under mixed models. Bioinformatics 19:1572–1574.

41. Röser M. 1989. Karyologische, systematische und chorologische Untersuchungen an der Gattung Helictotrichon Besser ex Schultes & Schultes im westlichen Mittelmeergebiet. Dissertationes Botanicae 145:1–250.

42. Röser M. 1994. Pathways of karyological differentiation in palms (Arecaceae). Plant Systematics and Evolution 189:83–122.
43. Röser M. 1996. Ecogeography of the grass genus Helictotrichon (Poaceae: Aveneae) in the Mediterranean and adjacent regions. Plant Systematics and Evolution 203:181–281.
44. Röser M, Winterfeld G, Grebenstein B, Hemleben V. 2001. Molecular diversity and physical mapping of 5S rDNA in wild and cultivated oat grasses (Poaceae: Aveneae). Molecular Phylogenetics and Evolution 21:198–217.
45. Sauer W. 1971. Zytologische Untersuchungen an Wildhafer-Sippen der Ostalpen. Helictotrichon Petzense, H. Parlatorei Und H. ×Krischae. Carinthia II 80:79–87.
46. Schönswetter P, Stehlik I, Holderegger R, Tribsch A. 2005. Molecular evidence for glacial refugia of mountain plants in the European Alps. Molecular Ecology 14:3547–3555.
47. Schwarzacher T, Ambros P, Schweizer D. 1980. Application of Giemsa banding to orchid karyotype analysis. Plant Systematics and Evolution 134:293–297.
48. Schweizer D. 1976. Reverse fluorescent chromosome banding with chromomycin and DAPI. Chromosoma 58:307–324.
49. Schweizer D, Ambros PF. 1994. Chromosome banding. In: Gosden JR, eds. Chromosome analysis protocols. Methods in molecular biology 29. Totowa: Humana Press, 97–112.
50. Swofford DL. 2002. PAUP*: Phylogenetic analysis using parsimony (*and other methods), version 4.0 beta. Sunderland, Massachusetts: Sinauer Associates.
51. Sumner AT. 1990. Chromosome banding. London: Unwin Hyman.
52. Vaio M, Speranza P, Vallis JF, Guerra M, Mazella C. 2005. Localization of the 5S and 45S rDNA sites and cpDNA sequence analysis in species of the Quadrifaria group of Paspalum (Poaceae, Paniceae). Annals of Botany 96:191–200.
53. Vollmann M. 1914. Flora von Bayern. Stuttgart: Eugen Ulmer.
54. Wendel JF, Schnabel A, Seelanan T. 1995. Bidirectional interlocus concerted evolution following allopolyploid speciation in cotton (Gossypium). Proceedings of the National Academy of Sciences of the USA 92:280–284.
55. Winterfeld G, Röser M. 2007a. Disposition of ribosomal DNAs in the chromosomes of perennial oats (Poaceae: Aveneae). Botanical Journal of the Linnean Society 155:193–210.
56. Winterfeld G, Röser M. 2007b. Chromosomal localization and evolution of satellite DNAs and heterochromatin in grasses (Poaceae), especially tribe Aveneae. Plant Systematics and Evolution 264:75–100.
57. Winterfeld G, Döring E, Röser M. 2009a. Chromosome evolution in wild oat grasses (Aveneae) revealed by molecular phylogeny. Genome 52:361–380.
58. Winterfeld G, Schneider J, Röser M. 2009b. Allopolyploid origin of Mediterranean species in Helictotrichon (Poaceae) and its consequences for karyotype repatterning and homogenization of rDNA repeat units. Systematics and Biodiversity 7:277–295.
59. Wölk A, Röser M. 2013. The new genus Trisetopsis and new combinations in oat-like grasses (Poaceae). Schlechtendalia 25:57–61.
60. Wölk A, Röser M. 2014. Polyploid evolution, intercontinental biogeographical relationships and morphology of the recently described African oat genus Trisetopsis (Poaceae). Taxon 63:773–788.
61. Wölk A, Winterfeld G, Röser M. 2015. Genome evolution in a Mediterranean species complex: phylogeny and cytogenetics of Helictotrichon (Poaceae) allopolyploids based on nuclear DNA sequences (rDNA, topoisomerase gene) and FISH. Systematics and Biodiversity 13:326–345.

CHAPTER 15

DIFFERENT AUTOSOMES EVOLVED INTO SEX CHROMOSOMES IN THE SISTER GENERA OF SALIX AND POPULUS

Jing Hou, Ning Ye, Defang Zhang, Yingnan Chen, Lecheng Fang, Xiaogang Dai & Tongming Yin

INTRODUCTION

Sex determination has the intriguing aspect in evolutionary and developmental biology. In contrast to the more completed sex chromosome evolution in animals, sex chromosomes in many plants are still at different evolutionary stages[1,2,3,4], and thus afford the opportunity of investigating the early stage of sex chromosome evolution. It is believed that sex chromosomes originate from a pair of autosomes[5,6], and the sex-determining systems in dioecious plants almost certainly evolved independently from ancestral hermaphrodites that lacked sex chromosomes[2,3]. Only about 4% of higher plants show full dioecism, with individuals of separate sexes[7,8,9].

Salicaceae is a family of dioecious woody plants, and the male or female flowers are arranged in morphologically different catkins (Figure 1) on the male or female trees[10]. *Salix* and *Populus* are the sister genera in*Salicaceae* (Figure 2). Genome analysis revealed that these two lineages originated from a common paleotetrapolyploid ancestor[11,12,13]. The chromosomes of *Salicaceae* are typically metacentric and small[14,15]. Based on cytological studies, there is no evidence of morphologically differentiated sex chromosomes in the *Salicaceae* species[16,17,18]. Various mechanisms have been proposed to explain the expression of gender in*Salicaceae*. Alström-Rapaport *et al.* proposed a multi-locus sex determination in the *Salix*, and that the presence of sex chromosomes was unlikely[19]. In contrast, Semerikov *et al.* reported that a single locus governed the sex determination and the female was the heterogametic gender in basket willow[20]. More recently, several studies revealed that there was a single locus located on chromosome XIX of *Populus*, with a peritelomeric localization in members of the *Aigeiros* subgenera[21,22] and a centromeric localization in subgenera of *Leuce*[23,24,25,26], appeared to contain a gene (genes) that controlled gender determination. Multiple lines of evidence suggested that poplar's chromosome XIX was in the process of evolving into an incipient sex chromosome[22,27]. The sex-determining region in poplar shows some interesting features, including location of gender locus, highly divergent haplotypes, severe recombination suppression, distinctive pattern of sRNA occurrence, and significantly fewer single nucleotide polymorphisms (SNP) than the rest of *Populus* genome[11,22,27]. Mapping studies on *P. deltoides*[22] and *P. alba*[26]revealed that sex determination occurred through a ZW system in which the female was the heterogametic gender[22]. However, the male sex was reported as the heterogametic gender in *P. nigra*[21] and *P. tremuloides*[23,24,25]. Thus, it is possible that both ZZ/ZW (female heterogamety) and XX/XY (male heterogamety) sex-determining systems could be present in some members of the family *Salicaceae*[27]. It has been confirmed that the sister genera of *Salix* and *Populus* originate from a common ancestor[11,12,13]. Since different gender determining systems probably evolve separately and quite recently in species of this family, *Salicaceae* is a desirable system to study the genetics and evolution of sex chromosomes in dioecious plants.

Citation: Jing Hou, Ning Ye, Defang Zhang, Yingnan Chen, Lecheng Fang, Xiaogang Dai & Tongming Yin, Different autosomes evolved into sex chromosomes in the sister genera of Salix and Populus, doi:10.1038/srep09076.

On willow and poplar, the male or female flowers are separately arranged in morphologically different catkins on the male or female trees. Photos taken by Tongming Yin and Jing Hou.

Figure 1: Flowers of the female and male trees in *Salicaceae*species.

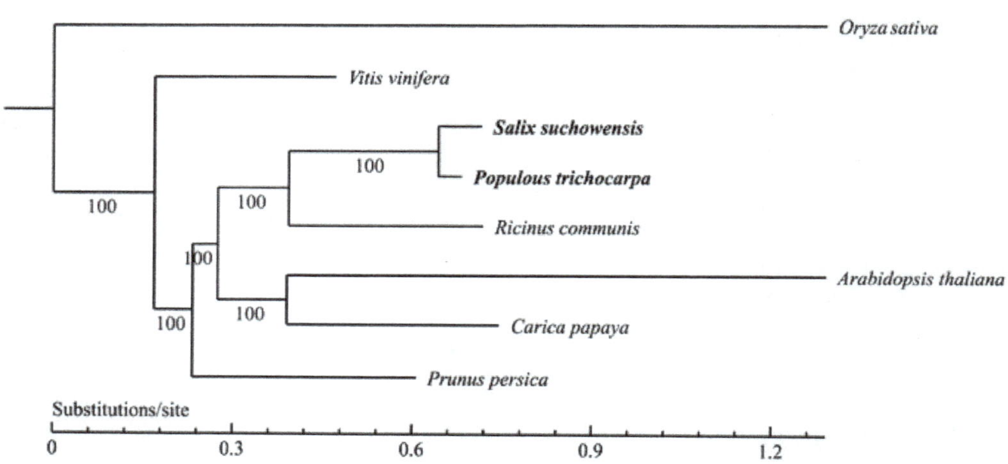

The phylogenetic tree was constructed with 1,881 single-copy genes on 4-fold degenerate sites. The branch length represents the neutral divergence rate. The posterior probabilities (credibility of the topology) for inner nodes are all 100%.

Figure 2: Phylogenetic tree of selected plant species.

S. suchowensis is an early flowering shrub willow that belongs to subgenus *Vetrix*28. Recently, its genome has been sequenced and publically available13. To gain further insight into the origin and evolution of sex chromosomes in *Salicaceae* species, we identified the sex chromosome in *S. suchowensis*, as well as compared the sex chromosomes between the sister genera of *Salix* and *Populus*.

RESULTS AND DISCUSSION

To map the gender locus of *S. suchowensis*, we established a large mapping pedigree in which the sequenced individual was the maternal parent. Among the progeny, 374 individuals were randomly selected for mapping and locating the gender locus in the willow genome. Investigation of sex phenotype confirmed constant gender for all of the mapping progeny across three locations in two continuous years. Among these individuals, 183 were female, 188 were male, and 3 remained sexually immature. Statistical analysis indicated that segregation of gender fit for the expected 1:1 Mendelian segregation ratio ($\chi^2 = 0.067$, where ($\chi^2 = 3.84$ for ≤ 0.05). Subsequently, genetic maps were separately constructed for the maternal and paternal parents by using amplified fragment length polymorphism (AFLP) markers and the pseudo-testcross strategy29. Totally, 1,137 1:1 segregating markers were generated by 90 primer combinations with the 374 mapping progeny. Among these markers, 494 testcross markers from the maternal parent were mapped into 266 bins distributed on 19 linkage groups, covering a genetic distance of 1924.5 cM (Supplementary Figure 1); Alternatively, 549 testcross loci from the male were assigned into 263 bins on 19 linkage groups, spanning a genetic distance of 2223.8 cM (Supplementary Figure 2). The coverage of the female and male maps was estimated to be 99.99% and 99.97% at 20 cM of a marker, respectively. Thus, the established maps achieved nearly complete genome coverage, with linkage group numbers equaled to the 19 haploid chromosome numbers in willow. Based on the established genetic maps, the gender locus was mapped as a 1:1 segregating morphological marker. Mapping results showed that gender locus could only be mapped on the maternal map, but was unmappable on the paternal map, indicating that the female was the heterogametic gender in willow, which was in agreement with the findings in *P. deltoides*22 and*P. alba*26 that revealed sex determination occurred through a ZW system, in which females are heterogametic22. It was also noteworthy that marker density in the vicinity of the gender locus (Supplementary Figure 1) was significantly higher than that expected by chance alone ($P \leq 0.0001$ with a Poisson calculator), which indicated severe recombination suppression around the gender locus. Recombination suppression has been recognized as a critical mechanism that triggered the divergence of the alternate sex chromosomes2·22·30·31.

To map the willow sequence scaffolds derived by Dai *et al.*13 along each chromosome of the willow genome, we carried out fine mapping with SNP markers that were generated by resequencing a subset of 80 progeny from the 374 mapping individuals. AFLP markers on the established maps were used as landmarks to anchor the SNP markers into the corresponding marker bins. In total, 2,548 testcross SNP markers from the maternal parent and 3,772 testcross SNP markers from the paternal parent were integrated into the female (Supplementary Figure 3) and male (Supplementary Figure 4) maps, respectively. Subsequently, SNP markers in each linkage group were used to map the willow sequence scaffolds derived by Dai *et al.*13 along the corresponding chromosome.

In total, 432 sequence scaffolds (covering a physical length of 194.1 Mb) and 630 sequence scaffolds (covering a physical length of 206.2 Mb) were mapped based on the female (Figure 3a) and male (Figure 3b) maps, respectively. There remained 7,009 unmapped sequence scaffolds larger than 2 kb, covering a total physical length of 38.4 Mb. Chromosome identities of the willow genome were designated by blasting willow sequences containing the mapped SNPs against the *P. trichocarpa* genome sequences. It was found that the linkage group containing the gender locus was chromosome XV in the willow genome (Figure 3a). This chromosome is an autosome in poplars. Thus, the willow›s sex chromosome corresponds to a poplar›s autosome. Previous studies revealed that chromosome XIX was the sex chromosome in poplars21·22·23·24·25·26, by contrast, chromosome XIX was identified as an autosome in willow.

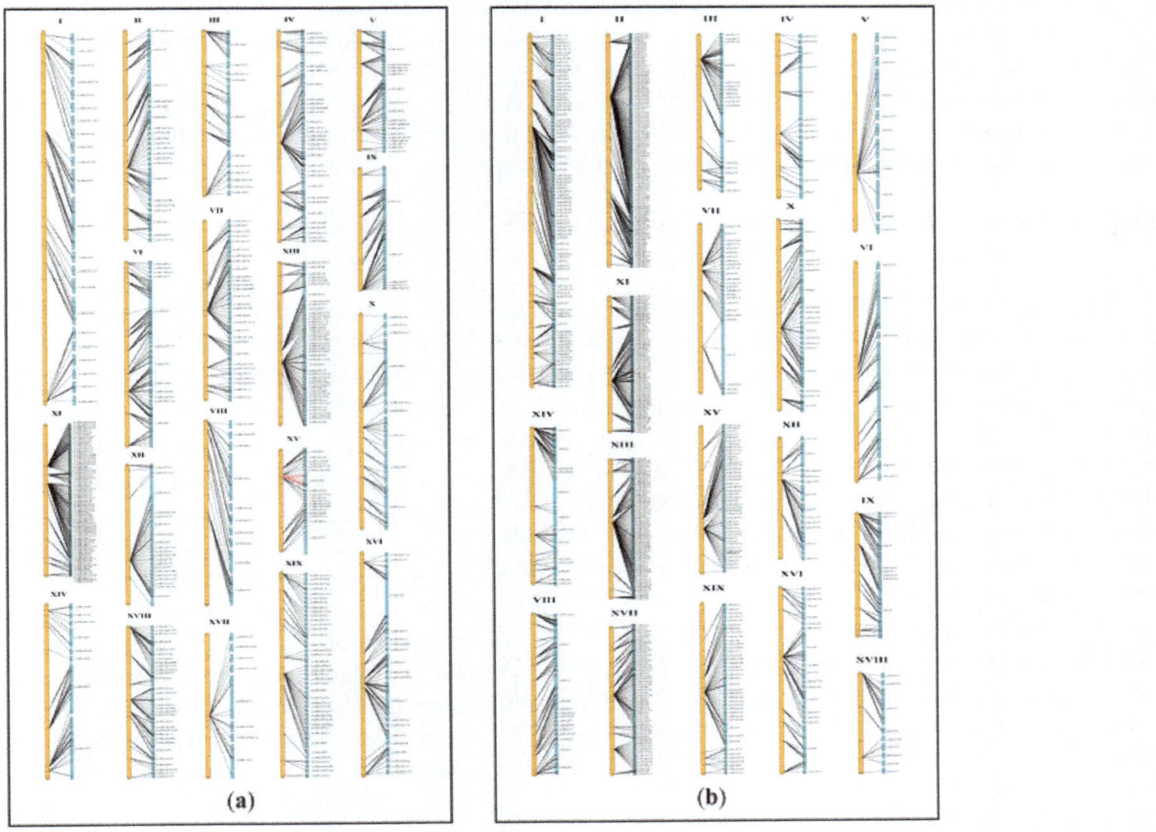

(a) anchor with the female map (b) anchor with the male map. Note: willow sequence scaffolds were obtained from Dai *et al.*'s study13. Sequence scaffolds were mapped according to the integrated SNP markers on the maternal genetic map of *S. suchowensis*. The orange bars on the left of each chromosome represented the linkage groups, and the discrete blue bars on the right represented the anchored sequence scaffolds. The mapped scaffolds were separated with evenly sized spaces, which did not represent the actual sizes of the uncaptured gaps. Chromosome identities were designated based on syntenic relationship between the willow's and poplar's chromosomes. As shown in Figure 3a, gender locus was mapped onto scaffold_64 on chromosome XV in the female *S. suchowensis*, and markers cosegregated with gender were indicated with red lines.

Figure 3: Anchoring the willow sequence scaffolds along each chromosome in the genomes of *Salix suchowensis*.

On willow's chromosome XV and chromosome XIX, 18 and 40 sequence scaffolds were mapped, respectively, along the corresponding chromosome based on the maternal map (Figure 3a). Because the sequenced individuals of both *P. trichocarpa* and *S. suchowensis* were females11·13, we first conducted syntenic analysis between sex chromosomes and the corresponding autosomes in these two lineages based on the willow's female map. Syntenic analysis revealed high collinearity on chromosome XIX (Figure 4a) and on chromosome XV (Figure 4b) between willow and poplar. In *P. trichocarpa*, complex haplotype divergence was observed between its alternate sex chromatids11·27. To explore whether the Z and W sex chromatids diverged with each other in willow, we examined the sequence scaffolds anchored on willow's chromosome XV based on the male map, which included 45 sequence scaffolds (Figure 3b). Contrary to the consistency revealed based on the female map (Figure 4b), syntenic analysis based on the male map revealed substantial chromosome rearrangements (Figure 4c), suggesting the occurrence of significant chromosome divergence between the Z and W chromatids in willow. We further compared willow's sex chromosome (XV) with poplar's sex chromosome (XIX), and found that sex chromosomes between these two lineages shared no syntenic chromosome segments (Figure 4d), indicating that sex chromosomes of poplar and willow originated from different ancestral chromosomes. Since sex chromosomes act as autosomes in the alternate genera of *Salicaceae*, we propose that turnover of autosomes into sex chromosomes occurred after the divergence of*Salix* and *Populus*.

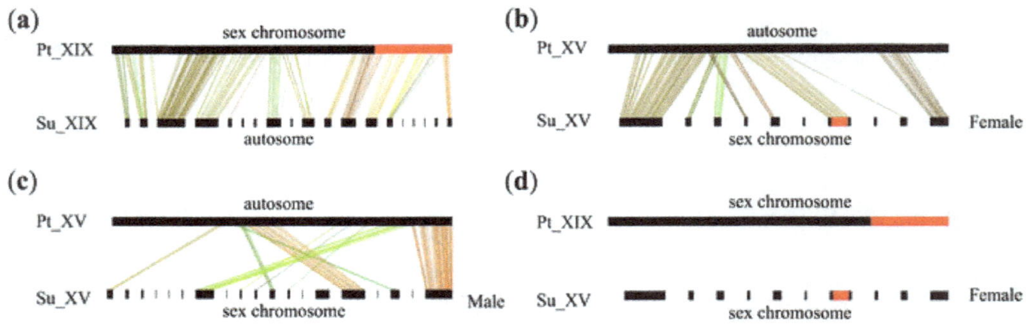

(a) synteny between poplar's and willow's chromosome XIX based on the female map. (b) synteny between poplar's and willow's chromosome XV based on the female map. (c) synteny between poplar's and willow's chromosome XV based on the male map. (d) synteny between poplar's chromosome XIX and willow's chromosome XV based on the female map. Note: Bar on the top of each chromosome pair represents poplar's chromosome. Bar at the bottom corresponds to the anchored sequence scaffolds on willow's chromosome. Only sequence scaffolds containing coding genes were included in this analysis. The red portions of the chromosome bars represent the sex-determining regions.

Figure 4: Syntenic analysis between sex chromosomes and the corresponding autosomes across *Populus trichocarpa* **and** *Salix suchowensis*.

Because evidence indicated that chromosome XV and chromosome XIX separately evolved into sex chromosomes after the divergence of willow and poplar, we proposed that turnover of different autosomes into sex chromosomes might be due to mobile of the sex-determining gene from chromosome XIX to chromosome XV in willow. Alternatively, the sex-determining gene may have translocated from chromosome XV to chromosome XIX in poplar. Therefore, there might be homologous sex-determining genes within the sex-determining regions between willow and poplar. Mapping of SSR markers revealed that the gender locus was in between SSR markers SSR151 (151,741 bp on scaffold 64) and SSR893 (893,817 bp on scaffold 64) (Figure 4b). In *P. nigra*21 and *P. deltoides*22, the close relatives to *P. trichocarpa* (the sequenced poplar species), gender locus was consistently mapped to the peritelomeric region upper the position of SSR marker O_206 (Figure 4a). We subsequently searched the homologous genes shared by these two regions between willow and poplar, and nine willow genes were found to have a total of 52 homologous genes in the sex-determining region of *P. trichocarpa*. However, all these poplar genes had homologous genes in other regions of the poplar genome, and none of these genes were found to be involved in sex determination based on gene annotation (Supplementary Table 1).

In recent studies, RNA-based sex-determining mechanisms have been found to play a significant role in higher plants27·32·33·34. For example, the tasselseed4 miRNA, i.e., miR172, was found to be involved in the sex determination of the male inflorescence in maize32·33. Interestingly, the miR172 family was conserved in *Populus*35 and located on the peritelomeric end of chromosome XIX27. We searched the homolgous sequences of miR172 in willow, and none of them appeared in the sex-determining region. More recently, Akagi *et al.* reported that a sex-determining candidate (*OGI*) encoded a small RNA targeting a feminizing gene (*MeGI*) in persimmons (*Diospyros lotus*), and the encoded small RNA acted as a sex determinant in this plant34. Search homologous genes of *OGI* and *MeGI* in poplar and willow showed that no homologous genes were found to be distributed on the sex chromosomes in either lineage (Supplementary Table 2). With all of these efforts, we have not identified the likely sex determinants in*Salicaceae*. The involvement of small RNAs in sex determination might explain why they have been difficult to identify to date34. RNA-based sex-determining mechanisms maybe present in *Salicaceae* species as well, thus small RNAs will be the primary focus in our future studies.

In plants, species with heterogametic females are less common36. Therefore, *Salix* species should represent a unique system for studying dioecy evolution in flowering plants. Although the sex-determining gene (genes) has not been identified with the current data, this work provides unique insight into the sex determination in *Salicaceae* and also sheds new light on understanding the genetics and evolution of sex chromosome in dioecious plants.

METHODS

Plant materials

An F_1 mapping pedigree with 2,546 progeny was established in 2010. The maternal parent of this pedigree was described by Dai *et al.*13, whereas the paternal parent was a two-year-old *S. suchowensis* collected from Linshu, Shandong Province of China. To map the gender locus, 374 offspring were randomly selected, and cuttings of the selected individuals were planted at three different locations in China: Sihong and Lishui in Jiangsu Province, as well as Shishou in Hubei Province. The field trials were established in 2012, and three replicates were planted for each progeny at each location.

Map construction

AFLP genotyping was conducted as described by Yin *et al.*37. GeneMapper Software (Version 3.7, Applied Biosystems) was used to score the amplicons in range of 50 ~ 500 bp. Map construction, genome length estimation, and map coverage calculation were performed following the description in Yin *et al.*37. A Poisson probability test was conducted following the description in Remington *et al.*38. The map charts were produced with MapChart 2.139.

Generating SNP markers

SNP markers were generated with a subset of 80 progeny from the 374 mapping individuals. The combination of two restriction enzymes, EcoRI and MseI (Takara Inc.), were used to reduce the complexity of the willow genome. For PCR amplification, the EcoRI primer had no selective nucleotide, and the MseI primer had a "G" selective nucleotide. The PCR amplification was conducted as described by Yin *et al.*37.

The sequencing library was constructed according to the manufacturer's standard protocol (Illumina, Inc.). Library quantification was carried out with a Quant-iTTM PicoGreen dsDNA Kit (Invitrogen, Inc.), and library qualify was evaluated with an Agilent 2100 Bioanalyzer (Agilent, Inc.). For each lane of a flow cell, 12 index-labeled samples were multiplexed, which enabled the sequencing of 84 samples (80 progeny and two replicates of the parents) in a full run, and the eighth lane was used for the control. Sequencing was performed on an Illumina HiSeq 2000 (Illumina, Inc.) at Nanjing Agricultural University following the manufacturer's protocols (Illumina, Inc.).

Raw sequencing data generated by the Illumina platform were analyzed with an integrated bioinformatics pipeline. The SAMTOOLS (http://samtools.sourceforge.net)40 and Varscan (http://varscan.source-forge.net) were combined to call the SNPs with default parameters41.

Anchoring the sequence scaffolds

AFLP markers on the established genetic maps were used to pull out tightly linked SNP markers with an LOD threshold ≥10, and these SNPs were assigned to the corresponding marker bins that showed the strongest linkage on each map. Subsequently, the mapped SNPs were used to anchor willow sequence scaffolds that were produced by Dai *et al.*13. Finally, willow sequences containing these SNPs were blasted against the *P. trichocarpa* genome sequences (ftp://ftp.jgi-psf.org/pub/compgen/phytozome/v9.0/Ptrichocarpa/) to designate the chromosome identities for the obtained linkage groups on the male and female maps.

Synteny and collinearity analyses

Genome sequences, Gff3 file, and the protein sequences of *P. trichocarpa* were retrieved from JGI website (ftp://ftp.jgi-psf.org/pub/compgen/phytozo- me/v9.0/Ptrichocarpa/). The corresponding information for willow was downloaded from the *S. suchowensis* website (115.29.234.170/willow). Synteny and collinearity analyses were performed with bioinformatics toolkit of MCScanX42. First, coding genes in willow and poplar genome were blasted by using BLASTP43 with an e-value cutoff of 1e−10, which produced a file in blast8 format. Then, an in-house perl script was used to filter the gff3 files and reformat the files to make them compatible with MCScanX. Finally, MCScanX produced the collinearity results, with the following parameters: −k 50, −m 25, −e 1e−5, where −k is the match score, −m is the maximum gaps, and −e is the e value cutoff. Dual_synteny_plotter in the MCScanX package was used to draw the collinearity charts.

ACKNOWLEDGEMENTS

This work was supported by the National Basic Research Project (2012CB114505), the Natural Science Foundation of China (31125008, 31400564) and Jiangsu Province (BK20130968). It was also enabled by the Program for Innovative Research Team (the Educational Department of China, the Universities of Jiangsu Province), the Doctorate Fellowship Foundation and PAPD (Priority Academic Program Development) program at Nanjing Forestry University.

REFERENCES

1. Jamsari, A., Nitz, I., Reamon-Büttner, S. M. & Jung, C. BAC-derived diagnostic markers for sex determination in asparagus. *Theor. Appl. Genet.* 108, 1140–1146 (2004).
2. Liu, Z. Y. *et al.* A primitive Y chromosome in papaya marks incipient sex chromosome evolution. *Nature* 427, 348–352 (2004).
3. Filatov, D. A. Evolutionary history of *Silene latifolia* sex chromosomes revealed by genetic mapping of four

genes.*Genetics* 170, 975–979 (2005).

4. Alexandrov, O. S., Divashuk, M. G., Yakovin, N. A. & Karlov, G. I.Sex chromosome differentiation in *Humulus japonicus* Siebold & Zuccarini, 1846 (Cannabaceae) revealed by fluorescence *in situ*hybridization of subtelomeric repeat. *Comp. Cytogenet.* 6, 39–247 (2012).
5. Smith, C. A. & Sinclair, A. H. Sex determination: insights from the chicken. *Bioessays* 26, 120–132 (2004).
6. Carvalho, A. B. & Clark, A. G. Y chromosome of *D. pseudoobscura* is not homologous to the ancestral *Drosophila* Y. *Science* 307, 108–110 (2005).
7. Lloyd, D. G & Webb, C. J. Secondary sex characters in plants. *Bot. Rev.* 43, 177–216 (1977).
8. Guttman, D. S. & Charlesworth, D. An X-linked gene with a degenerate Y-linked homologue in a dioecious plant. *Nature* 393, 263–266 (1998).
9. Heslop-Harrison, J. S. & Schwarzacher, T. Organisation of the plant genome in chromosomes. *Plant J.* 66, 18–33 (2011).
10. Rechinger, K. H. *Salix* taxonomy in Europe-problems, interpretations, observations. *Proceedings of the Royal Society of Edinburgh. Section B. Biological Sciences* 98, 1–12 (1992).
11. Tuskan, G. A. *et al.* The genome of black cottonwood, *Populus trichocarpa* (Torr. & Gray). *Science* 313, 1596–1604 (2006).
12. Berlin, S. *et al.* High-density linkage mapping and evolution of paralogs and orthologs in *Salix* and *Populus*. *BMC Genomics* 11, 129 (2010).
13. Dai, X. G. *et al.* The willow genome and divergent evolution from poplar after the common genome duplication. *Cell Res.* 24, 1274–1277 (2014).
14. Blackburn, K. B. & Harrison, J. W. H. A preliminary account of the chromosomes and chromosome behavior in the *Salicaceae*. *Ann. Bot.* 38, 361–378 (1924).
15. Islam-Faridi, M. N., Nelson, C. D., DiFazio, S. P., Gunter, L. E. &Tuskan, G. A. Cytogenetic analysis of *Populus trichocarpa*-ribosomal DNA, telomere repeat sequence, and marker-selected BACs. *Cytogenet. Genome Res.* 125, 74–80 (2009).
16. Peto, F. H. Cytology of poplar species and natural hybrids. *Can. J. Forest. Res.* 16, 445–455 (1938).
17. Van Dillewijn, C. Zytologische Studien in der Gattung *Populus* L.*Genetica* 22, 131–182 (1940).
18. Van Buijtenen, J. P. & Einspahr, D. W. Note on the presence of sex chromosomes in *Populus tremuloides*. *Bot. Gaz.* 121, 60–61 (1959).
19. Alström-Rapaport, C., Lascoux, M. & Gullberg, U. Sex determination and sex ratio in the dioecious shrub *Salix viminalis*L. *Theor. Appl. Genet.* 94, 493–497 (1997).
20. Semerikov, V. *et al.* Genetic mapping of sex-linked markers in*Salix viminalis* L. *Heredity* 91, 293–299 (2003).
21. Gaudet, M. *et al.* Genetic linkage maps of *Populus nigra* L. including AFLPs, SSRs, SNPs, and sex trait. *Tree Genet. Genomes* 4, 25–36 (2008).
22. Yin, T. M. *et al.* Genome structure and emerging evidence of an incipient sex chromosome in *Populus*. *Genome Res.* 18, 422–430 (2008).
23. Markussen, T., Pakull, B. & Fladung, M. Positioning of sex correlated markers for *Populus* in an AFLP- and SSR-marker based genetic map of *Populus tremula* × *tremuloides*. *Silvae Genet.*56, 180–184 (2007).
24. Pakull, B., Groppe, K., Meyer, M., Markussen, T. & Fladung, M.Genetic linkage mapping in aspen (*Populus tremula* L. and *Populus tremuloides* Michx.). *Tree Genet. Genomes* 5, 505–515 (2009).
25. Pakull, B. *et al.* Genetic mapping of linkage group XIX and identification of sex-linked SSR markers in a *Populus tremula*×*Populus tremuloides* cross. *Can. J. Forest. Res.* 41, 245–253 (2011).
26. Paolucci, I. *et al.* Genetic linkage maps of *Populus alba* L. and comparative mapping analysis of sex determination across*Populus* species. *Tree Genet. Genomes* 6, 863–875 (2010).
27. Tuskan, G. A. *et al.* The obscure events contributing to the evolution of an incipient sex chromosome in *Populus*: a retrospective working hypothesis. *Tree Genet. Genomes* 8, 559–571 (2012).
28. Smart, L. B. & Cameron, K. D. Genetic improvement of willow (*Salix* spp.) as a dedicated bioenergy crop. In: Vermerris W editors. *Genetic improvement of bioenergy crops*. New York: Springer (2008).
29. Grattapaglia, D. & Sederoff, R. R. Genetic-linkage maps of*Eucalyptus grandis* and *Eucalyptus urophylla* using a pseudotestcross-mapping strategy and RAPD markers. *Genetics*137, 1121–1137 (1994).
30. Ohno, S. Sex chromosomes and sex-linked genes. Berlin: Springer (1967).
31. Nei, M. Linkage modification and sex difference in recombination.*Genetics* 63, 681–699 (1969).
32. Chuck, G., Meeley, R., Irish, E., Sakai, H. & Hake, S. The maize*tasselseed4* microRNA controls sex determination and meristem cell fate by targeting *Tasselseed6/indeterminate spikelet1*. *Nat. Genet.* 39, 1517–1521 (2007).
33. Banks, J. MicroRNA, sex determination and floral meristem determinacy in maize. *Genome Biol.* 9, 204 (2008).
34. Akagi, T. *et al.* A Y-chromosome–encoded small RNA acts as a sex determinant in persimmons. *Science* 346, 646–650 (2014).
35. Barakat, A., Wall, P. K., Diloreto, S., Depamphilis, C. W. & Carlson, J. E. Conservation and divergence of microRNAs in *Populus*. *BMC Genomics* 8, 481–496 (2007).
36. Charlesworth, D. Plant sex chromosome evolution. *J. Exp. Bot.* 64, 405–420 (2013).
37. Yin, T. M., DiFazio, S. P., Gunter, L. E., Riemenschneider, D. &Tuskan, G. A. Large-scale heterospecific segregation distortion in*Populus* revealed by a dense genetic map. *Theor. Appl. Genet.* 109, 451–463 (2004).

38. Remington, D. L., Whetten, R. W., Liu, B. H. & O'Malley, D. M.Construction of an AFLP genetic map with nearly complete genome coverage in *Pinus taeda*. *Theor. Appl. Genet.* 98, 1279–1292 (1999).

39. Voorrips, R. E. MapChart: software for the graphical presentation of linkage maps and QTLs. *J. Hered.* 93, 77–78 (2002).

40. Li, H. *et al.* The sequence alignment/map format and SAMtools.*Bioinformatics* 25, 2078–2079 (2009).

41. Koboldt, D. C. *et al.* VarScan: variant detection in massively parallel sequencing of individual and pooled samples.*Bioinformatics* 25, 2283–2285 (2009).

42. Wang, Y. P. *et al.* MCScanX: a toolkit for detection and evolutionary analysis of gene synteny and collinearity. *Nucleic Acids Res* 40, e49–e49 (2012).

43. Altschul, S. F., Gish, W., Miller, W., Myers, E. W. & Lipman, D. J.Basic local alignment search tool. *J. Mol. Biol.* 215, 403–410 (1990).

CHAPTER 16

EVOLUTION BY GENE DUPLICATION OF MEDICAGO TRUNCATULA PISTILLATA-LIKE TRANSCRIPTION FACTORS

Edelín Roque, Mario A. Fares, Lynne Yenush, Mari Cruz Rochina, Jiangqi Wen, Kirankumar S. Mysore, Concepción Gómez-Mena, José Pío Beltrán and Luis A. Cañas

Keywords: Duplicated B-function genes, functional analyses, MADS-box transcription factors, Medicago truncatula, molecular evolution, PISTILLATA-like.

INTRODUCTION

Transcription factors (TFs) involved in developmental innovations are a paradigmatic case of evolution by gene duplication, with most TF duplications being concomitant with the emergence of major morphological innovations in angiosperms (Purugganan et al., 1995;Kramer et al., 1998; Aoki et al., 2004; Kim et al., 2004). A particularly attractive model system suited to address the mechanisms of gene diversification in plant evolution is provided by the large family of MADS-box TFs (Theissen et al., 2000; Airoldi and Davies, 2012).

Flower development is controlled by complex network interactions between TFs, most of them belonging to the MADS-box gene family (Schwarz-Sommer et al., 1990; Wellmer et al., 2014). These proteins form complexes that often interact with their own promoters and those of their targets to regulate their own and each other's expression. They contain an N-terminal DNA-binding MADS domain, followed by the intervening (I) and keratin-like (K) regions, the latter being essential for dimerization and formation of high-order complexes (Yang et al., 2003). Moreover, the highly variable C-terminal domain may have a role in protein complex formation and transcriptional regulation (reviewed by Kaufmann et al., 2005).

B-function MADS-box genes are involved in the specification of petal and stamen identity in probably all angiosperms (Litt and Kramer, 2010). InArabidopsis thaliana, this function is encoded by the combined activities of PISTILLATA (PI) and APETALA3 (AP3) (Bowman et al., 1989, 1991; Jack et al., 1992). PI and AP3 represent lineages that arose from a duplication event after the split between extant gymnosperms and angiosperms. Although several duplication events of the PI lineage have been documented throughout angiosperm evolution (Kramer et al., 1998, 2003;Kramer and Irish, 2000; Kim et al., 2004; Stellari et al., 2004; Viaene et al., 2009), few studies are available that combine functional and molecular analyses to understand the evolutionary forces that act on these duplicate gene lineages and the outcomes that ensure (Des Marais and Rausher, 2008).

Classic theory predicts that the genetic redundancy emerging from the duplication of genes relaxes the selection constraints on one gene copy allowing it to explore the genotype space and probe a wide range of phenotypes (Payne and Wagner, 2014). The evolutionary instability of genetic redundancy guarantees, however, the return of most duplicated genes to single-copy genes (Wolfe and Shields, 1997), which undermines the importance of the role of gene duplication in biological innovation. In contrast to this prediction, up to 30% of the genes in some organisms are duplicates (Blanc and Wolfe, 2004; Cui et al., 2006), sparking the idea that a number of factors allow duplicated genes to persist, and thus innovate, within the genomes. Determining these factors remains a major goal of evolutionary biology. Expression and functional diversification of TFs are essential to allow the persistence in duplicate (Singh and Hannenhalli, 2008). The link between gene expression and the rate of evolution (Drummond et al.,

Citation: Edelín Roque, Mario A. Fares, Lynne Yenush, Mari Cruz Rochina, Jiangqi Wen, Kirankumar S. Mysore, Concepción Gómez-Mena, José Pío Beltrán and Luis A. Cañas, Evolution by gene duplication of Medicago truncatula PISTILLATA-like transcription factors, doi: 10.1093/jxb/erv571.

2005; Gout et al., 2010), however plausible in TFs, remains a major unknown in the evolution of TFs after duplication.

Models have been proposed to explain how selection operates on duplicated genes (Innan and Kondrashov, 2010). Studies have identified many factors influencing the functional fates of duplicates, including selection for increased dosage (Conant and Wolfe, 2008), stoichiometric and gene balance (Birchler et al., 2005; Freeling and Thomas, 2006), the position of genes in the protein interaction network (Alvarez-Ponce and Fares, 2012), and whether duplicates are the result of whole-genome, small-scale duplications or retropositions (Carretero-Paulet and Fares, 2012; Fares et al., 2013; Keane et al., 2014).

Arabidopsis homeotic floral genes are conserved in model legumes (Hechtet al., 2005). Medicago truncatula contains four B-function MADS-box genes: two AP3-like (MtNMH7 and MtTM6) (Roque et al., 2013) and twoPI-like paralogs (MtPI and MtNGL9) (Benlloch et al., 2009), thus representing a good model system to study the effects of gene duplication and functional divergence within the B-function MADS-box lineages. Our previous functional characterization of the Mt AP3-like genes revealed that these paralogs have undergone a qualitative subfunctionalization process, concomitant with a complete partitioning of the expression pattern of the ancestral gene lineage (Roque et al., 2013). Previous functional characterization of MtPI, one member of the duplicated PI-like genes, suggested that MtPI plays a major role in the specification of petal and stamen identity in M. truncatula (Benlloch et al., 2009). However, the contribution of MtNGL9 to floral development was unclear and its functional fate after gene duplication remains unknown. Here, we have integrated both functional analyses and molecular evolution studies to assess comprehensively the divergence in the B-function of duplicated M. truncatula PI-like genes and determine their mode of evolution after duplication.

MATERIALS AND METHODS

Plant material and growth conditions

Medicago truncatula cv. Jemalong lines R108 and A17, Medicago sativa cv. Regen SY-27, Pisum sativum cv. Alaska plants, and Arabidopsis thalianacv. Landsber erecta (Ler) plants were used in this study. Leguminous plants were grown in the greenhouse at 22 °C (day) and 18 °C (night) with a 16h light/8h dark photoperiod, in a mixture of soil/sand (3:1). Arabidopsis plants were grown in growth chambers at 21 °C under long-day (16h light) conditions, in a mixture of 1:1 perlite:vermiculite. Plants were irrigated with Hoagland No. 1 solution supplemented with oligoelements (Hewitt, 1966). The mtpi-2 mutant allele was isolated in a previous screening (Cheng et al., 2014).

Isolation and sequence analysis of *PsNGL9* and *MsPI*

PsNGL9 and MsPI coding regions were isolated to use their sequences in the molecular evolution approaches. PsNGL9 (KJ470632) was obtained by reverse transcription–PCR (RT–PCR) from Pisum sativum floral cDNA samples using the primers MtNGL9-ATG/MtNGL9-638 (see Supplementary Table S4 at JXB online). MsPI (KJ470631) was obtained by RT–PCR from M. sativa RSY-27 floral cDNA samples using primers MtPI-ATG/MtPI-548 (seeSupplementary Table S4). The MsPI 3'-untranslated region (UTR) and C-terminal region were obtained using the 3' rapid amplification of cDNA ends (RACE) system (Invitrogen) with nested gene-specific primer MsPI-363 (Supplementary Table S4). Sequence alignment and similarity comparisons of the coding region and inferred proteins were performed using CLUSTALW. Sequence alignments and similarity comparisons of several PI-like proteins were performed using the UNIPROT website (http://www.uniprot.org). MADS-domains (positions 1–61) were defined by the Prosite database (http://prosite.expasy.org/) using the AtPI protein sequence as the query.

Phylogenetic tree

The evolutionary history was inferred by using the maximum likelihood method based on the JTT matrix-based model (Jones et al., 1992). The tree with the highest log-likelihood value (–5459.065) is shown. Initial tree(s) for the heuristic search were obtained automatically by applying Neighbor-Joining and BioNJ algorithms to a matrix of pairwise distances estimated using the JTT model, and then selecting the topology with the superior log-likelihood value. A discrete Gamma distribution was used to model evolutionary rate differences among sites (5 categories +G, parameter=1.447). The rate variation model allowed for some sites to be evolutionarily invariable ([+I], 0.895% sites). The tree is drawn to scale, with branch lengths measured as the number of substitutions per site. The analysis involved 70 amino acid sequences obtained from GenBank, and two PI-like sequences that we have isolated from M. sativa and P. sativum(Supplementary Table S1). All positions containing gaps and missing data were eliminated. There were a total of 91 positions in the final data set. The substitution rates of the PI-like genes compared among different paralogs were inferred from the phylogenetic tree using the relative rate test implemented in MEGA 5 (Tamura et al., 2011).

RNA *in situ* hybridization

RNA in situ hybridization with digoxigenin-labeled probes was performed on 8 μm longitudinal paraffin sections of M. truncatula inflorescences as described previously (Ferrandiz et al., 2000). A 298bp fragment of MtPI(504–801 from ATG) and a 243bp fragment of MtNGL9 (511–753 from ATG) were introduced into the pGEM-T Easy vector. Digoxigenin-labeled RNA antisense and sense probes were synthesized by in vitro transcription using T7 and SP6 RNA polymerases, respectively. Signal was detected as a purple precipitate when viewed under the light microscope.

Southern blot hybridization

Plant genomic DNA was extracted from leaves of M. truncatula cv. Jemalong, line A17 using standard procedures. A 10 μg aliquot of DNA was digested with EcoRI, BamHI, and HindIII, and separated on a 0.7% agarose gel overnight. Southern blot hybridization was performed by the standard method using two different conditions (52 °C and 65 °C). cDNA probes were isolated by PCR using the primer pairs MtNGL9-511/MtNGL9-753 and MtPI-504/MtPI-801 for the MtNGL9 and MtPI genes, respectively (see Supplementary Table S4).

Yeast two-hybrid analysis

For construction of the two-hybrid plasmids, the cDNAs of the entire coding region of the MtPI and MtNGL9 genes were subcloned into the pBTM116 two-hybrid vector (Vojtek et al., 1993) to generate in-frame fusions with the LexA DNA-binding domain (BD), and the IKC fragments ofMtNMH7 and MtTM6 genes were subcloned in pACT2 (Clontech) to generate in-frame fusions with the Gal4 transcriptional activation domain (AD). BamHI/SalI sites in the case of the pBTM116 constructs andBamHI/EcoRI or BamHI/XhoI sites in the case of the pACT2 constructs were added to primers used in the PCRs (see Supplementary Table S4). The yeast two-hybrid strain CTY10-5d was co-transformed with the appropriate constructs and transformants were selected on minimal medium (SD-Leu-Trp). Selection for interaction was performed on minimal medium (SD-Leu-Trp) containing X-Gal (80mg l^{-1}) and 1 × BU salts [$Na_2HPO_4 \cdot 7H_2O$ (7g l^{-1}); NaH_2PO4 (3g l^{-1})]. Growth of yeast for blue staining was scored after 60h of incubation at 30 °C (Fig. 4, bottom). Equivalent expression levels of MtPI and MtNGL9 LexA fusions were confirmed by western blot. In order to be able to quantify protein–protein interactions, we used a β-gal liquid assay with orthonitrophenyl-β-D-galactopyranoside (ONPG) as the substrate, measuring absorbance at 405nm. β-Gal activity was calculated essentially as described (Ludin et al., 1998).

Molecular characterization of the *Tnt1* insertion mutants *mtngl9-1* and *mtngl9-2*

The M. truncatula population used for the screening of mutants has been described in detail (Roque et al., 2013; Cheng et al., 2014; Serwatowska et al., 2014). The mtngl9 alleles were identified by PCR screening of a segregating population of ~10 000 independent lines, using primers annealing to the MtNGL9 sequence (NGL9-F, Supplementary Table S4) in combination with primers annealing to the LTR borders of the Tnt1retroelement (Tnt1-R; Supplementary Table S4). PCR products (seeSupplementary Fig. S3B) were obtained and cloned into the pGEM-T easy vector for sequencing. The Tnt1 insertion in mtngl9-1 is located at 360bp from the start codon, in the second exon (Supplementary Fig. S2A) and theTnt1 insertion in mtngl9-2 is located at 183bp from the start codon, at the end of the first exon (Supplementary Fig. S2A). The R1 plants were genotyped using the following primer pairs: NGL9-F/Tnt1-R which amplified the T-DNA insertion, and NGL9-F/NGL9-590G or NGL9-F/NGL9-612G for the wild-type fragment in mtngl9-1 and mtngl9-2, respectively (Supplementary Table S4; Supplementary Fig. S2B, C). Heterozygous lines NF14948.1 and NF14948.2 were self-pollinated. Approximately a quarter of the resultant progeny co-segregated with theTnt1 insertion. Gene expression of MtNGL9 in the homozygous Tnt1mutants was performed by RT–PCR (Supplementary Table S4;Supplementary Fig. S2D).

Arabidopsis thaliana transformation and genotyping

MtPI and MtNGL9 cDNAs were cloned into the SalI/BamHI sites of the pBINJIT60 vector (Guerineau and Mullineaux, 1993), a pBIN19 derivate (Clontech Laboratories). We used the primers MtPI-SalI/MtPI-BamHI forMtPI cloning, and MtNGL9-SalI/MtNGL9-BamHI for MtNGL9 cloning (Supplementary Table S4). The transcription of MtPI and MtNGL9 is under the control of a tandem repeat of the 35S promoter of Cauliflower mosaic virus. This vector was inserted into the Agrobacterium tumefaciens strain GV3101::pMP90(RK) and Arabidopsis plants were transformed according to standard procedures (Bechtold and Pelletier, 1998). For each construct, kanamycin-resistant lines were used for phenotypic and molecular characterization. Heterozygous transgenic lines 35S::MtPI (4) and35S::MtNGL9 (22) were used as the pollen donor for crosses to the homozygous mutant pi-1 (Bowman et al., 1989). The resulting progeny were allowed to self-fertilize, and plants containing the transgenes and homozygous for the pi-1 allele (pi-1/pi-1;35S::MtPI or pi-1/pi-1;35S::MtNGL9) were identified in the next generation. To genotype thepi-1 mutation, we designed a CAPS (cleaved amplified polymorphic sequences) marker using the primers CAPS-FOR/CAPS-REV (Supplementary Table S4) that amplify 729bp and the BseGI enzyme thats cut only the wild-type PI sequence (Supplementary Fig. S6).

RT–PCR analysis

Total RNA was isolated from floral apices from wild-type M. truncatulaR108, M. sativa cv. RSY-27, and P. sativum cv. Alaska, and from mtngl9and mtpi-2 mutant flowers. Also, we isolated total RNA from leaves or floral buds of 35S::MtPI, 35S::MtNGL9, and pi-1/pi-1;35S::MtPI/MtNGL9, and Arabidopsis (Ler) plants. We used the RNeasy Plant mini Kit (Qiagen) according to the manufacturer's instructions. Total RNA was treated with rDNaseI of the DNase Treatment and Removal Kit (Ambion). For first-strand synthesis, total RNA (1 µg) was reverse transcribed in a 20 µl reaction mixture using the PrimerScript 1st strand cDNA Synthesis Kit (Takara). Aliquots of each cDNA were used as a template for real-time PCR or semi-quantitative RT–PCR with gene-specific primers (Supplementary Table S4). For real-time RT–PCR analysis, 1 µl of the reverse transcription reaction was used with 300nM of each primer mixed with the Power SYBR®Green PCR Master Mix (Applied Biosystems) according to the manufacturer's instructions. The reaction was carried out in 96-well optical reaction plates using an ABI PRISM 7500 Sequence Detection System and appropriate software (Applied Biosystems). The relative levels were determined by the $2^{-\Delta\Delta}$Ct method. In all cases, the efficiency of the primers was analyzed. Each experiment was done with two biological replicates, each one with three technical replicates. To normalize the variance among samples, we used Secret Agent (O-linked N-acetyl glucosamine transferase: TC77416) (Hartweck et al., 2002) for M. truncatula, PsActin-11 (Zhang et al., 2013) for P. sativum samples, elongation factor MsEF-1 for M. sativa, and Ubiquitin10 (UBQ10, At4g05320) (Czechowski et al., 2005) for A. thaliana. All used primers are listed in Supplementary Table S4.

Light microscopy and cryo-SEM

Images of M. truncatula R108, the mtpi-2 and mtngl9 mutant, Arabidopsis (Ler) wild type, 35S::MtPI, 35S::MtNGL9, and the pi-1 complementation lines flowers were obtained as described previously (Roque et al., 2007). For cryo-scanning electron microscopy (SEM), samples were frozen in slush nitrogen and attached to the specimen holder of a CT-1000C cryo-transfer system interfaced with a JEOL JSM-5410 scanning electron microscope. The samples were then transferred from the cryostage to the microscope sample stage, where the condensed surface water was sublimed by controlled warming to –85 °C. Afterwards, the sample was transferred back to the cryostage for gold coating by sputtering. Finally, the sample was returned to the microscope sample stage and viewed at an accelerating voltage of 15 keV.

Analysis of adaptive evolution

We tested whether MtPI and MtNGL9 gene copies diverged through adaptive changes in their protein-coding gene sequences after the duplication of their ancestral gene. To this end, we ran four maximum likelihood-based models, all of which are implemented in the program CODEML from the package PAML version 4.7 (Yang, 2007). These models are based on the calculation of the log-likelihood value of a particular model that supports a given hypothesis and that allows determination of the strength of selection on the gene by estimating the non-synonymous to synonymous rates ratio ($\omega = d_N/d_S$). Values of $\omega=1$, $\omega>1$, and $\omega<1$ indicate neutral evolution, positive selection, and purifying selection, respectively. For each of the models, we calculate a likelihood value and then compare the log-likelihood values among models by the likelihood ratio test (LRT), such that twice the difference in the likelihood values between two models under comparison can be approached to a χ^2distribution, with the degrees of freedom (df) being the number of parameters estimated by the most complex model and which are not estimated in the simplest model. We calculated the likelihood value for four models. The Goldman and Young model (Goldman and Young, 1994) assumes a single ω value for the entire tree and alignment. The free-ratio model (FRM) assumes a single ω for the alignment, but allows this value to be inferred for each branch of the tree, so that branches with adaptive evolution ($\omega>1$) can be identified using this model. The branch site model (BSM) allows for ω values to change among tree branches and codon sites, and this model is compared with its null model in which ω is forced to be $\omega=1$. The model that better fits the data and phylogeny is identified comparing the Goldman and Young model with the FRM, and the BSM with its null model.

Analysis of functional divergence

We used two methods to identify functional divergence, one based on a maximum likelihood framework (Gu, 1999) and another based on a non-parametric approach (Caffrey et al., 2012). The maximum likelihood approach tests for significant changes in the rates of evolution after gene duplication (Gu, 1999; Wang and Gu, 2001). To do so, this approach calculates the log-likelihood value under two hypotheses, one in which the parameter of functional divergence is larger than 0 (H_1: $\theta > 0$) and one with $\theta = 0$ (H_0). The log-likelihood values for both hypotheses are compared through an LRT (LRT = $2\Delta l$), with twice the difference in the log-likelihood values being approximated to a χ^2 distribution with 1 df. If the null hypothesis is rejected, and thus we demonstrated the existence of functional divergence signatures at the sequence level, the posterior Bayesian probabilities at each of the amino acid sites in the alignment is calculated and those sites with probabilities >0.75 are assumed to have undergone changes responsible of the functional divergence between the two clades of paralogs. We tested two types of functional divergence, Type I and Type II. Type I functional divergence is given when at a particular amino acid site, the rate of evolution of one paralog is very slow while that of its sister clade is fast, indicating stronger functional constraints at that site in one paralog compared with the other, or a shift in the rates of evolution between the two paralogs (Gu, 1999). Alternatively, type II functional divergence measures a significant change in the amino acid properties at a site in the alignment between the two paralogous clades but with that site evolving at low rates in both of the paralogs (Gu, 2006). We used the software Diverge version 3.0 (Gu et al., 2013) to test for functional divergence Types I and II. The non-parametric approach ofCaffrey et al. (2012) uses a similar method to that of Gu but without specifying the phylogenetic clades under test, and thus allows identification of very strong signatures of functional divergence in particular branches of the phylogenetic tree. For each of the branches, the method estimates the parameter of functional divergence (FD) and tests whether this parameter is larger than that estimated from 1000 alignments that are simulated using the sequence and phylogenetic parameters of the real alignment.

RESULTS

Expression pattern of *Mt PI*-like paralogs during floral development

We previously isolated two M. truncatula B-function MADS-box genes (Benlloch et al., 2009) belonging to the PI/GLO subfamily (MtPI andMtNGL9; Fig. 1), which are present in the Medicago genome as single copies (Supplementary Fig. S1). MtPI expression was detected at high levels in floral buds, whereas MtNGL9 expression was very low (Fig. 2I). Similar differences in expression levels between the two PI paralogs were also observed in P. sativum and M. sativa floral buds (Fig. 2I). To determine if the Mt PI-like paralogs underwent spatial expression divergence during floral development, we compared the expression pattern of the two MtPI paralogs using in situ hybridization (Fig. 2A).MtNGL9 mRNA was first detected in floral meristems at about stage 2, in the cells that give rise to common petal–stamen primordia (Benlloch et al., 2003) (Fig. 2E). At stage 4, MtNGL9 expression was observed only in the developing common primordia (Fig. 2F). MtNGL9 expression was maintained during the entire process of common primordia compartmentalization and during the development of petals and stamens. (Fig. 2G, H). At late stages, MtNGL9 mRNA expression was also detected in ovules (Fig. 2H). MtPI and MtNGL9 had similar temporal and spatial expression patterns in petals and stamen during floral development (Fig. 2). While MtPI showed a higher uniformly distributed expression in petals and stamens during development, the MtNGL9 signal seemed to be mainly confined to their epidermal cells in late developmental stages (Fig. 2G, H). Moreover, MtNGL9 has a differential expression in ovules.

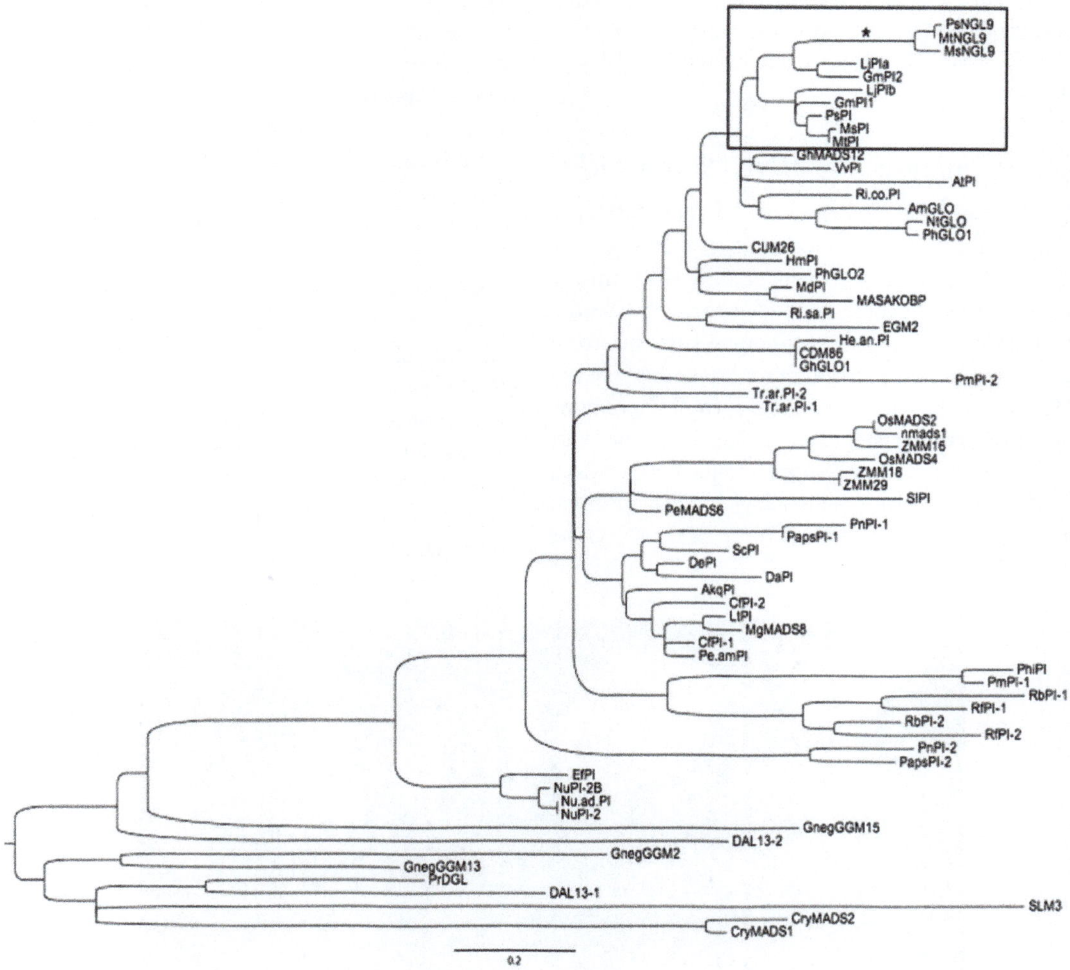

Figure 1: Phylogenetic tree of PI-like MADS-box genes. Sequences used in this analysis are listed inSupplementary Table S1. An accelerated branch incident on one of the Medicago truncatula PI-like gene paralogs is identified with an asterisk.

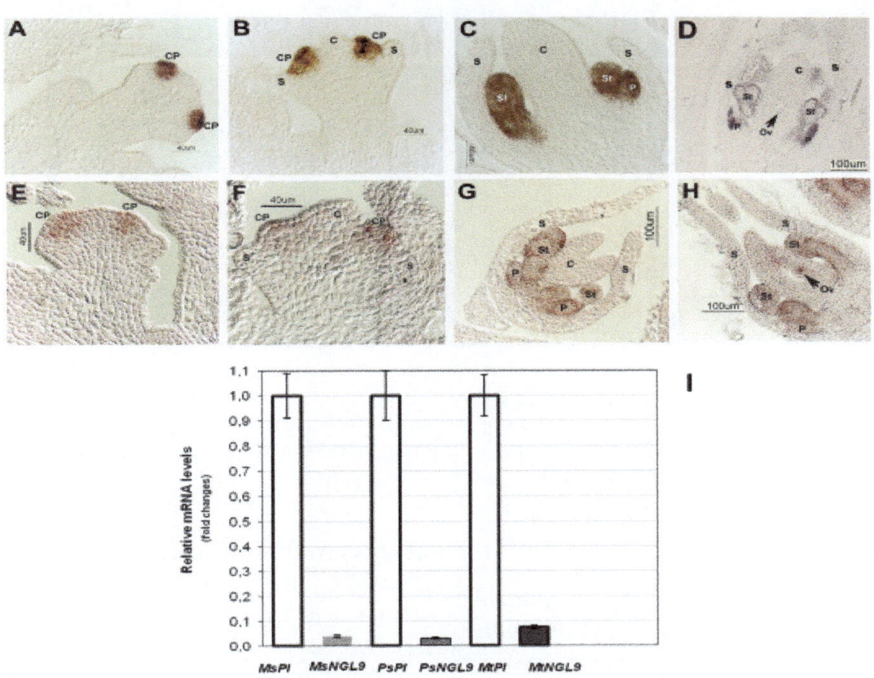

Figure 2: Expression patterns of theM. truncatula PI-like genes.In situ hybridization of theMtPI and MtNGL9 mRNAs inM. truncatula wild-type flower buds. (A, B) MtPIexpression was detected in cells of the common primordia (CP). (C, D) At late stages, MtPI was strongly expressed in the differentiated petals (P) and stamens (St). (E, F) MtNGL9 was weakly expressed at stages 2 and 4 in cells of the common primordia. (G) At stage 5, MtNGL9mRNA was weakly detected in the cells of differentiated petals and stamens. (H) At stage 7, MtNGL9 was also detected in ovules (Ov). (I) qRT–PCR analysis of MtPI and MtNGL9 mRNA levels in floral buds of three leguminous species. The

expression value ofMsPI, PsPI, and MtPI genes was set to 1.00 and the expression levels of MsNGL9, PsNGL9, and MtNGL9 were plotted relative to their respective PI values. (This figure is available in colour at JXBonline.)

Loss-of-function analyses of *MtNGL9* and *MtPI*

To investigate the specific contribution of the MtNGL9 gene in M. truncatula floral development, we looked for retrotransposon insertion mutants (Cheng et al., 2014). mtngl9 mutants were isolated in a reverse genetics approach (see the Materials and methods). mtngl9-1 contains aTnt1 insertion in the coding sequence 360bp from the start codon, in the second exon (Supplementary Fig. S2A) and mtngl9-2 contains a Tnt1insertion in the coding sequence 183bp from the start codon, at the end of the first exon (Supplementary Fig. S2A). Homozygous plants were genotyped using specific primers for both insertion lines (Supplementary Fig. S2B, C). Gene expression analysis indicated that no MtNGL9 transcript was detected in these plants (Supplementary Fig. S2D). We did not observe homeotic changes or any obvious mutant phenotype in floral organs (Fig. 3C). In contrast, mtpi-2 plants (Cheng et al., 2014) exhibited a complete conversion of petals to sepals and stamens to carpels (Fig. 3B). This phenotype resembles those of MtPI-RNAi flowers (Benlloch et al., 2009). Knockdown of MtPI resulted in a marked reduction in the expression of all B-function MADS box genes (Supplementary Fig. S3A), while the loss ofMtNGL9 caused a slight decrease in their expression (Supplementary Fig. S3B). The full loss-of-B-function phenotype in mtpi-2 and the pronounced effects of MtPI absence on the expression of the other B-function genes confirms MtPI as a master regulator in establishing the regulatory pathways for petal and stamen identity, while MtNGL9 does not seem to have a role in this regard. Taken together, these results suggest that pseudogenization might be the functional evolutionary fate for this gene.

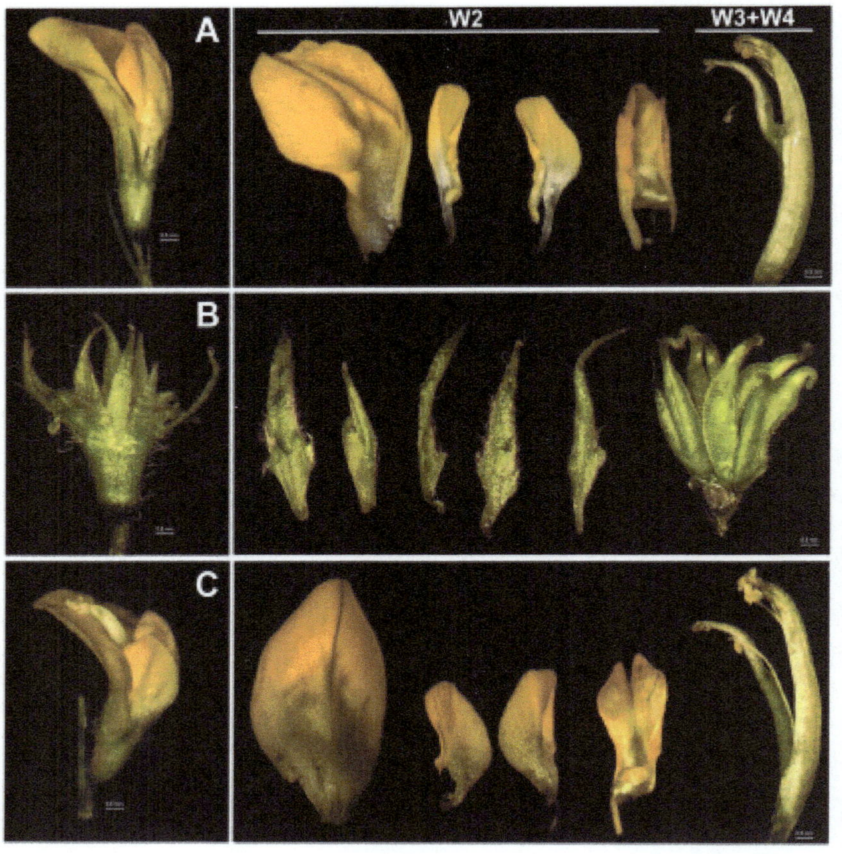

Figure 3: Phenotypes of mtpi-2 andmtngl9 mutants. (A) Wild-type Medicago truncatulaflower. Petals present in the second whorl (W2), and stamens and carpel in W3+W4, respectively. (B)mtpi-2 flower showing full conversion of petals into sepals and of stamens into carpels. (C) mtngl9 flowers do not show homeotic floral alterations.

Evolutionary analyses

To characterize the evolutionary forces acting on MtPI-like duplicated genes, we performed molecular evolution studies. Such studies rest on the assumption that functional genes are constrained by natural selection, while pseudogenes evolve under relaxed selective constraints (Petrov and Hartl, 2000; Zheng et al., 2007; Tutar, 2012).

The phylogeny of PI-like genes reveals different rates of evolution between MtPI and MtNGL9 from different species (Fig. 1), with the lineage leading to the MtNGL9 paralog (Fig.1, asterisk) being significantly longer (i.e. has a greater rate of amino acid evolution) than that leading to MtPI (Tajima's relative rate test: χ^2=5.45, df=1, P=0.019). This accelerated evolution in the branch leading to MtNGL9 could be the result of relaxed selective constraints or be indicative of the positive selection of adaptive amino acid substitutions in that lineage.

To examine these two possibilities, we tested the action of positive selection in the MtNGL9 lineage using the BSM as implemented in the program Codeml from the PAML package version 4.7 (Yang, 2007). Briefly, we compared the log-likelihood values for two hypotheses. The simplest hypothesis (i.e. null model or null hypothesis) assumes neutral evolution, hence relaxed constraints, in the branch of interest (the non-synonymous to synonymous rates ratio; $\omega = d_N/d_S = 1$). The alternative hypothesis, on the other hand, allows variable strengths of selection (ω) among codons in the lineage of interest (see the Material and methods). The value of ω is a good indicator of the strength of selection, with $\omega=1$, $\omega<1$, and $\omega>1$ indicating neutral evolution, purifying selection, and positive selection, respectively. The model testing for adaptive evolution (BSM) detected several amino acid sites in the MADS-box domain under positive selection (ω 3.71; Supplementary Table S2), although this model did not improve the log-likelihood value of the null model [likelihood ratio test: simple model (BSMN) $_0 = -10837.05$, BSM $_1 = -10836.37$; $\chi^2=1.36$, $0.10<P<0.25$). We also applied a model (the FRM) in which each branch of the tree was assumed to have a different ω value. This model showed, however, no evidence of positive selection in the branch leading to MtNGL9. The mean ω value for the entire tree and alignment, as estimated using the Goldman and Yang model (Goldman and Yang, 1994), was 0.14, indicative of strong purifying selection in this gene along the phylogenetic tree. The MtNGL9lineage showed an $\omega=0.56$, supporting relaxed selective constraints, but no neutral evolution, in this branch. While these relaxed constraints could be indicative of a process of non-functionalization, the strong conserved amino acid nature of important functional sites in this gene and the fact that this value is lower than 1 indicate that this gene copy has evolved under strong purifying selection. Overall, these analyses discard pseudogenization (i.e. non-functionalization) as a plausible fate for the gene copy exhibiting an accelerated rate of evolution.

We also tested for amino acid substitutions that may have led to functional changes between the two genes generated by gene duplication. To this end, we examined the substitution patterns in each amino acid site of the multiple sequence alignment containing these genes. Sites that are functionally important but that have diverged after gene duplication should be highly conserved within each gene clade that includes ortholog genes from other species but variable between the paralogs. Inspection of these patterns identifies a number of sites that present a conserved amino acid in one gene copy which is different from its sister gene copy.

No signatures of functional divergence between *MtNGL9* and *MtPI*

To test whether functional diversification events have followed the duplication giving rise to these two Mt PI-like gene copies, we searched for signatures of functional divergence in the amino acid sequences encoded by MtNGL9 and MtPI. We used two tests of functional divergence: (i) a maximum-likelihood approach (see the Materials and methods) (Gu, 1999); and (ii) a non-parametric probabilistic approach (Caffrey et al., 2012) that tests for functional divergence across the phylogenetic tree.

The maximum-likelihood test showed no evidence for either functional divergence Type I (parameter of functional divergence $\theta=0.087$, $P>0.1$) or Type II ($\theta=0.016$, $P>0.1$). Likewise, the non-parametric approach found no evidence of functional divergence within these groups (FD=0.01, $P>0.25$). Therefore, although some amino acid sites tended to be more conserved within paralogs than between them, formal statistical tests (Gu, 2003;Caffrey et al., 2012; Gu et al., 2013) could not find significant evidence supporting this observation.

Interaction strengths between *Medicago truncatula* PI-like proteins

It has been shown that Arabidopsis PI and AP3 proteins act jointly as heterodimers to perform the B-function (Goto and Meyerowitz, 1994; Jacket al., 1994; Riechmann et al., 1996; Krizek and Meyerowitz, 1996;McGonigle et al., 1996; Yang et al., 2003). The heterodimer formed by PI- and AP3-like proteins is the only functional DNA-binding dimer in the vast majority of angiosperms (Melzer et al., 2014).

To evaluate the effect of relaxed functional constraints on the MtNGL9function, we carried out yeast two-hybrid assays on pairwise combinations of B-class MADS-box proteins to determine if the dimerization capabilities were conserved in Mt PI-like gene products. Our results showed that both MtPI and MtNGL9 were able to interact with the Mt AP3-like proteins MtNMH7 and MtTM6 (Fig. 4). Also, we analyzed whether differences exist in the strengths with which PI-type MADS-box proteins interact. We have quantified the interaction of Mt PI-like proteins with MtNMH7 and MtTM6 by β-galactosidase activity assays. The MtPI/MtNMH7 and MtPI/MtTM6 pairs produced stronger interactions as compared with the MtNGL9/MtNMH7 and MtNGL9/MtTM6 pairs (Fig. 4).

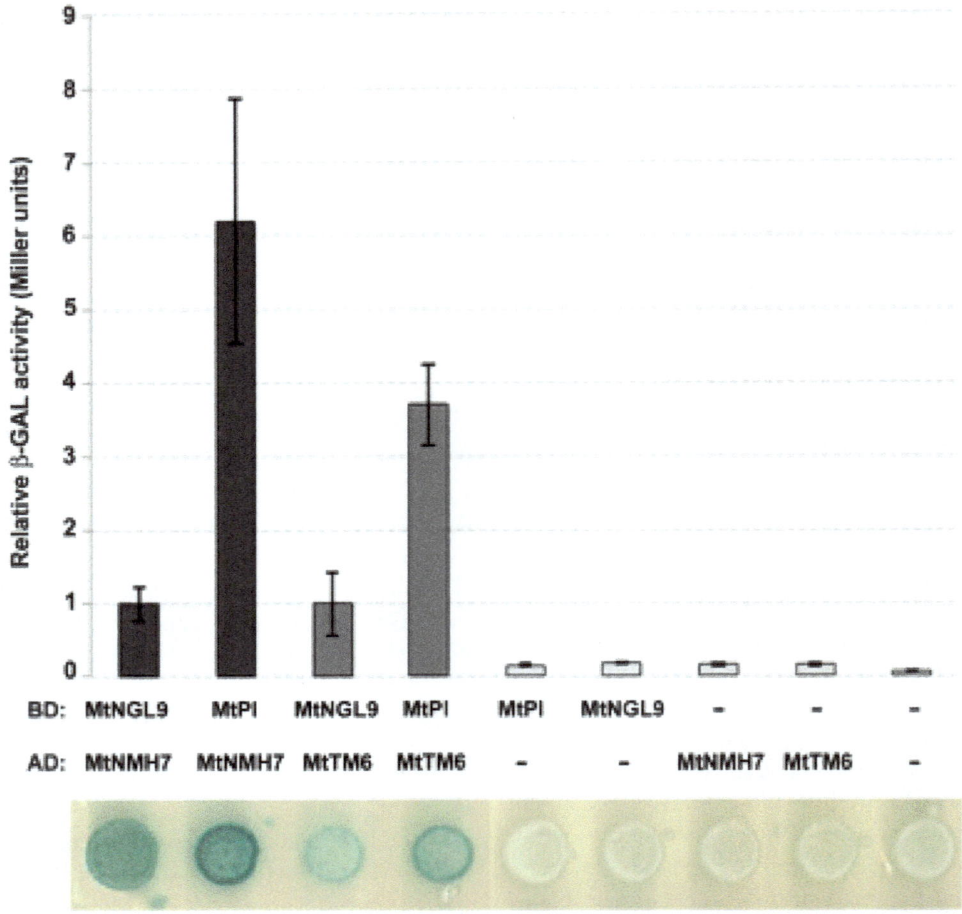

Figure 4: Quantification of the interaction between the M. truncatula B-function MADS-box proteins. BD, LexA-binding domain; AD, GAL4 activation domain. The β-GAL activity value of MtNGL9/MtNMH7 and MtNGL9/MtTM6 interactions was set to 1.00, and the MtPI/Mt-NMH7 and MtPI/MtTM6 interactions were plotted relative to these values, respectively. The values were calculated using 30 samples which correspond to 10 biological replicates. (This figure is available in colour at JXB online.)

Constitutive expression of *MtPI* and *MtNGL9* in Arabidopsis

To investigate the ability of MtPI and MtNGL9 proteins to induce B-function identity, we overexpressed these two genes in Arabidopsis plants. Constitutive expression of MtPI in Arabidopsis leads to homeotic conversion of sepals into petals (Fig. 5D). This phenotype is characteristic of the constitutive expression of most PI-like genes in Arabidopsis (Krizek and Meyerowitz, 1996; Berbel et al., 2005). The majority of 35S::MtPI(64.7%) plants displayed very open sepals transformed into petaloid organs (strong phenotype) (Fig. 5D). At the cellular level, these organs contained petal-like cells rounded and cobblestone-like in appearance with prominent cuticular thickenings (Fig. 5E). Occasionally, petal-like cells co-existed with cells retaining the typical pattern of the sepal cells, showing stomata (Fig. 5F, white arrow) and typical cuticular striation (Fig. 5F, yellow arrow). The remaining 35S::MtPI plants displayed flowers with partially opened sepals which occasionally showed a slight white color (medium phenotype) (Fig. 5G). In fact, the abaxial epidermal cells of the first whorl showed some cells with a small protrusion, which is reminiscent of cells typically found in petals (Fig. 5I; arrow). Most of the plants that overexpressed MtNGL9 exhibited the medium phenotype (81.8%) described above (Supplementary Fig. S5C). The remaining 35S::MtNGL9plants showed flowers with slightly opened sepals (weak phenotype; Fig. 5H) containing typical sepal-like cells (Fig. 5I).

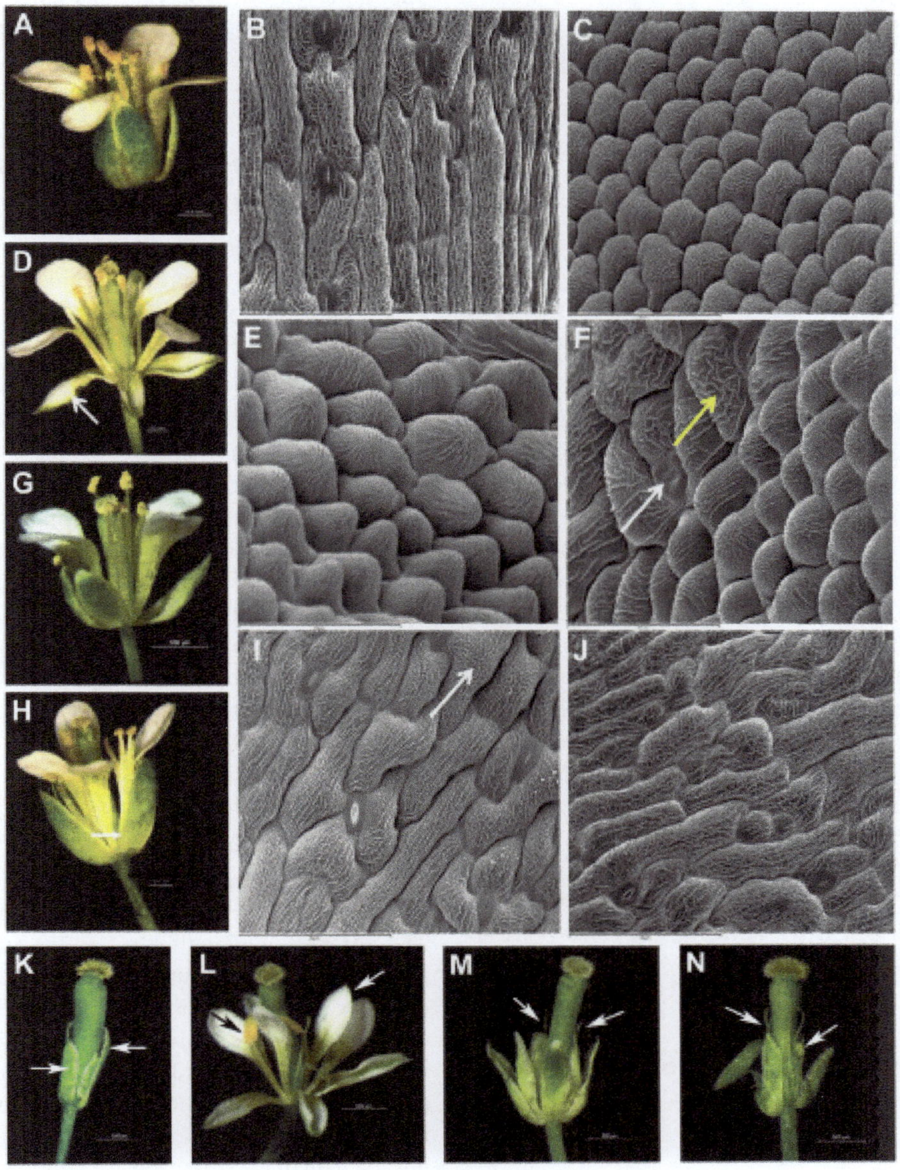

Figure 5: Constitutive expression ofMtPI and MtNGL9 in Arabidopsis and rescue of the pi-1 mutant floral phenotype. (A) Wild-type flower (Ler). (B) Abaxial epidermal cells of a wild-type sepal. (C) Abaxial epidermal cells of a wild-type petal. (D). Flower representative of the strong phenotype showing homeotic conversions of sepals into petals (arrow). (E, F) Abaxial epidermal cells of the first whorl organs of the 35S::MtPI line displaying the strong phenotype. (G) Flower representative of the medium phenotype. (H) Flower representative of the weak phenotype. Arrow: slightly opened sepals. (I) Abaxial epidermal cells of the first whorl organs of a medium 35S::MtNGL9 transgenic line. (J) Abaxial epidermal cells of the first whorl organs of a weak 35S::MtNGL9plant. (K) pi-1 mutant flower. (L) Flowers from 35S::MtPI;pi-1lines showing a near complete rescue of petal (white arrow) and stamen organs (black arrow). (M, N) 35S::MtNGL9;pi-1 flowers showing a weak rescue of petal and stamens (arrows).

We examined the expression levels of the MtPI and MtNGL9 transgenes in their respective overexpression lines by qRT–PCR (Supplementary Fig. S5A, B). Despite the high expression levels of MtNGL9 detected in 35S::MtNGL9lines, neither of these lines displayed the strong phenotype (Fig. 5G–J). We were able to observe differences in the relative effects of both constructs, therefore indicating that the MtPI protein was more active than MtNGL9.

To determine if Mt PI-like genes could rescue petal and stamen organ identity in the absence of endogenous PI activity, we crossed heterozygous35S::MtPI and 35S::MtNGL9 transgenic lines into a pi-1 mutant background. In the F_2 population of kanamycin-resistant plants, a number of independent transformants homozygous for pi-1 were identified (Supplementary Fig. S6), showing a range of phenotypes (Fig. 5L–N;Supplementary Table S3).

Only pi-1/pi-1;35S::MtPI plants showed near complete rescue of petal and stamen organs. Petals were indistinguishable from wild-type petals, and, in the third whorl, mosaic organs generally appeared, possessing both stamen and carpel characteristics. The restored stamens were not fully extended and did not produce fertile pollen (Fig. 5L; Supplementary Table S3). Meanwhile, pi-1/pi-1;35S::MtNGL9 lines (Supplementary Table S3) showed a weak pi-1 mutant rescue. The second whorl organs were larger than the second organs displayed in the pi-1 mutant flowers with a whitish color, but they never reached the size of the wild-type petals (Fig. 5M, N). The third whorl organs were generally filament-like structures and occasionally the stamens showed anther-like organs (Fig. 5M, N). Similar levels of MtPI or MtNGL9 transgenes were detected in floral buds of pi-1/pi-1;35S::MtPI or pi-1/pi-1;35S::MtNGL9 lines, respectively (Supplementary Fig. S5), suggesting that the degree of rescue in the complementation lines is not correlated with the transgene expression level, but with the ability of MtPI or MtNGL9 to replace PI in A. thaliana.

DISCUSSION

MtPI controls B-function in *Medicago truncatula*

Paralogous Mt PI-like genes arose from a duplication event that occurred prior to the speciation of legumes (Fig. 1). They may have originated during the whole-genome duplication (WGD) event that pre-dated speciation of Mt and other legumes ~50–60 Mya (Cannon et al., 2006). Two copies of both PI-like genes are also found in the legume subfamiliesP. sativum, L. japonicus, M. sativa, and G. max. Our phylogenetic analyses consistently placed one copy from each species in separate gene clades (Fig. 1). Our results showed that following this duplication event, only theMtPI duplicated copy has retained the B-function, based on its robust expression levels and the conserved expression pattern of the B-class genes, protein interaction capabilities, and the strong homeotic phenotype observed for the mtpi-2 mutant. The co-ordinated expression of MtPI withMtNMH7 in the inner cell layers of the petal and stamen organ primordia and with MtTM6 in the outer primordia cell layers is required to specify petal and stamen identity in M. truncatula (Roque et al., 2013). However,MtNGL9 does not appear to have a role in this regard, so could potentially be on the way to becoming a pseudogene.

MtNGL9 does not show evidence of pseudogenization

Duplicate genes that are stably preserved in genomes usually have divergent functions (He and Zhang, 2005). Indeed, duplicated genes destined to 'die' usually do so within a few million years after duplication, having acquired mutations that result in a non-functional gene (Lynch and Conery, 2000). Pseudogenes are expected to evolve neutrally; therefore, after a sufficient amount of time, the signature of purifying selection at the amino acid level will ultimately be erased (Zou et al., 2009).

MtNGL9 mutants do not present any homeotic change or any obvious floral mutant phenotype (Fig. 3C), suggesting that MtNGL9 does not have a role in the specification of petal and stamen fate. Our analyses discard pseudogenization as a plausible scenario for MtNGL9 evolution. First, the rate of evolution of MtNGL9 indicates strong purifying selection (i.e. $dN/dS < 1$). Secondly, MtNGL9 has persisted for >60 million years, as the duplication of the gene pre-dates the speciation of legumes (Cannon et al., 2006). Since the half-life for a duplicated gene is in the order of 20 million years (Lynch, 2007), then we can assume that MtNGL9 has not lost its function.

In the basal eudicot Papaver somniferum, one of the Paps PI-like proteins has lost dimerization capabilities with the Paps AP3-like proteins, thus indicating that this gene could be on the way to becoming a pseudogene in this species (Drea et al., 2007). In contrast, in support of a functional role for MtNGL9, this gene encodes a protein that conserves this competence to interact with the Mt AP3-like proteins, which it is an essential feature for the B-class MADS box proteins.

Mode of *Medicago truncatula PISTILLATA*-like gene evolution after duplication

Studies of MIKC^{C}-type MADS-box TFs in an array of species are contributing to a better understanding of how these genes may have changed their functions after duplication (Vandenbussche et al., 2004;Drea et al., 2007; Geuten and Irish, 2010; Pan et al., 2010; Fourquin et al., 2013; Roque et al., 2013; Serwatowska et al., 2014).

A number of theoretical models have been proposed to explain the persistence of duplicated genes in genomes. Paralogs may be selected for increased dosage or as a repository for gene conversion against deleterious changes in either copy and result in functional redundancy (Nadeau and Sankoff, 1997; Nowak et al., 1997; Gu, 2003; Gu et al., 2003). Alternatively, the paralogs may diverge either to generate new gene functions (neofunctionalization) (Taylor and Raes, 2004) or to partition multiple functions (subfunctionalization) through complementary degeneration (Force et al., 1999; Stoltzfus, 1999; Lynch and Force, 2000). These theoretical expectations are, however, only partially consistent with data (Lynch and Conery, 2000).

To describe the evolution of these duplicated genes, we have to invoke factors from different models of molecular evolution. The differential evolution of the two clades examined here is in agreement with the quantitative subfunctionalization model of Force and colleagues (Force et al., 1999): the unduplicated ancestor possessed elevated expression levels that were retained in only an Mt PI-like gene copy after the WGD pre-dating legumes speciation, the MtPI gene. Similar differences in expression levels between the two PI paralogs were also observed in P. sativum and M. sativa floral apices (Fig. 2I), suggesting ancestral changes in regulatory sequences. How can this differential expression between the two gene copies explain their evolutionary and functional patterns?

Whether gene expression can evolve independently from gene function remains an open question in evolutionary biology (Castillo-Davis et al., 2004). Many recent studies have explored molecular and population-genetic constraints on the rate of protein evolution and they have hypothesized that the best predictor of the evolutionary rates of proteins is the gene expression level (Pal et al., 2001; Herbeck et al., 2003; Rocha and Danchin, 2004; Subramanian and Kumar, 2004; Gout et al., 2010). Our results showed that the expression levels may have imposed additional constraints to the highly expressed MtPI protein. The lower expression of the MtNGL9 gene is likely to have relaxed the selective constraints on this gene, in agreement with the mistranslation model (Drummond et al., 2005), allowing it to evolve at a faster rate than MtPI.MtNGL9 has lost its ancestral expression level, and, as a consequence, its ancestral function seems to have decreased. In line with this, the constitutive expression of MtPI and MtNGL9 in Arabidopsis and the rescue of the pi-1 mutant by these genes revealed that MtPI had a greater ability than MtNGL9 to be a component of the transcriptional complex that specifies the second and the third whorl in a heterologous system.

TFs that function in the context of molecular networks are likely to be under stronger selective pressures (Luisi et al., 2015). Thus, the most parsimonious explanation for the preservation of both Mt PI-like duplicated gene copies is that both will be selected to achieve an optimal total dosage balance of their functional complex. In the case of ohnologs (duplicated genes generated by WGD), expression dosage is thought frequently to drive functional persistence, suggesting that the stoichiometry of expression among duplicated loci is important (Papp et al., 2003; Birchler and Veitia, 2007; Edger and Pires, 2009; Gout et al., 2010; Makino and McLysaght, 2012). MtNGL9 would have a far lower contribution to the total activity required to specify the B-function. The slight decrease of MtPI and the other B-class gene expression levels inmtngl9 floral buds suggests that MtNGL9 may be required to maintain the critical dosage for the B-function in M. truncatula.

On the other hand, we have detected MtNGL9 expression in ovules (Fig. 2H). The B-class gene MtNMH7 (Roque et al., 2013) and the euAP3 geneGmNMH7 in soybean (Wu et al., 2006) have also been detected in this floral tissue. Moreover, low levels of MtNGL9 transcript have been reported in vegetative tissues (Benlloch et al., 2009). The presence of transcripts in a given tissue does not necessarily mean that the gene activity is required for its proper development. However, it might be considered that changes in MtNGL9 regulation enabled its expression in different tissues and that the preservation of MtNGL9 and its partial functional relaxation subsequently provide the opportunity for the fixation of advantageous mutations, which can lead to new functions (He and Zhang, 2005). This makes neofunctionalization plausible as an alternative evolutionary fate for this gene, through the ability to form part of new complexes that control other development processes. The extent to which such divergences have led to a neofunctionalization of one of the duplicatedMtPI-like copies and possible acquisition of other roles in different developmental processes requires further investigation.

Our results provide evidence of the complex dynamics underlying the evolution by gene duplication of the B-class MADS-box subfamily. Notwithstanding this complexity, the B-function remains conserved in all eudicots, providing robustness to the specification of petal and stamen identities. Finally, our findings confirm MADS-box TFs as promising targets for novel studies aimed at a more complete understanding of the complexity of evolution by gene duplication.

ACKNOWLEDGEMENTS

This work was supported by grants from the Spanish Ministry of Economy and Competitiveness (MINECO; BIO2009-08134 and BIO2012-39849-CO2-01). MAF was supported by a grant from the MINECO (BFU2012-36346). We wish to thank Drs Santiago F. Elena Fito (IBMCP, CSIC-UPV), José V. Gimeno Alcáñiz (IATA, CSIC), Javier Paz-Ares (CNB, CSIC), and Carlos Alonso-Blanco (CNB, CSIC) for valuable suggestions and comments in the initial stages of this work. The technical assistance of Rafael Martínez-Pardo in the greenhouse is gratefully acknowledged.

REFERENCES

1. Airoldi CA Davies B . 2012. Gene duplication and the evolution of plant MADS-box transcription factors. Journal of Genetics and Genomics 39, 157–165.
2. Alvarez-Ponce D Fares MA . 2012. Evolutionary rate and duplicability in the Arabidopsis thaliana protein–protein interaction network. Genome Biology and Evolution 4, 1263–1274.
3. Aoki S Uehara K Imafuku M Hasebe M Ito M . 2004. Phylogeny and divergence of basal angiosperms inferred from APETALA3- and PISTILLATA-like MADS-box genes. Journal of Plant Research 117, 229–244.
4. Bechtold N Pelletier G . 1998. In planta Agrobacterium-mediated transformation of adult Arabidopsis thaliana plants by vacuum infiltration. Methods in Molecular Biology 82, 259–266.
5. Benlloch R Navarro C Beltrán J Cañas LA . 2003. Floral development of the model legume Medicago truncatula: ontogeny studies as a tool to better characterize homeotic mutations. Sexual Plant Reproduction 15, 231–241.
6. Benlloch R Roque E Ferrandiz C Cosson V Caballero T Penmetsa RV Beltran JP Canas LA Ratet P Madueno F . 2009. Analysis of B function in legumes: PISTILLATA proteins do not require the PI motif for floral organ development in Medicago truncatula . The Plant Journal 60, 102–111.
7. Berbel A Navarro C Ferrandiz C Canas LA Beltran JP Madueno F . 2005. Functional conservation of PISTILLATA activity in a pea homolog lacking the PI motif. Plant Physiology 139, 174–185.
8. Birchler JA Riddle NC Auger DL Veitia RA . 2005. Dosage balance in gene regulation: biological implications. Trends in Genetics 21, 219–226.
9. Birchler JA Veitia RA . 2007. The gene balance hypothesis: from classical genetics to modern genomics. The Plant Cell 19, 395–402.
10. Blanc G Wolfe KH . 2004. Widespread paleopolyploidy in model plant species inferred from age distributions of duplicate genes. The Plant Cell 16, 1667–1678.
11. Bowman JL Smyth DR Meyerowitz EM . 1989. Genes directing flower development in Arabidopsis . The Plant Cell 1, 37–52.
12. Bowman JL Smyth DR Meyerowitz EM . 1991. Genetic interactions among floral homeotic genes of Arabidopsis . Development 112, 1–20.
13. Caffrey BE Williams TA Jiang X Toft C Hokamp K Fares MA . 2012. Proteome-wide analysis of functional divergence in bacteria: exploring a host of ecological adaptations. PLoS One 7, e35659.
14. Cannon SB Sterck L Rombauts S et al. 2006. Legume genome evolution viewed through the Medicago truncatula and Lotus japonicus genomes. Proceedings of the National Academy of Sciences, USA 103, 14959–14964.
15. Carretero-Paulet L Fares MA . 2012. Evolutionary dynamics and functional specialization of plant paralogs formed by whole and small-scale genome duplications. Molecular Biology and Evolution 29, 3541–3551.
16. Castillo-Davis CI Hartl DL Achaz G . 2004. Cis-Regulatory and protein evolution in orthologous and duplicate genes. Genome Research 14, 1530–1536.

17. Cheng X Wang M Lee HK Tadege M Ratet P Udvardi M Mysore KS Wen J . 2014. An efficient reverse genetics platform in the model legume Medicago truncatula . New Phytologist 201, 1065–1076.

18. Conant GC Wolfe KH . 2008. Turning a hobby into a job: how duplicated genes find new functions. Nature Reviews Genetics 9, 938–950.

19. Cui L Wall PK Leebens-Mack JH et al. 2006. Widespread genome duplications throughout the history of flowering plants. Genome Research 16, 738–749.

20. Czechowski T Stitt M Altmann T Udvardi MK Scheible WR . 2005. Genome-wide identification and testing of superior reference genes for transcript normalization in Arabidopsis . Plant Physiology 139, 5–17.

21. Des Marais DL Rausher MD . 2008. Escape from adaptive conflict after duplication in an anthocyanin pathway gene. Nature 454, 762–765.

22. Drea S Hileman LC de Martino G Irish VF . 2007. Functional analyses of genetic pathways controlling petal specification in poppy. Development 134, 4157–4166.

23. Drummond DA Bloom JD Adami C Wilke CO Arnold FH . 2005. Why highly expressed proteins evolve slowly. Proceedings of the National Academy of Sciences, USA 102, 14338–14343.

24. Edger PP Pires JC . 2009. Gene and genome duplications: the impact of dosage-sensitivity on the fate of nuclear genes. Chromosome Research 17, 699–717.

25. Fares MA Keane OM Toft C Carretero-Paulet L Jones GW . 2013. The roles of whole-genome and small-scale duplications in the functional specialization of Saccharomyces cerevisiae genes. PLoS Genetics 9, e1003176.

26. Ferrandiz C Gu Q Martienssen R Yanofsky MF . 2000. Redundant regulation of meristem identity and plant architecture by FRUITFULL, APETALA1 and CAULIFLOWER . Development 127, 725–734.

27. Force A Lynch M Pickett FB Amores A Yan YL Postlethwait J . 1999. Preservation of duplicate genes by complementary, degenerative mutations. Genetics 151, 1531–1545.

28. Fourquin C del Cerro C Victoria FC Vialette-Guiraud A de Oliveira AC Ferrandiz C . 2013. A change in SHATTERPROOF protein lies at the origin of a fruit morphological novelty and a new strategy for seed dispersal in Medicago genus. Plant Physiology 162, 907–917.

29. Freeling M Thomas BC . 2006. Gene-balanced duplications, like tetraploidy, provide predictable drive to increase morphological complexity. Genome Research 16, 805–814.

30. Geuten K Irish V . 2010. Hidden variability of floral homeotic B genes in Solanaceae provides a molecular basis for the evolution of novel functions. The Plant Cell 22, 2562–2578.

31. Goldman N Yang Z . 1994. A codon-based model of nucleotide substitution for protein-coding DNA sequences. Molecular Biology and Evolution 11, 725–736.

32. Goto K Meyerowitz EM . 1994. Function and regulation of the Arabidopsis floral homeotic gene PISTILLATA . Genes and Development 8, 1548–1560.

33. Gout JF Kahn D Duret L , Paramecium Post-Genomics Consortium . 2010. The relationship among gene expression, the evolution of gene dosage, and the rate of protein evolution. PLoS Genetics 6, e1000944.

34. Gu X Zou Y Su Z Huang W Zhou Z Arendsee Z Zeng Y . 2013. An update of DIVERGE software for functional divergence analysis of protein family. Molecular Biology and Evolution 30, 1713–1719.

35. Gu X . 1999. Statistical methods for testing functional divergence after gene duplication. Molecular Biology and Evolution 16, 1664–1674.

36. Gu X . 2003. Evolution of duplicate genes versus genetic robustness against null mutations. Trends in Genetics 19, 354–356.

37. Gu X . 2006. A simple statistical method for estimating type-II (cluster-specific) functional divergence of protein sequences. Molecular Biology and Evolution 23, 1937–1945.

38. Gu Z Steinmetz LM Gu X Scharfe C Davis RW Li WH . 2003. Role of duplicate genes in genetic robustness against null mutations. Nature 421, 63–66.

39. Guerineau F Mulleneaux P . 1993. Plant transformation and expression vectors. In: Croy RRD , ed. Plant molecular biology Labfax. Oxford: BIOS Scientific Publishers, 121–148.

40. Hartweck LM Scott CL Olszewski NE . 2002. Two O-linked N-acetylglucosamine transferase genes of Arabidopsis thaliana L. Heynh. have overlapping functions necessary for gamete and seed development. Genetics 161, 1279–1291.

41. He X Zhang J . 2005. Rapid subfunctionalization accompanied by prolonged and substantial neofunctionalization in duplicate gene evolution. Genetics 169, 1157–1164.

42. Hecht V Foucher F Ferrandiz C et al. 2005. Conservation of Arabidopsis flowering genes in model legumes. Plant Physiology 137, 1420–1434.

43. Herbeck JT Wall DP Wernegreen JJ . 2003. Gene expression level influences amino acid usage, but not codon usage, in the tsetse fly endosymbiont Wigglesworthia . Microbiology 149, 2585–2596.

44. Hewitt EJ . 1966. Sand and water culture methods used in the study of plant nutrition . Farnham Royal, UK: Commonwealth Agricultural Bureau.

45. Innan H Kondrashov F . 2010. The evolution of gene duplications: classifying and distinguishing between models. Natural Reviews Genetics 11, 97–108.

46. Jack T Brockman LL Meyerowitz EM . 1992. The homeotic gene APETALA3 of Arabidopsis thaliana encodes a MADS-box and is expressed in petals and stamens. Cell 68, 683–697.

47. Jack T Fox GL Meyerowitz EM . 1994. Arabidopsis homeotic gene APETALA3 ectopic expression: transcriptional and posttranscriptional regulation

determines floral organ identity. Cell 76, 703–716.

48. Jones DT Taylor WR Thornton JM . 1992. The rapid generation of mutation data matrices from protein sequences. Computer Applications in the Biosciences 8, 275–282.

49. Kaufmann K Melzer R Theissen G . 2005. MIKC-type MADS-domain proteins: structural modularity, protein interactions and network evolution in land plants. Gene 347, 183–198.

50. Keane OM Toft C Carretero-Paulet L Jones GW Fares MA . 2014. Preservation of genetic and regulatory robustness in ancient gene duplicates of Saccharomyces cerevisiae . Genome Research 24, 1830–1841.

51. Kim S Yoo MJ Albert VA Farris JS Soltis PS Soltis DE . 2004. Phylogeny and diversification of B-function MADS-box genes in angiosperms: evolutionary and functional implications of a 260-million-year-old duplication. American Journal of Botany 91, 2102–2118.

52. Kramer E Di Stilio VS Schlulter PM . 2003. Complex patterns of gene duplication in the APETALA3 and PISTILLATA lineages of the Ranunculaceae. International Journal of Plant Sciences 164, 1–11.

53. Kramer EM Dorit RL Irish VF . 1998. Molecular evolution of genes controlling petal and stamen development: duplication and divergence within the APETALA3 and PISTILLATA MADS-box gene lineages. Genetics 149, 765–783.

54. Kramer EM Irish VF . 2000. Evolution of the petal and stamen developmental programs: evidence from comparative studies of the lower eudicots and basal angiosperms. International Journal of Plant Sciences 161, S29–S40.

55. Krizek BA Meyerowitz EM . 1996. The Arabidopsis homeotic genes APETALA3 and PISTILLATA are sufficient to provide the B class organ identity function. Development 122, 11–22.

56. Litt A Kramer EM . 2010. The ABC model and the diversification of floral organ identity. Seminars in Cell and Developmental Biology 21, 129–137.

57. Ludin K Jiang R Carlson M . 1998. Glucose-regulated interaction of a regulatory subunit of protein phosphatase 1 with the Snf1 protein kinase in Saccharomyces cerevisiae . Proceedings of the National Academy of Sciences, USA 95, 6245–6250.

58. Luisi P Alvarez-Ponce D Pybus M Fares MA Bertranpetit J Laayouni H . 2015. Recent positive selection has acted on genes encoding proteins with more interactions within the whole human interactome. Genome Biology and Evolution 7, 1141–1154.

59. Lynch M . 2007. The origins of genomic architecture . Sinauer Associates: Sunderland, MA.

60. Lynch M Conery JS . 2000. The evolutionary fate and consequences of duplicate genes. Science 290, 1151–1155.

61. Lynch M Force A . 2000. The probability of duplicate gene preservation by subfunctionalization. Genetics 154, 459–473.

62. Makino T McLysaght A . 2012. Ohnologs in the human genome are dosage balanced and frequently associated with disease. Proceedings of the National Academy of Sciences, USA 107, 9270–9274.

63. McGonigle B Bouhidel K Irish VF . 1996. Nuclear localization of the Arabidopsis APETALA3 and PISTILLATA homeotic gene products depends on their simultaneous expression. Genes and Development 10, 1812–1821.

64. Melzer R Harter A Rumpler F Kim S Soltis PS Soltis DE Theissen G . 2014. DEF- and GLO-like proteins may have lost most of their interaction partners during angiosperm evolution. Annals of Botany 114, 1431–1443.

65. Nadeau JH Sankoff D . 1997. Comparable rates of gene loss and functional divergence after genome duplications early in vertebrate evolution. Genetics 147, 1259–1266.

66. Nowak MA Boerlijst MC Cooke J Smith JM . 1997. Evolution of genetic redundancy. Nature 388, 167–171.

67. Pal C Papp B Hurst LD . 2001. Highly expressed genes in yeast evolve slowly. Genetics 158, 927–931.

68. Pan IL McQuinn R Giovannoni JJ Irish VF . 2010. Functional diversification of AGAMOUS lineage genes in regulating tomato flower and fruit development. Journal of Experimental Botany 61, 1795–1806.

69. Papp B Pal C Hurst LD . 2003. Dosage sensitivity and the evolution of gene families in yeast. Nature 424, 194–197.

70. Payne JL Wagner A . 2014. The robustness and evolvability of transcription factor binding sites. Science 343, 875–877.

71. Petrov DA Hartl DL . 2000. Pseudogene evolution and natural selection for a compact genome. Journal of Heredity 91, 221–227.

72. Purugganan MD Rounsley SD Schmidt RJ Yanofsky MF . 1995. Molecular evolution of flower development: diversification of the plant MADS-box regulatory gene family. Genetics 140, 345–356.

73. Riechmann JL Krizek BA Meyerowitz EM . 1996. Dimerization specificity of Arabidopsis MADS domain homeotic proteins APETALA1, APETALA3, PISTILLATA, and AGAMOUS. Proceedings of the National Academy of Sciences, USA 93, 4793–4798.

74. Rocha EP Danchin A . 2004. An analysis of determinants of amino acids substitution rates in bacterial proteins. Molecular Biology and Evolution 21, 108–116.

75. Roque E Gomez MD Ellul P Wallbraun M Madueno F Beltran JP Canas LA . 2007. The PsEND1 promoter: a novel tool to produce genetically engineered male-sterile plants by early anther ablation. Plant Cell Reports 26, 313–325.

76. Roque E Serwatowska J Cruz Rochina M Wen J Mysore KS Yenush L Beltran JP Canas LA . 2013. Functional specialization of duplicated AP3-like genes in Medicago truncatula . The Plant Journal 73, 663–675.

77. Schwarz-Sommer Z Huijser P Nacken W Saedler H Sommer H . 1990. Genetic control of flower development by homeotic genes in Antirrhinum majus . Science 250, 931–936.

78. Serwatowska J Roque E Gomez-Mena C Constantin GD Wen J Mysore KS Lund OS Johansen E Beltran JP Canas LA . 2014. Two euAGAMOUS genes control C-function in Medicago truncatula . PLoS One 9, e103770.

79. Singh LN Hannenhalli S . 2008. Functional diversification of paralogous transcription factors via divergence in DNA binding site motif and in expression. PLoS One 3, e2345.

80. Stellari GM Jaramillo MA Kramer EM . 2004. Evolution of the APETALA3 and PISTILLATA lineages of MADS-box-containing genes in the basal angiosperms. Molecular Biology and Evolution 21, 506–519.

81. Stoltzfus A . 1999. On the possibility of constructive neutral evolution. Journal of Molecular Evolution 49, 169–181.

82. Subramanian S Kumar S . 2004. Gene expression intensity shapes evolutionary rates of the proteins encoded by the vertebrate genome. Genetics 168, 373–381.

83. Tamura K Peterson D Peterson N Stecher G Nei M Kumar S . 2011. MEGA5: molecular evolutionary genetics analysis using maximum likelihood, evolutionary distance, and maximum parsimony methods. Molecular Biology and Evolution 28, 2731–2739.

84. Taylor JS Raes J . 2004. Duplication and divergence: the evolution of new genes and old ideas. Annual Review of Genetics 38, 615–643.

85. Theissen G Becker A Di Rosa A Kanno A Kim JT Munster T Winter KU Saedler H . 2000. A short history of MADS-box genes in plants. Plant Molecular Biology 42, 115–149.

86. Tutar Y . 2012. Pseudogenes. Comparative and Functional Genomics 2012, 424526.

87. Vandenbussche M Zethof J Royaert S Weterings K Gerats T . 2004. The duplicated B-class heterodimer model: whorl-specific effects and complex genetic interactions in Petunia hybrida flower development. The Plant Cell 16, 741–754.

88. Viaene T Vekemans D Irish VF Geeraerts A Huysmans S Janssens S Smets E Geuten K . 2009. Pistillata-duplications as a mode for floral diversification in (Basal) asterids. Molecular Biology and Evolution 26, 2627–2645.

89. Vojtek AB Hollenberg SM Cooper JA . 1993. Mammalian Ras interacts directly with the serine/threonine kinase Raf. Cell 74, 205–214.

90. Wang Y Gu X . 2001. Functional divergence in the caspase gene family and altered functional constraints: statistical analysis and prediction. Genetics 158, 1311–1320.

91. Wellmer F Bowman JL Davies B et al. 2014. Flower development: open questions and future directions. Methods in Molecular Biology 1110, 103–124.

92. Wolfe KH Shields DC . 1997. Molecular evidence for an ancient duplication of the entire yeast genome. Nature 387, 708–713.

93. Wu C Ma Q Yam KM Cheung MY Xu Y Han T Lam HM Chong K . 2006. In situ expression of the GmNMH7 gene is photoperiod-dependent in a unique soybean (Glycine max [L.] Merr.) flowering reversion system. Planta 223, 725–735.

94. Yang Y Fanning L Jack T . 2003. The K domain mediates heterodimerization of the Arabidopsis floral organ identity proteins, APETALA3 and PISTILLATA. The Plant Journal 33, 47–59.

95. Yang Z . 2007. PAML 4: phylogenetic analysis by maximum likelihood. Molecular Biology and Evolution 24, 1586–1591.

96. Zhang S Cheng F Wang C Zhang L An Y . 2013. Cloning and tissue-specific expression of predicted Pisum sativum actin isoform PEAc14-1. Biochemical Genetics 51, 722–727.

97. Zheng D Frankish A Baertsch R et al. 2007. Pseudogenes in the ENCODE regions: consensus annotation, analysis of transcription, and evolution. Genome Research 17, 839–851.

98. Zou C Lehti-Shiu MD Thibaud-Nissen F Prakash T Buell CR Shiu SH . 2009. Evolutionary and expression signatures of pseudogenes in Arabidopsis and rice. Plant Physiology 151, 3–15.

CHAPTER 17

THE ARABIDOPSIS THALIANA MOBILOME AND ITS IMPACT AT THE SPECIES LEVEL

Leandro Quadrana, Amanda Bortolini Silveira, George F Mayhew, Chantal LeBlanc, Robert A Martienssen, Jeffrey A Jeddeloh, Vincent Colot

INTRODUCTION

Transposable elements (TEs) are sequences that move and replicate around the genome. Depending on whether their mobilization relies on a RNA or DNA intermediate, they are classified as retrotransposons (class I) or DNA transposons (class II), respectively (Slotkin and Martienssen, 2007). TEs are further subdivided into distinct families, the prevalence of which differs between organisms because of a complex array of factors, including variable transposition activity and diverse selection pressures (Barrón et al., 2014). Given their mobile nature, TEs pose multiple threats to the physical and functional integrity of genomes. In particular, TEs can disrupt genes through insertion and also through excision in the case of DNA transposons. Thus, TE mobilization is a source of both germline and somatic mutations (Richardson et al., 2015; Barrón et al., 2014; Lisch, 2013). Although TEs are endogenous mutagens with potentially catastrophic effects, their mobilization might sometimes be beneficial. In fact, soon after their discovery, Barbara McClintock named TEs 'controlling elements' to emphasize their role in the control of gene action (McClintock, 1956). In mammals, transposition of some *LINE 1* retrotransposons occurs extensively during embryogenesis as well as in the adult brain, again suggesting functional relevance of somatic TE mobilization (Richardson et al., 2014). Nonetheless, TEs are under tight control to limit their mutational impact both within and across generations. In plants and mammals, a major control is through epigenetic silencing mechanisms, including DNA methylation and these mechanisms can in turn have 'epimutagenic' effects on adjacent genes (Slotkin and Martienssen, 2007; Weigel and Colot, 2012; Heard and Martienssen, 2014;Quadrana and Colot, 2016).

Despite the many documented short-term as well as evolutionary consequences of TE mobilization (Rebollo et al., 2012; Trono, 2016; Kazazian, 2004; Elbarbary et al., 2016; Bennetzen and Wang, 2014; Lisch, 2013), TEs are among the least investigated components of genomes, mainly because they are present in multiple, often degenerated copies, which complicate analysis. Thus, a species-wide view of the mobilome - i.e. of the set of TE families with transposition activity - is lacking for most organisms.

Studies in humans suggest that although at least half of the 3Gb genome is made up of TE sequences, mainly belonging to *LINE 1* and *SINE* retrotransposon families, a few of these only and none of the other TE families contain mobile copies (Richardson et al., 2015). In contrast, the number of TE families that have retained transposition activity is much larger in the mouse and these include several so-called endogenous retroviruses (ERVs) in addition to *LINEs* and *SINEs* (Richardson et al., 2015). In Drosophila, which has a much smaller genome (~120 Mb) characterized by a large repertoire of class I and class II TE families, the situation is again different with most TE families likely mobile (Mackay et al., 2012; Cridland et al., 2013;Robert et al., 2015; Rahman et al., 2015). However, in this species and even more so in mammals, the population genetics of the mobilome remains poorly characterized.

Citation: Leandro Quadrana, Amanda Bortolini Silveira, George F Mayhew, Chantal LeBlanc, Robert A Martienssen, Jeffrey A Jeddeloh, and Vincent Colot, The Arabidopsis thaliana mobilome and its impact at the species level, doi.org/10.7554/eLife.15716.001.

The flowering plant *A. thaliana* is particularly attractive for conducting a systematic survey of the mobilome and of its molecular as well as phenotypic impact at the species level. First, like Drosophila, *A. thaliana* has a compact genome and a large repertoire of class I and class II TE families (Arabidopsis Genome Initiative, 2000; Ahmed et al. 2011; Joly-Lopez and Bureau, 2014). Thus, most TE families are of relatively small size, which facilitates their study. Second, *A. thaliana* occupies a wide range of habitats across the globe and representative accessions have been extensively characterized both genetically and phenotypically (Weigel and Nordborg, 2015). Third, whole genome sequencing has been performed for >1000 *A. thaliana* accessions and DNA methylome as well as transcriptome data are also available for hundreds of these (Schmitz et al., 2013; Long et al., 2013; Dubin et al., 2015; Cao et al., 2011). Finally, genome-wide association studies (GWASs) are straightforward in this species (Weigel and Nordborg, 2015).

Here we present a comprehensive assessment of the *A. thaliana* mobilome, which radically changes the prevailing view of limited transposition potential in this species and provides important novel insights into the population genetics of TE mobilization (Hu et al., 2011; Maumus and Quesneville, 2014). Specifically, we show that the *A. thaliana* mobilome is composed of a very large number of class I and class II TE families overall, but differs extensively among accessions. We further show that TE mobilization is a complex trait and we have identified environmental as well as genetic factors that influence transposition in nature. These factors include the annual temperature range, the TE themselves and multiple gene loci, notably *MET2a*, which encodes a poorly characterized DNA methyltransferase. In addition, we present compelling evidence that TEs insert throughout the genome with no overt bias and that the mobilome has a pervasive impact on the expression and DNA methylation status of adjacent genes. These and other observations indicate that purifying selection is most probably the main factor responsible for the differential accumulation of TE sequences along the *A. thaliana* genome and notably their clustering in pericentromeric regions. Finally, we reveal the importance of the mobilome as a generator of large-effect alleles at loci underlying adaptive traits. Collectively, our approaches and findings provide a unique framework for detailed studies of the dynamics and impact of transposition in nature.

RESULTS

Composition of the *A. thaliana* Mobilome

The reference genome sequence of *A.thaliana* is 125 Mb long (TAIR 10) and contains ~32000, mostly degenerate TE copies that belong to 326 distinct families (Arabidopsis Genome Initiative, 2000; Ahmed et al., 2011). So far, transposition activity has been documented experimentally for nine TE families, mainly on the basis of studies carried out in the reference accession Col-0 (Ito and Kakutani, 2014; Tsay et al., 1993). To assess species-wide the composition of the *A.thaliana* mobilome, we used publically available Illumina short genome sequence reads (Schmitz et al., 2013; Schneeberger et al., 2011). First, we looked for TE copy number variation (CNV) between the reference accession Columbia (Col-0) and 211 accessions taken from across the globe. To limit the problem posed by the presence of TEs in multiple copies across the genome, with varying degrees of similarity to each other, we performed an aggregated CNV analysis based on the 11,851 annotated Col-0 TE sequences longer than 300 bp (see 'Materials and methods'). CNVs were detected for 263 TE families (Figure 1A and B; Figure 1—source data 1; see 'Materials and methods'), in keeping with the results of a previous study indicating that the vast majority of the TE sequences annotated in the Col-0 reference genome are absent from that of at least one of 80 accessions analyzed (Cao et al., 2011).

Since CNVs could reflect either recent TE mobilization or the gain or loss of TE copies through other types of chromosomal rearrangements, we then looked among the unmapped Illumina short reads for so-called 'split-reads' that contain TE extremities. Crucially, because most TE families generate short target site duplications (TSDs) of fixed size upon insertion, TSDs can serve as signatures of *bona fide* transposition events. We therefore developed a pipeline for the systematic identification of split-reads covering TE junctions that are absent from the reference genome and that produce, when mapped to the insertion site, a sequence overlap of the size of TSDs (3 to 15 bp, depending on the TE family, Figure 1C and D; see 'Materials and methods'). Our pipeline differs in that respect from that used in another study to detect the presence/absence of reference and non-reference TE insertions in the same set of accessions (Stuart et al., 2016). Results produced by our pipeline for the 292 annotated TE families that create TSDs upon transposition were verified visually to eliminate false positives (Figure 1—figure supplement 1; see 'Materials and methods'). Following this approach, non-reference TE insertions with TSDs were identified for 131 TE families in total (Figure 1—source data 2), which all also show CNV (Figure 1—source data 1).

Most (86%) non-reference TE insertions with TSDs are private or shared only by a few accessions and thus they typically correspond to recently derived alleles, as expected (Figure 1E). Moreover, recent transposition activity is only detected for between four and 66 TE families in any given accession, thus indicating large variations in the composition of the mobilome among accessions (Figure 1F). Nonetheless, we have probably identified most of the annotated TE families that compose the mobilome at the species level, because the number of TE families defined as mobile by the split-reads approach reaches a plateau after examining just 74 accessions (Figure 1G). The 53 class I *COPIA* families and the 40 class II Mutator-like (*MuDR*) families are the most mobile, as they account for 1408 and 729 of the 2835 non-reference TE insertions with TSDs identified in total, respectively (Figure 1H and Figure 1—source data 2 and 3). However, the number of non-reference insertions per accession is always small (<16) for any given family (Figure 1—source data 2), thus suggesting a lack of recent transposition bursts.

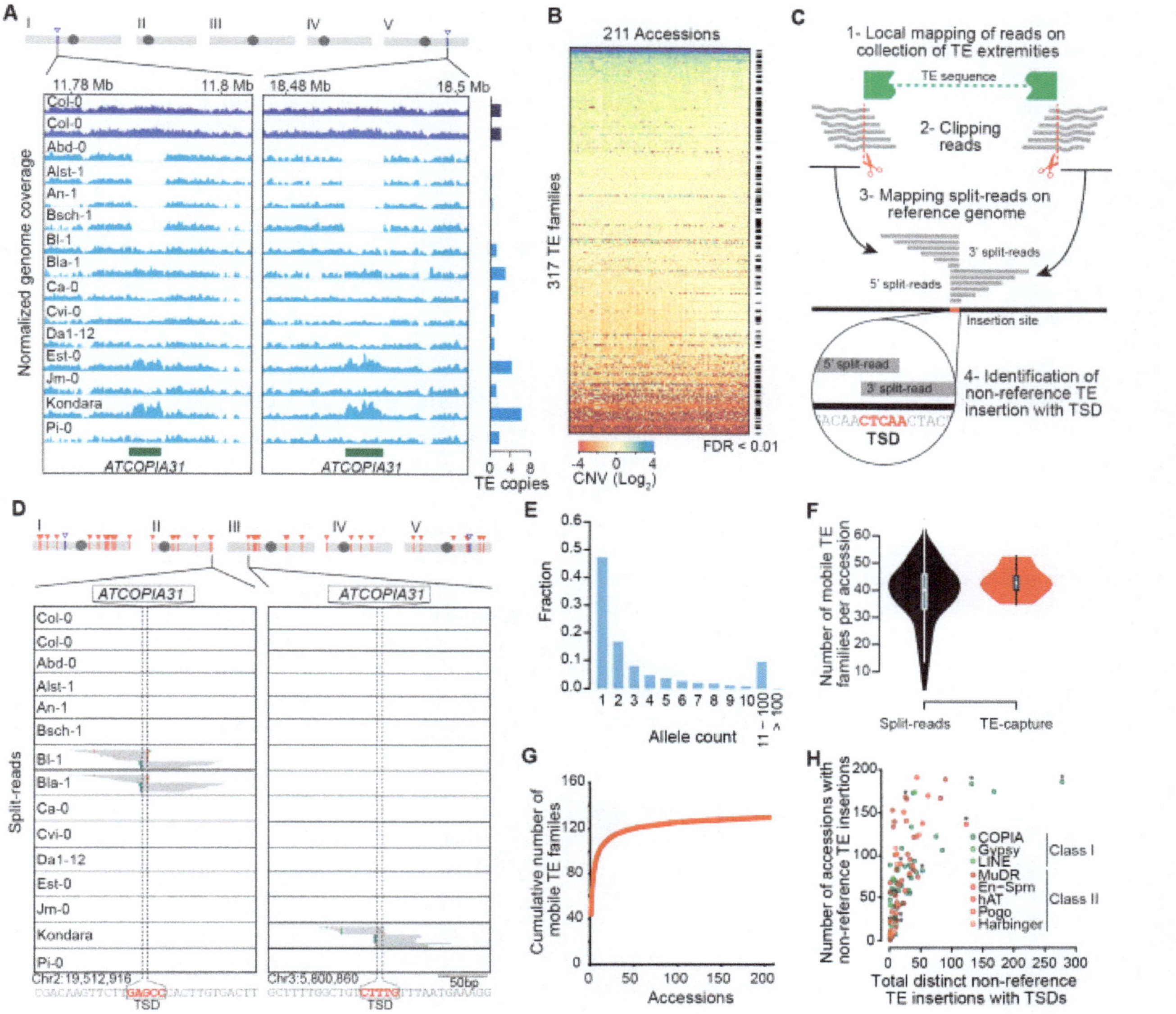

Figure 1: Overview of the ***A. thaliana*** **mobilome.** (A) Genome browser tracks showing normalized sequencing coverage over the two full-length ATCOPIA31 elements annotated in the reference genome (Col-0). CNV is detected as increased or decreased coverage in other accessions. Number of copies is indicated on the right. (B) Heat map representing CNVs (log2 ratio) for 317 TE families and 211 A. thaliana accessions. TE families with statistically significant CNV in at least one accession are indicated. Figure 1—source data 1 contains absolute copy number estimation of TE sequences. (C) Schematic representation of the bioinformatics pipeline to identify non-reference TE insertions with TSD using split-reads. 1- Reads are mapped on a collection of TE extremities from annotated TE sequences and reference sequences (Repbase update). 2- Reads aligning partially over TE extremities are extracted and clipped. 3- The unmapped portion of these split-reads are re-mapped on the Arabidopsis reference genome. 4- Non-reference TE insertions with TSDs are identified by searching for overlapping clusters of 5' and 3' split-reads. (D) Genome browser tracks showing split-reads for two non-reference ATCOPIA31 insertions and TSD reconstruction. Figure 1—source data 2 contains the coordinates of all non-reference TE insertions with TSDs. (E) Distribution frequency of allele counts for non-reference TE insertions with TSDs. (F) Number of mobile TE families per accession identified using split-read and TE-sequence capture. (G) Cumulative plot of the number of mobile TE families detected with increasing numbers of accessions. (H) The total number of non-reference TE insertions with TSDs is indicated in relation to the number of accessions with such insertions, for each of the 131 mobile TE families. Asterisks indicate the nine TE families with experimental evidence of transposition (Ito and Kakutani, 2014; Tsay et al., 1993). Figure 1—source data 3 contains the total number of distinct non-reference TE insertions with TSD for each TE family and super-family. Figure 1—figure supplement 2 shows TE-capture results. Figure 1—figure supplement 1 contains IGV screenshots showing the pattern of split-reads characteristic of true- and false-positive non-reference TE insertions with TSDs.

The ability to detect non-reference TE insertions with TSDs using split-reads is strongly dependent on read depth as well as sequence composition at the insertion site (Figure 1—figure supplement 2A; Hénaff et al., 2015). To assess the extent of this limitation, we used the assembled Ler-1 genome sequence recently obtained using PacBio long reads (see 'Materials and methods'). Although not annotated, this sequence assembly can serve to identify by whole genome comparison the Col-0 TEs flanked by TSDs that are absent from the corresponding position in the Ler-1 genome (see 'Materials and methods'). A total of 142 TEs belonging to 80 distinct families were identified in this way (Figure 1—source data 4), which is consistent with estimates obtained using other approaches (Ziolkowski et al., 2009; Hénaff et al., 2015). In contrast, we could detect only 78 Col-0-specific TEs with TSDs belonging to 49 TE families when using the split-reads pipeline to map Col-0 short reads onto the assembled Ler-1 genome. These results indicate therefore that the split-reads approach has a low sensitivity (Figure 1—figure supplement 1C; 45% false negatives and 10% false positives; False Discovery Rate: 15.3%).

To obtain an independent estimation of the composition of the mobilome, we also performed TE sequence capture (TE-capture; Baillie et al., 2011). Briefly, probes were designed to cover the 5' and 3' extremities of 310 TE elements belonging to 181 distinct families, including 117 of the 131 TE families identified as mobile with the split-reads approach (see 'Materials and methods'). Using genomic DNA extracted from 12 randomly chosen accessions (Figure 1—figure supplement 2D), we could validate by TE-capture most (87%) of the non-reference TE insertions with TSDs that were detected by the split-reads approach (Figure 1—figure supplement 2F; see 'Materials and methods'). As expected, TE-capture also uncovered many additional non-reference TE insertions with TSDs for the same TE families (Figure 1—figure supplement 2F). However, no such insertions were detected for the other TE families that could be captured but which were not identified as mobile by the split-reads approach in any of the 12 accessions. These results confirm that despite the low sensitivity of the latter, we have probably identified most of the TE families with TSDs that compose the *A. thaliana* mobilome at the species level. Finally, non-reference insertions were also identified for 30 TE families (including 15 *HELITRON* families) that could not be analyzed using our split-reads pipeline because they do not produce TSDs or have insertion sites located in low complexity regions (Figure 1—figure supplement 2B). Since most of the non-reference insertions for these 30 TE families are present in only one or two of the 12 accessions examined (Figure 1—figure supplement 2G), they likely reflect recent transposition events. Thus, there are altogether at least 165 TE families with recent transposition activity at the species level. Moreover, based on the TE-capture data, we can estimate that since they diverged from each other any two accessions have accumulated between ~200 and ~300 newly transposed TE copies (Figure 1—figure supplement 2H).

TE mobilization as a complex trait

The observation that the composition of the mobilome differs extensively between accessions (Figure 1F) suggests that it is influenced by environmental and genetic factors. To try to identify such factors, we first established that copy number (CN) correlates positively with the number of TE insertions with TSDs that are detected by TE-capture (Figure 2—figure supplement 1; see 'Materials and methods'). Thus, CNV is a reliable and quantitative estimator of differential TE mobilization between accessions, which we used to analyze the 113 TE families that were defined as mobile based both on the split-reads approach and TE-capture (Figure 2—source data 1).

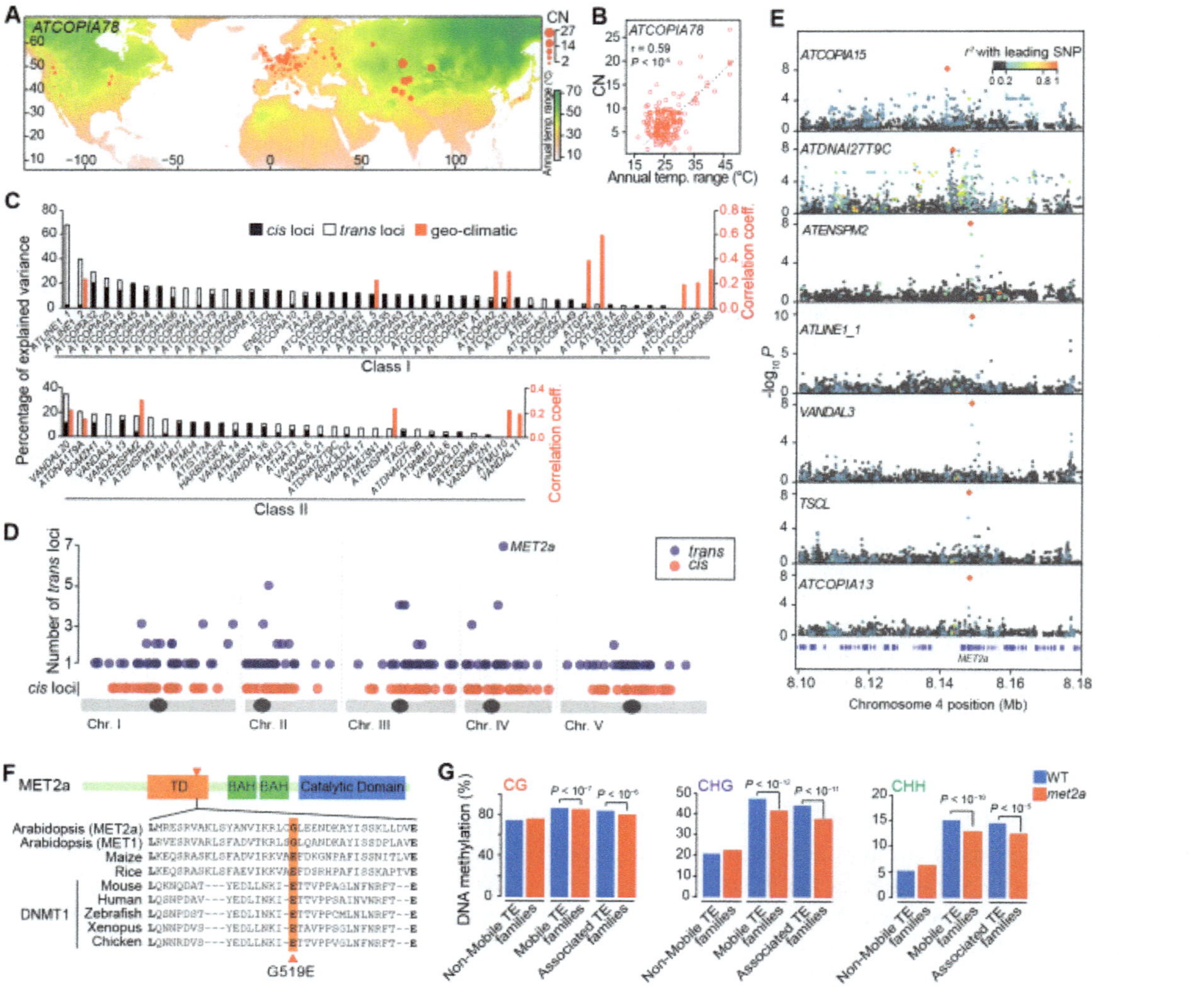

Figure 2: Environmental and genetic factors associated with differential mobilome activity. (A) Copy number (CN, red circles) of ATCOPIA78 in accessions distributed across the globe. Annual temperature range is also shown. (B) Partial Mantel correlation

between ATCOPIA78 CN and annual temperature range. (C) Fraction of CNV variance explained by SNPs (cis, and Trans) and partial Mantel correlation with geo-climatic variables. (D) Distribution of cis and Trans loci in the joined analysis (391 accessions) and number of TE families associated with a given Tran's locus. A complete list of the GWAS results is provided. (E) Manhattan plots displaying GWAS results for the seven TE families with a MET2a association. The leading SNP within each interval is indicated as a red diamond. Colors indicate the extent of linkage disequilibrium (r2) with the leading SNP. (F) Schematic view of the MET2a protein (TD: targeting domain; BAH: bromo adjacent homology domain) and sequence alignment of the TD. The amino acid substitution (G519E) that is present in some accessions is indicated (red arrow). (G) Average DNA methylation level over non-mobile, mobile and MET2a-associated TE families in WT and met2a Col-0 seedlings (Stroud et al., 2013). Statistically significant differences are indicated (MWU test). Figure 2—figure supplement 1 shows the positive correlation between CN and number of non-reference TE insertions with TSDs. Figure 2—figure supplement 2 shows climate association to CNVs. Figure 2—figure supplement 3 shows GWAS results for CNVs.

Controlling for population stratification and considering thirteen geo-climatic variables (Hancock et al., 2011), we uncovered robust correlations with CN for 15 class I and class II TE families. Among these, *ATCOPIA2* and *ATCOPIA78* share the highest number of geo-climatic variables correlated with CN (Figure 2—figure supplement 2). Moreover, the strongest correlation is between temperature annual range and CN for *ATCOPIA78* (Figure 2A and B). Given that at least one member of this TE family is transcriptionally induced by heat shock in the Col-0 accession (Ito et al., 2011; Cavrak et al., 2014), *ATCOPIA78* provides a compelling case of a causal link between climate and TE mobilization.

We next explored the possibility of using GWASs to identify genetic variants influencing TE mobilization (see 'Materials and methods'). For 33 TE families, a disproportionately large number of SNPs are associated with CNV, preventing further analysis. For the remaining 80 TE families, SNPs in linkage disequilibrium with each other and associated with CNVs delineate 230 loci. Moreover, 34% of these loci are also identified by GWAS using whole genome sequencing data obtained for another 180 accessions taken from Sweden (Long et al., 2013) (Figure 2—figure supplement 3A). This substantial overlap suggests a similar genetic architecture for the *A. thaliana* mobilome at both the local and global scales, which prompted us to perform joined GWASs using all 391 accessions in order to increase both sensitivity and specificity. Depending on the TE family, GWASs identified from 0 to 33 loci and collectively, associations explain between 2% and 67% of the variance in CN (Figure 2C).

Among the 334 loci detected in total, 130 encompass sites with reference or non-reference TE insertions (Figure 2C). Furthermore, each of these local (*cis*) genetic variants explains on average 5.2% of the total variance compared with 2.2% for distal (*Trans*) genetic variants. The higher explanatory power of *cis* variants is of course to be expected, as the TEs themselves are the primary determinants of the transposition process. Indeed, almost all *cis* SNPs that map to TE sequences in the reference genome are likely causal as they affect sequences involved in transposition, for example the long terminal repeats (LTRs) and the various open reading frames of LTR retrotransposons (Figure 2—figure supplement 3B and C).

While *cis* loci collectively explain most CN variance for class I TE families, this is not the case for class II TE families (Figure 2—figure supplement 3D). Given that class II TEs move by a cut and paste mechanism, some *trans* loci could in fact correspond to sites of excision. However, we could not find evidence of excision footprints, such as small insertions or deletions. Alternatively, the larger fraction of CN variance explained by *trans* loci for class II TE families may in part result from many of these families being non-autonomous, i.e. requiring factor(s) encoded by other TEs for their mobilization. Consistent with this possibility, the proportion of CN variance associated with *trans* loci as well as the number of TE annotations overlapping *trans* loci are higher for non-autonomous than autonomous class II TE families (Figure 2—figure supplement 3E). Although we did not investigate *trans* mobilization in depth, we readily identified one probable case, involving the non-autonomous and autonomous MuDR families *ATDNA1T9A*and *VANDAL16*, respectively (Figure 2—figure supplement 3F). Of note, CN do not co-vary between these two families, which could indicate that their transposition is differentially controlled. Finally, 'false' *trans* loci could also be caused by non-reference TE insertions that are sufficiently frequent to be in linkage disequilibrium with SNPs used for the GWASs but that we have failed to detect. However, such *trans* loci are not expected to be more prevalent for class II than class I TE families and should be very rare in any case given that the probability of missing moderately frequent (>5%) non-reference TE insertions by both the split-reads approach and TE-capture is low (Figure 2—figure supplement 3G).

Since transposition is controlled by multiple protein activities in addition to those encoded by the TEs themselves, we also examined genes located within or adjacent to *trans* loci. Overall, these genes do not appear to be enriched for any particular function and most of them are specific to a single TE family (Figure 2D and Figure 2—figure supplement 3H). These observations indicate either a complex genetic architecture of mobilome variation and/or spurious *trans* associations such as those considered above. Nonetheless, 22 *trans* loci stand out as they show association with CNV for two or more TE families (Figure 2D) and a causal link is evident in two cases. Indeed, the locus associated with CNV for respectively the retrotransposon and DNA transposon families *ATGP2* and *ATENSPM2* encodes the transcription factor ARF23, which recognizes motifs that are overrepresented in the sequence of these TEs. The second locus is associated with CNV for the largest number of TE families (four class I and three class II families, Figure 2E) and encodes the MET2a protein, a poorly characterized homolog of the main DNA maintenance methyltransferase MET1. Moreover, one of the *MET2a* SNPs is presumably causal as it leads to a non-synonymous amino-acid substitution (G519E) in a conserved domain of the protein (Figure 2F) that in the mammalian homolog Dnmt1 is required for the targeting to replication foci (Klein et al., 2011). A role for*MET2a* is also supported by the observation that *met2a* mutant plants (Stroud et al., 2013) lose some DNA methylation exclusively over mobile TE families. Furthermore, loss of methylation is more pronounced when only considering the seven TE families that show a *MET2a* association (Figure 2G). Intriguingly, CHG sites (where H=A, T or C), which are poor substrates for MET1 or Dnmt1 compared to CG sites (Law and Jacobsen, 2010), are the most affected in the *met2a* mutant. Whether or not this observation reflects an atypical recognition specificity for MET2a remains to be determined. Finally, we note that GWASs failed to detect any association with genes known to be involved in the epigenetic silencing of TEs (Ito and Kakutani, 2014) such as *MET1* and *DDM1*, presumably because of their essential function.

Genome localization of newly inserted TEs

In *A. thaliana* as in many other eukaryotes, TE sequences tend to cluster in pericentromeric regions (Arabidopsis Genome Initiative, 2000). Mechanistically, such clustering may result from insertion bias, selective constraints or differential elimination of TE copies through ectopic homologous recombination (Barrón et al., 2014). To distinguish between these possibilities, we looked at the genomic location of the 2835 non-reference TE insertions with TSDs detected with the split-reads approach and found that they are distributed almost evenly along chromosomes (Figure 3A). Since a similar distribution is observed for the non-reference TE insertions with TSDs detected exclusively using TE capture (Figure 3—figure supplement 1A), we can rule out an ascertainment bias of the split-reads approach towards non-reference TE insertions located along the chromosome arms. However, there is a clear trend towards a more pericentromeric localization when only considering non-reference TE insertions with TSDs that are shared by two or more accessions and that are thus presumably more ancestral (Figure 3B and Figure 3—figure supplement 1B). Moreover, the density of non-reference TE insertions with TSDs positively correlates with the recombination rate but negatively with gene density (Figure 3—figure supplement 1C, D and E). Finally, except for *COPIA* families, non-reference TE insertions with TSDs are globally under-represented within genes, where they are expected to be most detrimental (Figure 3C and Figure 3—figure supplement 1F and Figure 3—figure supplement 2A). Collectively, these observations provide strong evidence that TEs insert equally throughout the genome and are preferentially purged over time from the chromosome arms because of their deleterious effects on adjacent genes rather than as a consequence of ectopic homologous recombination.

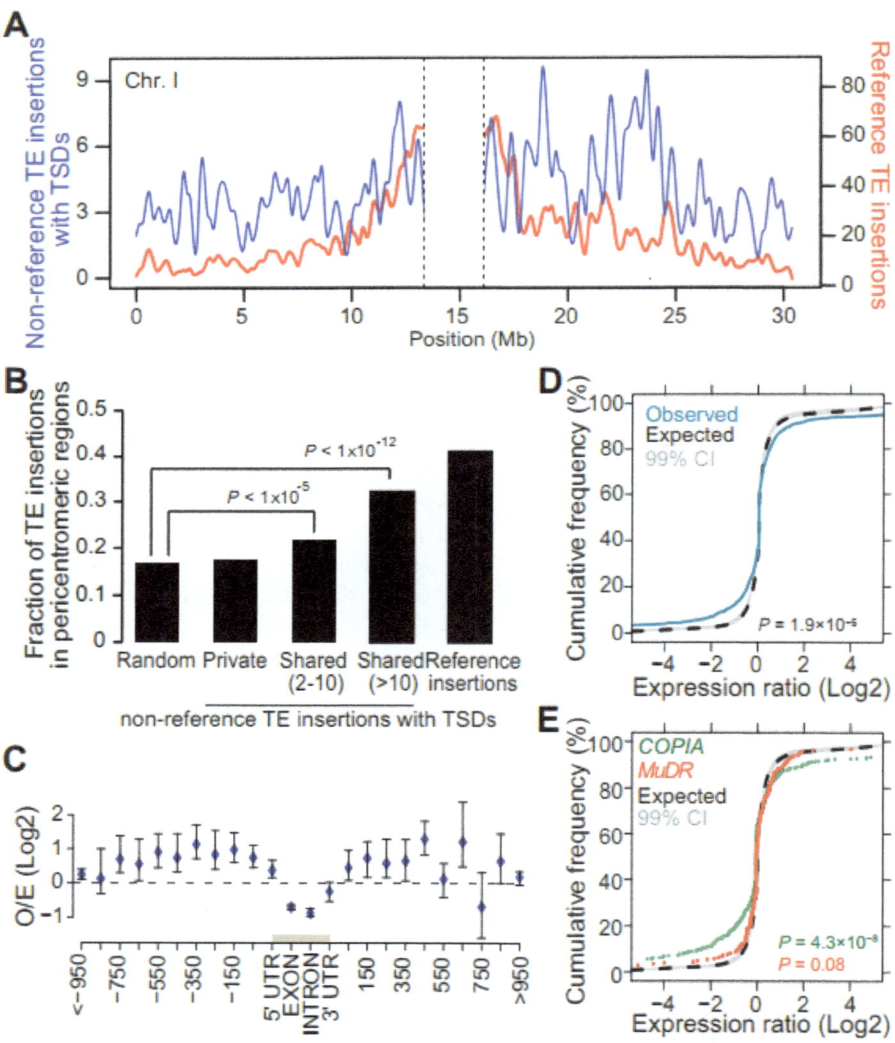

Figure 3: Genomic localization of non-reference TE insertions. (A) Density of non-reference TE insertions with TSDs (blue) and of annotated TE sequences (red) along the reference sequence of chromosome 1. Inner pericentromeric regions are masked. (B) Fraction of private and shared non-reference TE insertions with TSDs and of annotated TE sequences in outer pericentromeric

regions. Statistically significant differences are indicated (chi square test). (C) Observed/expected ratio (O/E) of private non-reference TE insertions with TSDs in and around genes. Errors bars are defined as 95% confidence intervals. (D) Cumulative distribution of gene expression ratios between alleles harboring and lacking non-reference TE insertions. Statistically significant differences were calculated using the KS test. (E) As D, but only for COPIA (green) or MuDR (red) non-reference TE insertions with TSDs. Figure 3—figure supplement 1 shows detailed analysis of the distribution of non-reference TE insertions with TSDs along the genome. Figure 3—figure supplement 2 shows local TE insertion preferences. Figure 3—figure supplement 3 shows global expression levels of gene affected by non-reference TE insertions. Figure 3—figure supplement 4 shows expression levels of genes affected in some accessions by a non-reference insertion with TSD in plants grown under control conditions or subjected to heat stress.

Although most TE families show no overt insertion bias at the genome scale, there are clear local insertion preferences. In agreement with previous observations (Fu et al., 2013; Miyao et al., 2003), private non-reference *COPIA* and *MuDR* insertions with TSDs are enriched at coding sequences and transcriptional start sites (TSS), respectively (Figure 3 figure supplement 2A). In addition, insertion sites for most TE superfamilies are enriched in specific DNA sequence motifs or exhibit biased sequence composition (Figure 3—figure supplement 2B and C). For example, *LINEs* tend to insert within poly(A) tracks, as expected for this superfamily of non-LTR retrotransposons, which integrate into the genome via poly(A)-dependent, target site–primed reverse transcription (TPRT; Beck et al., 2011).

Impact of Newly Inserted Tes on the Expression of Adjacent Genes

Transcriptome analyses in the reference accession Col-0 have revealed that *A.thaliana* genes nearest to TE sequences are expressed at lower levels compared with the genome-wide distribution of gene expression, suggesting that TE insertions tend to reduce the expression of neighboring genes (Hollister and Gaut, 2009). To investigate more directly the impact of TEs on the genes within or near which they insert, we examined RNA-seq data available for 144 accessions (Schmitz et al., 2013). Specifically, we considered all non-reference TE insertions with TSDs and calculated for each gene located within 1 kb of them (1616 genes in total), the ratio between the expression level in the accession(s) harboring the insertion and the median expression level in the accessions devoid of the insertion. Expression ratios expected under the null hypothesis (no effect of the TE insertions) were calculated by taking 10^6 randomly chosen sets of 1616 genes and assigning for each set the TE insertion 'presence/absence' label randomly among the 144 accessions (see 'Materials and methods'). Comparison of the distribution of the observed and expected expression ratios indicates that for a large fraction of genes, expression is indeed significantly altered when TEs insert within or near them (Figure 3D, $p<1.9\times10^{-5}$). These alterations are most pronounced for the *COPIA* insertions, which are overrepresented in genes and less pronounced for the *MuDR* insertions, despite the latter being overrepresented around the TSS of genes (Figure 3E and Figure 3—figure supplement 2A). Although other TE superfamilies show similar trends, we could not draw firm conclusions in these cases because of insufficient statistical power (Figure 3—figure supplement 3). This notwithstanding, it is clear that TE insertions induce both increases and decreases in gene expression with equal frequency (Figure 3D and E). Thus our findings contradict the prevailing view of a dampening effect of TE insertions on the expression of adjacent genes (Hollister and Gaut, 2009) and suggest instead a stronger selection against TE insertions when they occur close to highly expressed genes.

To complement the re-analysis of transcriptome data, we also measured by RT-qPCR the expression level of 19 genes with recent *COPIA* or *MuDR* Insertions, using nine different accessions grown under control conditions or subjected to heat shock. *COPIA* insertions were found to have more dramatic and systematic effects on gene expression in stressed plants (Figure 3—figure supplement 4) which in the case of*ATCOPIA78* can be related to its transcriptional sensitivity to heat shock (Ito et al., 2011; Cavrak et al., 2014). On the other hand, we could not detect any effect of the MuDR insertions under the two conditions tested (Figure 3—figure supplement 4). These findings are in agreement with those of the transcriptome analysis and indicate in addition that the effect of TE insertions on the expression of adjacent genes can vary substantially in relation to the environment.

Impact of Newly Inserted TEs on the DNA Methylation Status of Adjacent Sequences

TE sequences are typically targeted by the RNA-directed DNA methylation (RdDM) machinery in *A. thaliana* (Lippman et al., 2004; Lister et al., 2008; Cokus et al., 2008) and we have previously provided genome-wide evidence that DNA methylation can spread from RdDM targets to flanking sequences, with possible consequences on gene expression (Ahmed et al., 2011). To investigate the effect of new TE insertions on the DNA methylation status of adjacent sequences, we used MethylC-Seq data available for 140 accessions (Schmitz et al., 2013). Analysis of this data set first indicated that mobile TE families have on average higher CG, CHG and CHH methylation than non-mobile TE families (Figure 4A). Furthermore, DNA methylation is also higher for most mobile TE families in the accessions with evidence of recent transposition activity (Figure 4—figure supplement 1A). These observations prompted us to examine in addition methylome data obtained for several mutation accumulation (MA) lines (Becker et al., 2011; Schmitz et al., 2011). Mobile TE families suffer less sporadic DNA methylation loss than non-mobile families (Figure 4B). These findings are entirely consistent with DNA methylation playing an important role in the control of TE mobility and they suggest in turn that most of the recent TE insertions we have identified are present in the methylated state. Moreover, given that DNA methylation is likely established over newly inserted TE copies by RdDM in a progressive manner across multiple generations (Teixeira et al., 2009; Marí-Ordóñez et al., 2013), unmethylated non-reference TE insertions should be mainly private and reflect very recent transposition events.

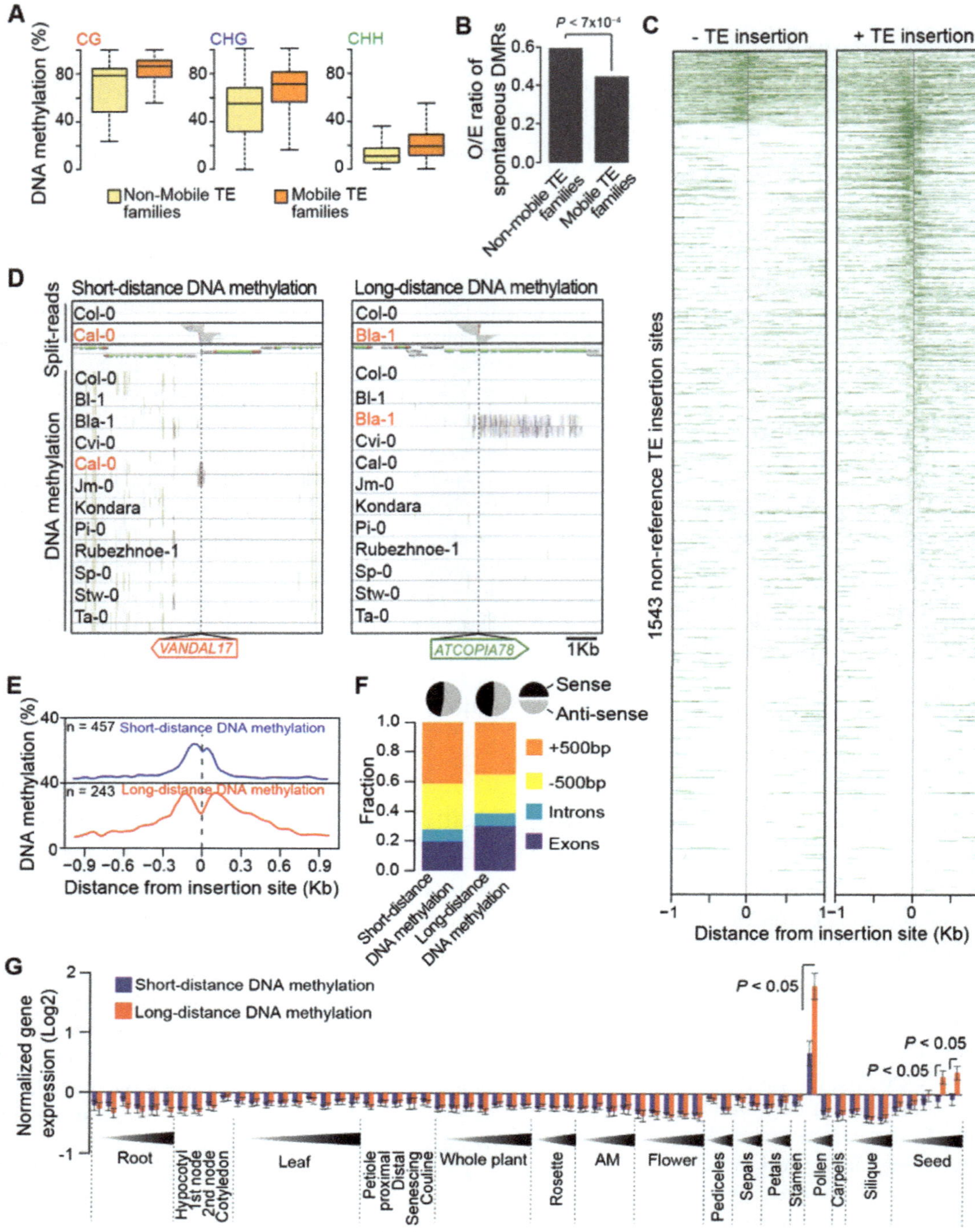

Figure 4: DNA methylation of non-reference TE insertion sites. (A) Boxplot representation of average DNA methylation level for mobile and non-mobile TE families across all accessions. (B) O/E ratio of spontaneous DMRs identified in mutation accumulation lines (Becker et al., 2011; Schmitz et al., 2011) for non-mobile and mobile TE families. Statistically significant differences were calculated using a chi square test. (C) Average DNA methylation level in 50bp windows upstream and downstream of 1543 insertions sites for accessions lacking or containing a given non-reference TE insertion with TSD. (D) Genome browser tracks showing examples of insertion sites respectively associated with short- and long-distance DNA methylation. (E) Meta-analysis of DNA methylation around non-reference TE insertions sites. (F) Distribution of non-reference TE insertions associated with short- or long-distance DNA methylation according to their position relative to genes (stacked bar plot) and proportion of insertions in the two possible orientations relative to the closest gene (pie charts). (G) Average expression level in different organs and at different developmental time points (in Col-0) of genes with non-reference TE insertions with TSDs and affected by short- (blue) or long-distance (red) DNA methylation. Error bars are s.e.m. Statistical significance of differences was calculated using a MWU test. Figure 4—figure supplement 1 shows DNA methylation of TE families and impact on sequences flanking non-reference TE insertions with TSDs.

Based on these considerations, we next analyzed the DNA methylation status of 1543 TE insertion sites for which reliable data could be extracted across all 140 accessions (Figure 4C). Approximately 10% of sites are methylated in most accessions, including systematically in the one(s) containing the TE. As expected, these sites are preferentially located within TE-rich, pericentromeric regions. In contrast, another 40% of sites are devoid of methylation in the accession(s) containing the TE insertion as well as in most of the other accessions. This absence of adjacent DNA methylation could indicate either that the TE insertions themselves are unmethylated or else that DNA methylation does not spread from them. Finally, 50% of sites are methylated exclusively or almost exclusively in the accession(s) with the TE insertion (Figure 4—figure supplement 1B), thus suggesting that at these sites TEs are methylated and that DNA methylation did spread into adjacent sequences. Why some sites may be refractory to DNA methylation spreading when others are not is unclear, as we did not identify any feature that could distinguish them, such as the identity of the TE or the sequence composition at the insertion site.

Further analysis of DNA methylation associated with TE insertions indicates that it affects all three sequence contexts and that it generally extends for up to 300 bp on both sides of the insertions (Figure 4D and E; Figure 4—figure supplement 1B), a distance that closely matches that previously reported for the spreading of DNA methylation from RdDM targets (Ahmed et al., 2011). For 243 insertion sites however, DNA methylation extends over much longer distances (up to 3.5 kb) on one or the other side of the insertion (Figure 4D and E). While most of these sites lie within or close to genes, the TE insertions are not preferentially orientated with respect to gene transcription (Figure 4F), which rules out sense-antisense transcription as the likely trigger for this long-distance DNA methylation. Proximity to pericentromeric heterochromatin can also be ruled out, because most of the genes with long-distance DNA methylation are located on the chromosome arms (Figure 4—figure supplement 1C). To explore potential mechanisms further, we made used of the wealth of epigenomic data available in Col-0 to examine the 142 Col-0 TE insertions with TSDs that are absent from the assembled Ler-1 genome sequence (Figure 1—source data 4). Methylome data (Stroud et al., 2013) indicate that 121 of these 142 Col-0 TE copies are methylated and that DNA methylation tends to extend into flanking sequences, predominantly over short distances, but occasionally over much longer distances (<300 pb: 63 TE insertions; >1 kb: 36 TE insertions; Figure 1—source data 4). These results confirm those obtained for the non-reference TE insertions. In addition, analysis of Col-0 small RNA-seq data (Fahlgren et al., 2009) indicates that in contrast to short-distance DNA methylation, long-distance DNA methylation aligns with 24-nt siRNAs (Figure 4—figure supplement 1D). Thus, genes affected by the latter type of DNA methylation have presumably become secondary targets of RdDM, as was shown for a transgene (Kanno et al., 2008; Daxinger et al., 2009). Moreover, genes affected by long-distance DNA methylation in accessions other than Col-0 tend in the latter accession, where they are by definition in the ancestral state, to be most highly expressed in both pollen and seeds and more highly expressed in these two organs than genes affected by short-distance DNA methylation (Figure 4G). Given that RdDM activity is also maximal in these organs (Teixeira and Colot, 2010), our observations suggest that secondary RdDM results from the concomitance of strong transcription and strong RdDM at target loci.

Finally, our analysis of TE-associated DNA methylation indicates that it accounts for at least 7% of the so-called gene C-DMRs (i.e. regions of differential methylation at CG, CHG and CHH sites) identified in nature, which are typically low frequency gain of DNA methylation variants (Schmitz et al., 2013). These and similar findings reported recently (Stuart et al., 2016) confirm and extend previous results that first indicated that many natural gene C-DMRs are not *bona fide* epialleles but rather new alleles caused by TE insertions (Schmitz et al., 2013). Nonetheless, examination of one TE-insertion allele shared among 13 accessions indicates that it is present in the unmethylated state in one accession and thus possibly subjected to epigenetic variation in nature (Figure 4—figure supplement 1B).

TE Insertions as Motors of Adaptive Changes

Although TEs tend to insert with no overt bias at the genome scale (Figure 3A), we detected nineteen 10 kb windows with a high load of non-reference TE insertions (Figure 5A). Such enrichment could result from insertion preferences or reflect an absence of strong negative selection. In fact, three of these 10 kb windows span genes encoding nucleotide-binding domain and leucine-rich repeat containing (NLR) proteins, which function as immune receptors in plants and are known to be under diversifying selection (Chae et al., 2014). Moreover, a fourth 10 kb window spans the gene *FLC*, which encodes a key repressor of flowering and is one of the main genetic factors causing natural variation in the onset of flowering, another key adaptive trait (Ietswaart et al., 2012). Remarkably, the *FLC* locus has the highest number of non-reference TE insertions (seven in total) across the genome. These insertions belong to several COPIA families and affect four distinct *FLC* haplotypes in total (Figure 5A and Figure 5—figure supplement 1A). Moreover, five insertions are located within the first intron (Figure 5B), which plays an important role in the epigenetic regulation of *FLC*in response to cold (Ietswaart et al., 2012). Although four of these insertions as well as another intronic insertion were previously identified among early flowering accessions (Liu et al., 2004; Lempe et al., 2005), causality could not be established unequivocally because of numerous other sequence polymorphisms in complete linkage disequilibrium. To obtain direct proof of a causal role for the seven TE insertions we identified, we used publically available transcriptomic (Schmitz et al., 2013) as well as phenotypic data (Li et al., 2010; Lempe et al., 2005) and compared *FLC* expression as well as flowering time among accessions that have the same *FLC* haplotype but differ by the presence or absence of a TE insertion (see 'Material and methods'). Results of these comparisons indicate that the TE-containing accessions have systematically much reduced *FLC* expression and flower much earlier than their TE-free counterparts (Figure 5C and D; Figure 5—figure supplement 1B and C). Thus, we can conclude that TEs are recurrent generators of major effect *FLC* alleles, which in turn suggests that they contribute significantly to the high level of allelic heterogeneity observed at this locus (Li et al., 2014)

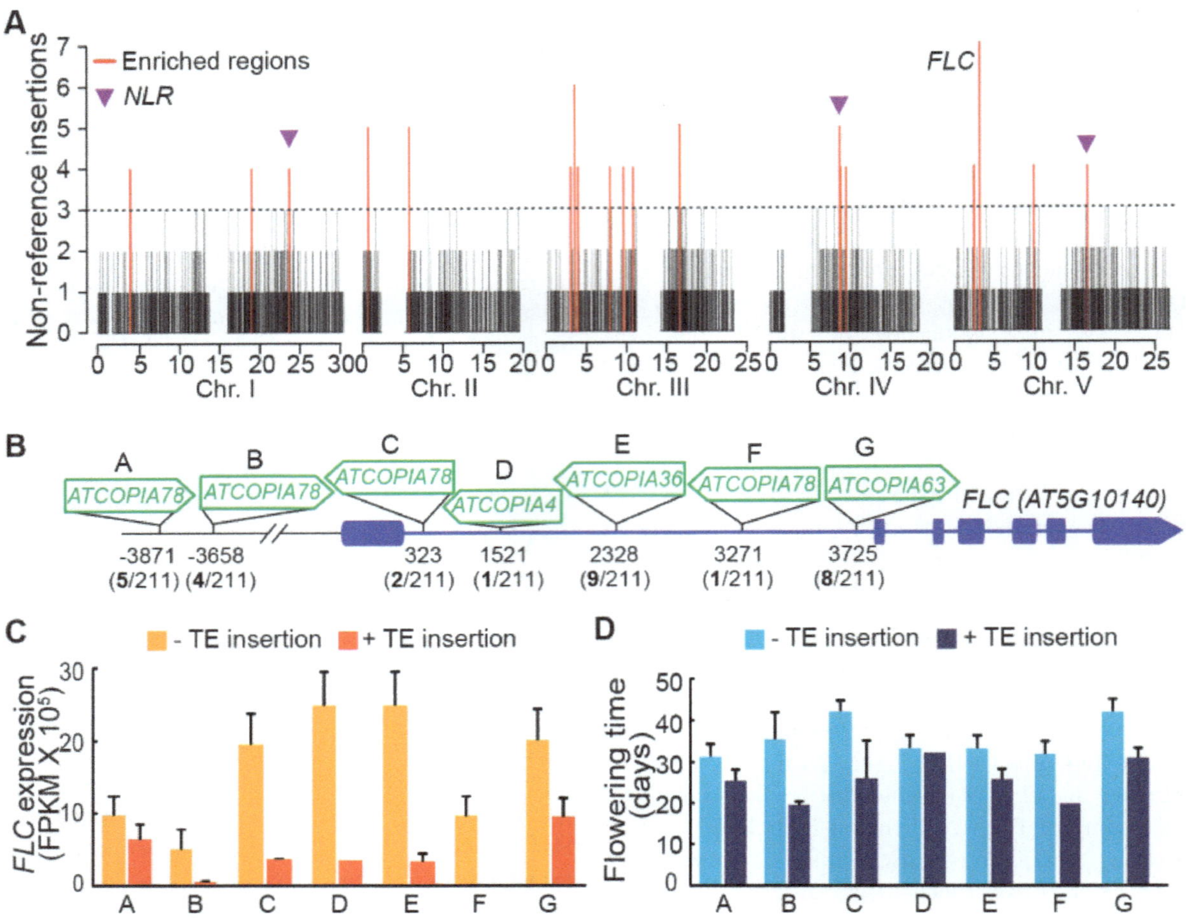

Figure 5: **Local enrichment of non-reference TE insertions with TSDs.** (A) Density of non-reference TE insertions with TSDs in 10 kb windows. The 19 regions statistically enriched in such insertions are indicated by red bars. (B) Position and identity of the seven non-reference TE insertions with TSDs spanning the FLC locus. (D) and (E) Level of FLC expression (D) and flowering time (E) for accessions of same FLC haplotype but differing by the presence or absence of the relevant TE insertion. Errors bars are s.e.m. Figure 5—figure supplement 1 shows the reconstruction of the FLC haplotypes and additional analyses of the effect on flowering timeof non-reference TE insertions with TSDs at the locus.

DISCUSSION

We have shown that the *A. thaliana* mobilome is particularly rich at the species level, being composed of at least half of the 326 TE families that are annotated in the reference genome. This finding is at odds with the prevailing view that most TE families are mere molecular fossils in *A. thaliana,* since they contain a much lower proportion of 'young', i.e. non-degenerated TE copies than in the close relative *A. lyrata* (Hu et al., 2011; Maumus and Quesneville, 2014). Furthermore, we provide definitive evidence that TEs insert throughout the genome, with no overt bias towards pericentromeric regions, which contrasts with the observed clustering of annotated TE sequences around centromeres. However, these discrepancies are easily resolved, since we have also shown that despite the richness of the *A. thaliana* mobilome, most TEs tend to be rapidly purged by natural selection in this species when they insert in the chromosome arms, which are gene-rich. Indeed, our systematic survey indicates that TEs have pervasive effects on the expression and DNA methylation status of genes near or within which they insert. Incidentally, the deleterious effects associated with most transposition events in *A. thaliana* may also explain in part the fact that we did not detect any recent transposition bursts, as these should be strongly counter-selected. Furthermore, because *A. thaliana* is predominantly self-fertilizing, the purging of deleterious TE insertions should be accelerated in this species compared to *A. lyrata*, which is an obligated out-crosser. Given this difference in mating systems, the TE population dynamics of *A. thaliana* and *A. lyrata* are expected to differ significantly (Lockton and Gaut, 2010). Thus, homologous recombination could play a more prominent role in the elimination of TE insertions in *A. lyrata,* as is thought to be the case in *D. melanogaster* (Barrón et al., 2014). However, comprehensive studies similar to those presented here remain to be performed for *A. lyrata* in order to identify conclusively the forces that shape the TE landscape in this species.

We have also shown that the composition and activity of the mobilome vary greatly between accessions. GWASs revealed that this variation is caused in part by sequence polymorphisms within the TEs themselves (*cis* variation), which is in agreement with empirical data and theoretical models indicating that TE families contain only one or a few active, autonomous (i.e. master) TE copies at any one time (Becker et al., 2011). The fact that we could readily detect such *cis* variants may again be linked to the mating system of *A. thaliana*, which on the one hand should increase the probability that disabling mutations accumulate within the few active TE copies that are present within a given lineage, before these copies could transpose further; and on the other hand should decrease the probability of acquiring new active copies through crosses.

Another important result of the GWASs is that natural variation at the *MET2a* locus, which encodes a poorly characterized DNA methyltransferase, has a significant impact on mobilome composition and activity across accessions, being associated with differential transposition activity for seven class I and class II TE families. While the role of MET2a in transposition control remains to be determined experimentally, it is noteworthy that none of the epigenetic repressors of TE activity identified through genetic screens, such as MET1 or DDM1, are associated with natural variation of the mobile me, presumably because of their essential function. Altogether, these findings illustrate the power of GWASs in identifying the genetic factors affecting transposition in nature.

Being a complex trait, TE mobilization is also modulated by environmental factors, and we have identified temperature annual range as a clear contributor to the variation in *ATCOPIA78* mobilization across accessions. Remarkably, this TE family has generated several rare alleles with large effects at the *FLC* locus, which is a major genetic determinant of the onset of flowering in nature. It is therefore tempting to speculate that *ATCOPIA78* may endow *A. thaliana* with a unique ability to adapt to global warming and the associated increase in droughts by facilitating the creation of early flowering *FLC* alleles. Additionally, these observations may provide insights into how *A. thaliana* has been able to colonize efficiently the entire northern hemisphere from a few glacial refugia located in Southern Europe (François et al., 2008).

In summary, our findings have far reaching implications, as they indicate that part of the missing heritability that plagues many GWASs may be accounted for by recent and thus rare TE insertion alleles with large effects (Vinkhuyzen et al., 2013; Brachi et al., 2011). More generally, our study highlights the need for similar species-wide explorations of the mobilome in a variety of organisms in order to assess the true mutational and epimutational impact of transposition as well as its contribution to natural phenotypic variation. In this respect, it can be anticipated that the advent of long read sequencing technologies will greatly facilitate such studies, especially in organisms with large, repeat-rich genomes.

MATERIALS AND METHODS

Data Sources

Whole genome sequencing data (Illumina short paired-end reads) for *A. thaliana* accessions were obtained from the NCBI SRA archive for 210 accessions collected worldwide (Schmitz et al., 2013) SRA012474 as well as for the 180 Swedish accessions (Long et al., 2013; SRA052536). Illumina short paired-end reads from Ler-1 (Schneeberger et al., 2011) and the assembled Ler-1genome obtained using PacBio long reads (unpublished) were retrieved fromhttp://1001genomes.org/data/MPI/MPISchneeberger2011/releases/current/Ler-1/Reads/ and https://github.com/PacificBiosciences/DevNet/wiki/Arabidopsis-P5C3, respectively. High quality SNP imputations for these accessions were obtained from the 1001 Genomes Project (http://1001genomes.org/). Processed MethylC-seq and RNA-seq data from accessions collected worldwide (Schmitz et al., 2013) were obtained from NCBI GEO (GSE43857 and GSE43858, respectively). Processed BS-seq data for the *met2a* mutant (Stroud et al., 2013) were obtained from NCBI GEO (GSE39901). Flowering time data were retrieved from the phenotypic information deposited at https://www.arabidopsis.org/servlets/TairObject?type=germplasm&id=6530472136 as well as (Li et al., 2010; Lempe et al., 2005).

Local climate data and coordinates for the 211 accessions analyzed using CNV were retrieved from (Hancock et al., 2011) (http://bergelson.uchicago.edu/wp-content/uploads/2015/04/allvars948_notnormd_011311.txt.zip). TE-capture sequencing data have been deposited in the ENA short read archive under project number PRJEB11706 and accessions codes ERR1121179 to ERR1121190.

CNV Analysis

We performed an aggregated CNV analysis for each TE family by considering only annotated TE sequences longer than 300bp as a joined pseudo-annotation. Families composed in the Col-0 reference genome exclusively of TE sequences shorter than 350bp (seven families, *AR12, AR3, ATCLUST1, ATDNATA1, ATMSAT1, ATSINE1* and *ATSINE3*), that contain too many (>1000) copies or that overlap with genomic regions with abnormal coverage (four families, *ATREP15, ATREP10D, ATREP3* and *HELITRONY3*; see next section) were not considered. Illumina pair-end short reads were mapped onto the TAIR10 reference genome using Bowtie2 (using the arguments --mp 13 --rdg 8,5 --rfg 8,5 --very-sensitive) and PCR-duplicates were removed using Picard. Read coverage (RC) was computed in non-overlapping windows of 100bp spanning the joined pseudo-annotation for each TE family. RC over each bin was corrected for GC content (Yoon et al., 2009) and normalized to the genome-wide RD. Normalized RC across joined pseudo-annotations were compared between each accession and two independent re-sequenced Col-0 reference genomes (Schmitz et al., 2013; Long et al., 2013). CNVs were detected by performing a distribution-independent permutation test for each of the two references. In order to be as stringent as possible, the maximum *P*-value from these two comparisons were considered. These p-values were compared with an empirical null-distribution of p-values constructed by randomizing a million times the TE annotation labels over the 100bp windows. We defined the false discovery rate (FDR) of an observed *P*-value as the fraction of significant (that is, below the observed *P*-value in question) hits found in the randomized set. We considered statistically significant TE-CNVs when FDR was below 10^{-5}.

Filtering out Genomic Regions with Aberrant Coverage or Low Sequence Complexity

Read mapping to the reference *A. thaliana* genome sequence revealed several regions with very high coverage, which correspond mainly to the 180 bp centromeric repeat unit, the 45S and 5S rDNA repeat units, *ATHILA* TE sequences, telomeres and plastid DNA-like sequences. These as well as the few regions not suitable for sequence alignment because of low complexity were identified by mapping genome sequencing reads obtained for two independent lines of the Col-0 accession (Schmitz et al., 2013; Long et al., 2013) onto the TAIR10 genome sequence. RC was calculated on consecutive non-overlapping windows of 100 bp across the entire genome. After correction for GC content (Yoon et al., 2009), consecutive windows (allowing one window gap) with a RC greater or lower than three median absolute deviations from the median RC signal were merged to define larger segments. These sequences spanned 1,125,487 bp (0.94% of the TAIR10 genome sequence) and were excluded from all analyses.

Identification of non-reference TE insertions with TSDs using split-reads

The split-read analysis was performed in four steps: (i) extraction of reads not mapping to the reference genome; (ii) forced mapping to a collection of TE sequence extremities and soft-clipping of mapped reads; (iii) mapping of clipped reads to the reference genome; (iv) identification of clipped reads that reveal target site duplications (TSDs). Briefly, for each accession we retrieved reads that do not map to the TAIR10 reference genome (containing the SAM flag 4), but we did not make use of information about discordant paired-end reads, unlike in the Jitterbug approach (Hénaff et al., 2015). Unmapped reads were then aligned (using Bowtie2 in --local mode to allow for soft clip alignments) to a collection of 5' and 3' TE sequence extremities (300 bp) obtained from the Col-0 for each TE family or from Repbase Update (Jurka et al., 2005) in the case of the *ARNOLDY2, ATCOPIA62, ATCOPIA95, TA12, TAG1* families, which do not contain copies with intact extremities in the Col-0 genome. Next, we selected all reads with one end (≥20 nt) mapping to a TE extremity (by locating reads where the CIGAR string contained only one 'S' character with a value equal or greater to 20). These reads were recursively soft clipped by 1 nt and mapped to the TAIR10 reference genome using Bowtie2 (using the arguments --end-to-end --mp 13 --rdg 8,5 --rfg 8,5 --very-sensitive) until the soft-clipped read length reached 20 nt. Read clusters composed of five or more reads clipped from the same extremity and overlapping with read clusters composed of reads clipped from the other extremity were taken to indicate the presence of a bona fide TE insertion only if the size of the overlap was equal or less than two-fold larger than that reported for TSDs for the corresponding TE family. Putative non-reference TE insertions overlapping genomic regions with aberrant coverage (as defined above) or located within inner pericentromeres (Clark et al., 2007) or spanning the corresponding donor TE sequence were filtered out. Presence or absence of these putative non-reference TE insertions was verified in other accessions by relaxing the parameters used to detect them in the first place. Specifically, we asked for the presence of a minimum of two rather than ten soft-clipped reads spanning the corresponding TSD coordinates. This improved the discovery of putative TE-insertions that are shared by more than one accession.

Estimation of Sensitivity

The rate of false negatives was estimated indirectly by first searching for TE insertions with TSDs that are present in the Col-0 reference genome but absent from the Ler-1 assembled genome sequence recently obtained using PacBio long-reads (unpublished). Specifically, annotated TE sequences together with 1 kb of upstream and downstream sequences were aligned using BLAT to the Ler-1 genome sequence. Heavily truncated TE sequences (covering less than 50% of full-length copies) were not considered, as most of these are unlikely to have been mobilized recently. Out of the 7931 TE sequences analyzed, 200 are not present in the Ler-1 genome sequence and among these, 142 have TSDs (Figure 1—source data 4). This value was then compared to that obtained by mapping Col-0 Illumina short split-reads to the Ler-1 genome sequence. A total of 64 TE insertions with TSDs could not be detected using that approach, thus giving a false negative rate of 45% (Figure 1 figure supplement 2C). Furthermore, 15 TE insertions were called that are in fact not present in the Col-0 genome sequence, thus giving a false positive rate of 10%; (i.e. FDR of 15.3%; Figure 1—figure supplement 2C).

Manual Filtering of Residual False Positives

We analyzed two re-sequenced Col-0 genomes (Schmitz et al., 2013; Long et al., 2013) using our split-read pipeline. Non-reference TE insertions with TSDs detected in other accessions and also identified in at least one of the two Col-0 genomes were filtered out, as they are most likely errors. Additionally, after manual inspection using the Integrative Genomics Viewer (IGV; Robinson et al., 2011), we eliminated all putative non-reference TE insertions with TSDs for which read-clusters comprise a disproportionately high number of reads (more than three MADs over the median whole genome coverage) as well as all putative non-reference TE insertions with TSDs that are present in 50% of accessions, as these are frequently the result of sequencing errors or mapping artifacts. Indeed, the pattern of 5' and 3' split-reads at the insertion site differs radically between false and *bona fide* non-reference TE insertions, which enable their discrimination by visual inspection (Figure 1 figure supplement 1). We also observed a high rate of non-reference TE insertions with TSDs within TE sequences, mostly in the case of the *ATHILA* and *GYPSY* families. Visual inspection of these insertions indicates that they most likely result from mapping artifacts and they were therefore excluded. Finally, we examined 12 non-reference TE insertions with TSDs by PCR and all were validated

TE Sequence Capture

A complete description of our TE-capture design and protocol will be published elsewhere. Briefly 2.1M biotinylated capture probes (Roche-NimbleGen) were designed to cover 200bp at each end of 310 potentially mobile TEs, belonging to 181 of the 326 TE families annotated in TAIR10. These 181 TE families include 117 of the 131 mobile TE families identified using the split-read approach as well as most other TE families for which non-degenerate and thus potentially mobile copies are present in the Col-0 genome. Mobile or potentially mobile TE families with skewed sequence composition or very high copy number (including all ATHILA families) were excluded from the design in order to avoid overt biases in the capture of TE sequences. Twelve DNA sequence libraries for the accessions Pi-0, Bl-1, Sp-0, Ta-0, Lip-0, Bla-1, Stw-0, Cvi-0, Kondara, Jm-0, Gre-0 and Rubezhnoe-1 were prepared following the Illumina TruSeq paired-end kit, with some modifications. One µg of genomic DNA was sheared to a median fragment size of 400-450 bp. Fragments were subjected to end repair, A-tailing and index adapter ligation following the manufacturer's instructions. Libraries were then amplified through 7 cycles of ligation-mediated PCR using the KAPA HiFi Hot Start Ready Mix and primers AATGATACGGCGACCACCGAGA and CAAGCAGAAGACGGCATACGAG at a final concentration of 2 μM. PCR products were cleaned-up using the Qiaquick PCR Purification kit following the manufacturer's instructions, except that DNA was eluted in PCR grade water. Amplified DNA libraries were then pooled and one μg of the multiplex pool was used in the first hybridization step to capture probes (72 hr). Captured DNA was recovered using Streptavidin Dynabeads, washed and PCR amplified (5 cycles) as above. Amplified captured DNA was then subjected to a second round of hybridization (12 hr), recovery and amplification (14 cycles). Hybridization, captured DNA recovery and washing were performed as described in the NimbleGen SeqCapture EZ protocol. Pre-capture and post-second capture PCR products were run on an Agilent Bioanalyser DNA 1000 chip. Enrichment for capture TE sequences was confirmed by qPCR and estimated at 50–100 fold depending on the TE. Pair-end sequencing was performed with Illumina HiSeq 2000 and 100 bp reads. Between 10.7 and 16.1 million pairs were sequenced per library. Pairs were mapped to the TAIR10 reference genome using Bowtie2 with the arguments --mp 13 --rdg 8, 5 --rfg 8, 5 --very-sensitive -X 1000. . Discordant pairs were remapped using Bowtie2 with the parameters --mp 13 --rdg 8, 5 --rfg 8, 5 -D 20 -R 10 -N 0 -L 15 -i S, 1, 0.50 -k 100. TE insertions were detected in each accession using discordantly mapped reads and the algorithm Hydra (Quinlan et al., 2010) using the arguments: pairDiscordants: -d 50000 –z 1000 –x0 –r 100; dedupDiscordants: -s3; hydra: -ms 50 –li –use all –mld (10*mad) –mno (median+ (20*mad). Enrichment factor was ~250 fold on average (Figure 1—figure supplement 2E). Comparing the results of TE-capture with those obtained using our split-read pipeline gives respectively false negative and positive rates of 56% and 7% (i.e. FDR of 14%; Figure 1—figure supplement 2E). These values are similar to those obtained using the assembled genome sequence of Ler-1 and confirm the low sensitivity of the split-read pipeline.

GWAS of CNVs

Unlike for the initial CNV analysis, we considered this time only the annotated TE sequences that are at least half the size of the corresponding reference TE sequence (obtained from the *A. thaliana* RepeatMasker repeat library, http://repeatmasker.org) as this improved significantly the correlation between CN and the number of non-reference TE insertions. For each of the 113 families identified as mobile by both the split-reads approach and TE-capture, aggregated CN was obtained for each accession by summing the CNs estimated for each annotated copy. GWAS on these CN values was carried out with imputed SNPs (MAF > 5%) from the 211 accessions collected worldwide (Schneeberger et al., 2011; Schmitz et al., 2013), the 180 Swedish accessions (Long et al., 2013) and the joined dataset using a linear mixed model (LMM) (Kang et al., 2010). Kinship matrix was included in the model as a random effect to control population structure. Out of the 113 TE families, 33 displayed a genome inflation factor (GIF) greater than 1.10 and were excluded from subsequent analysis. A conservative threshold value (p-value $< 1\times10^{-8}$) was set to call statistically associated SNPs. Proximal associated SNPs in linkage disequilibrium ($r^2 > 0.2$) were identified using Plink (Purcell et al., 2007) and combined in blocks to build statistical associated intervals, which were expanded by 1 or 5 kb on either side. Variance explained by the leading SNP within each locus was calculated using the following equation: $\sigma^2_{SNP} = 2(MAF)(1-MAF)\frac{\beta^2}{\sigma^2_y},$ in which MAF is the minor allele frequency, β is the SNP effect estimated by the LMM and σ_y^2 is the variance of the phenotype Y. Total variance explained by*cis* and *trans* loci was computed as the sum of the single-locus explained variances, under the assumption of additive contributions. The probability P(f) of missing a non-reference TE insertion with TSD underlying a 'false' *trans* locus through both the split reads and TE-capture approaches was calculated using the following equation:

$$P(f) = FN \times \binom{n}{12-k} \times f^{12-k} \times (1-f)^k$$

where f is the frequency of the minor allele for the *trans* locus under consideration (reported in the), FN is the false negative rate for the split-reads bioinformatic pipeline (0.56), n is the number of accessions analyzed by TE-capture (12) and k is the number of accessions without the non-reference TE insertion with TSD among the 12 accessions analyzed by TE-capture (k ranges from 0 to 12 and calculations were all performed using k=12).

Climate Analysis

We selected 12 geo-climatic variables representing different ecological layers: Aridity, number of frosty days, number of consecutive frost-free days, day length in the spring, maximum temperature in the warmest month, minimum temperature in the coldest month, temperature annual range, photosynthetically active radiation, precipitation in the wettest month, precipitation in the driest month, relative humidity in the summer, landscape slope and thermal (Hancock et al., 2011). CNs for the 113 families confirmed as mobile by the split-read approach and TE-capture were used to calculate a partial Mantel correlation with the 13 geo-climatic variables. Population structure kinship was included in the test to control population stratification. Partial Mantel tests were conducted using the 'ecodist' package in R. A threshold of $p<0.01$, corrected for multiple testing (8.33×10^{-4}), was set to call statistically correlated variables. In addition to using the partial Mantel test, we also applied linear models to regress CN as a function of the climatic variables. Although this method does not control for population structure, it largely confirmed the associations found by the partial Mantel test.

Characterization of *cis* and *Trans* Associations

Protein coding genes, miRNA genes, ncRNAs and TE annotations were retrieved from (ftp://ftp.arabidopsis.org/home/tair/Genes/TAIR10_genome_release/TAIR10_gff3/TAIR10_GFF3_genes.gff). GWAS intervals were defined as *cis* associations if they overlap with a TE annotation or a non-reference TE insertion with TSDs of the same family. All other GWAS intervals were defined as *Tran's* associations. The *cis* associations are over-represented 34 times (p-value $<1 \times 10^{-16}$) when compared to randomly chosen genomic intervals of the same size, which is consistent with TE activity being primarily determined by the TE sequence itself. All genomic annotations overlapping with *Trans* intervals (within 1 or 5 kb) were considered as being putatively causal.

Analysis of the localization of non-reference TE insertions with TSDs along chromosomes

To assess if non-reference TE insertions with TSDs are enriched in pericentromeric regions, their number within these regions was compared with that expected from a random distribution. Insertions with different allele counts (private, shared by 2–10 accessions, shared by >10 accessions) were considered separately. TE distribution in the reference genome (TAIR10) was obtained by counting the number of TE sequences located within pericentromeres (minus genomic regions showing aberrant coverage and inner pericentromeres). The expected distribution for the 2835 non-reference TE insertions with TSDs was calculated by randomizing 10^6 times their position across the chromosomes (genomic regions showing coverage deviation, the inner pericentromeres, or coordinates spanning the corresponding reference TE sequence were excluded). This set of random positions was used as a control for all subsequent analyses. Insertion distribution over genes and neighboring sequences was performed using a metagene. Briefly, protein coding gene features were extracted from the TAIR10 annotation and coordinates of non-reference TE insertions with TSDs were crossed with the set of genic features according to the following stepwise hierarchy: 5' UTR > 3' UTR> exon > intron > intergenic regions. For insertions that do not overlap protein-coding genes, the distance to the closest gene was calculated and reported as negative or positive distance according to the gene orientation. Expected insertion distribution under the null hypothesis was retrieved by applying this procedure for each of the 10^6 randomized sets of insertions. To assess if non-reference TE insertions with TSDs are enriched within small clusters, we divided the genome into 10 kb non-overlapping windows and we counted the number of insertions events within them. The observed and expected (random) densities of non-reference TE insertions with TSDs per window were compared and significant enrichment was declared when the number of insertions found within a window was in the upper 0.005% tail of the random distribution.

Reconstruction of the Historical Recombination Landscape

Historical recombination was estimated using LDhat (McVean et al., 2004) as described before (Choi et al., 2013). Briefly, biallelic SNPs (MAF ≥ 0.1) from the 210 accessions collected worldwide were selected and split into blocks of 5000 SNPs with overlap of 500 SNPs. SNPs located in inner pericentromeres were excluded from the analysis. Blocks of SNPs were formatted using the 'convert' program. A likelihood lookup table was generated for 210 individuals with program 'complete' using the following parameters: –n 210 –rhomax 100 –n_pts 100 –theta 0.001. Population-scaled recombination rate (ρ/kb, $\rho = 4Ner$, where Ne is the effective population size and r is the per-generation recombination rate) was calculated using the 'interval' program with the following parameters: -its 60000000 –bpen 5 –sam 40000. Recombination rates for contiguous blocks were joined at overlap position 250. Population-scaled recombination rate map is provided

Identification of DNA Sequence Motifs Overrepresented At Non-Reference TE Insertion Sites with TSDS

Sequence spanning non-reference insertion sites were analyzed using Bioprospector Release 2 (Liu et al., 2001) using the following parameters: -r 1 -n 200 -a 1 -W 'TSD size'. Background sequence distribution for the reference genome was obtained using the 'genomebg' program. Sequence logo was produced using the seqLogo package version 1.36.0.

Assessing the Impact of Non-Reference TE Insertions with TSDS on Gene Expression

All genes with a non-reference TE insertion with TSD within 1 kb were retrieved and their expression analyzed using transcriptome data available for 144 accessions (Schmitz et al., 2013). For each gene, we calculated the ratio between the median gene expression level for the accessions harboring the TE insertion and the median gene expression level for accessions lacking that insertion. Distribution plots of observed gene expression ratios were compared to the expected distribution under the null hypothesis (random effect). This expected distribution was obtained by calculating the gene expression ratio for 10^6 randomly chosen sets of genes for which the TE insertion presence/absence 'label' was randomly assigned between the accessions. Statistical significance of differences between the observed and expected distributions was determined using the Kolmogorov-Smirnov test.

Expression levels of genes affected by non-reference insertions with TSDs

RNA was extracted using the RNeasy plant mini kit (Qiagen) from plants grown under normal conditions (10 days old seedlings grown in liquid medium) or subjected to heat shock treatment (Ito et al., 2011). RT-qPCR was performed as described previously (Silveira et al., 2013). Primers details are given RT-qPCR results (one biological replicate only) are indicated relative to those obtained for a gene (*AT5G13440*) that shows invariant expression under multiple conditions.

Gene expression Profile of Genes Affected by Long-Distance DNA Methylation

Expression data in the Col-0 accession were obtained fromhttp://www.weigelworld.org/resources/ microarray/AtGenExpress/AtGE_dev_gcRMA.txt.zip/at_download/file for the 224 and 162 genes affected respectively by short- and long-distance DNA methylation in accessions with non-reference TE insertions with TSDs (by definition, these TE insertions are absent in Col-0). Triplicate data for each developmental time point was averaged and then normalized across the developmental time-point series. Average expression level was then calculated for each time point for all genes affected by short-distance DNA methylation and compared to the average calculated for all genes affected by long-distance DNA methylation. Statistical significance of differences between these two averages was calculated using the non-parametric Mann-Whitney U test.

Determination of *FLC* Haplotypes

Haplotype analysis was performed as described previously (Li et al., 2014). Briefly, SNPs within 100 kb of *FLC* were retrieved for the 211 worldwide accessions and used as input into fast PHASE version 1.4.0 (Scheet and Stephens, 2006). Default parameters were kept, except for the -Pzp option. For each SNP, haplotype membership with the highest likelihood was assigned.

Code Availability

Source code for the split-read pipeline can be accessed at https://github.com/LeanQ/SPLITREADER

ACKNOWLEDGEMENTS

We thank members of the Colot lab and especially Mathilde Etcheverry for discussions. We thank Edith Heard and Pierre Capy for critical reading of an earlier version of the manuscript. This work was supported by the European Union Seventh Framework Programme Network of Excellence EpiGeneSys (Award 257082, to VC), the Investissements d'Avenir ANR-10-LABX-54 MEMO LIFE, ANR-11-IDEX-0001-02 PSL* Research University and ANR-12-ADAP-0020-01 (to VC) and the Chaire Blaise Pascal (to RAM). LQ was the recipient of postdoctoral fellowships from the ANR-10-LABX-54 MEMO LIFE and ANR-11-IDEX-0001-02 PSL* Research University. ABS was the recipient of postdoctoral fellowships from the ANR-12-ADAP-0020-01 and from the Brazilian National Council for Scientific and Technological Development (CNPq). LQ and VC conceived and designed the study. LQ performed all of the bioinformatic analyses. ABS, GFM and JAJ designed the TE-sequence capture, ABS performed the experiment and CL and RAM performed the sequencing of captured fragments. ABS performed all of the other experimental analyses. LQ and VC wrote the paper, with additional input from all authors. Correspondence and requests for materials should be addressed to colot@biologie.ens.fr.

REFERENCES

1. Ahmed I, Sarazin A, Bowler C, Colot V, Quesneville H. Genome-wide evidence for local DNA methylation spreading from small RNA-targeted sequences in Arabidopsis. Nucleic Acids Research.2011; 39:6919–6931. doi: 10.1093/nar/gkr324.
2. Arabidopsis Genome Initiative Analysis of the genome sequence of the flowering plant *Arabidopsis thaliana.* Nature. 2000; 408:796–815. doi: 10.1038/35048692.
3. Baillie JK, Barnett MW, Upton KR, Gerhardt DJ, Richmond TA, De Sapio F, Brennan PM, Rizzu P, Smith S, Fell M, Talbot RT, Gustincich S, Freeman TC, Mattick JS, Hume DA, Heutink P, Carninci P, Jeddeloh JA, Faulkner GJ. Somatic retrotransposition alters the genetic landscape of the human brain. Nature. 2011; 479:534–537. doi: 10.1038/nature10531.
4. Barrón MG, Fiston-Lavier AS, Petrov DA, González J, Petrov DA. Population genomics of transposable elements in Drosophila. Annual Review of Genetics. 2014; 48:561–581. doi: 10.1146/annurev-genet-120213-092359.
5. Beck CR, Garcia-Perez JL, Badge RM, Moran JV. LINE-1 elements in structural variation and disease. Annual Review of Genomics and Human Genetics. 2011; 12:187–215. doi: 10.1146/annurev-genom-082509-141802.
6. Becker C, Hagmann J, Müller J, Koenig D, Stegle O, Borgwardt K, Weigel D. Spontaneous epigenetic variation in the *Arabidopsis thaliana* methylome. Nature. 2011; 480:245–252. doi: 10.1038/nature10555.
7. Bennetzen JL, Wang H. The contributions of transposable elements to the structure, function, and evolution of plant genomes. Annual Review of Plant Biology. 2014; 65:505–530. doi: 10.1146/annurev-arplant-050213-035811.
8. Brachi B, Morris GP, Borevitz JO. Genome-wide association studies in plants: the missing heritability is in the field. Genome Biology. 2011; 12:232. doi: 10.1186/gb-2011-12-10-232.
9. Cao J, Schneeberger K, Ossowski S, Günther T, Bender S, Fitz J, Koenig D, Lanz C, Stegle O, Lippert C, Wang X, Ott F, Müller J, Alonso-Blanco C, Borgwardt K, Schmid KJ, Weigel D. Whole-genome sequencing of multiple *Arabidopsis thaliana* populations. Nature Genetics. 2011; 43:956–963. doi: 10.1038/ng.911.
10. Cavrak VV, Lettner N, Jamge S, Kosarewicz A, Bayer LM, Mittelsten Scheid O, Scheid OM. How a retrotransposon exploits the plant's heat stress response for its activation. PLoS Genetics.2014; 10:e15716 doi: 10.1371/journal.pgen.1004115.
11. Chae E, Bomblies K, Kim ST, Karelina D, Zaidem M, Ossowski S, Martín-Pizarro C, Laitinen RA, Rowan BA, Tenenboim H, Lechner S, Demar M, Habring-Müller A, Lanz C, Rätsch G, Weigel D. Species-wide genetic incompatibility analysis identifies immune genes as hot spots of deleterious epistasis. Cell. 2014; 159:1341–1351. doi: 10.1016/j.cell.2014.10.049.
12. Choi K, Zhao X, Kelly KA, Venn O, Higgins JD, Yelina NE, Hardcastle TJ, Ziolkowski PA, Copenhaver GP, Franklin FC, McVean G, Henderson IR. Arabidopsis meiotic crossover hot spots overlap with H2A.*Z nucleosomes at gene* promoters. Nature Genetics. 2013; 45:1327–1336. doi: 10.1038/ng.2766.
13. Clark RM, Schweikert G, Toomajian C, Ossowski S, Zeller G, Shinn P, Warthmann N, Hu TT, Fu G, Hinds DA, Chen H, Frazer KA, Huson DH, Schölkopf B, Nordborg M, Rätsch G, Ecker JR, Weigel D. Common sequence polymorphisms shaping genetic diversity in *Arabidopsis thaliana*. Science.2007; 317:338–342. doi: 10.1126/science.1138632.
14. Cokus SJ, Feng S, Zhang X, Chen Z, Merriman B, Haudenschild CD, Pradhan S, Nelson SF, Pellegrini M, Jacobsen SE. Shotgun bisulphite sequencing of the Arabidopsis genome reveals DNA methylation patterning. Nature. 2008; 452:215–219. doi: 10.1038/nature06745.
15. Cridland JM, Macdonald SJ, Long AD, Thornton KR. Abundance and distribution of transposable elements in two Drosophila QTL mapping resources. Molecular Biology and Evolution.2013; 30:2311–2327. doi: 10.1093/molbev/mst129.

16. Daxinger L, Kanno T, Bucher E, van der Winden J, Naumann U, Matzke AJ, Matzke M, Winden Johannes van der. A stepwise pathway for biogenesis of 24-nt secondary siRNAs and spreading of DNA methylation. The EMBO Journal. 2009; 28:48–57. doi: 10.1038/emboj.2008.260.
17. Dubin MJ, Zhang P, Meng D, Remigereau MS, Osborne EJ, Paolo Casale F, Drewe P, Kahles A, Jean G, Vilhjálmsson B, Jagoda J, Irez S, Voronin V, Song Q, Long Q, Rätsch G, Stegle O, Clark RM, Nordborg M, Casale FP. DNA methylation in Arabidopsis has a genetic basis and shows evidence of local adaptation. eLife. 2015; 4:e15716 doi: 10.7554/eLife.05255.
18. Elbarbary RA, Lucas BA, Maquat LE. Retrotransposons as regulators of gene expression. Science.2016; 351 doi: 10.1126/science.aac7247.
19. Fahlgren N, Sullivan CM, Kasschau KD, Chapman EJ, Cumbie JS, Montgomery TA, Gilbert SD, Dasenko M, Backman TW, Givan SA, Carrington JC. Computational and analytical framework for small RNA profiling by high-throughput sequencing. RNA. 2009; 15:992–1002. doi: 10.1261/rna.1473809.
20. François O, Blum MG, Jakobsson M, Rosenberg NA, Michael GB. Demographic history of european populations of *Arabidopsis thaliana*. PLoS Genetics. 2008; 4:e15716 doi: 10.1371/journal.pgen.1000075.
21. Fu Y, Kawabe A, Etcheverry M, Ito T, Toyoda A, Fujiyama A, Colot V, Tarutani Y, Kakutani T. Mobilization of a plant transposon by expression of the transposon-encoded anti-silencing factor.EMBO Journal. 2013; 32:2407–2417. doi: 10.1038/emboj.2013.169.
22. Hancock AM, Brachi B, Faure N, Horton MW, Jarymowycz LB, Sperone FG, Toomajian C, Roux F, Bergelson J, Gianluca Sperone F. Adaptation to climate across the *Arabidopsis thaliana* genome.Science. 2011; 334:83–86. doi: 10.1126/science.1209244.
23. Heard E, Martienssen RA. Transgenerational epigenetic inheritance: myths and mechanisms. Cell.2014; 157:95–109. doi: 10.1016/j.cell.2014.02.045.
24. Hollister JD, Gaut BS. Epigenetic silencing of transposable elements: a trade-off between reduced transposition and deleterious effects on neighboring gene expression. Genome Research.2009; 19:1419–1428. doi: 10.1101/gr.091678.109.
25. Hu TT, Pattyn P, Bakker EG, Cao J, Cheng JF, Clark RM, Fahlgren N, Fawcett JA, Grimwood J, Gundlach H, Haberer G, Hollister JD, Ossowski S, Ottilar RP, Salamov AA, Schneeberger K, Spannagl M, Wang X, Yang L, Nasrallah ME, Bergelson J, Carrington JC, Gaut BS, Schmutz J, Mayer KF, Van de Peer Y, Grigoriev IV, Nordborg M, Weigel D, Guo YL. The *Arabidopsis lyrata*genome sequence and the basis of rapid genome size change. Nature Genetics. 2011; 43:476–481. doi: 10.1038/ng.807.
26. Hénaff E, Zapata L, Casacuberta JM, Ossowski S. Jitterbug: somatic and germline transposon insertion detection at single-nucleotide resolution. BMC Genomics. 2015; 16:768. doi: 10.1186/s12864-015-1975-5.
27. Ietswaart R, Wu Z, Dean C. Flowering time control: another window to the connection between antisense RNA and chromatin. Trends in Genetics. 2012; 28:445–453. doi: 10.1016/j.tig.2012.06.002.
28. Ito H, Gaubert H, Bucher E, Mirouze M, Vaillant I, Paszkowski J. An siRNA pathway prevents transgenerational retrotransposition in plants subjected to stress. Nature. 2011; 472:115–119. doi: 10.1038/nature09861.
29. Ito H, Kakutani T. Control of transposable elements in *Arabidopsis thaliana*. Chromosome Research.2014; 22:217–223. doi: 10.1007/s10577-014-9417-9.
30. Joly-Lopez Z, Bureau TE. Diversity and evolution of transposable elements in Arabidopsis.Chromosome Research. 2014; 22:203–216. doi: 10.1007/s10577-014-9418-8.
31. Jurka J, Kapitonov VV, Pavlicek A, Klonowski P, Kohany O, Walichiewicz J. Repbase Update, a database of eukaryotic repetitive elements. Cytogenetic and Genome Research. 2005; 110:462–467. doi: 10.1159/000084979.
32. Kang HM, Sul JH, Service SK, Zaitlen NA, Kong SY, Freimer NB, Sabatti C, Eskin E, Susan K, Zaitlen Sit-yee Kong, Nelson B. Variance component model to account for sample structure in genome-wide association studies. Nature Genetics. 2010; 42:348–354. doi: 10.1038/ng.548.
33. Kanno T, Bucher E, Daxinger L, Huettel B, Böhmdorfer G, Gregor W, Kreil DP, Matzke M, Matzke AJ. A structural-maintenance-of-chromosomes hinge domain-containing protein is required for RNA-directed DNA methylation. Nature Genetics. 2008; 40:670–675. doi: 10.1038/ng.119.
34. Kazazian HH. Mobile elements: drivers of genome evolution. Science. 2004; 303:1626–1632. doi: 10.1126/science.1089670.
35. Klein CJ, Botuyan MV, Wu Y, Ward CJ, Nicholson GA, Hammans S, Hojo K, Yamanishi H, Karpf AR, Wallace DC, Simon M, Lander C, Boardman LA, Cunningham JM, Smith GE, Litchy WJ, Boes B, Atkinson EJ, Middha S, B Dyck PJ, Parisi JE, Mer G, Smith DI, Dyck PJ. Mutations in DNMT1 cause hereditary sensory neuropathy with dementia and hearing loss. Nature Genetics. 2011; 43:595–600. doi: 10.1038/ng.830.
36. Kofler R, Nolte V, Schlötterer C. Tempo and Mode of Transposable Element Activity in Drosophila.PLoS Genetics. 2015; 11:e15716 doi: 10.1371/journal.pgen.1005406.
37. Law JA, Jacobsen SE. Establishing, maintaining and modifying DNA methylation patterns in plants and animals. Nature Reviews. Genetics. 2010; 11:204–220. doi: 10.1038/nrg2719.
38. Lempe J, Balasubramanian S, Sureshkumar S, Singh A, Schmid M, Weigel D. Diversity of flowering responses in wild *Arabidopsis thaliana* strains. PLoS Genetics. 2005;1 doi: 10.1371/journal.pgen.0010006.
39. Li Y, Huang Y, Bergelson J, Nordborg M, Borevitz JO. Association mapping of local climate-sensitive quantitative trait loci in *Arabidopsis thaliana*. Proceedings of the National Academy of Sciences of the

United States of America. 2010; 107:21199–21204. doi: 10.1073/pnas.1007431107.

40. Li P, Filiault D, Box MS, Kerdaffrec E, van Oosterhout C, Wilczek AM, Schmitt J, McMullan M, Bergelson J, Nordborg M, Dean C. Multiple FLC haplotypes defined by independent cis-regulatory variation underpin life history diversity in *Arabidopsis thaliana*. Genes & Development.2014; 28:1635–1640. doi: 10.1101/gad.245993.114.

41. Lippman Z, Gendrel AV, Black M, Vaughn MW, Dedhia N, McCombie WR, Lavine K, Mittal V, May B, Kasschau KD, Carrington JC, Doerge RW, Colot V, Martienssen R, Richard McCombie W. Role of transposable elements in heterochromatin and epigenetic control. Nature. 2004; 430:471–476. doi: 10.1038/nature02651.

42. Lisch D. How important are transposons for plant evolution? Nature Reviews. Genetics. 2013; 14:49–61. doi: 10.1038/nrg3374.

43. Lister R, O'Malley RC, Tonti-Filippini J, Gregory BD, Berry CC, Millar AH, Ecker JR, Harvey Millar A. Highly integrated single-base resolution maps of the epigenome in Arabidopsis. Cell.2008; 133:523–536. doi: 10.1016/j.cell.2008.03.029.

44. Liu X, Brutlag DL, Liu JS. BioProspector: discovering conserved DNA motifs in upstream regulatory regions of co-expressed genes. Pacific Symposium on Biocomputing. 2001; 6:127–138.

45. Liu J, He Y, Amasino R, Chen X. siRNAs targeting an intronic transposon in the regulation of natural flowering behavior in Arabidopsis. Genes & Development. 2004; 18:2873–2878. doi: 10.1101/gad.1217304.

46. Lockton S, Gaut BS. The evolution of transposable elements in natural populations of self-fertilizing*Arabidopsis thaliana* and its outcrossing relative *Arabidopsis lyrata*. BMC Evolutionary Biology.2010; 10:10. doi: 10.1186/1471-2148-10-10.

47. Long Q, Rabanal FA, Meng D, Huber CD, Farlow A, Platzer A, Zhang Q, Vilhjálmsson BJ, Korte A, Nizhynska V, Voronin V, Korte P, Sedman L, Mandáková T, Lysak MA, Seren Ü, Hellmann I, Nordborg M. Massive genomic variation and strong selection in *Arabidopsis thaliana* lines from Sweden. Nature Genetics. 2013; 45:884–890. doi: 10.1038/ng.2678.

48. Mackay TF, Richards S, Stone EA, Barbadilla A, Ayroles JF, Zhu D, Casillas S, Han Y, Magwire MM, Cridland JM, Richardson MF, Anholt RR, Barrón M, Bess C, Blankenburg KP, Carbone MA, Castellano D, Chaboub L, Duncan L, Harris Z, Javaid M, Jayaseelan JC, Jhangiani SN, Jordan KW, Lara F, Lawrence F, Lee SL, Librado P, Linheiro RS, Lyman RF, Mackey AJ, Munidasa M, Muzny DM, Nazareth L, Newsham I, Perales L, Pu LL, Qu C, Ràmia M, Reid JG, Rollmann SM, Rozas J, Saada N, Turlapati L, Worley KC, Wu YQ, Yamamoto A, Zhu Y, Bergman CM, Thornton KR, Mittelman D, Gibbs RA. The *Drosophila melanogaster* Genetic Reference Panel. Nature.2012; 482:173–178. doi: 10.1038/nature10811.

49. Marí-Ordóñez A, Marchais A, Etcheverry M, Martin A, Colot V, Voinnet O. Reconstructing de novo silencing of an active plant retrotransposon. Nature Genetics. 2013; 45:1029–1039. doi: 10.1038/ng.2703.

50. Maumus F, Quesneville H. Ancestral repeats have shaped epigenome and genome composition for millions of years in *Arabidopsis thaliana*. Nature Communications. 2014; 5:4104. doi: 10.1038/ncomms5104.

51. McClintock B. Intranuclear systems controlling gene action and mutation. Brookhaven Symposia in Biology. 1956; 8:58–74.

52. McVean GA, Myers SR, Hunt S, Deloukas P, Bentley DR, Donnelly P. The fine-scale structure of recombination rate variation in the human genome. Science. 2004; 304:581–584. doi: 10.1126/science.1092500.

53. Miyao A, Tanaka K, Murata K, Sawaki H, Takeda S, Abe K, Shinozuka Y, Onosato K, Hirochika H. Target site specificity of the Tos17 retrotransposon shows a preference for insertion within genes and against insertion in retrotransposon-rich regions of the genome. Plant Cell. 2003; 15:1771–1780. doi: 10.1105/tpc.012559.

54. Purcell S, Neale B, Todd-Brown K, Thomas L, Ferreira MA, Bender D, Maller J, Sklar P, de Bakker PI, Daly MJ, Sham PC. PLINK: a tool set for whole-genome association and population-based linkage analyses. American Journal of Human Genetics. 2007; 81:559–575. doi: 10.1086/519795.

55. Quadrana L, Colot V. Plant Transgenerational Epigenetics. Annual Review of Genetics 2016

56. Quinlan AR, Clark RA, Sokolova S, Leibowitz ML, Zhang Y, Hurles ME, Mell JC, Hall IM. Genome-wide mapping and assembly of structural variant breakpoints in the mouse genome. Genome Research. 2010; 20:623–635. doi: 10.1101/gr.102970.109.

57. Rahman R, Chirn GW, Kanodia A, Sytnikova YA, Brembs B, Bergman CM, Lau NC. Unique transposon landscapes are pervasive across *Drosophila melanogaster* genomes. Nucleic Acids Research. 2015; 43:10655–10672. doi: 10.1093/nar/gkv1193.

58. Rebollo R, Romanish MT, Mager DL. Transposable elements: an abundant and natural source of regulatory sequences for host genes. Annual Review of Genetics. 2012; 46:21–42. doi: 10.1146/annurev-genet-110711-155621.

59. Richardson SR, Morell S, Faulkner GJ. L1 retrotransposons and somatic mosaicism in the brain. Annual Review of Genetics. 2014; 48:1–27. doi: 10.1146/annurev-genet-120213-092412.

60. Richardson SR, Doucet AJ, Kopera HC, Moldovan JB, Garcia-Perez JL, Moran JV. The influence of LINE-1 and SINE retrotransposons on mammalian genomes. Microbiology Spectrum. 2015; 3:e15716 doi: 10.1128/microbiolspec.MDNA3-0061-2014.

61. Robinson JT, Thorvaldsdóttir H, Winckler W, Guttman M, Lander ES, Getz G, Mesirov JP. Integrative genomics viewer. Nature Biotechnology. 2011; 29:24–26. doi: 10.1038/nbt.1754.

62. Scheet P, Stephens M. A fast and flexible statistical model for large-scale population genotype data: applications to inferring missing genotypes and haplotypic phase. American Journal of Human Genetics. 2006; 78:629–

644. doi: 10.1086/502802.

63. Schmitz RJ, Schultz MD, Lewsey MG, O'Malley RC, Urich MA, Libiger O, Schork NJ, Ecker JR. Transgenerational epigenetic instability is a source of novel methylation variants. Science.2011; 334:369–373. doi: 10.1126/science.1212959.

64. Schmitz RJ, Schultz MD, Urich MA, Nery JR, Pelizzola M, Libiger O, Alix A, McCosh RB, Chen H, Schork NJ, Ecker JR, Matthew D, Nery MP. Patterns of population epigenomic diversity. Nature.2013; 495:193–198. doi: 10.1038/nature11968.

65. Schneeberger K, Ossowski S, Ott F, Klein JD, Wang X, Lanz C, Smith LM, Cao J, Fitz J, Warthmann N, Henz SR, Huson DH, Weigel D. Reference-guided assembly of four diverse*Arabidopsis thaliana* genomes. Proceedings of the National Academy of Sciences of the United States of America. 2011; 108:10249–10254. doi: 10.1073/pnas.1107739108.

66. Silveira AB, Trontin C, Cortijo S, Barau J, Del Bem LE, Loudet O, Colot V, Vincentz M. Extensive natural epigenetic variation at a de novo originated gene. PLoS Genetics. 2013; 9:e15716 doi: 10.1371/journal.pgen.1003437.

67. Slotkin RK, Martienssen R. Transposable elements and the epigenetic regulation of the genome.Nature Reviews. Genetics. 2007; 8:272–285. doi: 10.1038/nrg2072.

68. Stroud H, Greenberg MV, Feng S, Bernatavichute YV, Jacobsen SE, Maxim VC. Comprehensive analysis of silencing mutants reveals complex regulation of the Arabidopsis methylome. Cell.2013; 152:352–364. doi: 10.1016/j.cell.2012.10.054.

69. Stuart T, Eichten SR, Cahn J, Borevitz J, Lister R. Population scale mapping of novel transposable element diversity reveals links to gene regulation and epigenomic variation. bioRxiv. 2016:039511. doi: 10.1101/039511.

70. Teixeira FK, Heredia F, Sarazin A, Roudier F, Boccara M, Ciaudo C, Cruaud C, Poulain J, Berdasco M, Fraga MF, Voinnet O, Wincker P, Esteller M, Colot V. A role for RNAi in the selective correction of DNA methylation defects. Science. 2009; 323:1600–1604. doi: 10.1126/science.1165313.

71. Teixeira FK, Colot V. Repeat elements and the Arabidopsis DNA methylation landscape. Heredity.2010; 105:14–23. doi: 10.1038/hdy.2010.52.

72. Trono D. Transposable Elements, Polydactyl Proteins, and the Genesis of Human-Specific Transcription Networks. Cold Spring Harbor Symposia on Quantitative Biology. 2016:027573. doi: 10.1101/sqb.2015.80.027573.

73. Tsay YF, Frank MJ, Page T, Dean C, Crawford NM, Mary J. Identification of a mobile endogenous transposon in *Arabidopsis thaliana*. Science. 1993; 260:342–344. doi: 10.1126/science.8385803.

74. Vinkhuyzen AA, Wray NR, Yang J, Goddard ME, Visscher PM, Naomi R, Wray JY. Estimation and partition of heritability in human populations using whole-genome analysis methods. Annual Review of Genetics. 2013; 47:75–95. doi: 10.1146/annurev-genet-111212-133258.

75. Weigel D, Colot V. Epialleles in plant evolution. Genome Biology. 2012; 13:249. doi: 10.1186/gb-2012-13-10-249.

76. Weigel D, Nordborg M. Population Genomics for Understanding Adaptation in Wild Plant Species.Annual Review of Genetics. 2015; 49:315–338. doi: 10.1146/annurev-genet-120213-092110.

77. Yoon S, Xuan Z, Makarov V, Ye K, Sebat J, Kenny Ye. Sensitive and accurate detection of copy number variants using read depth of coverage. Genome Research. 2009; 19:1586–1592. doi: 10.1101/gr.092981.109.

78. Ziolkowski PA, Koczyk G, Galganski L, Sadowski J. Genome sequence comparison of Col and Ler lines reveals the dynamic nature of Arabidopsis chromosomes. Nucleic Acids Research.2009; 37:3189–3201. doi: 10.1093/nar/gkp183.

CHAPTER 18 ATYPICAL CENTROMERES IN PLANTS

Maria Cuacos, F. Chris H. Franklin and Stefan Heckmann

Keywords: cenH3, centromere, holocentric chromosomes, kinetochore, meiosis, mitosis, neocentromeres, plants

CENTROMERE TYPES

Centromeres are chromosomal loci where kinetochores assemble. Kinetochore proteins mediate cell cycle regulation, sister chromatid cohesion, spindle microtubule attachment and chromosome movements (Lermontova et al., 2014). These functions are essential for genome stability by mediating faithful mitotic and meiotic chromosome segregation. Any failure leads to chromosome missegregation and ultimately genome instability.

Kinetochore establishment and centromere maintenance in active eukaryotic centromeres generally depends on the presence of the centromeric histone H3 variant cenH3 (also called CENP-A in mammals; De Rop et al., 2012). Although essential for genome integrity, contrary to expectation centromeric cenH3 localization is not specified by centromere specific DNA sequence(s) except in budding yeast (Clarke and Carbon, 1985). It is rather determined epigenetically. Centromere loci and centromeric DNAs are highly diverse varying dramatically in size and sequence composition between species. Centromeres can range in size from the 125 bp "point" centromeres in budding yeast up to mega bp-sized "regional" centromeres in humans and plants. In the most extreme case, poly- or holocentromeres can even extend over the entire chromosome length.

Citation: Cuacos M, Franklin FC and Heckmann S (2015) Atypical centromeres in plants. Front. Plant Sci. 6:913. doi: 10.3389/fpls.2015.00913.

Although centromeric DNAs are not conserved often plant centromeres contain distinct satellite DNA sequences and families of long terminal repeat (LTR) retrotransposons (Houben and Schubert, 2003; Neumann et al., 2011). However, these repeats are neither necessary nor sufficient for centromere activity since gain of new centromeric activity over unique DNA sequences can occur (Nasuda et al., 2005; Han et al., 2006).

Centromeric DNAs are one of the fastest evolving sequences in eukaryotic genomes (Bensasson, 2011; Melters et al., 2013). It is interesting that such an essential and functionally conserved chromosomal locus has so rapidly evolved with regards to its structure, extension and DNA sequence composition. For instance, (i) the sequence composition and centromere extension vary dramatically between closely-related species, e.g., *Solanum* (Zhang et al., 2014) or *Oryza* species (Yi et al., 2013), and even between centromeres within one species, e.g., *Pisum sativum* (Macas et al., 2007), (ii) different centromere types, e.g., mono- and holocentromeres, have evolved between different insect lineages (Drinnenberg et al., 2014) and even between closely-related dodder species (Pazy and Plitmann, 1994, 1995), or (iii) albeit functionally similar, unconventional centromeres, e.g., cenH3- and CENP-C-independent insect holocentromeres (Drinnenberg et al., 2014) or "meta-polycentric" centromeres in *Pisum* or *Lathyrus* (Neumann et al., 2012, 2015), and unconventional kinetochores, e.g., kinetoplastid kinetochores devoid of any conventional components (Akiyoshi and Gull, 2014), have evolved.

Studies on "atypical" plant centromeres such as neocentromeres or holocentromeres have contributed to our general knowledge of the structure, regulation and function of centromeres. In this review, we focus on such unusual centromere types in plants, highlight recent discoveries and discuss their implications.

HOLOCENTROMERES

Most studied organisms possess one size-restricted centromere (monocentromere) per chromosome (Figure 1A). However, in various species so-called holocentromeres ("holo-" from Greek: entire) initially described by Schrader (1935), characterized by an almost chromosome-wide extension occur (Figure 1B). They are also called diffuse centromeres or polycentromeres—for the rest of the review we will use the terms holocentromere or holocentric chromosome. Holocentromeres evolved by convergent evolution in diverse eukaryotic lineages including green algae, invertebrates, and plants (Melters et al., 2012). Around 800 species as diverse as nematodes, spiders, and sedges are reported to possess holocentromeres.

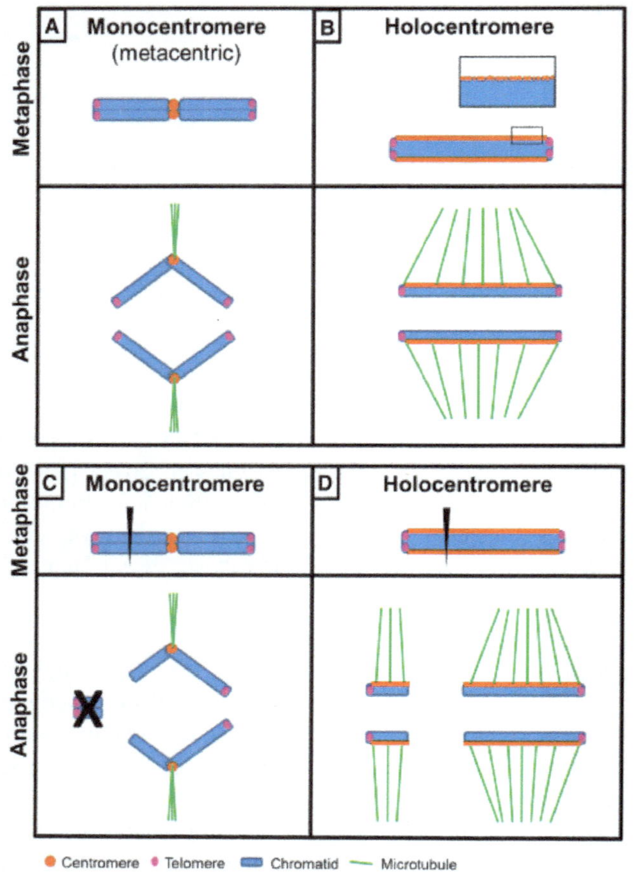

Figure 1: Structure and behavior of a monocentric and a holocentric chromosome. (A) A metacentric chromosome shows a primary constriction during metaphase. During anaphase chromatids move as V-shaped structures due to microtubule attachment to the size-restricted centromere. **(B)** A holocentric chromosome shows an almost chromosome-wide centromere extension and no primary constriction during metaphase. Sister chromatids are not discernible. During anaphase spindle microtubule attachment to the holocentromere results in chromatids moving as linear bars parallel to the spindle. Inset, various centromeric subdomains fuse to one functional composite linear holocentromere at metaphase. **(C)** Breakage of a monocentric chromosome results in loss of the acentric chromosome fragment during anaphase, whereas **(D)** after chromosome breakage of a holocentric chromosome both fragments retain kinetic activity due to the almost chromosome-wide centromere extension and thus can be transmitted. Note absence of telomeric repeats at broken chromosome ends. In case of holocentric chromosomes of *Luzula elegans*, rapid telomere-mediated "chromosome healing" occurs (Jankowska et al., 2015).

CHROMOSOME CLASSIFICATION: MONOCENTROMERE VS. HOLOCENTROMERE

If a given chromosome is comparatively large, classical cytogenetic techniques are applicable for the classification as holo- or monocentric. In the case of a holocentromere, a mitotic metaphase chromosome lacks a primary constriction and during anaphase chromatids move as linear bars parallel to the spindle due to almost chromosome-wide spindle attachment (holokinetic behavior; Figure 1B). If a chromosome is monocentric, it bears a primary constriction and the chromatids either move as a v-shaped structure (metacentric), as a linear bar perpendicular to the spindle pole (acrocentric) or as a configuration in between both extreme cases due to microtubule attachment to the defined size-restricted centromere (Figure 1A).

A more direct approach to classify the chromosome structure is through immunolocalization of centromere components. Although many components are functionally conserved, they are often divergent in sequence composition even between closely-related species. Thus, centromere-related antibodies are not universally available. The discovery of phosphorylation of threonine 120 of histone H2A (H2AThr120ph) as universal mark for active centromeres in plant species with mono- and holocentromeres will allow the classification of (comparatively large) chromosomes to be refined (Demidov et al., 2014).

HOLOCENTROMERES IN PLANTS

In flowering plants, holocentromeres are found among the monocots *Cyperaceae* (sedges),*Juncaceae* (rushes; Malheiros et al., 1947; Hakansson, 1958) and *Chionographis* (string flowers;Tanaka, 1977) as well as in dicots such as *Cuscuta* subgenus *Cuscuta* (dodders; Pazy and Plitmann, 1995), *Drosera* (sundews; Sheikh et al., 1995), or in the nutmeg tree *Myristica fragrans* (Flach, 1966). There is evidence for 228 plant species with holocentromeres (Melters et al., 2012). This number is likely an underestimate since for instance many species possess comparatively small chromosomes and thus chromosome classification is cytologically challenging. In addition, all*Juncaceae* and *Cyperaceae* are predicted to be holocentric (the genus *Carex* within *Cyperaceae*consists of around 2000 species, Reznicek, 1990), which actually suggests an even higher number of species with holocentromeres within only these two families. However, there are contradictory reports such as in the genus *Drosera*, where species are reported to possess holocentromeres (Kondo and Lavarack, 1984; Sheikh and Kondo, 1995; Sheikh et al., 1995) and monocentromeres (Shirakawa et al., 2011; Demidov et al., 2014).

KARYOTYPE EVOLUTION IN SPECIES WITH HOLOCENTROMERES

A chromosome-wide centromere organization allows, in theory, rapid karyotype evolution. Unlike monocentric chromosomes, basically each part of a given holocentric chromosome has centromere activity and thus theoretically a high probability of being transmitted during nuclear divisions after chromosome breakage (Figures 1C,D). This karyotype flexibility conferred by a holocentromere is reflected in (i) the extremely wide and almost continuous chromosome number found among related holocentric species, e.g., *Carex* 2n = 12–124, *Eleocharis* 2n = 6–196, or *Juncus* 2n = 18–170 (Bureš et al., 2013), (ii) interspecies chromosome number variation, e.g., *Eleocharis kamtschatica*with 2n = 41–47 (Yano and Hoshino, 2006), or (iii) the negative correlation between chromosome number and chromosome size in *Luzula* (e.g., Nordenskiold, 1951; Jankowska et al., 2015).

Typically double strand breaks (DSBs) in DNA are resolved by non-homologous end-joining or homologous recombination (Knoll et al., 2014). However, telomerase can add telomeric repeats at break sites leading to "chromosome healing" (Tsujimoto, 1993; Tsujimoto et al., 1999; Nelson et al., 2011). Monocentric chromosome healing at DSB sites results in deletion or loss of the distal acentric chromosome fragment and thus needs to be tightly regulated. Unlike monocentric chromosomes, stable transmission of (artificially induced) holocentric chromosome fragments during mitosis and meiosis suggests that chromosome healing is a more or less common phenomenon in holocentric chromosome species (e.g., Hakansson, 1954; Nordenskiold, 1964). Recently, Jankowska et al. (2015) showed that irradiation of holocentric *Luzula elegans*chromosomes results in a range of heteromorphic derived karyotypes. Independent of their size all chromosomes/fragments showed centromere activity and gradual telomere-mediated "chromosome healing." Newly formed telomere repeats were cytologically detectable 21 days after irradiation in ~50% of cases, increasing to >95% after 3 months. In the progenies of the irradiated plants all the chromosomes/fragments possessed telomeric repeats. This rapid and efficient *de novo* telomere formation is likely conferred by a telomerase-mediated healing process and important for fragment stabilization/karyotype fixation.

The combination of holokinetic chromosomes and rapid telomere formation at DSBs allows stable transmission of chromosome fragments and thus rapid karyotype evolution. Additionally, polyploidy and proliferation/removal of high copy sequences are involved in rapid genome evolution (e.g., Kuta et al., 2004; Zedek et al., 2010; Bozek et al., 2012). However, how holocentric chromosome species can deal with this cytological "chromosomal chaos" is largely unknown. In holocentric *Lepidoptera* with a high intraspecific cytogenetic variation, a high degree of synteny at fine scales is found, suggesting an adaptive mechanism (d'Alencon et al., 2010). Similar studies in plants are missing.

STRUCTURE AND BEHAVIOR OF HOLOCENTRIC CHROMOSOMES

In plant holocentromeres, cenH3 is found along mitotic chromosomes representing active centromeres as in species with monocentromeres (Nagaki et al., 2005). In some *Luzula* and*Rhynchospora* species with comparatively large chromosomes cenH3-positive chromosome regions form a groove-like structure except at chromosome ends during mitotic metaphase (Nagaki et al., 2005; Heckmann et al., 2011; Cabral et al., 2014; Wanner et al., 2015). It seems likely that a centromeric groove is a structural adaptation of relatively large holocentric chromosomes or a distinct evolutionary accommodation within certain genera. Ultrastructural analysis of mitotic *L. elegans* chromosomes showed that cenH3 containing chromatin is found at the periphery of each individual chromatid and that microtubules attach to cenH3- and not H2AThr120ph-chromatin during mitosis (Wanner et al., 2015). H2AThr120ph is enriched in the centromeric groove and completely absent along the axis where chromatids are in close contact, suggesting that H2AThr120ph is not involved in holocentric sister chromatid cohesion (Wanner et al., 2015). No differentiation between holocentric chromatids is found microscopically in *L. elegans* probably owing to almost chromosome-wide centromeric cohesion (Heckmann et al., 2011; Wanner et al., 2015). Mitotic sister chromatids are only discernible after staining sister-chromatid exchanges (Heckmann and Houben, 2013). Microtubule attachment regions are concentrated on the pericentromeric rims, possibly increasing attachment stability during separation of sister chromatids. Bundles of 2–4 individual microtubules are distributed along the entire centromere length with a mean distance between individual bundles of 300–400 nm during mitosis and 350–500 nm during meiosis (Heckmann et al., 2014a; Wanner et al., 2015). In *C. elegans* with much smaller holocentromeres the number of microtubules attachments is ~85 genome-wide or ~15 per chromosome (O'Toole et al., 2003). In summary, H2AThr120ph and cenH3 are found within the longitudinal centromeric groove of (large) holocentric plant metaphase chromosomes and microtubule attachment is enriched along the groove rim.

The almost chromosome-wide holocentromere architecture is also reflected in the distribution of epigenetic marks. The cell cycle-dependent phosphorylation of serine 10 or serine 28 of H3 typically enriched in pericentromere regions of monocentric plant chromosomes (Houben et al., 1999; Gernand et al., 2003) occurs in *Luzula* and *Rhynchospora* uniformly along the chromosomes (Gernand et al., 2003; Nagaki et al., 2005; Guerra et al., 2006). Similarly typical eu- and heterochromatin epigenetic marks or early/late DNA replicating chromatin domains are detected uniformly along *L. elegans* mitotic chromosomes at normal resolution (Heckmann et al., 2013). However, using super-high-resolution light microscopy interspersed units of various chromatin types were distinguished. Intermingling of different chromatin domains throughout the *L. elegans*genome is correlated with the distribution of highly repetitive DNA and likely reflects interplay between scattered chromosome-wide centromere organization and overall genome organization (Heckmann et al., 2013).

A scattered polycentric centromere arrangement is microscopically reflected in cenH3 dynamics during the cell cycle. During interphase cenH3 is found dispersed, in prophase as small foci along chromosomes and during metaphase as composite linear axial line along each sister chromatid (Buchwitz et al., 1999; Nagaki et al., 2005; Heckmann et al., 2011). In *C. elegans* a polycentric chromosome arrangement is revealed at fine scale resolution (Gassmann et al., 2012; Steiner and Henikoff, 2014). About 700 individual centromeric sites -single cenH3 nucleosomes flanked by well-spaced canonical nucleosomes- are preferentially found at dispersed sites of permissive chromatin (Steiner and Henikoff, 2015).

CenH3 is also found with low density in roughly 2900 broad chromosome domains of low transcriptional activity and low nucleosome turnover that put together represent roughly half the genome (Gassmann et al., 2012; Steiner and Henikoff, 2014). Thus, *C. elegans* holocentromeres are polycentromeres consisting of individual point centromeres as the basic units of assembly.

A holo-/polycentric chromosome, composed of various centromeric subdomains, should have a high risk of misorientation during anaphase due to potential merotelic spindle attachment to individual subdomains. In *C. elegans*, chromokinesin KLP-19 counteracts persistent merotelic attachments (Powers et al., 2004). However, it is unclear how holocentric plants circumvent this holocentromere-associated challenge. In dicentric chromosomes, when a critical distance between two active centromeres is reached, the chromosome can break during anaphase due to not forming one functional centromere at metaphase (see below). In case of a holocentric chromosome, the distance between individual centromere subunits must be likewise restricted. In *C. elegans*, with comparatively small holocentric chromosomes, the genomic cenH3 distribution indicates a distance between individual centromere subunits of maximally 1.9 Mb ranging from 290 bp to 1.9 Mb with a median of 83 kb (Steiner and Henikoff, 2014). Notably, during divisions probably not all centromeric regions (only ~15%) are kinetically active in *C. elegans*, thus the maximum distance might be even higher. Similar studies in plants with (larger) holocentric chromosomes are, to date, lacking. Thus, it is unclear what the maximum functional inter-subunit distance tolerated between individual centromere units of a given holocentric plant chromosome is and additionally, it is unclear whether as in *C. elegans*, only a subset of these centromeric domains are kinetically active.

HOLOCENTROMERE IDENTITY

Recently, holocentromere-enriched satellite DNA sequences and retrotransposons preferentially bound by cenH3 were found in *R. pubera* (Marques et al., 2015) similar to most plant monocentromeres (Houben and Schubert, 2003; Neumann et al., 2011). Thus, also in species with holocentric chromosomes centromere-specific repetitive DNAs can occur. Stretched *Pisum*chromosomes show multiple centromeres consisting of satellite DNAs (Neumann et al., 2012,2015). This "meta-polycentricity" may be an evolutionary link toward the development of holocentromeres in species such as *Rhynchospora*.

Unlike *Rhynchospora*, in *L. elegans* neither typical centromere-associated retrotransposons nor any holocentromere-associated satellite DNAs are found (Heckmann et al., 2013). Thus, cenH3 may be associated with a centromere-specific chromatin status rather than with specific centromeric DNA sequences. In *L. nivea* the 178-bp tandem repeat sequence LCS1 (Haizel et al., 2005) which shares some similarity with the centromeric tandem repeat RCS2 of rice (Dong et al., 1998; Nonomura and Kurata, 2001) has been described. Whether LCS1 plays a centromeric role is uncertain.

In *C. elegans* a short DNA motif is enriched at individual centromeric sites, however, it is likely not a direct target for cenH3 (Steiner and Henikoff, 2014). Accordingly, basically any DNA sequence can acquire centromere activity and extrachromosomal arrays are even segregated after few cycles in *C. elegans* (Stinchcomb et al., 1985; Yuen et al., 2011). Thus, in *C. elegans* centromeric nucleosomes are inherited epigenetically rather than being DNA sequence-dependent.

Limited available data suggest that there are different ways of defining holocentromeres with regards to centromeric sequences. In *R. pubera* there are centromere-specific satellite DNAs and retrotransposons, whereas in *L. elegans* and *C. elegans* no centromere-specific sequences are found. Thus, possibly different evolutionary scenarios with regards to centromere-specific DNA sequences led to the formation of holocentromeres. Further studies will clarify, e.g., how in *L. elegans* individual centromeric subunits are defined or whether between closely-related species with holo- and/or monocentromeres such as *Cuscuta* or *Drosera* different or similar centromeric DNA sequences occur.

MEIOTIC ADAPTATIONS OF HOLOCENTRIC CHROMOSOMES

Sexual reproduction is characterized by the process of meiosis, during which two consecutive rounds of chromosome segregation follow one single round of DNA replication generating haploid gametes. Cytologically, during the first meiotic division homologous chromosomes (homologs) are separated and during the second meiotic division chromatids disjoin. To allow faithful transmission typically homologs pair and perform reciprocal genetic exchange, termed crossover, physically linking homologs and thus ensuring balanced chromosome segregation during meiosis I. Proper chromosome segregation further depends on mono-orientation of fused sister kinetochores during meiosis I and on bi-orientation of sister kinetochores during meiosis II. In monocentromere species this is realized by a two-step loss of cohesion, i.e., along chromosome arms during meiosis I and at sister centromeres during meiosis II.

The two-step loss of cohesion is hampered in a holocentromere due to the lack of defined broad chromosomal centromere and arm domains allowing their spatial distinction. As adaptation, species with holocentromeres evolved different strategies to conduct faithful meiotic chromosome segregation, such as "chromosome remodeling" in *C. elegans* (Schvarzstein et al., 2010) or "functional monocentricity" in *Heteroptera* (Viera et al., 2009). Another alternative meiotic process is found in plants with holocentromeres characterized by separation of sister chromatids already during meiosis I (Cabral et al., 2014; Heckmann et al., 2014a). *Luzula* and *Rhynchospora*display a functional holocentromere throughout meiosis. Prophase I events are cytologically similar to those found in species with monocentric chromosomes including meiotic DSB induction and progression even in achiasmatic chromosomes of *R. tenuis*. At metaphase I, contrary to a monopolar orientation of sister monocentromeres, sister holocentromeres are unfused and interact, individually bi-orientated, with the meiotic spindle, resulting in a separation of sister chromatids already during meiosis I. The homologous non-sister chromatids are kept usually terminally linked by chromatin threads until anaphase II when they are separated ensuring haploidization. These chromatin threads are heterochromatic, enriched in satellite DNAs in *L. elegans* and possibly formed in a crossover-independent manner as their occurrence in achiasmatic chromosomes of *R. tenuis* suggests. Notably, monocentric chromosomes can also associate via chromatin threads (e.g., crane flies or Drosophila, LaFountain et al., 2002; Hughes et al., 2009), suggesting that terminal heterochromatin can be "sticky" enabling an achiasmatic association of homologs and the link of homologous non-sister chromatids in *Luzula* or *Rhynchospora* species. Whether an achiasmatic cohesin-mediated or another unknown mechanism is involved is unclear. Alternatively, these chromatin threads may be formed in a crossover-dependent manner in such a way that only distinct terminal chiasmata persist, being released later than those in interstitial regions. Although *R. tenuis* chromosomes are apparently achiasmatic, DSB and synaptonemal complex formation -prerequisites for crossover formation- suggest that distinct terminal crossovers mediating chromatin thread formation might occur in *R. tenuis*. Additionally, these threads are most likely not caused by catenated late replicating DNA, as homologs and non-sister chromatids are connected as, e.g., in *L. elegans*. Chromatin threads are not described for species with holocentromeres using a different meiotic mode. Therefore, chromatin threads are likely conducive for an inverted meiotic chromatid segregation process in species with holocentromeres (Heckmann et al., 2014b). Interestingly, reminiscent of the situation in monocentric species where centromeres are typically recombination cold spots (Yelina et al., 2012), in species with holocentromeres crossovers mainly occur in non-centromeric (i.e., terminal) chromosome regions. Thus, centromeric regions are crossover cold spots in both mono- and holocentromere species.

EVOLUTION OF HOLOCENTROMERES

An intriguing question is whether a holocentric or a monocentric chromosome structure appeared first during evolution. Nagaki et al. (2005) proposed for *Luzula* that a 90°direction turn of a monocentromere in an ancestral *Luzula* species together with subsequent centromere extension could be the basis of holocentricity. Neumann et al. (2012) proposed that spreading of centromere-competent satellite(s) was the cause of a transition from a monocentric to a polycentric chromosome structure in *L. nivea*. Villasante et al. (2007) proposed the "telomere to centromere" model that predicts an origin of holocentromeres from monocentromeres. The "centromere drive" hypothesis suggests a transition from mono- to holocentric chromosomes in order to suppress centromere drive (Malik and Henikoff, 2002). Melters et al. (2012) proposed that monocentricity was the ancestral chromosome configuration and that holocentricity evolved multiple times independently. Currently it is generally accepted that an independent transition from mono- into holocentromeres occurred in total on at least 13 occasions in eukaryotic lineages with the exception of vertebrates (four times in plants and nine times in animals; Melters et al., 2012).

In four holocentric insect lineages cenH3 and CENP-C were independently lost while other outer and inner kinetochore components such as NDC80 or MIS12 remain (Drinnenberg et al., 2014). In*Luzula*, *Rhynchospora*, and *C. elegans* this transition occurred without a loss of cenH3 (Buchwitz et al., 1999; Nagaki et al., 2005; Cabral et al., 2014). Two cenH3s are even found in *C. elegans*, Hcp-3 and Cpar-1 (Monen et al., 2005), and in *L. nivea*, LnCENH3-A and LnCENH3-B (Moraes et al., 2011). CenH3 is essential for mitosis but dispensable for male meiosis in *C. elegans* (Monen et al., 2005). Interestingly, the weakly expressed isoform, Cpar-1, specifically localizes to meiotic chromosomes and is cleaved in its N-terminal tail by separase at anaphase I (Monen et al., 2015). Whether these dynamics reflect a meiotic adaptation compensating for a holocentric chromosome structure during meiosis is unknown. In *L. nivea* two cenH3s are transcribed and at least LnCENH3-B is found in centromeric nucleosomes (Moraes et al., 2011). Whether both cenH3s are essential for centromere activity, are functionally diverged or redundant, and whether this duplication is related to holocentromere occurrence, is at present unclear.

Apparently, a transition from mono- into holocentromeres can evolve differently with regards to cenH3: either entire cenH3 loss in insects, "partial functional loss" being dispensable for male meiosis in *C. elegans*, or complete mitotic and meiotic retention of cenH3 in holocentric plants.Drinnenberg et al. (2014) proposed that an event early in the evolution of insects, e.g., a lineage-specific evolution of a centromeric protein similar to *Umbrea* in flies (Ross et al., 2013), enabled a cenH3 loss in holocentric insects. Unlike holocentric insects, in *L. nivea* and *C. elegans*holocentromeres coincide with two cenH3 variants. Also highly dynamic centromere architectures found in some *Fabeae* species positively correlate with the presence of two active cenH3 variants (Neumann et al., 2015). Possibly two cenH3 variants, albeit also found in monocentric species (e.g.,Sanei et al., 2011), either enable or are the consequence of structural centromere changes. Thus, it is tempting to speculate that either any variation in the level of cenH3, loss or duplication, might enable centromere plasticity and thus structural centromere evolution or that structural changes in centromere architecture may render centromere-dependency on cenH3. In a nutshell, transitions from mono- into holocentromeres are likely based on distinct evolutionary scenarios rather than on one common pivotal event.

In *Cuscuta*, plants of the subgenera *Monogyna* and *Grammica* are reported to have monocentromeres while members of the subgenus *Cuscuta* are reported to have holocentromeres (Pazy and Plitmann, 1994, 1995). Similarly, some *Drosera* species have monocentromeres (Shirakawa et al., 2011; Demidov et al., 2014) while others are reported to have holocentromeres (Kondo and Lavarack, 1984; Sheikh and Kondo, 1995; Sheikh et al., 1995). These species potentially offer a great opportunity to gain further insight into the evolution of differing centromere types between closely-related species and of the mechanisms involved. Additionally, occurrence of holo- and monocentromeres between closely related *Cuscuta* or *Drosera* species might offer a possibility to generate hybrids between species with holo- and with monocentromeres. This would be an attractive model system to study two different centromere types within a hybrid.

Another striking question is why holocentricity has arisen multiple times during evolution in diverse eukaryotic lineages but not in all? One common explanation is their advantage in relation to DSBs when compared to monocentric chromosomes. However, widespread occurrence of monocentromeres and the fact that holocentromeres are not described (so far) in vertebrates suggest that potential advantages conferred by a holocentromere are counteracted by certain disadvantages; possibly the potential of merotelic chromosome attachments during nuclear divisions or the faithful segregation of meiotic chromosomes. This may explain why species with holocentromeres evolved diverse meiotic chromosome segregation processes and why manifold mechanisms for sexual and/or asexual reproduction are found in species with holocentromeres (e.g., in aphids, Manicardi et al., 2015).

DI-, TRI- AND META-POLYCENTRIC CHROMOSOMES

Dicentric chromosomes (chromosomes with two active centromeres) are typically unstable (Figure2A). They form anaphase bridges which lead to chromosome breakage, as already observed by Barbara McClintock (1939) in maize. Dicentric chromosomes are reported in plant and non-plant species such as *Drosophila*, yeast and human (Stimpson et al., 2012). There are even reports about tricentric (with three centromeres; Zhang et al., 2010) and meta-polycentric (with up to five centromeres) chromosomes (Neumann et al., 2012, 2015). Typically, di- or tricentric chromosomes arise as a consequence of profound genome rearrangements (Stimpson et al., 2012) although naturally occurring di- and meta-polycentric chromosomes do exist.

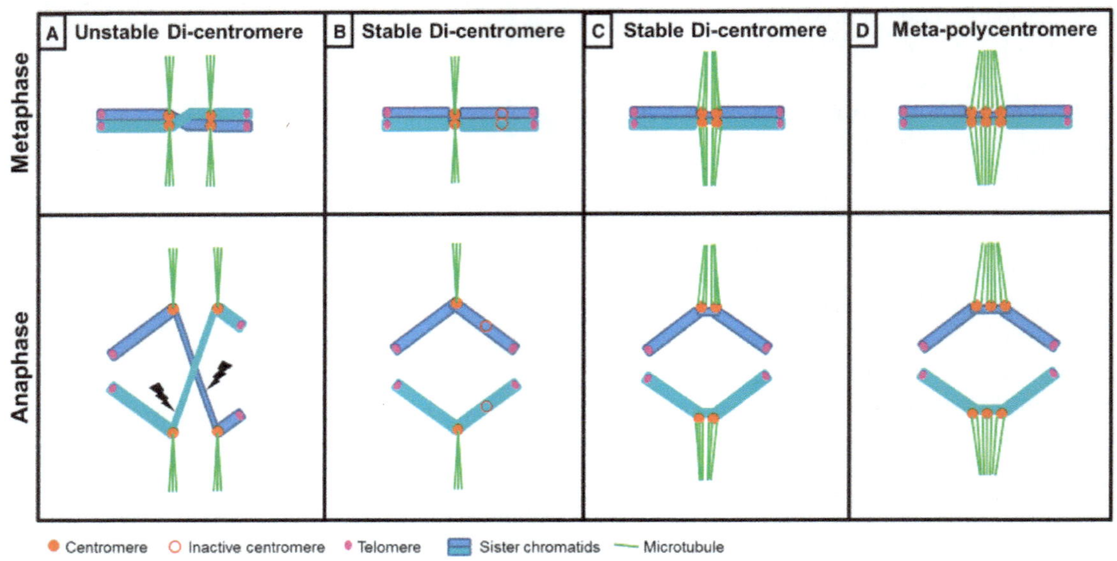

Figure 2: Schematic representation of chromosomes with multiple centromeres. (A) A chromosome with two active centromeres (di-centromere or dicentric chromosome) is typically unstable. A twist between sister chromatids within the region between both centromeres leads to merotelic spindle attachment to two kinetochores on the same chromatid resulting in an anaphase bridge and subsequent chromosome breakage.**(B,C)** Stabilization of a dicentric chromosome can occur by **(B)** inactivation of one centromere, so that the chromosome behaves as monocentric or **(C)** when the close proximity between two active centromeres enables them to behave as one functional unit. **(D)** Meta-polycentric chromosome with three functional centromeres within one constriction.

Stabilization of di- and tricentric chromosomes can occur via different mechanisms. One mechanism is the epigenetic inactivation of one of the centromeres leading to a functional monocentric chromosome (Figure 2B). If two or three centromeres are different in size, the small centromere(s) is/are inactivated (Han et al., 2009a; Zhang et al., 2010). An inactivated centromere can be reactivated when detached from the active centromere, demonstrating that it retains centromeric capability (Schubert et al., 1995; Han et al., 2009a). Stabilization can also occur if one centromere exhibits functional dominance over the other(s). A tricentric wheat chromosome (Zhang et al., 2010) was found with one large and two small centromeres that due to being in close proximity, function as one unit. All three centromeres contain centromeric sequences, cenH3 (the small centromeres containing 30% the amount of the large centromere), H3S10ph and bind spindle microtubules. This tricentric chromosome exhibits features of dicentrics: the two smaller centromeres can be inactivated, positively correlating with increased amounts of H3K27me2/3 or, when the two smaller centromeres are active, chromosome breakage occurs. However, in 70% of progenies the intact tricentric chromosomes was transmitted, possibly due to dominant pulling forces of the large over the small centromeres during anaphase (Zhang et al., 2010).

Stabilization is further dependent on proximity between the two centromeres: if they are "close enough," both active centromeres can behave as a functional unit and orientate to the same pole (Figure 2C), whereas when a critical distance is reached between the two active centromeres the chromosome can break due to merotelic spindle attachments. The critical distance is estimated to be around 10 Mb in a human X chromosome (Sullivan and Willard, 1998) and up to 20 Mb in an engineered human dicentric chromosome (Higgins et al., 2005). A naturally occurring stable "dicentric" chromosome is found in rice (Wang et al., 2013). Two cenH3-binding domains composed of typical centromeric repeats are separated by ~400 kb 5S rDNA sequences that do not associate with cenH3. Also canonical centromeres contain blocks of H3- instead of cenH3-containing chromatin of considerable size, e.g., rice Cen8 (Nagaki et al., 2004; Yan et al., 2008), potato Cen9 and Cen11 (Gong et al., 2012) or "dicentric" maize chromosome 5 with an estimated "gap" of 2.8 Mb (Wolfgruber et al., 2009). In the most extreme case, meta-polycentric chromosomes in pea and closely related *Lathyrus* species contain three to five functional centromeres within stretched primary constrictions (Neumann et al., 2012, 2015; Figure 2D). In these meta-polycentric chromosomes of pea up to several megabases are estimated to lie between cenH3-containing domains

(Neumann et al., 2012). Thus, it seems likely that as long as the distance between multiple centromeres per chromosome is limited, these centromeres can function together, similar to the situation found in holocentric chromosomes.

NEOCENTROMERES

A neocentromere is a chromosomal locus outside the endogenous centromere that acquires kinetic activity. They are described in various organisms including plants, yeast, flies, chicken and humans (Guerra et al., 2010; Burrack and Berman, 2012; Scott and Sullivan, 2014). The term "neocentromere" has been traditionally used to define two different phenomena: (i) a *de novo*centromere formation occurring after chromosome breakage or endogenous centromere inactivation, which allows transmission of the (re-arranged) chromosome (Figure 3), and (ii) the kinetic motility of terminal or subterminal heterochromatin, which is pulled to the cell poles during meiosis in plants (Figure 4).

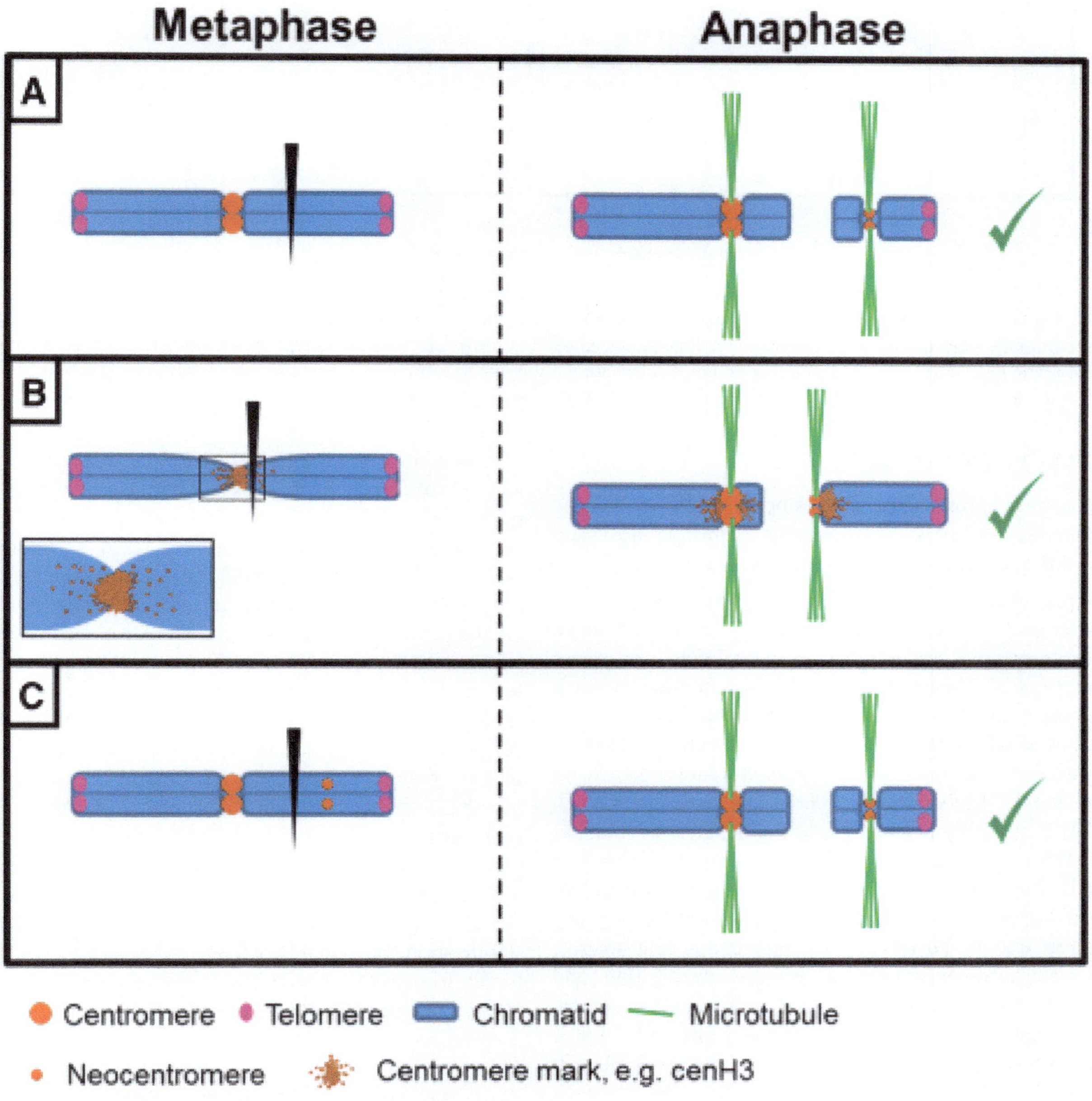

Figure 3. Formation and behavior of *de novo* centromeres. (A) Following chromosome breakage, an acentric fragment can form a neocentromere allowing its proper transmission. **(B)** Chromosome breakage close to the endogenous centromere may lead to neocentromere formation due to presence or spreading of centromeric marks (e.g., cenH3) to pericentromeric regions. **(C)** Neocentromere formation in an intact chromosome leads to a dicentric chromosome structure. If this results in chromosome breakage, a centromere- and a neocentromere-containing fragment will result.

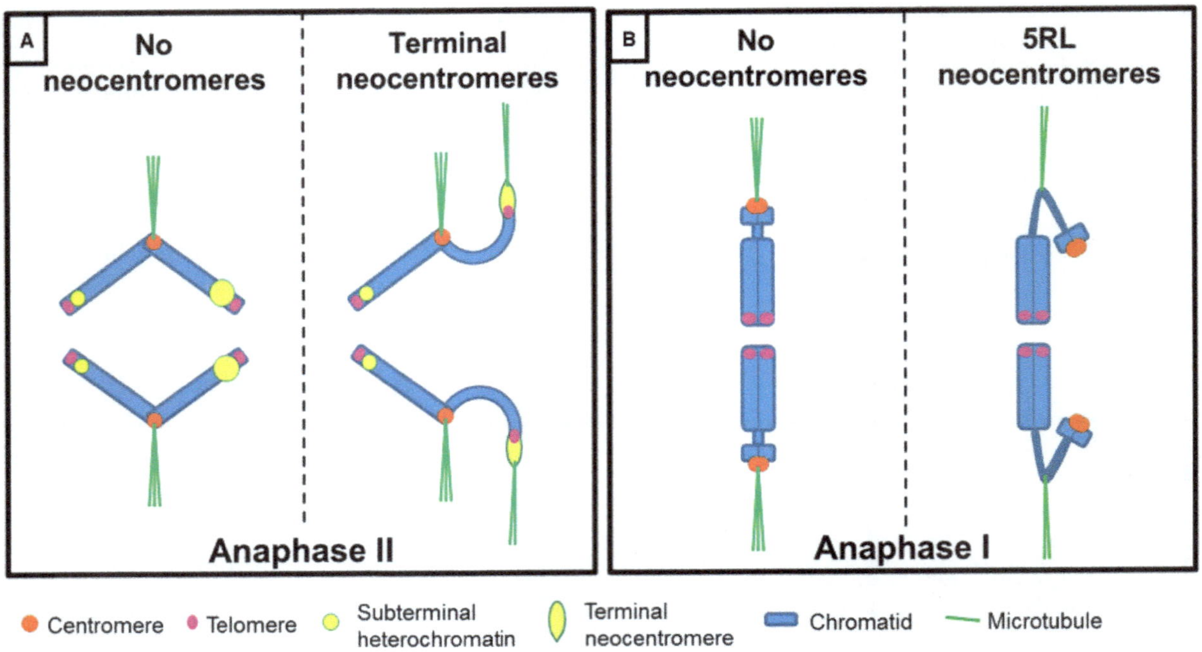

Figure 4. Schematic representation of meiosis-specific neocentromeres in plants. (A) During anaphase II, terminal neocentromeres are visible as heterochromatic stretches directed toward the cell poles ahead of the centromere. Telomeric regions are not stretched to the poles. Large heterochromatic regions (represented as larger yellow circles) are more prone to form neocentromeres than small ones. **(B)** Neocentromere in rye 5RL arises at an interstitial heterochromatic constriction and can substitute for the centromere during anaphase I.

DE NOVO CENTROMERES

In monocentric chromosome species occasionally an acentric chromosome fragment is stably transmitted during mitosis and meiosis due to neocentromere formation (Figures 3A,B).

In non-plant species, human neocentromeres are the best described to date. They usually appear in rearranged chromosomes associated with developmental delays or cancer and are typically isolated from clinical samples (Burrack and Berman, 2012). The first human neocentromere was reported by Voullaire et al. (1993) in chromosome 10. To date, neocentromeres are found in all human chromosomes except chromosome 19 (Marshall et al., 2008; Liehr et al., 2010; Klein et al., 2012), typically in euchromatic regions devoid of alpha-satellite DNAs characteristic of human centromeres (Alonso et al., 2010). These initial findings suggested that (alpha) satellite DNAs and/or heterochromatin are not necessary for (human) centromere activity. However, like endogenous centromeres, human neocentromeres form a primary constriction and contain all tested centromere-associated proteins except CENP-B whose localization is sequence-specific and requires a CENP-B box found in endogenous centromeres (Saffery et al., 2000).

In plants, *de novo* centromere formation has been documented in barley and maize. They appear in rearranged acentric chromosomes devoid of typical centromeric DNA sequences but contain centromeric proteins including cenH3 (Nasuda et al., 2005; Topp et al., 2009; Fu et al., 2013;Zhang et al., 2013; Liu et al., 2015).

In a barley telocentric chromosome derived from a 7HS isochromosome a neocentromere occurred close to the endogenous centromere (Nasuda et al., 2005). In maize, two *de novo* centromeres were reported in chromosome 3, one in the short arm (3S, Topp et al., 2009) and another one in a derivative of the long arm called Dp3a (Fu et al., 2013). In maize 3S cenH3 amounts varied between lines and low amounts of cenH3 positively correlated with low transmission rates. CenH3 levels increased over generations and the neocentromere became more stable, thus accumulation of cenH3 over time likely stabilizes the neocentromere (Topp et al., 2009). The neocentromere in Dp3a formed within protein-coding genes (Fu et al., 2013), similar to, e.g., rice centromere 8 which contains actively transcribed genes (Nagaki et al., 2004; Yan et al., 2008). Although Dp3a was occasionally transmitted during meiosis, it got frequently lost during somatic divisions suggesting that the neocentromere was unstable. The cenH3-binding region was ~350 kb, considerable smaller than the megabase-sized cenH3 binding domains of canonical maize centromeres (Wolfgruber et al., 2009). Possibly the amount of cenH3 within the 350 kb is not sufficient for proper centromere activity and larger cenH3 amounts, potentially acquired over successive generations (Topp et al., 2009), are required for stabilization of neocentromeres. Neocentromeres are also found in supernumerary maize B chromosomes (Zhang et al., 2013; Liu et al., 2015) indicating that they are not exclusive to A chromosomes.

ORIGIN OF *DE NOVO* CENTROMERES

Why these neocentromeres appear and why specifically in a given chromosome region is an intriguing question. *De novo* neocentromeres might represent "latent" centromeres (Voullaire et al., 1993; Choo, 1997), locations of ancestral centromeres following centromere repositioning events (Rocchi et al., 2012). Alternatively, spreading of centromeric chromatin "marks" to adjacent chromosomal regions may explain neocentromeres arising close to an endogenous centromere (Maggert and Karpen, 2001; Nasuda et al., 2005; Shang et al., 2013; Thakur and Sanyal, 2013;Zhang et al., 2013; Figure 3B). In chicken, for instance, cenH3 is found in pericentromeric regions and that could trigger neocentromere formation when the endogenous centromere is removed (Shang et al., 2013).

For neocentromeres located far away from the canonical centromere, an interesting point arises: are they a *consequence*, or conversely a *cause*, of chromosome breakage? Generally a neocentromere is proposed to "rescue" acentric chromosome fragments allowing their transmission during cell divisions (hence, they would be a *consequence*). However, neocentromere formation could also be the *cause*. If a new region acquires the ability to assemble an active kinetochore, this could lead to a dicentric structure and subsequent chromosome breakage (Figure3C). Possibly, both options exist: whereas experimentally centromere excision results in neocentromere formation (Ishii et al., 2008; Ketel et al., 2009; Shang et al., 2013; Thakur and Sanyal, 2013), artificially tethering of centromeric components to non-centromeric loci or overexpression of cenH3 can lead to ectopic kinetochore formation and chromosome instability (Heun et al., 2006;Barnhart et al., 2011; Mendiburo et al., 2011; Olszak et al., 2011; Teo et al., 2013). CenH3 is also found in non-centromeric chromatin in, e.g., human and chicken cells (Shang et al., 2013; Bodor et al., 2014) but whether these ectopic cenH3 domains can nucleate under certain conditions a *de novo* centromere is unclear. It has been proposed that small neocentromeres could be nucleated with relatively high frequency. While presence of the endogenous larger centromere prevents them from becoming active, in fragments detached from the main centromere this activation could occur (Liu et al., 2015).

TERMINAL NEOCENTROMERES IN PLANTS

Terminal neocentromeres were first described by Kattermann (1939) in rye who called them "T-chromosomes" for "terminal chromosomes." He described an activity in chromosomes from inbred lines resembling terminal centromeres. Prakken and Müntzing (1942) and Rhoades and Vilkomerson (1942) described in rye and maize respectively, that chromosome ends were *attracted*to the poles. The term "neo-centric" was used for the first time by Rhoades (1942). Later reports described similar chromosome activities, primarily in grasses but also in lilies and a moss (Dawe and Hiatt, 2004; Figure 4A).

Maize terminal neocentromeres appear in heterochromatic DNA domains known as knobs (Rhoades and Vilkomerson, 1942). Under standard conditions, knobs are inert. However, presence of a chromosome 10 carrying a distinct large knob called abnormal 10 (Ab10) renders them active as neocentromeres during meiosis (Rhoades and Vilkomerson, 1942). Ab10-dependent terminal neocentromere activity causes the preferential transmission - meiotic drive- of knobbed chromosomes to the egg cell during female meiosis (Rhoades, 1942). Terminal neocentromeres are genetically regulated by a locus called *Smd1* (suppressor of meiotic drive 1; Dawe and Cande, 1996) and further unknown genes (Hiatt et al., 2002; Hiatt and Dawe, 2003). Smd1 regulates the activation of terminal neocentromeres and their subsequent preferential segregation (Dawe and Cande, 1996). Interestingly, terminal neocentromeres are necessary but not sufficient for the preferential transmission of Ab10 and knobbed chromosomes since deletion of *smd1* led to neocentromere activity without meiotic drive (Dawe and Cande, 1996). Moreover, the two major components of maize knobs, i.e., a 180-bp satellite repeat and a 350-bp tandemly repeated sequence (TR-1; Peacock et al., 1981; Ananiev et al., 1998; Gonzalez-Sanchez et al., 2007) are independently regulated in such a way that both repeats "compete" genetically to be preferentially transmitted (Hiatt et al., 2002; Kanizay et al., 2013). Thus, maize neocentromeres might constitute a mechanism by which knobs and associated loci are preferentially transmitted in a process resembling the behavior of "selfish" B chromosomes (Houben et al., 2014).

Maize terminal neocentromeres lack, based on immunological analysis, cenH3 and CENP-C (Dawe et al., 1999; Dawe and Hiatt, 2004) but interact with spindle microtubules albeit in a lateral way and not the canonical end-on fashion (Yu et al., 1997). The poleward motility of chromosome arms is not affected by the microtubule-stabilizing drug taxol suggesting that a meiosis specific microtubule-based motor protein could be responsible for their kinetic activity (Hiatt et al., 2002). Identification of such protein remains elusive.

Rye terminal neocentromeres were initially observed in inbred populations (Kattermann, 1939;Prakken and Müntzing, 1942) and later also in open pollinated rye varieties (Kavander and Viinikka, 1987; Manzanero and Puertas, 2003). During meiosis, all rye chromosomes can show terminal neocentromeres that interact with microtubules (Puertas et al., 2005) but they preferentially occur in chromosomes with large C-banding positive heterochromatic blocks (Manzanero and Puertas, 2003; Figure 4A). The repetitive sequences pSc200 and pSc250 (Vershinin et al., 1995), but not centromeric or telomeric sequences, are stretched at the neocentromeres (Manzanero and Puertas, 2003). Terminal heterochromatin domains of rye B chromosomes show no neocentromeric activity (Manzanero and Puertas, 2003). Thus, only a subset of terminal heterochromatin can acquire kinetic activity. Similar to maize, rye terminal neocentromeres are genetically regulated. Hayward (1962) reported that rye neocentromeres are controlled by a polygenic system and Puertas et al. (2005) proposed a model based on two *trans*-acting genes. Contrary to maize, no examples of meiotic drive are reported for rye terminal neocentromeres.

Terminal neocentromeres are active together with the endogenous centromere stretching the chromosome arms to the same, or the opposite pole of the centromere (Manzanero and Puertas, 2003). Thus, chromosomes behave as di- or polycentric during meiotic divisions but chromosome breakage does not occur and the endogenous kinetochore leads the chromosome movement to the poles (Yu et al., 1997; Manzanero and Puertas, 2003).

Initial observations in maize suggested that terminal neocentromeres could enable migration of acentric fragments to the cell poles (Rhoades and Dempsey, 1966; Dawe and Cande, 1996). However, it was later demonstrated that only when an acentric fragment was brought to the equator plate (possibly by interaction with other bivalents) could it then migrate to the pole during anaphase I (Yu et al., 1997). Similarly, rye terminal neocentromeres fail to move acentric fragments unless they are physically linked (even by chromatin threads) to a centromere (Puertas et al., 2005). Therefore, a *cis*-acting centromere is essential for terminal neocentromere activity.

ORIGIN OF TERMINAL NEOCENTROMERES

Why would the chromosome ends acquire kinetic activity during meiosis? Frequently terminal neocentromeres appear in situations of genetic instability, such as inbred populations, interspecific hybrids, heat stress, X-ray irradiated plants or presence of abnormal chromosomes (Dawe and Hiatt, 2004). However, they also appear in normal cultivars of diploid rye. Neocentromeres in fission yeast (Ishii et al., 2008), *Drosophila* (Platero et al., 1999) and human (Marshall et al., 2008) preferentially occur in subtelomeric regions. This could indicate that subtelomeres are preferred domains to acquire centromeric activity or, in line with the "centromeres from telomeres" model (Villasante et al., 2007), that subtelomeric regions might have retained some of the ancestral features which (especially under unusual genomic conditions) allow them to recover centromeric activity. Recently, an interesting functional exchange of roles between centromeres and telomeres during yeast meiosis was found (Fennell et al., 2015): in mutants that fail to form a meiotic bouquet (attachment of telomeres to the nuclear envelope during meiotic prophase, essential for proper meiotic progression in some species) centromeres can functionally replace telomeres and (at least partially) rescue the mutant phenotype. This indicates that centromeres and telomeres are not completely independent functional entities.

Kinetochore-independent chromosome movements similar to terminal neocentromeres are also found in other organisms. For example during mitotic metaphase in the African blood lily*Haemanthus katherinae* a poleward force acts on chromosome arms facilitating movement of acentric chromosome fragments (Khodjakov et al., 1996), or in the germ line of *Parascaris*, large amounts of satellite DNAs (heterochromatin) are enriched at its chromosome termini functionally acting as centromeres but getting eliminated rapidly in somatic tissues (Pimpinelli and Goday, 1989). Whether similar mechanisms are involved in the regulation of terminal neocentromere activity is unclear.

OTHER TYPES OF NEOCENTROMERES

Key differences between *de novo* and terminal neocentromeres are that the latter do not mediate sister chromatid cohesion nor lead chromosome movement in the absence of an active endogenous centromere (Table 1). An intermediate situation is found in the long arm of chromosome 5 of rye (5RL; Figure 4B). A neocentromere in 5RL is found that is active only during meiosis, can coincide with the centromere(s) and is associated with heterochromatin (Manzanero et al., 2000, 2002), similar to terminal neocentromeres. However, unlike terminal neocentromeres, it mediates sister chromatid cohesion at anaphase I and leads chromosome movement when the endogenous centromere is inactive, similar to the situation found in *de novo* centromeres (Manzanero et al., 2000, 2002; Cuacos et al., 2011).

Table 1: Comparison of the three types of neocentromeres described with regards to their main features.

Feature	*De novo* centromere	Terminal neocentromere	5RL neocentromere
Replaces the centromere	Yes	No	Occasionally
Sister-chromatid cohesion	Yes	No	Yes
Visible constriction	Yes	No	Yes
Genomic location	Eu- and heterochromatin	Heterochromatin	Heterochromatin
Centromeric proteins	Yes	No	No
Interaction with spindle microtubules	Yes (end-on)	Yes (lateral in maize, end-on in rye)	Yes (end-on)
Species	Human, chicken, yeast, Drosophila, barley, maize	Plants (e.g., maize, rye)	Rye, wheat-rye addition lines

This neocentromere arises in an interstitial constriction in haploid rye, wheat-rye and wheat-Triticale hybrids and wheat-rye addition lines involving chromosome 5R, both in mono(telo)somic and di(telo)somic conditions (Schlegel, 1987; Manzanero et al., 2000, 2002; Cuacos et al., 2011). When the 5RL neocentromere is active, the constriction is cytologically stretched up to several times the chromosome length, reaching the cell poles before the onset of anaphase I (Manzanero et al., 2000, 2002; Cuacos et al., 2011). Interestingly, it can lead chromosome movement together with the centromere or alone (Manzanero et al., 2000, 2002; Cuacos et al., 2011) suggesting that this neocentromere, unlike terminal neocentromeres, does not require a *cis*-acting centromere to be active (Figure 4B).

The heterochromatic 5RL interstitial constriction lacks typical centromeric and telomeric sequences but contains the repetitive sequences pSc119.2 (Bedbrook et al., 1980; McIntyre et al., 1990) and UCM600 (Manzanero et al., 2002; Gonzalez-Garcia et al., 2011; Cuacos et al., 2011). Proteins accumulate at the constriction and a thin bundle of microtubules is end-on attached (Manzanero et al., 2002; Cuacos et al., 2011), showing that interaction between spindle microtubules and the constriction occurs. Notably, kinetic activity is cenH3- and CENP-C-independent (Cuacos, 2013) similar to maize terminal neocentromeres.

Varying frequencies of active rye 5RL neocentromeres in wheat-rye addition lines (Manzanero et al., 2000, 2002) suggested the influence of an environmental factor in its activity. Indeed, treating plants with an organophosphate pesticide significantly increased a basal frequency of ~10% neocentromeres up to ~50% (Cuacos et al., 2011). Concomitantly, an alteration of the meiotic spindle was found possibly facilitating the interaction of the spindle microtubules with the constriction (Cuacos et al., 2011). However, why this region acquires under certain conditions such ability is not well understood.

In *Bromus marginatus* and *B. pseudolaevipes* hybrids (Walters, 1952), *Aegilops markgrafii*(Schubert, 2011), and wheat chromosomes in the progeny from Triticale × tritordeum hybrids (Carvalho et al., 2008) a stretched constriction during meiosis, similar to the one found in rye 5RL, was described. In *A. markgrafii* this constriction is not related to neocentromeric activity since cenH3 and microtubules are exclusively found at the endogenous centromere (Schubert, 2011). Whether constrictions in *Bromus* or wheat acquire kinetic activity is not documented.

Heterochromatin within the 5RL neocentromere could explain its ability to maintain sister chromatid cohesion, as heterochromatin within pericentromeric regions does in native centromeres (Bernard et al., 2001; Nonaka et al., 2002). Moreover, a heterochromatic environment is likely conducive for neocentromere formation, as fission yeast (Ishii et al., 2008) and *Drosophila* (Platero et al., 1999; Olszak et al., 2011) neocentromeres are commonly found in heterochromatic regions. Also human neocentromeres arising at euchromatic domains recruit heterochromatic proteins (Saffery et al., 2000) and plant terminal neocentromeres are heterochromatin-associated. However, the latter do not maintain sister chromatid cohesion suggesting that heterochromatin is necessary but not sufficient to maintain sister chromatid cohesion.

NEO-CENTROMERE FORMATION AND EVOLUTION

Different models have been proposed to explain the origin of the centromere. Villasante et al. (2007) proposed that centromeres were derived from telomeres, in such a way that subtelomeric regions were the first proto-centromeres recognized by spindle microtubules. This model is, e.g., supported by telomere-like sequences found within centromeric regions of several species (Villasante et al., 2007 and references therein). The model assumes that centromeres were, from their origin, associated with repetitive DNA elements.

Another model proposes the opposite: centromeres originally formed on single-copy non-repetitive DNA loci and subsequently acquired highly repeated sequences (Gong et al., 2012; Rocchi et al., 2012). In support of this model, repeat-free centromeres have been found. For example in rice, the centromere of chromosome 8 contains low amounts of satellite DNAs (Nagaki et al., 2004) and some species lack the centromeric satellite CentO (Lee et al., 2005). Also, in potato five centromeres contain unique- or low-copy repetitive DNA (Gong et al., 2012). In non-plant species, repeat-less centromeres are found in chicken (Shang et al., 2010) and the genus Equus (Wade et al., 2009). This model is further supported by the fact that centromere repositioning has occurred in mammals and plants, i.e., a new centromere formed in a different chromosome region while the "old" one became inactive (Han et al., 2009b; Rocchi et al., 2012). Consistently, shattered remnants of centromeric satellites are found throughout plant and animal genomes (Rocchi et al., 2012;Lysak, 2014).

In potato, together with repeat-free centromeres there are centromeres containing megabases of repetitive satellite DNA with an extremely large monomer size up to 5 kb (Gong et al., 2012) which are not, or only rarely found in closely related species (Gong et al., 2012; Wang et al., 2014b; Zhang et al., 2014) suggesting a relatively recent species-specific amplification. The coexistence of centromeres with and without repetitive DNA suggests that accumulation of repeats could be a sudden, rather than progressive, process. These sequences could be incorporated from another centromere or constitute new repeats, for which retrotransposons might play a relevant role (Gong et al., 2012; Gao et al., 2015). Later accumulation or expansion of repeats could occur by several mechanisms such as replication slippage, rolling circle replication followed by reinsertion or transposition. Moreover, maize centromeric satellites are highly diverged from ancient maize centromere satellites and chromosome-specific satellites are not found, suggesting existence of a homogenization mechanism (Bilinski et al., 2015). Finally, maize chromosomes transferred to a species with larger genome such as oat experience a rapid expansion of centromere size (cenH3-binding domain; Wang et al., 2014a) agreeing with previous observations that correlated centromere and genome size in grasses (Zhang and Dawe, 2012). Altogether, this demonstrates that neo-centromere evolution can be highly dynamic.

CONCLUDING REMARKS

Despite striking progress made in plant centromere biology, many questions remain to be answered. How could such an essential chromosomal locus evolve structurally so diverse, from a point centromere up to a holocentromere? The more data on centromeres becomes available, the more challenging it is to define the centromere. The centromere could be best described as a "conserved function, but distinct structure, organization and features between organisms." In future, a better understanding in different organisms and in different centromere types of the precise nature of epigenetic mechanisms specifying centromere location, formation and maintenance as well as of the unsolved connection between epigenetics and genetics behind centromere biology is needed.

We anticipate that further studies of (plant) species with atypical centromeres will broaden our mainly monocentric chromosome-biased knowledge on centromere definition and will help in establishing more accurate models of centromere biology and evolution. Future research on plant centromere biology will also help plant genetic engineering (e.g., artificial chromosome or double haploid plant production) and thus ultimately help plant breeding.

CONFLICT OF INTEREST STATEMENT

The authors declare that the research was conducted in the absence of any commercial or financial relationships that could be construed as a potential conflict of interest.

ACKNOWLEDGMENT

This work was supported by the BBSRC (grant numbers BB/M004902/1 and BB/J007633/1).

REFERENCES

1. Akiyoshi, B., and Gull, K. (2014). Discovery of unconventional kinetochores in kinetoplastids. *Cell* 156, 1247–1258. doi: 10.1016/j.cell.2014.01.049
2. Alonso, A., Hasson, D., Cheung, F., and Warburton, P. E. (2010). A paucity of heterochromatin at functional human neocentromeres. *Epigenetics Chromatin* 3, 6. doi: 10.1186/1756-8935-3-6
3. Ananiev, E. V., Phillips, R. L., and Rines, H. W. (1998). A knob-associated tandem repeat in maize capable of forming fold-back DNA segments: are chromosome knobs megatransposons? *Proc. Natl. Acad. Sci. U.S.A.* 95, 10785–10790. doi: 10.1073/pnas.95.18.10785

4. Barnhart, M. C., Kuich, P. H., Stellfox, M. E., Ward, J. A., Bassett, E. A., Black, B. E., et al. (2011). HJURP is a CENP-A chromatin assembly factor sufficient to form a functional *de novo* kinetochore. *J. Cell Biol.* 194, 229–243. doi: 10.1083/jcb.201012017

5. Bedbrook, J. R., Jones, J., O'dell, M., Thompson, R. D., and Flavell, R. B. (1980). A molecular description of telometic heterochromatin in secale species. *Cell* 19, 545–560. doi: 10.1016/0092-8674(80)90529-2

6. Bensasson, D. (2011). Evidence for a high mutation rate at rapidly evolving yeast centromeres. *BMC Evol. Biol.* 11:211. doi: 10.1186/1471-2148-11-211

7. Bernard, P., Maure, J. F., Partridge, J. F., Genier, S., Javerzat, J. P., and Allshire, R. C. (2001). Requirement of heterochromatin for cohesion at centromeres. *Science* 294, 2539–2542. doi: 10.1126/science.1064027

8. Bilinski, P., Distor, K., Gutierrez-Lopez, J., Mendoza, G. M., Shi, J., Dawe, R. K., et al. (2015). Diversity and evolution of centromere repeats in the maize genome. *Chromosoma* 124, 57–65. doi: 10.1007/s00412-014-0483-8

9. Bodor, D. L., Mata, J. F., Sergeev, M., David, A. F., Salimian, K. J., Panchenko, T., et al. (2014). The quantitative architecture of centromeric chromatin. *Elife* 3, e02137. doi: 10.7554/eLife.02137

10. Bozek, M., Leitch, A. R., Leitch, I. J., Záveská Drábková, L., and Kuta, E. (2012). Chromosome and genome size variation in*Luzula* (*Juncaceae*), a genus with holocentric chromosomes. *Bot. J. Linn. Soc.* 170, 529–541. doi: 10.1111/j.1095-8339.2012.01314.x

11. Buchwitz, B. J., Ahmad, K., Moore, L. L., Roth, M. B., and Henikoff, S. (1999). A histone-H3-like protein in *C. elegans*. *Nature*401, 547–548. doi: 10.1038/44062

12. Bureš, P., Zedek, F., and Marková, M. (2013). "Holocentric Chromosomes," in *Plant Genome Diversity*, Vol. 2, eds J. Greilhuber, J. Dolezel, and J. F. Wendel (Vienna: Springer), 187–208.

13. Burrack, L. S., and Berman, J. (2012). Neocentromeres and epigenetically inherited features of centromeres. *Chromosome Res.*20, 607–619. doi: 10.1007/s10577-012-9296-x

14. Cabral, G., Marques, A., Schubert, V., Pedrosa-Harand, A., and Schlögelhofer, P. (2014). Chiasmatic and achiasmatic inverted meiosis of plants with holocentric chromosomes. *Nat. Commun.* 5, 5070. doi: 10.1038/ncomms6070

15. Carvalho, A., Guedes-Pinto, H., Heslop-Harrison, J. S., and Lima-Brito, J. (2008). Wheat neocentromeres found in F1 Triticale × Tritordeum hybrids (AABBRHch) After 5-Azacytidine Treatment. *Plant Mol. Biol. Rep.* 26, 46–52. doi: 10.1007/s11105-008-0021-z

16. Choo, K. H. (1997). Centromere DNA dynamics: latent centromeres and neocentromere formation. *Am. J. Hum. Genet.* 61, 1225–1233. doi: 10.1086/301657

17. Clarke, L., and Carbon, J. (1985). The structure and function of yeast centromeres. *Annu. Rev. Genet.* 19, 29–55. doi: 10.1146/annurev.ge.19.120185.000333

18. Cuacos, M. (2013). *Analysis of a Secondary Constriction in Chromosome 5R of Rye and its Neocentromeric Activity*. Ph.D. thesis, Universidad Complutense de Madrid, Madrid.

19. Cuacos, M., Gonzalez-Garcia, M., Gonzalez-Sanchez, M., Puertas, M. J., and Vega, J. M. (2011). Activation of rye 5RL neocentromere by an organophosphate pesticide. *Cytogenet. Genome Res.* 134, 151–162. doi: 10.1159/000325744

20. d'Alencon, E., Sezutsu, H., Legeai, F., Permal, E., Bernard-Samain, S., Gimenez, S., et al. (2010). Extensive synteny conservation of holocentric chromosomes in *Lepidoptera* despite high rates of local genome rearrangements. *Proc. Natl Acad. Sci. U.S.A.*107, 7680–7685. doi: 10.1073/pnas.0910413107

21. Dawe, R. K., and Cande, W. Z. (1996). Induction of centromeric activity in maize by suppressor of meiotic drive 1. *Proc. Natl. Acad. Sci. U.S.A.* 93, 8512–8517. doi: 10.1073/pnas.93.16.8512

22. Dawe, R. K., and Hiatt, E. N. (2004). Plant neocentromeres: fast, focused, and driven. *Chromosome Res.* 12, 655–669. doi: 10.1023/B:CHRO.0000036607.74671.db

23. Dawe, R. K., Reed, L. M., Yu, H. G., Muszynski, M. G., and Hiatt, E. N. (1999). A maize homolog of mammalian CENPC is a constitutive component of the inner kinetochore. *Plant Cell* 11, 1227–1238. doi: 10.1105/tpc.11.7.1227

24. De Rop, V., Padeganeh, A., and Maddox, P. (2012). CENP-A: the key player behind centromere identity, propagation, and kinetochore assembly. *Chromosoma* 121, 527–538. doi: 10.1007/s00412-012-0386-5

25. Demidov, D., Schubert, V., Kumke, K., Weiss, O., Karimi-Ashtiyani, R., Buttlar, J., et al. (2014). Anti-phosphorylated histone H2AThr120: a universal microscopic marker for centromeric chromatin of mono- and holocentric plant species. *Cytogenet. Genome Res.* 143, 150–156. doi: 10.1159/000360018

26. Dong, F. G., Miller, J. T., Jackson, S. A., Wang, G. L., Ronald, P. C., and Jiang, J. M. (1998). Rice (*Oryza sativa*) centromeric regions consist of complex DNA. *Proc. Natl Acad. Sci. U.S.A.* 95, 8135–8140. doi: 10.1073/pnas.95.14.8135

27. Drinnenberg, I. A., Deyoung, D., Henikoff, S., and Malik, H. S. (2014). Recurrent loss of CenH3 is associated with independent transitions to holocentricity in insects. *eLife* 3, e03676. doi: 10.7554/eLife.03676

28. Fennell, A., Fernandez-Alvarez, A., Tomita, K., and Cooper, J. P. (2015). Telomeres and centromeres have interchangeable roles in promoting meiotic spindle formation. *J. Cell Biol.* 208, 415–428. doi: 10.1083/jcb.201409058

29. Flach, M. (1966). Diffuse centromeres in a dicotyledonous plant. *Nature* 209, 1369–1370. doi: 10.1038/2091369b0

30. Fu, S., Lv, Z., Gao, Z., Wu, H., Pang, J., Zhang, B., et al. (2013). *De novo* centromere formation on a chromosome fragment in maize. *Proc. Natl. Acad. Sci. U.S.A.* 110, 6033–6036. doi: 10.1073/pnas.1303944110

31. Gao, D., Jiang, N., Wing, R. A., Jiang, J., and Jackson, S. A. (2015). Transposons play an important role in the

evolution and diversification of centromeres among closely related species. *Front. Plant Sci.* 6:216. doi: 10.3389/fpls.2015.00216

32. Gassmann, R., Rechtsteiner, A., Yuen, K. W., Muroyama, A., Egelhofer, T., Gaydos, L., et al. (2012). An inverse relationship to germline transcription defines centromeric chromatin in *C. elegans*. *Nature* 484, 534–537. doi: 10.1038/nature10973

33. Gernand, D., Demidov, D., and Houben, A. (2003). The temporal and spatial pattern of histone H3 phosphorylation at serine 28 and serine 10 is similar in plants but differs between mono- and polycentric chromosomes. *Cytogenet. Genome Res.* 101, 172–176. doi: 10.1159/000074175

34. Gong, Z., Wu, Y., Koblizkova, A., Torres, G. A., Wang, K., Iovene, M., et al. (2012). Repeatless and repeat-based centromeres in potato: implications for centromere evolution. *Plant Cell* 24, 3559–3574. doi: 10.1105/tpc.112.100511

35. Gonzalez-Garcia, M., Cuacos, M., Gonzalez-Sanchez, M., Puertas, M. J., and Vega, J. M. (2011). Painting the rye genome with genome-specific sequences. *Genome* 54, 555–564. doi: 10.1139/g11-003

36. Gonzalez-Sanchez, M., Gonzalez-Garcia, M., Vega, J. M., Rosato, M., Cuacos, M., and Puertas, M. J. (2007). Meiotic loss of the B chromosomes of maize is influenced by the B univalent co-orientation and the TR-1 knob constitution of the A chromosomes.*Cytogenet. Genome Res.* 119, 282–290. doi: 10.1159/000112075

37. Guerra, M., Brasileiro-Vidal, A. C., Arana, P., and Puertas, M. J. (2006). Mitotic microtubule development and histone H3 phosphorylation in the holocentric chromosomes of *Rhynchospora tenuis* (*Cyperaceae*). *Genetica* 126, 33–41. doi: 10.1007/s10709-005-1430-7

38. Guerra, M., Cabral, G., Cuacos, M., Gonzalez-Garcia, M., Gonzalez-Sanchez, M., Vega, J., et al. (2010). Neocentrics and holokinetics (holocentrics): chromosomes out of the centromeric rules. *Cytogenet. Genome Res.* 129, 82–96. doi: 10.1159/000314289

39. Haizel, T., Lim, Y. K., Leitch, A. R., and Moore, G. (2005). Molecular analysis of holocentric centromeres of *Luzula* species.*Cytogenet. Genome Res.* 109, 134–143. doi: 10.1159/000082392

40. Hakansson, A. (1954). Meiosis and pollen mitosis in x-rayed and untreated spikelets of *Eleocharis palustris*. *Hereditas* 15, 325–345. doi: 10.1111/j.1601-5223.1954.tb02976.x

41. Hakansson, A. (1958). Holocentric chromosomes in *Eleocharis*. *Hereditas* 44, 531–540. doi: 10.1111/j.1601-5223.1958.tb03498.x

42. Han, F., Gao, Z., and Birchler, J. A. (2009a). Reactivation of an inactive centromere reveals epigenetic and structural components for centromere specification in maize. *Plant Cell* 21, 1929–1939. doi: 10.1105/tpc.109.066662

43. Han, Y., Zhang, Z., Liu, C., Liu, J., Huang, S., Jiang, J., et al. (2009b). Centromere repositioning in cucurbit species: implication of the genomic impact from centromere activation and inactivation. *Proc. Natl. Acad. Sci. U.S.A.* 106, 14937–14941. doi: 10.1073/pnas.0904833106

44. Han, F., Lamb, J. C., and Birchler, J. A. (2006). High frequency of centromere inactivation resulting in stable dicentric chromosomes of maize. *Proc. Natl. Acad. Sci. U.S.A.* 103, 3238–3243. doi: 10.1073/pnas.0509650103

45. Hayward, M. D. (1962). Genetic control of neocentric activity in rye. *Heredity* 17, 439–441. doi: 10.1038/hdy.1962.43

46. Heckmann, S., and Houben, A. (2013). "Holokinetic centromeres," in *Plant Centromere Biology*, eds J. Jiang and J. A. Birchler (Oxford: Wiley-Blackwell), 83–94. doi: 10.1002/9781118525715.ch7

47. Heckmann, S., Jankowska, M., Schubert, V., Kumke, K., Ma, W., and Houben, A. (2014a). Alternative meiotic chromatid segregation in the holocentric plant *Luzula elegans*. *Nat. Commun.* 5:4979. doi: 10.1038/ncomms5979

48. Heckmann, S., Schubert, V., and Houben, A. (2014b). Holocentric plant meiosis: first sisters, then homologues. *Cell Cycle* 13, 3623–3624. doi: 10.4161/15384101.2014.986628

49. Heckmann, S., Macas, J., Kumke, K., Fuchs, J., Schubert, V., Ma, L., et al. (2013). The holocentric species *Luzula elegans* shows interplay between centromere and large-scale genome organization. *Plant J.* 73, 555–565. doi: 10.1111/tpj.12054

50. Heckmann, S., Schroeder-Reiter, E., Kumke, K., Ma, L., Nagaki, K., Murata, M., et al. (2011). Holocentric chromosomes of*Luzula elegans* are characterized by a longitudinal centromere groove, chromosome bending, and a terminal nucleolus organizer region. *Cytogenet. Genome Res.* 134, 220–228. doi: 10.1159/000327713

51. Heun, P., Erhardt, S., Blower, M. D., Weiss, S., Skora, A. D., and Karpen, G. H. (2006). Mislocalization of the Drosophila centromere-specific histone CID promotes formation of functional ectopic kinetochores. *Dev. Cell* 10, 303–315. doi: 10.1016/j.devcel.2006.01.014

52. Hiatt, E. N., and Dawe, R. K. (2003). Four loci on abnormal chromosome 10 contribute to meiotic drive in maize. *Genetics* 164, 699–709.

53. Hiatt, E. N., Kentner, E. K., and Dawe, R. K. (2002). Independently regulated neocentromere activity of two classes of tandem repeat arrays. *Plant Cell* 14, 407–420. doi: 10.1105/tpc.010373

54. Higgins, A. W., Gustashaw, K. M., and Willard, H. F. (2005). Engineered human dicentric chromosomes show centromere plasticity. *Chromosome Res.* 13, 745–762. doi: 10.1007/s10577-005-1009-2

55. Houben, A., Banaei-Moghaddam, A. M., Klemme, S., and Timmis, J. N. (2014). Evolution and biology of supernumerary B chromosomes. *Cell. Mol. Life Sci.* 71, 467–478. doi: 10.1007/s00018-013-1437-7

56. Houben, A., and Schubert, I. (2003). DNA and proteins of plant centromeres. *Curr. Opin. Plant Biol.* 6, 554–560. doi: 10.1016/j.pbi.2003.09.007

57. Houben, A., Wako, T., Furushima-Shimogawara, R., Presting, G., Künzel, G., Schubert, I., et al. (1999). The cell cycle dependent phosphorylation of histone

H3 is correlated with the condensation of plant mitotic chromosomes. *Plant J.* 18, 675–679. doi: 10.1046/j.1365-313x.1999.00496.x

58. Hughes, S. E., Gilliland, W. D., Cotitta, J. L., Takeo, S., Collins, K. A., and Hawley, R. S. (2009). Heterochromatic threads connect oscillating chromosomes during prometaphase i in drosophila oocytes. *PLoS Genet.* 5:e1000348. doi: 10.1371/journal.pgen.1000348

59. Ishii, K., Ogiyama, Y., Chikashige, Y., Soejima, S., Masuda, F., Kakuma, T., et al. (2008). Heterochromatin integrity affects chromosome reorganization after centromere dysfunction. *Science* 321, 1088–1091. doi: 10.1126/science.1158699

60. Jankowska, M., Fuchs, J., Klocke, E., Fojtová, M., Polanská, P., Fajkus, J., et al. (2015). Holokinetic centromeres and efficient telomere healing enable rapid karyotype evolution. *Chromosoma* doi: 10.1007/s00412-015-0524-y [Epub ahead of print].

61. Kanizay, L. B., Albert, P. S., Birchler, J. A., and Dawe, R. K. (2013). Intragenomic conflict between the two major knob repeats of maize. *Genetics* 194, 81–89. doi: 10.1534/genetics.112.148882

62. Kattermann, G. (1939). Ein neuer karyotyp bei roggen. *Chromosoma* 1, 284–299. doi: 10.1007/BF01271637

63. Kavander, T., and Viinikka, Y. (1987). Neocentric activity in open-pollinated cultivars of rye. *Hereditas* 107, 229–233. doi: 10.1111/j.1601-5223.1987.tb00289.x

64. Ketel, C., Wang, H. S., Mcclellan, M., Bouchonville, K., Selmecki, A., Lahav, T., et al. (2009). Neocentromeres form efficiently at multiple possible loci in *Candida albicans*. *PLoS Genet.* 5:e1000400. doi: 10.1371/journal.pgen.1000400

65. Khodjakov, A., Cole, R. W., Bajer, A. S., and Rieder, C. L. (1996). The force for poleward chromosome motion in *Haemanthus*cells acts along the length of the chromosome during metaphase but only at the kinetochore during anaphase. *J. Cell Biol.* 132, 1093–1104. doi: 10.1083/jcb.132.6.1093

66. Klein, E., Rocchi, M., Ovens-Raeder, A., Kosyakova, N., Weise, A., Ziegler, M., et al. (2012). Five novel locations of Neocentromeres in human: 18q22.1, Xq27.1 approximately 27.2, Acro p13, Acro p12, and heterochromatin of unknown origin.*Cytogenet. Genome Res.* 136, 163–166. doi: 10.1159/000336648

67. Knoll, A., Fauser, F., and Puchta, H. (2014). DNA recombination in somatic plant cells: mechanisms and evolutionary consequences. *Chromosome Res.* 22, 191–201. doi: 10.1007/s10577-014-9415-y

68. Kondo, K., and Lavarack, P. S. (1984). Study of some Australian species of *Drosera* L. (Droseraceae). *Bot. J. Linn. Soc.* 88, 317–333. doi: 10.1111/j.1095-8339.1984.tb01579.x

69. Kuta, E., Bohanec, B., Dubas, E., Vizintin, L., and Przywara, L. (2004). Chromosome and nuclear DNA study on *Luzula*—a genus with holokinetic chromosomes. *Genome* 47, 246–256. doi: 10.1139/g03-121

70. LaFountain, J. R., Cole, R. W., and Rieder, C. L. (2002). Partner telomeres during anaphase in crane-fly spermatocytes are connected by an elastic tether that exerts a backward force and resists poleward motion. *J. Cell Sci.* 115, 1541–1549.

71. Lee, H. R., Zhang, W., Langdon, T., Jin, W., Yan, H., Cheng, Z., et al. (2005). Chromatin immunoprecipitation cloning reveals rapid evolutionary patterns of centromeric DNA in *Oryza* species. *Proc. Natl. Acad. Sci. U.S.A.* 102, 11793–11798. doi: 10.1073/pnas.0503863102

72. Lermontova, I., Sandmann, M., and Demidov, D. (2014). Centromeres and kinetochores of Brassicaceae. *Chromosome Res.* 22, 135–152. doi: 10.1007/s10577-014-9422-z

73. Liehr, T., Kosyakova, N., Weise, A., Ziegler, M., and Raabe-Meyer, G. (2010). First case of a neocentromere formation in an otherwise normal chromosome 7. *Cytogenet. Genome Res.* 128, 189–191. doi: 10.1159/000271471

74. Liu, Y., Su, H., Pang, J., Gao, Z., Wang, X. J., Birchler, J. A., et al. (2015). Sequential *de novo* centromere formation and inactivation on a chromosomal fragment in maize. *Proc. Natl. Acad. Sci. U.S.A.* 112, E1263–E1271. doi: 10.1073/pnas.1418248112

75. Lysak, M. A. (2014). Live and let die: centromere loss during evolution of plant chromosomes. *New Phytol.* 203, 1082–1089. doi: 10.1111/nph.12885

76. Macas, J., Neumann, P., and Navratilova, A. (2007). Repetitive DNA in the pea (*Pisum sativum* L.) genome: comprehensive characterization using 454 sequencing and comparison to soybean and *Medicago truncatula*. *BMC Genom.* 8:427. doi: 10.1186/1471-2164-8-427

77. Maggert, K. A., and Karpen, G. H. (2001). The activation of a neocentromere in Drosophila requires proximity to an endogenous centromere. *Genetics* 158, 1615–1628.

78. Malheiros, N., De Castro, K., and Camara, A. (1947). Chromosomas sem centromero localizado O caso da *Luzula* purpurea Link.*Agronomia Lusitana* 9, 51–74.

79. Malik, H. S., and Henikoff, S. (2002). Conflict begets complexity: the evolution of centromeres. *Curr. Opin. Genet. Dev.* 12, 711–718. doi: 10.1016/S0959-437X(02)00351-9

80. Manicardi, G. C., Mandrioli, M., and Blackman, R. L. (2015). The cytogenetic architecture of the aphid genome. *Biol. Rev. Camb. Philos. Soc.* 90, 112–125. doi: 10.1111/brv.12096

81. Manzanero, S., and Puertas, M. J. (2003). Rye terminal neocentromeres: characterisation of the underlying DNA and chromatin structure. *Chromosoma* 111, 408–415. doi: 10.1007/s00412-002-0224-2

82. Manzanero, S., Puertas, M. J., Jimenez, G., and Vega, J. M. (2000). Neocentric activity of rye 5RL chromosome in wheat.*Chromosome Res.* 8, 543–554. doi: 10.1023/A:1009275807397

83. Manzanero, S., Vega, J. M., Houben, A., and Puertas, M. J. (2002). Characterization of the constriction with neocentric activity of 5RL chromosome in wheat. *Chromosoma* 111, 228–235. doi: 10.1007/s00412-002-0211-7

84. Marques, A., Ribeiro, T., Neumann, P., Macas, J.,

Novak, P., Schubert, V., et al. (2015). Holocentromeres in *Rhynchospora* are associated with genome-wide centromere-specific repeat arrays interspersed amongst euchromatin. *Proc. Natl. Acad. Sci. U.S.A.* (in press).

85. Marshall, O. J., Chueh, A. C., Wong, L. H., and Choo, K. H. (2008). Neocentromeres: new insights into centromere structure, disease development, and karyotype evolution. *Am. J. Hum. Genet.* 82, 261–282. doi: 10.1016/j.ajhg.2007.11.009

86. McClintock, B. (1939). The behavior in successive nuclear divisions of a chromosome broken at meiosis. *Proc. Natl. Acad. Sci. U.S.A.* 25, 405–416. doi: 10.1073/pnas.25.8.405

87. McIntyre, C. L., Pereira, S., Moran, L. B., and Appels, R. (1990). New *Secale cereale* (rye) DNA derivatives for the detection of rye chromosome segments in wheat. *Genome* 33, 635–640. doi: 10.1139/g90-094

88. Melters, D. P., Bradnam, K. R., Young, H. A., Telis, N., May, M. R., Ruby, J. G., et al. (2013). Comparative analysis of tandem repeats from hundreds of species reveals unique insights into centromere evolution. *Genome Biol.* 14, R10–R10. doi: 10.1186/gb-2013-14-1-r10

89. Melters, D. P., Paliulis, L. V., Korf, I. F., and Chan, S. W. (2012). Holocentric chromosomes: convergent evolution, meiotic adaptations, and genomic analysis. *Chromosome Res.* 20, 579–593. doi: 10.1007/s10577-012-9292-1

90. Mendiburo, M. J., Padeken, J., Fulop, S., Schepers, A., and Heun, P. (2011). Drosophila CENH3 is sufficient for centromere formation. *Science* 334, 686–690. doi: 10.1126/science.1206880

91. Monen, J., Hattersley, N., Muroyama, A., Stevens, D., Oegema, K., and Desai, A. (2015). Separase cleaves the N-Tail of the CENP-A related protein CPAR-1 at the meiosis I metaphase-anaphase transition in *C. elegans*. *PLoS ONE* 10:e0125382. doi: 10.1371/journal.pone.0125382

92. Monen, J., Maddox, P. S., Hyndman, F., Oegema, K., and Desai, A. (2005). Differential role of CENP-A in the segregation of holocentric *C. elegans* chromosomes during meiosis and mitosis. *Nat. Cell Biol.* 7, 1248–1255. doi: 10.1038/ncb1331

93. Moraes, I. C., Lermontova, I., and Schubert, I. (2011). Recognition of *A. thaliana* centromeres by heterologous CENH3 requires high similarity to the endogenous protein. *Plant Mol. Biol.* 75, 253–261. doi: 10.1007/s11103-010-9723-3

94. Nagaki, K., Cheng, Z., Ouyang, S., Talbert, P. B., Kim, M., Jones, K. M., et al. (2004). Sequencing of a rice centromere uncovers active genes. *Nat. Genet.* 36, 138–145. doi: 10.1038/ng1289

95. Nagaki, K., Kashihara, K., and Murata, M. (2005). Visualization of diffuse centromeres with centromere-specific histone H3 in the holocentric plant *Luzula nivea*. *Plant Cell* 17, 1886–1893. doi: 10.1105/tpc.105.032961

96. Nasuda, S., Hudakova, S., Schubert, I., Houben, A., and Endo, T. R. (2005). Stable barley chromosomes without centromeric repeats. *Proc. Natl. Acad. Sci. U.S.A.* 102, 9842–9847. doi: 10.1073/pnas.0504235102

97. Nelson, A. D., Lamb, J. C., Kobrossly, P. S., and Shippen, D. E. (2011). Parameters affecting telomere-mediated chromosomal truncation in *Arabidopsis*. *Plant Cell* 23, 2263–2272. doi: 10.1105/tpc.111.086017

98. Neumann, P., Navratilova, A., Koblizkova, A., Kejnovsky, E., Hribova, E., Hobza, R., et al. (2011). Plant centromeric retrotransposons: a structural and cytogenetic perspective. *Mob. DNA* 2, 4. doi: 10.1186/1759-8753-2-4

99. Neumann, P., Navratilova, A., Schroeder-Reiter, E., Koblizkova, A., Steinbauerova, V., Chocholova, E., et al. (2012). Stretching the rules: monocentric chromosomes with multiple centromere domains. *PLoS Genet.* 8:e1002777. doi: 10.1371/journal.pgen.1002777

100. Neumann, P., Pavlíková, Z., Koblížková, A., Fuková, I., Jedličková, V., Novák, P., et al. (2015). Centromeres off the hook: massive changes in centromere size and structure following duplication of CenH3 gene in fabeae species. *Mol. Biol. Evol.* 32, 1862–1879. doi: 10.1093/molbev/msv070

101. Nonaka, N., Kitajima, T., Yokobayashi, S., Xiao, G., Yamamoto, M., Grewal, S. I., et al. (2002). Recruitment of cohesin to heterochromatic regions by Swi6/HP1 in fission yeast. *Nat. Cell Biol.* 4, 89–93. doi: 10.1038/ncb739

102. Nonomura, K. I., and Kurata, N. (2001). The centromere composition of multiple repetitive sequences on rice chromosome 5.*Chromosoma* 110, 284–291. doi: 10.1007/s004120100148

103. Nordenskiold, H. (1951). Cyto-taxonomical studies in the genus *Luzula*.1. somatic chromosomes and chromosome numbers.*Hereditas* 37, 325–355. doi: 10.1111/j.1601-5223.1951.tb02898.x

104. Nordenskiold, H. (1964). Effect of X-Irradiation on Diploid + Polyploid *Luzula*. *Hereditas* 51, 344. doi: 10.1111/j.1601-5223.1964.tb01940.x

105. O'Toole, E. T., Mcdonald, K. L., Mäntler, J., Mcintosh, J. R., Hyman, A. A., and Müller-Reichert, T. (2003). Morphologically distinct microtubule ends in the mitotic centrosome of *Caenorhabditis elegans*. *J. Cell Biol.* 163, 451–456. doi: 10.1083/jcb.200304035

106. Olszak, A. M., Van Essen, D., Pereira, A. J., Diehl, S., Manke, T., Maiato, H., et al. (2011). Heterochromatin boundaries are hotspots for *de novo* kinetochore formation. *Nat. Cell Biol.* 13, 799–808. doi: 10.1038/ncb2272

107. Pazy, B., and Plitmann, U. (1994). Holocentric chromosome behavior in *Cuscuta* (*Cuscutaceae*). *Plant Syst. Evol.* 191, 105–109. doi: 10.1007/BF00985345

108. Pazy, B., and Plitmann, U. (1995). Chromosome divergence in the genus *Cuscuta* and its systematic implications. *Caryologia* 48, 173–180. doi: 10.1080/00087114.1995.10797327

109. Peacock, W. J., Dennis, E. S., Rhoades, M. M., and Pryor, A. J. (1981). Highly repeated DNA sequence limited to knob heterochromatin in maize. *Proc. Natl. Acad. Sci. U.S.A.* 78, 4490–4494. doi: 10.1073/pnas.78.7.4490

110. Pimpinelli, S., and Goday, C. (1989). Unusual kinetochores and chromatin diminution in *Parascaris. Trends Genet.* 5, 310–315. doi: 10.1016/0168-9525(89)90114-5

111. Platero, J. S., Ahmad, K., and Henikoff, S. (1999). A distal heterochromatic block displays centromeric activity when detached from a natural centromere. *Mol. Cell* 4, 995–1004. doi: 10.1016/S1097-2765(00)80228-2

112. Powers, J., Rose, D. J., Saunders, A., Dunkelbarger, S., Strome, S., and Saxton, W. M. (2004). Loss of KLP-19 polar ejection force causes misorientation and missegregation of holocentric chromosomes. *J. Cell Biol.* 166, 991–1001. doi: 10.1083/jcb.200403036

113. Prakken, R., and Müntzing, A. (1942). A meiotic peculiarity in rye, simulating a terminal centromere. *Hereditas* 28, 441–482. doi: 10.1111/j.1601-5223.1942.tb03291.x

114. Puertas, M. J., Garcia-Chico, R., Sotillo, E., Gonzalez-Sanchez, M., and Manzanero, S. (2005). Movement ability of rye terminal neocentromeres. *Cytogenet. Genome Res.* 109, 120–127. doi: 10.1159/000082390

115. Reznicek, A. A. (1990). Evolution in sedges (*Carex, Cyperaceae*). *Can J. Bot.* 68, 1409–1432. doi: 10.1139/b90-180

116. Rhoades, M. M. (1942). Preferential segregation in maize. *Genetics* 27, 395–407.

117. Rhoades, M. M., and Dempsey, E. (1966). The effect of abnormal chromosome 10 on preferential segregation and crossing over in maize. *Genetics* 53, 989–1020.

118. Rhoades, M. M., and Vilkomerson, H. (1942). On the anaphase movement of chromosomes. *Proc. Natl. Acad. Sci. U.S.A.* 28, 433–436. doi: 10.1073/pnas.28.10.433

119. Rocchi, M., Archidiacono, N., Schempp, W., Capozzi, O., and Stanyon, R. (2012). Centromere repositioning in mammals.*Heredity (Edinb)* 108, 59–67. doi: 10.1038/hdy.2011.101

120. Ross, B. D., Rosin, L., Thomae, A. W., Hiatt, M. A., Vermaak, D., De La Cruz, A. F. A., et al. (2013). Stepwise evolution of essential centromere function in a *Drosophila neogene*. *Science* 340, 1211–1214. doi: 10.1126/science.1234393

121. Saffery, R., Irvine, D. V., Griffiths, B., Kalitsis, P., Wordeman, L., and Choo, K. H. (2000). Human centromeres and neocentromeres show identical distribution patterns of >20 functionally important kinetochore-associated proteins. *Hum. Mol. Genet.* 9, 175–185. doi: 10.1093/hmg/9.2.175

122. Sanei, M., Pickering, R., Kumke, K., Nasuda, S., and Houben, A. (2011). Loss of centromeric histone H3 (CENH3) from centromeres precedes uniparental chromosome elimination in interspecific barley hybrids. *Proc. Natl. Acad. Sci. U.S.A.* 108, E498–E505. doi: 10.1073/pnas.1103190108

123. Schlegel, R. (1987). Neocentric activity in chromosome 5R of rye revealed by haploidy. *Hereditas* 107, 1–6. doi: 10.1111/j.1601-5223.1987.tb00264.x

124. Schrader, F. (1935). Notes an the mitotic behavior of long chromosomes. *Cytologia* 6, 422–430. doi: 10.1508/cytologia.6.422

125. Schubert, I., Rieger, R., and Fuchs, J. (1995). Alteration of basic chromosome number by fusion-fission cycles. *Genome* 38, 1289–1292. doi: 10.1139/g95-170

126. Schubert, V. (2011). No neocentric activity on *Aegilops markgrafii* chromosome E. *Cytogenet. Genome Res.* 132, 100–104. doi: 10.1159/000318996

127. Schvarzstein, M., Wignall, S. M., and Villeneuve, A. M. (2010). Coordinating cohesion, co-orientation, and congression during meiosis: lessons from holocentric chromosomes. *Genes Dev.* 24, 219–228. doi: 10.1101/gad.1863610

128. Scott, K. C., and Sullivan, B. A. (2014). Neocentromeres: a place for everything and everything in its place. *Trends Genet.* 30, 66–74. doi: 10.1016/j.tig.2013.11.003

129. Shang, W. H., Hori, T., Martins, N. M., Toyoda, A., Misu, S., Monma, N., et al. (2013). Chromosome engineering allows the efficient isolation of vertebrate neocentromeres. *Dev. Cell* 24, 635-648. doi: 10.1016/j.devcel.2013.02.009

130. Shang, W. H., Hori, T., Toyoda, A., Kato, J., Popendorf, K., Sakakibara, Y., et al. (2010). Chickens possess centromeres with both extended tandem repeats and short non-tandem-repetitive sequences. *Genome Res.* 20, 1219–1228. doi: 10.1101/gr.106245.110

131. Sheikh, S. A., and Kondo, K. (1995). Differential staining with orcein, giemsa, cma, and dapi for comparative chromosome study of 12 species of australian *Drosera* (*Droseraceae*). *Am. J. Bot.* 82, 1278–1286. doi: 10.2307/2446251

132. Sheikh, S. A., Kondo, K., and Hoshi, Y. (1995). Study of diffused centromeric nature of *Drosera chromosomes. Cytologia (Tokyo)* 60, 43–47. doi: 10.1508/cytologia.60.43

133. Shirakawa, J., Nagano, K., and Hoshi, Y. (2011). A chromosome study of two centromere differentiating *Drosera* species, *D. arcturi* and *D. regia*. *Caryologia* 64, 453–463.

134. Steiner, F. A., and Henikoff, S. (2014). Holocentromeres are dispersed point centromeres localized at transcription factor hotspots. *Elife* 3, e02025. doi: 10.7554/eLife.02025

135. Steiner, F. A., and Henikoff, S. (2015). Diversity in the organization of centromeric chromatin. *Curr. Opin. Genet. Dev.* 31, 28–35. doi: 10.1016/j.gde.2015.03.010

136. Stimpson, K. M., Matheny, J. E., and Sullivan, B. A. (2012). Dicentric chromosomes: unique models to study centromere function and inactivation. *Chromosome Res.* 20, 595–605. doi: 10.1007/s10577-012-9302-3

137. Stinchcomb, D. T., Shaw, J. E., Carr, S. H., and Hirsh, D. (1985). Extrachromosomal DNA transformation of *Caenorhabditis elegans*. *Mol. Cell. Biol.* 5, 3484–3496. doi: 10.1128/MCB.5.12.3484

138. Sullivan, B. A., and Willard, H. F. (1998). Stable dicentric × chromosomes with two functional centromeres. *Nat. Genet.* 20, 227–228. doi: 10.1038/3024

139. Tanaka, N. (1977). Chromosome studies in *Chionographis* (*Liliaceae*). 1. holokinetic nature of chromosomes in *Chionographis*-japonica-maxim.

Cytologia 42, 753–763. doi: 10.1508/cytologia.42.753

140. Teo, C. H., Lermontova, I., Houben, A., Mette, M. F., and Schubert, I. (2013). *De novo* generation of plant centromeres at tandem repeats. *Chromosoma* 122, 233–241. doi: 10.1007/s00412-013-0406-0

141. Thakur, J., and Sanyal, K. (2013). Efficient neocentromere formation is suppressed by gene conversion to maintain centromere function at native physical chromosomal loci in *Candida albicans*. *Genome Res.* 23, 638–652. doi: 10.1101/gr.141614.112

142. Topp, C. N., Okagaki, R. J., Melo, J. R., Kynast, R. G., Phillips, R. L., and Dawe, R. K. (2009). Identification of a maize neocentromere in an oat-maize addition line. *Cytogenet. Genome Res.* 124, 228–238. doi: 10.1159/000218128

143. Tsujimoto, H. (1993). Molecular cytological evidence for gradual telomere synthesis at the broken chromosome ends in wheat. *J. Plant Res.* 106, 239–244. doi: 10.1007/BF02344591

144. Tsujimoto, H., Usami, N., Hasegawa, K., Yamada, T., Nagaki, K., and Sasakuma, T. (1999). *De novo* synthesis of telomere sequences at the healed breakpoints of wheat deletion chromosomes. *Mol. Gen. Genet.* 262, 851–856. doi: 10.1007/s004380051150

145. Vershinin, A. V., Schwarzacher, T., and Heslop-Harrison, J. S. (1995). The large-scale genomic organization of repetitive DNA families at the telomeres of rye chromosomes. *Plant Cell* 7, 1823–1833. doi: 10.1105/tpc.7.11.1823

146. Viera, A., Page, J., and Rufas, J. S. (2009). Inverted meiosis: the true bugs as a model to study. *Genome Dyn* 5, 137–156. doi: 10.1159/000166639

147. Villasante, A., Abad, J. P., and Mendez-Lago, M. (2007). Centromeres were derived from telomeres during the evolution of the eukaryotic chromosome. *Proc. Natl. Acad. Sci. U.S.A.* 104, 10542–10547. doi: 10.1073/pnas.0703808104

148. Voullaire, L. E., Slater, H. R., Petrovic, V., and Choo, K. H. (1993). A functional marker centromere with no detectable alpha-satellite, satellite III, or CENP-B protein: activation of a latent centromere? *Am. J. Hum. Genet.* 52, 1153–1163.

149. Wade, C. M., Giulotto, E., Sigurdsson, S., Zoli, M., Gnerre, S., Imsland, F., et al. (2009). Genome sequence, comparative analysis, and population genetics of the domestic horse. *Science* 326, 865–867. doi: 10.1126/science.1178158

150. Walters, M. S. (1952). Spontaneous chromosome breakage and atypical chromosome movement in meiosis of the hybrid*Bromus marginatus* × *B. Pseudolaevipes*. *Genetics* 37, 8–25.

151. Wang, G., Li, H., Cheng, Z., and Jin, W. (2013). A novel translocation event leads to a recombinant stable chromosome with interrupted centromeric domains in rice. *Chromosoma* 122, 295–303. doi: 10.1007/s00412-013-0413-1

152. Wang, K., Wu, Y., Zhang, W., Dawe, R. K., and Jiang, J. (2014a). Maize centromeres expand and adopt a uniform size in the genetic background of oat. *Genome Res.* 24, 107–116. doi: 10.1101/gr.160887.113

153. Wang, L., Zeng, Z., Zhang, W., and Jiang, J. (2014b). Three potato centromeres are associated with distinct haplotypes with or without megabase-sized satellite repeat arrays. *Genetics* 196, 397–401. doi: 10.1534/genetics.113.160135

154. Wanner, G., Schroeder-Reiter, E., Ma, W., Houben, A., and Schubert, V. (2015). The ultrastructure of mono- and holocentric plant centromeres: an immunological investigation by structured illumination microscopy and scanning electron microscopy.*Chromosoma* doi: 10.1007/s00412-015-0521-1 [Epub ahead of print].

155. Wolfgruber, T. K., Sharma, A., Schneider, K. L., Albert, P. S., Koo, D. H., Shi, J., et al. (2009). Maize centromere structure and evolution: sequence analysis of centromeres 2 and 5 reveals dynamic Loci shaped primarily by retrotransposons. *PLoS Genet.*5:e1000743. doi: 10.1371/journal.pgen.1000743

156. Yan, H., Talbert, P. B., Lee, H. R., Jett, J., Henikoff, S., Chen, F., et al. (2008). Intergenic locations of rice centromeric chromatin. *PLoS Biol.* 6:e286. doi: 10.1371/journal.pbio.0060286

157. Yano, O., and Hoshino, A. T. (2006). Cytological studies of aneuploidy in *Eleocharis kamtschatica* (*Cyperaceae*). *Cytologia* 71, 141–147. doi: 10.1508/cytologia.71.141

158. Yelina, N. E., Choi, K., Chelysheva, L., Macaulay, M., De Snoo, B., Wijnker, E., et al. (2012). Epigenetic remodeling of meiotic crossover frequency in *Arabidopsis thaliana* DNA methyltransferase mutants. *PLoS Genet.* 8:e1002844. doi: 10.1371/journal.pgen.1002844

159. Yi, C., Zhang, W., Dai, X., Li, X., Gong, Z., Zhou, Y., et al. (2013). Identification and diversity of functional centromere satellites in the wild rice species *Oryza brachyantha*. *Chromosome Res.* 21, 725–737. doi: 10.1007/s10577-013-9374-8

160. Yu, H. G., Hiatt, E. N., Chan, A., Sweeney, M., and Dawe, R. K. (1997). Neocentromere-mediated chromosome movement in maize. *J. Cell Biol.* 139, 831–840. doi: 10.1083/jcb.139.4.831

161. Yuen, K. W. Y., Nabeshima, K., Oegema, K., and Desai, A. (2011). Rapid *de novo* centromere formation occurs independently of heterochromatin protein 1 in *C. elegans* Embryos. *Curr. Biol.* 21, 1800–1807. doi: 10.1016/j.cub.2011.09.016

162. Zedek, F., Smerda, J., Smarda, P., and Bures, P. (2010). Correlated evolution of LTR retrotransposons and genome size in the genus *Eleocharis*. *BMC Plant Biol.* 10:265. doi: 10.1186/1471-2229-10-265

163. Zhang, B., Lv, Z., Pang, J., Liu, Y., Guo, X., Fu, S., et al. (2013). Formation of a functional maize centromere after loss of centromeric sequences and gain of ectopic sequences. *Plant Cell* 25, 1979–1989. doi: 10.1105/tpc.113.110015

164. Zhang, H., and Dawe, R. K. (2012). Total centromere size and genome size are strongly correlated in ten grass species.*Chromosome Res.* 20, 403–412. doi: 10.1007/s10577-012-9284-1

165. Zhang, H., Koblizkova, A., Wang, K., Gong, Z., Oliveira, L., Torres, G. A., et al. (2014). Boom-bust turnovers of megabase-sized centromeric DNA in *Solanum* species: rapid evolution of DNA sequences associated with centromeres. *Plant Cell* 26, 1436–1447. doi: 10.1105/tpc.114.123877

166. Zhang, W., Friebe, B., Gill, B. S., and Jiang, J. (2010). Centromere inactivation and epigenetic modifications of a plant chromosome with three functional centromeres. *Chromosoma* 119, 553–563. doi: 10.1007/s00412-010-0278-5

INDEX

N

O

R

S

T

W

Y

Z